About Island Press

Since 1984, the nonprofit organization Island Press has been stimulating, shaping, and communicating ideas that are essential for solving environmental problems worldwide. With more than 800 titles in print and some 40 new releases each year, we are the nation's leading publisher on environmental issues. We identify innovative thinkers and emerging trends in the environmental field. We work with world-renowned experts and authors to develop cross-disciplinary solutions to environmental challenges.

Island Press designs and executes educational campaigns in conjunction with our authors to communicate their critical messages in print, in person, and online using the latest technologies, innovative programs, and the media. Our goal is to reach targeted audiences—scientists, policymakers, environmental advocates, urban planners, the media, and concerned citizens—with information that can be used to create the framework for long-term ecological health and human well-being.

Island Press gratefully acknowledges major support of our work by The Agua Fund, The Andrew W. Mellon Foundation, Betsy & Jesse Fink Foundation, The Bobolink Foundation, The Curtis and Edith Munson Foundation, Forrest C. and Frances H. Lattner Foundation, G.O. Forward Fund of the Saint Paul Foundation, Gordon and Betty Moore Foundation, The Kresge Foundation, The Margaret A. Cargill Foundation, The Overbrook Foundation, The S.D. Bechtel, Jr. Foundation, The Summit Charitable Foundation, Inc., V. Kann Rasmussen Foundation, The Wallace Alexander Gerbode Foundation, and other generous supporters.

The opinions expressed in this book are those of the author(s) and do not necessarily reflect the views of our supporters.

About the Pacific Institute for Studies in Development, Environment, and Security

The Pacific Institute is one of the world's leading research and policy nonprofits working to create a healthier planet and sustainable communities. Based in Oakland, California, we conduct interdisciplinary research and partner with stakeholders to produce solutions that advance environmental protection, economic development, international security, and social equity—nationally and internationally. Since the Institute's founding in 1987, we have worked to change policy and find real-world solutions to problems like water shortages, habitat destruction, global warming, and environmental injustice, with the fundamental idea that adequate, safe, clean water is closely connected to every vital resource issue of our time. We have moved the focus of water thinking away from narrow approaches and toward more integrated and sustainable water practices and concepts through rigorous independent research, extensive policy engagement, and intensive outreach to the public. The Pacific Institute has formulated a new vision for long-term water planning in California and internationally; developed a new approach for valuing well-being in local communities; worked on transborder environment and trade issues in North America and beyond; analyzed standards in global environmental protection; clarified key concepts and criteria for sustainable water use; offered recommendations for reducing conflicts over water in the Middle East, Latin America, and Central Asia; championed the human right to water; assessed the impacts of global warming on freshwater resources; and created programs to address environmental justice concerns in low-income communities and communities of color. Our research has reached tens of millions of people through reports, projects, speeches, testimony, and media; more than 100 research papers, reports, and testimonies are available free on the Pacific Institute websites at www.pacinst.org and www.worldwater.org.

THE WORLD'S WATER

Volume 8

The Biennial Report on Freshwater Resources

Peter H. Gleick
with Newsha Ajami, Juliet Christian-Smith, Heather Cooley,
Kristina Donnelly, Julian Fulton, Mai-Lan Ha, Matthew Heberger,
Eli Moore, Jason Morrison, Stewart Orr, Peter Schulte, and
Veena Srinivasan

Washington | Covelo | London

Copyright © 2014 Pacific Institute for Studies in Development, Environment, and Security

All rights reserved under International and Pan-American Copyright Conventions. No part of this book may be reproduced in any form or by any means without permission in writing from the publisher: Island Press, 2000 M Street NW, Suite 650, Washington, DC 20009.

ISLAND PRESS is a trademark of the Center for Resource Economics.

Library of Congress Card Catalog Number 98024877
ISBN 10: 1-61091-481-3 (cloth)
ISBN 13: 978-1-61091-481-9 (cloth)
ISBN 10: 1-61091-482-1 (paper)
ISBN 13: 978-1-61091-482-6 (paper)

ISSN 15287-7165

Printed on recycled, acid-free paper ♻

Manufactured in the United States of America
10 9 8 7 6 5 4 3 2 1

Keywords: water management, water quality, fracking, desalination, virtual water, water conflict, green jobs, the Dead Sea, Syria, corporate water use

---

Island Press would like to thank the New Mexico Water Initiative, a project of the Hanuman Foundation, for generously supporting the production of this book.

Contents

Foreword by Ismail Serageldin xi

Introduction xiii

ONE   Global Water Governance in the Twenty-First Century   1

    Heather Cooley, Newsha Ajami, Mai-Lan Ha, Veena Srinivasan, Jason Morrison, Kristina Donnelly, and Juliet Christian-Smith

    Global Water Challenges   2
    The Emergence of Global Water Governance   6
    Conclusions   15

TWO   Shared Risks and Interests: The Case for Private Sector Engagement in Water Policy and Management   19

    Peter Schulte, Stuart Orr, and Jason Morrison

    The Business Case for Investing in Sustainable Water Management   21
    Utilizing Corporate Resources While Ensuring Public Interest Outcomes and Preventing Policy Capture   28
    Moving Forward: Unlocking Mutually Beneficial Corporate Action on Water   31

THREE   Sustainable Water Jobs   35

    Eli Moore, Heather Cooley, Juliet Christian-Smith, and Kristina Donnelly

    Water Challenges in Today's Economy   36
    Job Quality and Growth in Sustainable Water Occupations   54
    Conclusions   57
    Recommendations   58

FOUR   Hydraulic Fracturing and Water Resources: What Do We Know and Need to Know?   63

    Heather Cooley and Kristina Donnelly

    Overview of Hydraulic Fracturing   64
    Concerns Associated with Hydraulic Fracturing Operations   65
    Water Challenges   67
    Conclusions   77

FIVE   Water Footprint   83
Julian Fulton, Heather Cooley, and Peter H. Gleick

     The Water Footprint Concept   83
     Water, Carbon, and Ecological Footprints and Nexus Thinking   85
     Water Footprint Findings   87
     Conclusion   90

SIX   Key Issues for Seawater Desalination in California: Cost and Financing   93
Heather Cooley and Newsha Ajami

     How Much Does Seawater Desalination Cost?   93
     Desalination Projects and Risk   101
     Case Studies   106
     Conclusions   117

SEVEN   Zombie Water Projects   123
Peter H. Gleick, Matthew Heberger, and Kristina Donnelly

     The North American Water and Power Alliance—NAWAPA   124
     The Reber Plan   131
     Alaskan Water Shipments   133
     Las Vegas Valley Pipeline Project   136
     Diverting the Missouri River to the West   140
     Conclusions   144

WATER BRIEF

One   The Syrian Conflict and the Role of Water   147
     Peter H. Gleick

Two   The Red Sea–Dead Sea Project Update   153
     Kristina Donnelly

Three   Water and Conflict: Events, Trends, and Analysis (2011–2012)   159
     Peter H. Gleick and Matthew Heberger

Four   Water Conflict Chronology   173
     Peter H. Gleick and Matthew Heberger

DATA SECTION

Data Table 1:   Total Renewable Freshwater Supply by Country (2013 Update)   221

Data Table 2:   Freshwater Withdrawal by Country and Sector (2013 Update)   227

Data Table 3A: Access to Improved Drinking Water by Country, 1970–2008   236

Data Table 3B: Access to Improved Drinking Water by Country, 2011 Update   247
Data Table 4A: Access to Improved Sanitation by Country, 1970–2008   252
Data Table 4B: Access to Improved Sanitation by Country, 2011 Update   263
Data Table 5:   MDG Progress on Access to Safe Drinking Water by Region   268
Data Table 6:   MDG Progress on Access to Sanitation by Region   271
Data Table 7:   Monthly Natural Runoff for the World's Major River Basins, by Flow Volume   274
Data Table 8:   Monthly Natural Runoff for the World's Major River Basins, by Basin Name   292
Data Table 9:   Area Equipped for Irrigation Actually Irrigated   310
Data Table 10:   Overseas Development Assistance for Water Supply and Sanitation, by Donating Country, 2004–2011   317
Data Table 11:   Overseas Development Assistance for Water Supply and Sanitation, by Subsector, 2007–2011   320
Data Table 12:   Per Capita Water Footprint of National Consumption, by Country, 1996–2005   323
Data Table 13:   Per Capita Water Footprint of National Consumption, by Sector and Country, 1996–2005   332
Data Table 14:   Total Water Footprint of National Consumption, by Country, 1996–2005   341
Data Table 15:   Total Water Footprint of National Consumption, by Sector and Country, 1996–2005   350
Data Table 16A: Global Cholera Cases Reported to the World Health Organization, by Country, 1949–1979   359
Data Table 16B: Global Cholera Cases Reported to the World Health Organization, by Country, 1980–2011   367
Data Table 17A: Global Cholera Deaths Reported to the World Health Organization, by Country, 1949–1979   373
Data Table 17B: Global Cholera Deaths Reported to the World Health Organization, by Country, 1980–2011   381
Data Table 18A: Perceived Satisfaction with Water Quality in Sub-Saharan Africa   387
Data Table 18B: Regional Assessment of Satisfaction with Water (and Air) Quality   389
Data Table 18C: Countries Most and Least Satisfied with Water Quality   392

WATER UNITS, DATA CONVERSIONS, AND CONSTANTS   395

COMPREHENSIVE TABLE OF CONTENTS   405

Volume 1: The World's Water 1998–1999: The Biennial Report on Freshwater Resources   405

Volume 2: The World's Water 2000–2001: The Biennial Report on Freshwater Resources   408

Volume 3: The World's Water 2002–2003: The Biennial Report on Freshwater Resources   411

Volume 4: The World's Water 2004–2005: The Biennial Report on Freshwater Resources   414

Volume 5: The World's Water 2006–2007: The Biennial Report on Freshwater Resources   418

Volume 6: The World's Water 2008–2009: The Biennial Report on Freshwater Resources   422

Volume 7: The World's Water, Volume 7: The Biennial Report on Freshwater Resources   426

Volume 8: The World's Water, Volume 8: The Biennial Report on Freshwater Resources   430

COMPREHENSIVE INDEX   435

Foreword

It is a privilege to write a foreword to volume 8 of *The World's Water*, an indispensable biennial report on the condition of the earth's water resource. Water is life. Without water, there is no life as we know it. Yet humans have abused it and misused it and polluted, wasted, and destroyed it.

In the 1990s, when I was vice president of environmentally and socially sustainable development for the World Bank, I was appalled at the lack of international engagement on the issue of water. In a speech I gave in Stockholm in August 1995, I said: "The wars of this century have been for oil, but the wars of the next century shall be for water"—a statement that echoed around the news media of the world. Many accused me of being an alarmist, but gradually, within a few years, the issue of water and water management started taking its rightful place among the national, regional, and global priorities of various actors around the planet. What was lacking was a proper global perspective on the issue of water that would go way beyond chapter 18 of Rio's Agenda 21.

And here, like a beacon unto the multitudes, the careful and meticulous scientific work of the Pacific Institute and its president, Peter Gleick, came to the fore in the form of *The World's Water* series. It provided the most thoughtful, comprehensive, and interesting coverage of the available knowledge and the insights needed to deal with the myriad problems that water represents. Sixteen years later, the eighth volume of *The World's Water* continues that distinguished tradition. Although many notable agencies of the United Nations, the World Water Council, the Global Water Partnership, the World Resources Institute, and others have been producing excellent reports on water issues, *The World's Water* continues its tradition of excellence, and I heartily recommend it to anyone interested in dispassionate and thoughtful discussion of the complex and important issues involved.

As we have come to expect, this year's report, beyond providing a host of useful data, also marshals many experts to address important themes of water and water use. Starting with global water governance, it deals with the issue that most concerned me so many years ago. Today, as I sit in Egypt and watch the unfolding debates about the management of the Nile's waters as Ethiopia proceeds to build its Grand Ethiopian Renaissance Dam (GERD), and as I watch the agonies of populations in places enduring droughts and the general mismanagement of our most precious resources in so many countries, I can think of no more important topic that should elicit an international focus on cooperation rather than conflict. We must see that nature does not recognize political boundaries and that, just as water flows in rivers shared by countries, climate change affects all countries, not only those whose economic conduct plays a major role in causing it.

Beyond global water governance (addressed in chapter 1), the importance of engaging the private sector in smart water management is addressed in chapter 2. Here, if public and private partnerships are to thrive, transparency must be the order of the day. Such collaboration will help us attain the goal of ensuring that all people have access to clean water and proper sanitation and that water is properly managed in agriculture, still the sector accounting for the largest water withdrawals.

Economic issues are widely discussed in this report. There are great opportunities to generate employment and jobs by smart investments in water systems (discussed in chapter 3). The rapidly expanding technique of hydraulic fracturing (or "fracking") raises problems and poses risks to water systems from inappropriately conducted fossil-fuel extraction, all of which is discussed in chapter 4. The cost and financing of seawater desalination projects are discussed in chapter 6. And recurrently revived large-scale infrastructure schemes and projects—here named "zombie water projects"—are discussed in chapter 7.

Chapter 5 of this report presents background and detail on a tool of enormous value, the "water footprint," and discusses its application worldwide. If only the concerned parties would all listen and make use of the insights generated by this tool.

But why should I take up more of the readers' time? Let them delve into the riches between these covers. It is time to read, learn, and reflect . . . and to enjoy.

Ismail Serageldin
Librarian of Alexandria
Former Vice President, World Bank

Introduction

When the first volume of *The World's Water* was published, the United Nations' Millennium Development Goals had not yet been established. Today, as we finish volume 8, we are racing toward the 2015 deadline for those goals, but basic human needs for water and sanitation remain unsatisfied. Deadlines aside, if we are ever to achieve the objectives of global health and sustainability, water security is critical.

In this volume, we continue to provide data on access (by country and by urban-rural split) to water and sanitation. We also provide data on the ongoing, unchecked severe cholera outbreaks that are a direct consequence of our failure to meet basic human needs for safe water. Human-caused climate changes are also increasingly apparent, with growing evidence around the world in the form of alterations of natural hydrology and impacts on our built water infrastructure. Ecosystems that rely on freshwater continue to deteriorate in many parts of the world as human competition for water grows. And conflicts over water allocations and use are growing, not diminishing, as reflected in the Water Conflict Chronology—a regular feature of these volumes.

The World's Water, however, has always been about more than water "problems." My colleagues and I believe strongly that there are successful and expandable solutions out there, and we try to describe and discuss them in every volume. New thinking about solutions and the "soft path for water" are both needed and available, if we look around. As a result, *The World's Water* has always tried to explore new ideas, successful case studies, and innovative communications efforts.

There is no shortage of topics to address, and as always it is a challenge to try to choose among them for inclusion in the books. In this latest volume we tackle some new topics and revisit and update some older ones.

Chapter 1 offers an overview of the complex institutions and ideas that constitute "global water governance" and offers some thoughts about how to improve and manage water systems at local, continental, and global scales while still protecting the critical needs of communities.

Chapter 2 revisits previous Pacific Institute efforts on how to engage the corporate sector in more sustainable water management, reporting, and use. Some innovative efforts are under way under the auspices of the United Nations Global Compact office and the United Nations' CEO Water Mandate, bringing new voices into the water discussion. The hope, still to be fully realized, is that responsible and sustainable corporate water management will play a role alongside the efforts of governments and nongovernmental organizations in moving toward more sustainable water policy.

The third chapter in this volume summarizes recent work completed at the Pacific Institute on water-related jobs. Jobs, in different sectors of the economy, are required to manage, deliver, and treat the freshwater required to satisfy our commercial, institutional, industrial, agricultural, and domestic needs for goods and services. In addition, growing attention is being given to sustainable measures such as low-impact development, water reuse, watershed restoration, water conservation and efficiency, and many other proven and promising practices. As the country shifts to more sustainable water management, new services, occupations, and markets are emerging. Our research suggests that water is a worthy arena for exploring green jobs.

Chapter 4 expands on the work we presented in the previous volume focused on energy and water, especially the increasingly controversial issue of hydraulic fracturing, or "fracking." This practice has been particularly controversial in the United States but is expanding globally, wherever unconventional natural gas development is under way. To better identify and understand the key issues, the Pacific Institute conducted extensive interviews with a diverse group of stakeholders, including representatives from state and federal agencies, academia, industry, environmental groups, and community-based organizations from across the United States. This chapter provides a short summary of the key issues identified in the interviews and an assessment and synthesis of existing research.

Chapter 5 digs into the useful concept of the "water footprint." We use water for many purposes: drinking, bathing, washing our clothes, watering our gardens, and more. In addition to these direct uses, however, water is needed to produce nearly everything we use and consume, from the food we eat and the clothes we wear to the technological devices that are integral to our modern society. A full measure of the water footprint of an individual, industrial sector, or society is the combination of direct and indirect water needed to provide our basket of goods and services. This chapter discusses the tool of "footprinting," how it is related to other footprint indicators, such as the carbon footprint and the ecological footprint, how it is calculated, and the findings from some recent research.

Chapter 6 returns to the issue of desalination. In this chapter, we provide a short summary of the economics of various major desalination projects and the spatial and temporal variability associated with these costs. We also describe project-related risks associated with desalination plants, and we provide two case studies of seawater desalination projects in the United States to highlight some of the issues around cost and financing of these projects.

The final chapter, chapter 7, tackles the issue of water "zombies." As any consumer of popular culture today knows, a zombie is an undead creature, something that was killed but came back to "life" in some form to terrify, entertain, and gross people out. Zombie books, movies, costumes, makeup, computer games, and more are big business. But not all zombies are fictional, and some are potentially really dangerous—at least to our pocketbooks and environment. These include "zombie water projects," which we define as large, costly water projects that are proposed, killed off for one reason or another, and brought back to life, even if the project itself is socially, politically, economically, or environmentally unjustified. In chapter 7 we offer—in an only partly tongue-in-cheek presentation—a description of some major zombie water projects from the past century: serious proposals for massive projects to use, divert, or transfer water that died only to come back to life in one form or another. In short, this chapter argues that modern civilization must learn that our ability to do something doesn't mean that we actually should do it, especially in the field of large-scale geoengineering.

The chapters are supplemented with shorter In Brief reports on items of interest. The current volume includes an update on the Red Sea–Dead Sea proposals, the role of water in the current Syrian conflict, and a detailed update of our popular Water Conflict Chronology, with entries through the end of 2012. The chronology offers historical examples of conflicts related to water going back millennia, and it is also available in a flexible and useful format online at http://www.worldwater.org, where readers can sort water conflicts by time, location, type of conflict, and more and see the results in active maps.

We provide in the back of volume 8 a complete table of contents and integrated index across all previous volumes, to help readers find information in the earlier editions that might be useful or relevant in their research or other efforts.

Finally, *The World's Water* again offers a wide variety of important, useful, and popular data on water in a series of data tables. In this volume we present updated data on water supply and use by country, the latest progress (or lack thereof) toward the Millennium Development Goals for water and sanitation, runoff data for the world's largest and most important rivers, irrigation data, estimates of funding for water from overseas development agencies, tables on water footprints of consumption, and more.

If there is any good news, it is that water is receiving more attention from policy makers, scientists, the media, and members of the public. Education and awareness are key steps in moving toward implementing smart and effective solutions. We are happy to think we are playing even a small role.

Special thanks to all of my coauthors, both within the Pacific Institute and outside with partner organizations. This project continues to benefit from the support of Island Press, including Emily Davis, my new(ish) editor. Thank you, Emily, for your help getting this latest volume out the door!

Peter H. Gleick
Oakland, California, 2013

CHAPTER 1

Global Water Governance in the Twenty-First Century

Heather Cooley, Newsha Ajami, Mai-Lan Ha, Veena Srinivasan, Jason Morrison, Kristina Donnelly, and Juliet Christian-Smith

Growing pressure on the world's water resources is having major impacts on our social and economic well-being. Even as the planet's endowment of water is expected to remain constant, human appropriation of water, already at 50 percent by some measures, is expected to increase further (Postel et al. 1996). Pressures on water resources are likely to worsen in response to population growth, shifts toward more meat-based diets, climate change, and other challenges. Moreover, the world's water is increasingly becoming degraded in quality, raising the cost of treatment and threatening human and ecosystem health (Palaniappan et al. 2010). Furthermore, the physical availability of freshwater resources does not guarantee that a safe, affordable water supply is available to all. At least 780 million people do not have access to clean drinking water, some 2.5 billion people lack access to safe sanitation systems, and 2–5 million people—mainly children—die as a result of preventable water-related diseases every year (Gleick 2002; UN 2009; WHO and UNICEF 2012).

There is growing recognition that the scope and complexity of water-related challenges extend beyond national and regional boundaries and therefore cannot be adequately addressed solely by national or regional policies. In a recent report, the United Nations notes that "water has long ceased to be solely a local issue" (UN 2012a, 40). In particular, widespread water scarcity and lack of access to water supply and sanitation threaten socioeconomic development and national security for countries around the world. Additionally, people around the world share and exchange water directly and indirectly through natural hydrologic units and systems and through global trade (i.e., "virtual water," discussed below). Furthermore, climate change and the growing presence of multinational companies within the water sector play a role in globalizing water issues (Hoekstra 2006).

Over the past sixty years, a number of efforts have sought to address the many challenges facing the water sector. Early efforts to address these challenges were almost entirely based on developing large-scale physical infrastructure, such as dams and reservoirs, to produce new water supplies. Amid a growing recognition that technology and infrastructure alone were not sufficient to address persistent water management concerns, discourse about water governance began to emerge in the early 1990s. In its first World Water Development Report, the United Nations strongly stated that the "water crisis is essentially a crisis of governance and societies are facing a number of social, economic and political challenges on how to govern water more effectively" (UN 2003b, 370). In this chapter, we describe some of the major global water challenges and identify key deficiencies in global water governance in addressing these challenges. We conclude with several recommendations for improving global water governance in order to better address major water concerns in the twenty-first century.

Global Water Challenges

As described below, the scope and complexity of water-related challenges extend beyond traditional national and regional boundaries. Such challenges require broader thinking and more comprehensive solutions.

Water Scarcity

Water scarcity is a major challenge, affecting every continent around the world. Water scarcity occurs when water demand nears (or exceeds) the available water supply. Several groups, including the World Resources Institute and the International Water Management Institute (IWMI), have developed tools to promote a better understanding of where and how water risks are emerging around the world. The IWMI, for example, estimates that 1.2 billion people—nearly 20 percent of the world's population—live in areas of physical water scarcity, where water withdrawals for agriculture, industry, and domestic purposes exceed 75 percent of river flows. An additional 500 million people live in areas approaching physical scarcity. Another 1.6 billion people live in areas of economic water scarcity, where water is available but human capacity or financial resources limit access. In these areas, adequate infrastructure may not be available or, if water is available, its distribution may be inequitable (IWMI 2007).

But water scarcity isn't solely a natural phenomenon; it's also a human one. Numerous human activities—such as untimely water use, pollution, insufficient or poorly maintained infrastructure, and inadequate management systems—can result in or exacerbate water scarcity. As noted by the United Nations, there are adequate water resources to meet our needs, but water "is distributed unevenly and too much of it is wasted, polluted and unsustainably managed" (UN 2012b).

Widespread declines in groundwater levels are one symptom of water scarcity. Groundwater is an important source of freshwater in many parts of the world. Some areas, however, have become overly dependent on groundwater supplies. In the past two decades, advances in well-drilling techniques have significantly reduced the cost of extracting groundwater. Driven, in part, by these technological advancements, groundwater withdrawals have tripled over the past fifty years (UN 2012a). In some areas, the rate of groundwater extraction now consistently exceeds natural recharge rates, causing widespread depletion and declining groundwater levels. A recent analysis of groundwater extraction by hydrologist Yoshihide Wada and colleagues (2010) finds that depletion rates doubled between 1960 and 2000 and are especially high in parts of China, India, and the United States. Much of the groundwater extracted supports agriculture (67 percent), although it is also used for domestic (22 percent) and industrial (11 percent) purposes.

Water Quality

While most water assessments emphasize water quantity, water quality is also critical for satisfying basic human and environmental needs. The quality of the world's water is under increasing threat as a result of population growth, expanding industrial and agricultural activities, and climate change. Poor water quality threatens human and ecosystem health, increases water treatment costs, and reduces the availability of safe water for drinking and other uses (Palaniappan et al. 2010). It also limits economic productivity and development opportunities. Indeed, the United Nations finds that "water quality is a global concern as risks of degradation translate directly into social and economic impacts" (UN 2012a, 403).

Water quality concerns are widespread, although the true extent of the problem remains unknown. In developing countries, an estimated 90 percent of sewage and 70 percent of industrial waste is discharged into waterways without any treatment at all (UN 2003a). Asian rivers are the most polluted in the world, and bacteria levels from human waste in these rivers are three times higher than the global average. Moreover, lead levels in these rivers are twenty times more than in rivers in industrialized countries (UNESCO 2005).

Drinking Water and Sanitation Access

The failure to provide safe drinking water and adequate sanitation services to all people is perhaps the greatest development failure of the twentieth century. In an attempt to remedy this failure, the United Nations established the Millennium Development Goals (MDGs), eight targets designed to tackle extreme poverty. At the direction of United Nations member countries, UN organizations and multilateral and bilateral development agencies have been working to achieve these goals by the year 2015. While many of the MDGs are widely acknowledged to be associated with water, including those related to improving gender equality and reducing child mortality, Target 7.C specifically aims to reduce by half the proportion of the population without sustainable access to safe drinking water and basic sanitation by 2015. Although not without their critics, the MDGs have served to highlight the importance of water, sanitation, and hygiene in improving health and economic opportunities (UN 2012a).

By UN measures (which are acknowledged to have important limitations), significant progress has been made in improving access to drinking water. In 1990, 76 percent of the global population had access to an "improved drinking water source"—defined as one that, by nature of its construction or through active intervention, is likely to be protected from outside contamination, in particular from contamination with fecal matter—whereas by 2010, this number had grown to 89 percent (WHO and UNICEF 2012). The global population as a whole is on track to meet the MDG drinking water target; however, global aggregates hide large regional disparities. For example, while India and China have made significant progress, sub-Saharan Africa, where only 61 percent of the population has access to an improved water source, is unlikely to achieve the MDG drinking water target. Additionally, coverage in the least developed countries is worse than in other developing countries. Finally, even within countries, there are disparities between urban and rural communities and between the rich and the poor (WHO and UNICEF 2012).

Despite this progress, access to an improved drinking water source remains out of reach for many people. An estimated 780 million people do not have access to basic water service (WHO and UNICEF 2012). Additionally, the MDG drinking water target is based on access to an improved supply of water with little or no consideration of whether the water is affordable, whether the water is safe for consumption, or whether that access is being maintained over time. For example, naturally occurring arsenic pollution in groundwater affects nearly 140 million people in seventy countries on all continents (UN 2009). In Bangladesh alone, nearly 70 million people are exposed to groundwater contaminated with arsenic beyond the recommended limits of the World Health Organization (UN 2009).

Far less progress has been made in achieving the MDG sanitation targets. In 1990, nearly half of the global population had access to improved sanitation. By 2010, the percentage of people with access to improved sanitation had increased to 63 percent. An estimated 2.5 billion people still lack access to improved sanitation (WHO and UNICEF 2012). The global population is not on track to meet the sanitation target, and coverage is especially low in sub-Saharan Africa and in southern Asia.

Water and Ecosystems

Freshwater ecosystems are among the most extensively altered systems on Earth. Rivers, streams, and lakes have been subjected to chemical, physical, and biological alteration as a result of large-scale water diversions, introduction of invasive species, overharvesting, pollution, and climate change (Carpenter et al. 2011). An estimated 20–35 percent of freshwater fish are vulnerable or endangered, mostly because of habitat alteration, although pollution, invasive species, and overharvesting are also to blame (Cosgrove and Rijsberman 2000). About half of the world's wetlands have been lost since 1900, and much of the remaining wetland area is degraded (Zedler and Kercher 2005). Freshwater ecosystem conditions are likely to continue to decline unless action is taken to address acute threats and better manage freshwater resources.

Globalization and Virtual Water Flows

Globalization is characterized by the production and movement of goods and services around the world, and water is a key ingredient, either directly or indirectly, in almost every good produced. Consequently, the movement of goods effectively results in the movement of water around the world. Existing patterns of trade, however, are not necessarily water efficient. Many factors are at play when global trade decisions are made, and water is rarely one of them. The concept of "virtual water"—the water embedded in the production of food and other products—has been introduced as a way to evaluate the role of trade in distributing water resources. Some have argued that by allowing those living in water-scarce regions to meet some of their water needs through the import of water-intensive goods, international trade can provide a mechanism to improve global water-use efficiency (Allan 1993). Others, however, have posited that it simply externalizes the environmental burden of producing a particular product. In any case, the facts suggest that countries' relative water endowments are not dictating global trade patterns. Indeed, three of the world's top ten food exporters are considered water scarce, and three of the top ten food importers are water rich (World Economic Forum Water Initiative 2011). Furthermore, globalization increases dependence on others for essential goods and increases vulnerability to external water scarcity (Hoekstra and Mekonnen 2012).

Climate Change

Rising concentrations of greenhouse gases resulting from human activities are causing large-scale changes to Earth's climate. These climatic changes will have major implications for global water resources. As temperatures rise, the flows of water in the hydrologic cycle will accelerate. In short, climate change will intensify the water cycle, altering water availability, timing, quality, and demand. Indeed, all of the major international and national assessments of climate change have concluded that freshwater systems are among the most vulnerable, presenting risk for all sectors of society (Compagnucci et al. 2001; SEG 2007; Kundzewicz et al. 2007; Bates et al. 2008; USGCRP 2013). A technical report on freshwater resources released in 2008 by the Intergovernmental Panel on Climate Change (IPCC) concludes that "water and its availability and quality will be the main pressures on, and issues for, societies and the environment under climate change" (Bates et al. 2008).

A community's vulnerability to climate change will depend upon the magnitude of the impact and the community's sensitivity and adaptive capacity. As noted by Kenneth D. Frederick of Resources for the Future and Peter H. Gleick of the Pacific Institute (1999), "the socioeconomic impacts of floods, droughts, and climate and nonclimate factors affecting the supply and demand for water will depend in large part on how society adapts." The poor and those living in developing countries are the most vulnerable because they have fewer social, technological, and financial resources to enable them to adapt (UNFCCC 2007).

Water-Energy-Food Nexus

Throughout the twentieth century, the close connections between water, energy, and food were largely unknown or were ignored in policy decision making. Water, energy, and food systems, and the governance institutions set up to manage them, were often separated by well-defined silos, and managers rarely communicated with one another. Water systems were often designed and constructed with the assumption that energy would be cheap and abundant, and energy systems were designed and constructed with the assumption that water would be cheap and abundant. Likewise, food systems have been operated as though neither the cost nor the availability of water and energy would constrain production. We now understand that this is no longer true: these critical resources are closely interconnected, and a growing interest in the water-energy-food nexus highlights the need to better understand and manage these interdependencies:

- Agriculture is a major user of water, accounting for 70 percent of all freshwater withdrawals. Agriculture is also a major user of energy, and food prices are sensitive to energy prices and policies on fertilizers, pesticides, and transportation to distribute products. Meeting the food and fiber demands of a growing population that is simultaneously shifting toward a more water-intensive diet will require a rethinking of how water is used.

- Energy is a major user of water. In the United States, for example, thermoelectric power plants account for nearly 50 percent of all freshwater withdrawals (Kenny et al. 2009). Newly proposed energy sources, such as biofuels, are placing additional strains on local water resources and global food systems.

- Large amounts of energy are required to capture, treat, distribute, and use water. Population growth and climate change are prompting some to consider importing water over longer distances, accessing groundwater from greater depths, or developing more marginal, lower-quality supplies that require extensive treatment.

Failure to consider these linkages in policy and decision making can lead to unintended consequences. Biofuels, for example, have emerged as an alternative to traditional, fossil-fuel-based energy sources, and many governments have instituted mandates and incentives to promote biofuel development. The European Union has committed to converting 10 percent of its transportation fuel to biofuels by 2020 (UN 2012a). In 2009–2010, nearly 40 percent of domestic corn use in the United States was for fuel (USDA 2010). However, first-generation biofuels, which represent the vast majority of biofuels produced today, are water and chemical intensive, and their development increases pollution of and competition for limited water resources. Additionally, biofuels compete with food crops for land and water resources, contributing to increased food prices and threats to food security. The impacts of increasing biofuel production make it clear that national decision making is linked to global agricultural output, food prices, and water availability.

The Emergence of Global Water Governance

The importance of governance as a key factor in addressing water-related challenges began to emerge in the late twentieth century amid a growing recognition that technology and infrastructure alone were not sufficient to address persistent water management concerns. Indeed, in its first World Water Development Report, the United Nations issued a strongly worded statement that the "water crisis is essentially a crisis of governance and societies are facing a number of social, economic and political challenges on how to govern water more effectively" (UN 2003b). Early water governance efforts emphasized the local and regional scales, in part because water challenges were largely perceived as local issues. But the scope and complexity of water challenges, as described above, highlight the need for a more comprehensive and coordinated global effort.

Despite the need, discussions about global water governance have been limited. One of the few definitions of global water governance comes from a 2008 study that defines it as "the development and implementation of norms, principles, rules, incentives, informative tools, and infrastructure to promote a change in the behavior of actors at the global level in the area of water governance" (Pahl-Wostl et al. 2008, 422). Thus, global water governance focuses on the processes of international cooperation and multilateralism. It comprises formal and informal instruments—including global governmental and nongovernmental organizations, regimes, actors, frameworks, and agreements—created to balance interests and meet global water challenges that span national and regional boundaries. It informs the way challenges are tackled (or not) at the regional and international levels by various players (from governmental bodies to civil society organizations) and suggests opportunities for, and barriers to, meeting global objectives. Global water governance also facilitates interaction and dialogue among key players to inform the development of solutions to problems at local, national, and regional levels to ease global pressures.

Evaluating the Effectiveness of Global Water Governance Today

Current global water governance systems were established during a time when approaches to water resource development and management differed from those encountered today (Jury and Vaux 2007). Persistent and emerging water challenges suggest that an assessment is needed to determine how global water governance efforts can be improved to more effectively address twenty-first-century water challenges and to leverage opportunities afforded by new thinking and innovative technologies. We describe below some key deficiencies and recommend ways in which governance can be improved to better address major freshwater concerns.

Intergovernmental Organizations Lack Clear Leadership and Coordination

A large number of organizations exist to address water challenges at various scales—particularly the United Nations system, multilateral lending institutions, and regional basin organizations—all working on different aspects of water management and service delivery. While global summits and forums have helped to identify major challenges and issue areas, implementation of coherent action is hampered by differing agendas among organizations and agencies that overlap in some areas but not in others.

At the international level, leadership and coordinated action within the water sector could emerge from the United Nations' system of agencies and programs. UN-Water was created in 2003 to serve as the interagency coordinating mechanism to promote coherence and coordination of UN system actions and other nontraditional partners and stakeholders (e.g., public and private sectors and civil society) related to the implementation of the international agenda defined by the Millennium Declaration and the World Summit on Sustainable Development. UN-Water, however, has several deficiencies. In particular, it "does not have a strong mandate," nor does it make centralized policies (Pahl-Wostl et al. 2008, 427). UN-Water also has its own areas of focus (water and climate change, water quality, water supply and sanitation, and transboundary water), which fail to address the full range of water-related challenges. Additionally, inadequate personnel and funding hamper UN-Water's goal of promoting collaboration among the various agencies and programs that focus on different water-related issues and challenges, among them the United Nations Educational, Scientific, and Cultural Organization (UNESCO), the United Nations Environment Programme (UNEP), the United Nations Children's Fund (UNICEF), the United Nations Development Programme (UNDP), the Food and Agriculture Organization of the United Nations (FAO), the World Health Organization (WHO), and the World Meteorological Organization (WMO).

The lack of clear leadership manifests itself in several ways. In particular, bilateral funding agencies are more likely to focus their efforts on their own priorities. For example, the German development agency GIZ has spent considerable resources on addressing the food-water-energy nexus, climate change, and access to water and sanitation. The US Agency for International Development (USAID), on the other hand, is focused on biodiversity, food security, climate change, and water access and sanitation. While all of these efforts are aligned with global priorities, lack of coordination can hinder their effectiveness.

Recommendation: Secure a Sustainable Funding Source and a Stronger Mandate for Coordinating Intergovernmental Organizations

The global nature of water-related challenges requires clear leadership and coordination. Intergovernmental agreements produced at world summits and forums require effective intergovernmental organizations to play the leading role in coordinating action. The United Nations system, as the sole global governance organization with the legitimacy and authority of member governments, must lead. UN-Water offers a potential starting point, given its existing mandate to coordinate action. To fulfill its mission, however, it (or any other intergovernmental mechanism established to coordinate action) must be given the resources and an empowered mandate to do so. This requires governments to fulfill pledges made at previous UN summits (such as the 2002 World Summit on Sustainable Development, or Johannesburg Summit) to ensure that financial resources are made available. It also requires political will from the United Nations to provide a stronger mandate for the organization and the ability to overcome traditional interagency rivalry that hampers cooperation.

Recommendation: Promote Greater Collaboration to Build Understanding and Coordinate Action

To effectively address the interlinked nature of the problems, it is imperative that water-related action be led not from within a silo but rather with a deep understanding of the cross-sector issues—for example, taking into consideration development, energy, biodiversity, climate change, food security, and more. Building this understanding requires close, continuous collaboration among the different organizations and individuals involved. UN-Water's 2013 theme of international cooperation is a positive step in that direction. Government-led efforts to encourage participation by actors through multistakeholder processes (such as the parallel meetings at Rio+20, the 2012 United Nations Conference on Sustainable Development) are key to promoting this collaboration. Likewise, the United Nations' current approach to developing the new Sustainable Development Goals is an encouraging development. By instituting a process that brings together development agencies, civil society groups, and the private sector to define water-related goals and potential actions, the UN approach promotes better understanding, which can lead to more coordinated action and better outcomes.

The Role of Nongovernmental Actors Is Expanding

Today, a study of global governance cannot be limited to merely governmental or intergovernmental processes. The rise and influence of a broad range of new actors, with their own sources of authority and power, are indicative of a more complicated global governance structure. These actors, who come from the private, nongovernmental, academic, and media sectors, act independently or, increasingly, in networks to bring about new thinking and solutions. These new global actors have fostered innovation and performed important functions, such as serving as watchdogs of governmental and private sector activities within the water sector. However, concerns have been raised about some who are engaged in public policy, particularly regarding their legitimacy, accountability, and relationship with existing public governance structures. For example, some of these new initiatives may be undermining government-led efforts, operating outside of local priorities, or, in the case of some privately led initiatives, engaging in policy capture (see, e.g., the discussion of corporate actions regarding water in chapter 2 of this volume). Their

centers of authority or the constituencies for whom they speak has also been a subject of debate. Although it is clear that these actors will continue to play an important role in global water governance, efforts should be made to understand what their role should be and their relationship with government-led efforts.

Recommendation: Explore and Develop Guidelines and Principles to Help Govern Nongovernmental Processes

As more parties become involved, effort is needed to better understand and define the roles and responsibilities of each in order to leverage unique capabilities. For entities that are actively engaging in areas that are in the traditional realm of governments, clear guidance as to how these new processes should interact with existing processes is needed. Realizing that these processes can potentially undermine one another, some organizations, such as the United Nations' CEO Water Mandate, have developed guidelines and principles to govern how the private sector engages in water policy (see, e.g., Morrison et al. 2010). More efforts like these are needed to ensure that civil society and private sector efforts and initiatives complement existing government-led processes where possible.

Some potential overarching principles developed for sustainability standards systems (many of which are global action networks, or GANs) that could serve as starting points for further exploration include the following (Ward and Ha 2012):

> *Respect* the unique roles of governments and states.
>
> *Engage* public sector actors.
>
> *Support* sharing of information and resources with public sector actors.
>
> *Build on* existing public sector and international norms.
>
> *Assess and review* the range of public sector implications and relationships.

Water Sector Funding Is Inadequate and Too Narrowly Focused

The international community, including the major economies and international organizations, has played a significant role in funding water sector improvements, especially in developing countries. Yet funding remains limited and too narrowly focused. Funding commitments made by major economies at the 2002 Johannesburg Summit and among the Group of Eight countries have thus far not materialized. Additionally, a recent survey conducted by the World Health Organization (2012) finds that overall funding for the water sector is low—and is skewed toward capital expenditures for drinking water systems in urban areas. Expenditures for sanitation, operation and maintenance costs, and rural systems are much lower.

Recommendation: Develop Financing Mechanisms to Support Ongoing Operation and Maintenance Costs

Funding is needed to support ongoing operation and maintenance costs of water infrastructure. Available funding is insufficient to operate and maintain the existing infrastructure or to support the people and institutions needed to manage it effectively. As a result, systems are poorly managed or fall into disrepair, increasing the long-term costs. Additional funding is needed to support the ongoing operation and maintenance of new and existing water-related infrastructure.

New Funders Often Fail to Abide by Environmental and Social Lending Standards

For much of the twentieth century, the World Bank, the Asian Development Bank, intergovernmental agencies, and bilateral donors were the main funders of large-scale infrastructure in the developing world. In recent years, new economic realities and players have emerged. Commercial banks and energy and construction companies in the global South are playing an increasingly important role and are fundamentally changing water resource management. For instance, Pacific Environment's China program director, Kristen McDonald, and her colleagues reported in 2009 that Chinese financial institutions, state-owned enterprises, and private firms were involved in at least ninety-three major dam projects overseas. These and other new players—predominantly energy and construction companies from Thailand, Vietnam, China, Russia, and Malaysia—had not adopted internationally accepted environmental and social lending standards and norms. Furthermore, these new funders forced the World Bank and the Asian Development Bank to reconfigure their own lending practices to further dilute their environmental and social safeguards (Molle et al. 2009).

Recommendation: Establish New Lending Standards and Compliance Strategies

Commercial banks and energy and construction companies play an increasingly important role in financing water resource development projects. In the case of dam construction, for example, these new players do not meet even the World Bank's standards—which are already weaker than the recommendations of the World Commission on Dams. The failure to abide by social and environmental lending standards poses a threat to local environmental and social systems. New environmental and social lending standards are needed to ensure that lending promotes sustainable development objectives. The new players, along with civil society organizations, should be included in crafting and designing these new standards in order to ensure compliance.

Knowledge and Technology Transfer Efforts Remain Largely Top-Down

Over the past several decades, water-related knowledge and technological innovation have grown tremendously, with new techniques and ideas emerging from governmental bodies, independent research institutions, and academic bodies around the world. The challenge lies in getting this knowledge and technology to places that can implement them. Intergovernmental processes to foster technology and knowledge transfers—mainly through forums such as the annual Water Environment Federation Technical Exhibition and Conference (WEFTEC) and the like—have predominantly been in a top-down manner. There is growing recognition, however, that even innovative technologies that are thought to be highly effective may not be appropriate everywhere. Each technology is developed and crafted according to local circumstances, which can differ dramatically from one region to another. As a result, an off-the-shelf approach to technology and knowledge transfers may not lead to the desired outcome or may lead to unintended consequences. Implementation of Green Revolution concepts to industrialize agriculture in the Punjab region of India provides an example of a top-down, single-focus transfer of knowledge and technology that has led to several unintended consequences, including groundwater overdraft in some areas. Today, the state of Punjab is trying to manage these problems by revisiting and reforming state agricultural policy and regulations using a more bottom-up technology and knowledge transfer approach (Tiwana et al. 2007).

Recommendation: Promote Open-Access Knowledge Transfer

Over the past few decades, there has been tremendous growth in the technologies available for transferring knowledge and information. Geospatial technologies, the Internet, and mobile devices are just a few of the technologies available to improve communication. Although reliance on such technologies must be carefully considered, given the global variations in their application and use, they can provide tremendous opportunity for new ways of getting information to water users and of connecting water stakeholders and researchers with one another and with decision makers. Global institutions can play an important role in facilitating the use and distribution of these new technologies. Extending access to new and emerging scientific findings can enable and empower the local research community to better understand and identify local problems and design or demand specific solutions to improve local water governance (Jury and Vaux 2005; Hutchings et al. 2012). There is also a need for better communication of complicated scientific knowledge to policy makers and decision makers in order to influence development of comprehensive management strategies and inform the policy-making process.

Recommendation: Facilitate Effective Technology Transfer by Engaging Local Communities in the Decision-Making Process

Empowering local communities to identify their water issues and solutions allows them to select an approach that more closely aligns with their social and cultural realities. On-site education and capacity building play a major role in facilitating successful and effective bottom-up or horizontal technology and knowledge transfer. Especially in regions with very limited access to and understanding of state-of-the-art technological solutions, or with limited institutional capacity to provide local technological training, international institutions such as the UNESCO-IHE Institute for Water Education can foster capacity building and educational efforts to facilitate implementation and operational learning of imported technologies. Also, continuous monitoring and performance assessment of a transferred technology can provide an opportunity to adjust and calibrate implementation and operational processes to prevent undesirable outcomes. Global institutions can also facilitate focused research and development investment, especially by those in the developed world with financial resources, to advance technologies and make them more accessible to the developing world.

Recommendation: Improve Understanding and Communication of Risk and Uncertainty

Some uncertainty inherent in hydrologic and water resource management systems is unavoidable. Yet the development of management practices and strategies relies heavily on future supply and demand predictions, which are fraught with uncertainty. Water resource managers around the world use various supply and demand predictions in their decision-making processes. A better understanding of the uncertainties and risks associated with them can lead to the development of more effective planning and management strategies that reflect these limitations. New decision support tools should include an uncertainty assessment component, which would offer an array of decisions and the uncertainties and risks associated with them in order to provide an opportunity for adaptive and flexible management approaches. Effective communication of these uncertainties and risks to policy makers and the general public is also an important element of adaptive and flexible water resource management practice (UN 2012a).

Data Collection Efforts Are Inadequate

Good data and ongoing monitoring activities are the cornerstones of effective water management and governance. We live in an information era, and vast amounts of water data are collected in different ways and at a variety of temporal and spatial scales, from local stream gauges to global satellites. Current attempts at information sharing, such as UN-Water's Activity Information System, Documentation Center, and Key Water Indicator Portal, provide key data necessary to tackle the water challenges identified earlier. Despite these improvements, there are still regions lacking basic water data and information. Even when the data are collected, they are often not widely available or their quality is poor. Efforts are needed to improve the collection, compilation, and reporting of comprehensive water-related data.

Recommendation: Develop a Centralized Global Water Data Portal

The rational management of water is predicated on the availability of comprehensive data. Capacity needs to be developed in all countries to collect, manage, and analyze water information. Some of the key data needed include precipitation, runoff, virtual water flows, groundwater levels, and overall water demand and supply. Where resources are inadequate to collect and compile these data, they should be provided through international aid or other mechanisms. Also, as developing countries undergo economic transitions, monitoring and reporting need to be integrated into new laws. These efforts would benefit from a centralized global water portal in which to assemble the reported data, especially where local governments lack the financial or technological capacity to provide such services. Finally, international data protocols, standard data formats, and sharing arrangements are needed in order to increase comparability of data worldwide.

Recommendation: Leverage New Data Collection Technologies

New local data collection and monitoring efforts are emerging that engage stakeholders through crowdsourcing, or reporting of information through electronic devices. Mobile connectivity is outpacing fixed landline phones and access to computers, especially in many developing countries that lack telephone network infrastructures. New monitoring efforts that use cell phones and other RSS technologies, such as the WASH SMS Project, capitalize on the widespread and rapidly growing use of mobile devices throughout the world to facilitate the flow of information between communities, governmental entities, and service providers (Hutchings et al. 2012). These data can provide timely information on local water systems, including the availability and quality of water. Small-scale, local data collection and reporting efforts such as these should be encouraged.

Lack of Transparency and Accountability Limits the Effectiveness of Water Sector Investments and Fosters Corruption

The water sector lacks transparency and adequate participation from key stakeholders, especially in marginalized communities, and this in turn leads to an accountability deficit and can result in ineffective or inefficient management strategies and investments. A 2008 report by Transparency International and the Water Integrity Network finds that a lack of transparency and participation contributes to rampant corruption across the water sector, including in water management, drinking water and sanitation service provision, irrigation, and hydropower development. The water sector is especially prone to corruption because of the complex system of agencies responsible for its management and delivery; the growing presence of private actors and informal providers that operate in legal gray zones (where the actors are the de facto water service providers allowed to operate by governments but who may not have official license); and the large sums of money required for infrastructure investments. Addressing the issue is especially challenging because of the general focus within the sector on technological solutions rather than governance. The report further finds that the poor and most vulnerable are the most likely victims because they are more exposed to the informal sector (where corruption is more prevalent) and have limited resources and avenues to voice their concerns. This, in turn, exacerbates corruption because those most affected by it are unable to call for greater accountability (Transparency International and Water Integrity Network 2008).

Recommendation: Adopt New Standards, Codes, and Best Practices for Water Resource Development and Management to Promote Greater Transparency and Participation

Water resource development and management are guided by a series of standards, codes, and best practices. These standards, codes, and practices, which include both mandatory and voluntary initiatives, must provide a regulatory framework that brings about greater transparency, promotes participation and oversight to tackle corrupt practices, and develops best-practice guidance where regulatory frameworks are weak or poorly implemented. Both governments and GANs can play a key role in their formulation. For example, Kenya has adopted a human rights–based approach to the water sector that places an emphasis on transparency and participation. Likewise, the United Nations' CEO Water Mandate released its *Corporate Water Disclosure Guidelines: Public Exposure Draft* in an effort to promote greater transparency in the private sector's water use and allow stakeholders to better evaluate this use. These efforts are encouraging; however, more can and should be done.

Recommendation: Promote Capacity Building and Increase Participation in Water Management

To bring about greater participation in water management and better implementation of frameworks that promote transparency, serious effort is needed to build the capacity of governmental officials and civil society groups, especially community-based organizations. Governments and GANs can provide technical know-how and financial resources to ensure that local governmental officials and community-based organizations, two groups with an intimate knowledge of local problems, can be key advocates for change. For example, the Freshwater Action Network focuses much of its effort on providing capacity building to its civil society members in order for them to engage in decision-making processes, call for greater transparency, and hold governmental and private sector actors accountable.

Recommendation: Empower Communities through Long-Term and Short-Term Education and Outreach Efforts

Education and outreach promote greater understanding about a particular issue and can help facilitate change by redefining acceptable behaviors and social norms. Knowledge is power; hence, it can empower communities, especially the poor and most vulnerable, to demand change and accountability. While education and outreach efforts often occur at the local level, global efforts can provide educational tools, platforms, and strategies for planning effective educational programs. For example, the UNESCO-IHE Institute for Water Education, established in the Netherlands in 2003, was developed to educate and train professionals and build the capacity of sector organizations, knowledge centers, and other institutions in developing countries and countries in transition. These efforts are needed at every scale. Household- and community-scale efforts can promote behavioral changes, facilitate grassroots support and demand for better regulations and enforcement, and bring about transparency and accountability. Education and capacity building at larger scales can promote effective interventions at the watershed, national, and international levels to develop better standards, regulation, and enforcement.

There Has Been a Failure to Adopt Broad-Based Agreements on Transboundary Watercourses

Many rivers, lakes, and groundwater aquifers are shared by two or more nations, and most of the planet's available freshwater crosses political borders, ensuring that politics inevitably intrude on water policy. Indeed, international river basins cover about half of Earth's land surface, and about 40 percent of the world's population relies on these shared water sources. Since transboundary watersheds traverse political and jurisdictional lines, heterogeneous and sometimes conflicting national laws and regulatory frameworks make management a major challenge, particularly when no single national government has authority over another. As such, transboundary water management often requires the creation of international guidelines or specific agreements between riparian states.

While the value of transboundary watershed treaties has regularly been demonstrated, there are political and financial constraints that make their adoption difficult in many parts of the world. In 1997, the General Assembly of the United Nations adopted the Convention on the Law of the Non-Navigational Uses of International Watercourses. This UN convention sets forth principles for equitable and reasonable utilization of international watercourses and for equitable participation. More than a decade after its adoption by the vast majority of the General Assembly, however, the convention has not yet obtained enough signatures to enable it to enter into force and effect. As of February 22, 2013, thirty countries had ratified or acceded to the convention; thirty-five signatures are needed for the convention to enter into force.[1]

Recommendation: Bring into Force the United Nations Convention on the Law of the Non-Navigational Uses of International Watercourses

As much as we hope that treaties will be developed in all transboundary watersheds to foster cooperation and collaboration among all riparian states, political and financial constraints make this difficult in many areas of the world. Therefore, adopting an effective international legal framework is a critical step in addressing future challenges. The 1997 United Nations Convention on the Law of the Non-Navigational Uses of International Watercourses represents an important contribution to the strengthening of the rule of law regarding the protection and preservation of international watercourses, and it should be brought into force.

---

1. See International Water Law Project, Status of the Watercourse Convention as of 22 February 2013, http://www.internationalwaterlaw.org/documents/intldocs/watercourse_status.html.

Existing Interbasin Agreements Lack Flexibility

Global climate change will pose a wide range of challenges to freshwater resources, altering water quantity, water quality, and system operations and imposing new governance complications. For countries whose watersheds and river basins lie wholly within their own political boundaries, adapting to increasingly severe climatic variability and changes will be difficult enough. When those water resources cross borders and implicate multiple political entities and actors, sustainable management of shared water resources in a changing climate will be especially difficult and will require active coordination, engagement, and participation of all the actors sharing the basin. In particular, most transboundary water agreements are based on the assumption that future water supply and quality will not change. Moreover, most treaties and international agreements fail to include adequate mechanisms for addressing changing social, economic, or climate conditions (for an early analysis of this problem, see Goldenman 1990 and Gleick 2000).

Recommendation: Improve Flexibility of Existing Interbasin Agreements

No two water treaties are the same. Each is developed under unique circumstances, addresses different concerns, and has a particular set of constraints. Additionally, climate change will affect each basin differently. As a result, each treaty must be evaluated to determine what flexibility mechanisms currently exist and where significant vulnerabilities remain. This process should be started before a problem arises so as to improve the atmosphere for cooperation and negotiation. Additionally, transboundary watershed countries should consider incorporating provisions into existing treaties to allow for greater flexibility in the face of change, including (1) creation of flexible allocation strategies and water quality criteria; (2) agreement on response strategies for extreme events, such as floods and drought; (3) development of clear amendment and review procedures to allow for changing hydrologic, social, and climatic conditions or in response to new scientific knowledge; and (4) establishment of joint management institutions that can, for example, facilitate a climate vulnerability and adaptation assessment (Cooley and Gleick 2011).

Conclusions

Throughout the twentieth century, water governance efforts emphasized the local and regional scales, in part because water challenges were largely perceived as local issues. However, there is growing recognition that the scope and complexity of water-related challenges extend beyond national and regional boundaries and therefore cannot be adequately addressed solely by national or regional policies. Discussions about global water governance, however, have been limited. Water governance studies that have taken a broader perspective have largely focused on transboundary water resources. Global water governance has also been discussed within the context of other, more prominent global governance challenges (notably climate change and energy) and within discussions of global development objectives. However, there has been little to no discussion about global water governance that looks more holistically at global water challenges and the structures and approaches needed to meet these challenges.

In this chapter, we have defined global water governance, identified key deficiencies in global water governance, and offered recommendations for how it can be improved to better address major water concerns in the twenty-first century. We noted that the global dimensions of water governance are difficult and complex issues. Such governance and the institutional structures that accompany it are complicated by local, regional, and national factors. Indeed, there is no single practice or policy that will "solve" the water challenges facing the world today. This chapter, however, provides several paths forward to more efficient and effective water governance in an effort to promote a more robust and sustainable approach to solving water problems in the twenty-first century.

References

Allan, J. A. 1993. Fortunately there are substitutes for water: Otherwise our hydro-political futures would be impossible. In *Priorities for Water Resources Allocation and Management*, 13–26. London: Overseas Development Administration.

Bates, B. C., Z. W. Kundzewicz, S. Wu, and J. P. Palutikof, eds. 2008. Climate Change and Water. IPCC Technical Paper VI of the Intergovernmental Panel on Climate Change. Geneva, Switzerland: IPCC Secretariat.

Carpenter, S. R., E. H. Stanley, and M. J. Vander Zanden. 2011. State of the world's freshwater ecosystems: Physical, chemical, and biological changes. *Annual Review of Environment and Resources* 36:75–99. doi:10.1146/annurev-environ-021810-094524.

Compagnucci, R., L. da Cunha, K. Hanaki, C. Howe, G. Mailu, I. Shiklomanov, E. Stakhiv, and P. Döll. 2001. Hydrology and water resources. In *Climate Change 2001: Impacts, Adaptation, and Vulnerability*, edited by J. J. McCarthy, O. F. Canziani, N. A. Leary, D. J. Dokken, and K. S. White, 191–233. Contribution of Working Group II to the Third Assessment Report of the Intergovernmental Panel on Climate Change. Cambridge, England: Cambridge University Press. http://www.grida.no/publications/other/ipcc_tar/.

Cooley, H., and P. H. Gleick. 2011. Climate-proofing transboundary water agreements. *Hydrological Sciences Journal* 56 (4): 711–718.

Cosgrove, W. J., and F. R. Rijsberman. 2000. *World Water Vision: Making Water Everybody's Business*. World Water Council. London: Earthscan.

Frederick, K. D., and P. H. Gleick. 1999. Water and Global Climate Change: Potential Impacts on U.S. Water Resources. Washington, DC: Pew Center on Global Climate Change. http://www.c2es.org/docUploads/clim_change.pdf.

Gleick, P. H. 2000. How much water is there and whose is it? The world's stocks and flows of water and international river basins. In *The World's Water 2000–2001: The Biennial Report on Freshwater Resources*, 19–38. Washington, DC: Island Press.

———. 2002. Dirty Water: Estimated Deaths from Water-Related Diseases 2000–2020. Pacific Institute Research Report. Oakland, CA: Pacific Institute.

Goldenman, G. 1990. Adapting to climate change: A study of international rivers and their legal arrangements. *Ecology Law Quarterly* 17 (4): 741–802.

Hoekstra, A. Y. 2006. The Global Dimension of Water Governance: Nine Reasons for Global Arrangements in Order to Cope with Local Water Problems. Value of Water Research Report Series No. 20. Delft, Netherlands: UNESCO-IHE Institute for Water Education.

Hoekstra, A. Y., and M. M. Mekonnen. 2012. The water footprint of humanity. *Proceedings of the National Academy of Sciences* 109 (9): 3232–3237. doi:10.1073/pnas.1109936109.

Hutchings, M. T., A. Dev, M. Palaniappan, V. Srinivasan, N. Ramanathan, and J. Taylor. 2012. mWASH: Mobile Phone Applications for the Water, Sanitation, and Hygiene Sector. Oakland, CA: Pacific Institute and Nexleaf Analytics. http://www.pacinst.org/wp-content/uploads/2013/02/full_report36.pdf.

International Water Management Institute (IWMI). 2007. *Water for Food, Water for Life: A Comprehensive Assessment of Water Management in Agriculture*. London: Earthscan; Colombo, Sri Lanka: International Water Management Institute.

Jury, W. A., and H. Vaux. 2005. The role of science in solving the world's emerging water problems. *Proceedings of the National Academy of Sciences* 102 (44): 15715–15720.

———. 2007. The emerging global water crisis: Managing scarcity and conflict between water users. *Advances in Agronomy* 95:1–76.

Kenny, J. F., N. L. Barber, S. S. Hutson, K. S. Linsey, J. K. Lovelace, and M. A. Maupin. 2009. Estimated Use of Water in the United States in 2005. US Geological Survey Circular 1344. Reston, VA: US Geological Survey. http://pubs.usgs.gov/circ/1344/pdf/c1344.pdf.

Kundzewicz, Z. W., L. J. Mata, N. W. Arnell, P. Döll, P. Kabat, B. Jiménez, K. A. Miller, T. Oki, Z. Sen, and I. A. Shiklomanov. 2007. Freshwater resources and their management. In *Climate Change 2007: Impacts, Adaptation, and Vulnerability*, edited by M. L. Parry, O. F. Canziani, J. P. Palutikof, P. J. van der Linden, and C. E. Hanson, 173–210. Contribution of Working Group II to the Fourth Assessment Report of the Intergovernmental Panel on Climate Change. Cambridge, England: Cambridge University Press. http://www.ipcc.ch/pdf/assessment-report/ar4/wg2/ar4-wg2-chapter3.pdf.

McDonald, K., P. Bosshard, and N. Brewer. 2009. Exporting dams: China's hydropower industry goes global. *Journal of Environmental Management* 90:S294–S302.

Molle, F., T. Foran, and M. Käkönen. 2009. *Contested Waterscapes in the Mekong Region: Hydropower, Livelihoods, and Governance*. London: Earthscan.

Morrison, J., P. Schulte, J. Christian-Smith, S. Orr, N. Hepworth, and G. Pegram. 2010. The CEO Water Mandate Guide to Responsible Business Engagement with Water Policy. United Nations Global Compact and Pacific Institute. http://www.unglobalcompact.org/docs/issues_doc/Environment/ceo_water_mandate/Guide_Responsible_Business_Engagement_Water_Policy.pdf.

Pahl-Wostl, C., J. Gupta, and D. Petry. 2008. Governance and the global water system: Towards a theoretical exploration. *Global Governance* 14:419–436.

Palaniappan, M., P. H. Gleick, L. Allen, M. J. Cohen, J. Christian-Smith, and C. Smith. 2010. Clearing the Waters: A Focus on Water Quality Solutions. Report prepared for the United Nations Environment Programme. Oakland, CA: Pacific Institute.

Postel, S. L., G. C. Daily, and P. R. Ehrlich. 1996. Human appropriation of renewable fresh water. *Science* 271 (5250): 785–788. doi:10.1126/science.271.5250.785.

Scientific Expert Group on Climate Change (SEG). 2007. Confronting Climate Change: Avoiding the Unmanageable and Managing the Unavoidable, edited by R. M. Bierbaum, J. P. Holdren, M. C. MacCracken, R. H. Moss, and P. H. Raven. Report prepared for the United Nations Commission on Sustainable Development. Research Triangle Park, NC: Sigma Xi; Washington, DC: United Nations Foundation.

Tiwana, N. S., N. Jerath, S. S. Ladhar, G. Singh, R. Paul, D. K. Dua, and H. K. Parwana. 2007. State of Environment: Punjab—2007. Chandigarh, Punjab: Punjab State Council for Science and Technology. http://www.soeatlas.org/PDF_Map%20Gallery/SoE%20report%20of%20Punjab.pdf.

Transparency International and Water Integrity Network. 2008. *Global Corruption Report 2008: Corruption in the Water Sector*. Cambridge, UK: Cambridge University Press. http://www.transparency.org/whatwedo/pub/global_corruption_report_2008_corruption_in_the_water_sector.

United Nations (UN). 2003a. Water: A Matter of Life and Death. Fact Sheet. International Year of Freshwater 2003. http://www.un.org/events/water/factsheet.pdf.

———. 2003b. *Water for People, Water for Life*. World Water Development Report 1. Paris: UNESCO Publishing; New York: Berghahn Books. http://unesdoc.unesco.org/images/0012/001297/129726e.pdf.

———. 2009. *Water in a Changing World*. World Water Development Report 3. Paris: UNESCO Publishing; London: Earthscan. http://unesdoc.unesco.org/images/0018/001819/181993e.pdf.

———. 2012a. *Managing Water under Uncertainty and Risk*. World Water Development Report 4. Paris: UNESCO Publishing. http://unesdoc.unesco.org/images/0021/002156/215644e.pdf.

———. 2012b. Water Scarcity. http://www.un.org/waterforlifedecade/scarcity.shtml (accessed May 5).

United Nations Educational, Scientific, and Cultural Organization (UNESCO). 2005. Sound of Our Water: Water Problems. http://unesco.uiah.fi/water/material/05_water_problem_html.

United Nations Framework Convention on Climate Change (UNFCCC). 2007. Climate Change: Impacts, Vulnerabilities, and Adaptation in Developing Countries. Bonn, Germany: United Nations Framework Convention on Climate Change. http://unfccc.int/resource/docs/publications/impacts.pdf.

United States Department of Agriculture (USDA). 2010. USDA Agricultural Projections to 2019. Long-Term Projections Report OCE-2010-1. Washington, DC: United States Department of Agriculture, Economic Research Service.

United States Global Change Research Program (USGCRP). 2013. Draft National Climate Assessment Report. Washington, DC: United States Global Change Research Program.

Wada, Y., L. P. H. van Beek, C. M. van Kempen, J. W. T. M. Reckman, S. Vasak, and M. F. P. Bierkens. 2010. Global depletion of groundwater resources. *Geophysical Research Letters* 37 (20): 1–5. doi:10.1029/2010GL044571.

Ward, H., and M. Ha. 2012. Voluntary Social and Environmental Standards and Public Governance: Reviewing the Evidence and Setting Principles for Standards-Setters. London: Foundation for Democracy and Sustainable Development; Oakland, CA: Pacific Institute. http://www.sustainabilitystandards101.org/report/project/public-governance/.

World Economic Forum Water Initiative. 2011. *Water Security: The Water-Food-Energy-Climate Nexus*, edited by Dominic Waughray. Washington, DC: Island Press.

World Health Organization (WHO). 2012. UN-Water Global Analysis and Assessment of Sanitation and Drinking-Water: The Challenge of Extending and Sustaining Services. GLAAS 2012 Report. http://www.un.org/waterforlifedecade/pdf/glaas_report_2012_eng.pdf (accessed May 11, 2012).

World Health Organization (WHO) and United Nations Children's Fund (UNICEF). 2012. Progress on Drinking Water and Sanitation: 2012 Update. New York: United Nations Children's Fund; Geneva, Switzerland: World Health Organization. http://whqlibdoc.who.int/publications/2012/9789280646320_eng_full_text.pdf.

Zedler, J. B., and S. Kercher. 2005. Wetland resources: Status, trends, ecosystem services, and restorability. *Annual Review of Environment and Resources* 30:39–74.

CHAPTER 2

Shared Risks and Interests

The Case for Private Sector Engagement in Water Policy and Management

Peter Schulte, Stuart Orr, and Jason Morrison

Some practitioners have suggested there is a strong business case for companies to advance sustainable water management in their operations, supply chain, and society more broadly. The concept of "shared risk" articulates that the conditions that drive water risks for industry also create risk for other segments of society and that there is a shared interest across sectors in more sustainable water management and robust water governance processes. Others have challenged the legitimacy of this concept, citing inherent conflicts between corporate and other societal uses of water and a history of corporate policies that prioritize profit over the public interest. This chapter explores this debate. In doing so, it strives to demonstrate shared interest among sectors and an opportunity for companies to contribute to sustainable water management, but it also acknowledges that there are indeed conflicting interests that pose real threats to the public interest. Though there is significant potential for a more progressive and enlightened corporate stance in the ways companies view and advance public interest goals, we are surely only at the beginning of this paradigm shift. Ultimately, this chapter aims to foster a nuanced debate of how the private sector can meaningfully support and facilitate strong water governance in ways that simultaneously improve business viability and advance the public interest.

Industry relies on freshwater both directly, to manufacture goods, and indirectly, in the production of supplies that feed industrial processes. Water is used as a solvent; to cool industrial processes, dilute contaminants, irrigate crops, and extract fossil fuels; and as a key ingredient in many products, among other uses. In response to water scarcity and pollution, inadequate management systems, and associated challenges in a growing number of regions around the world, businesses are increasingly making the strategic decision to promote and invest in sustainable water management (CDP 2012; Ceres 2009). Their decisions are based on their growing understanding that water risks are caused not only by a company's own water use and pollution but also by the watershed context in which the company operates.

Many water-intensive companies seek to improve water-use efficiency and ensure adequate wastewater treatment in their operations, while others are going beyond their "factory fence lines" to encourage and facilitate more sustainable water management throughout their supply chain and to engage with the watersheds in which they operate

(Pegram et al. 2009; Newborne and Mason 2012; Morrison et al. 2010). They do so through a variety of means, including but not limited to the following (Morrison et al. 2010):

- Facilitating water-use efficiency and pollution reduction measures of other actors in their watersheds
- Advocating for efficient, equitable, and ecologically sustainable water policies and practices at the local, national, and international scales
- Sharing data and information to improve public water management
- Investing in public water infrastructure expansions or upgrades
- Using internal facilities to meet local water supply and treatment needs
- Using financial and technical resources to support local water institutions in catchment planning and management
- Supplementing infrastructure to ensure local supply to communities and industry

The stated objective of these "beyond the fence line" engagement strategies, often collectively referred to as "policy engagement" (Morrison et al. 2010), is to reduce business risk by supporting a stable business environment and ensuring consistent access to water supplies, thereby strengthening a company's license to operate and building a company's standing among its stakeholders. These efforts also seek to identify and reduce the company's adverse impacts on a region's water-related challenges. Such strategies are grounded in the premise that they advance the public interest and are mutually beneficial to companies, their stakeholders, and other actors in the watershed. This, in turn, stems from the idea that many water-related risks, such as scarcity, pollution, inadequate infrastructure, insufficient management capacity, and climate change, affect a wide range of actors and are shared among companies, governments, civil society, communities, and other actors (Pegram et al. 2009).

While many companies are beginning to engage in water policy as a key element of their water management strategies, some nongovernmental organizations (NGOs) and academics are calling into question whether such strategies truly advance the public interest. Some, such as the Public Services International Research Unit (PSIRU), as detailed in its article "Conflicts, Companies, Human Rights, and Water—A Critical Review of Local Corporate Practices and Global Corporate Initiatives," argue that such strategies in reality perpetuate a history of undue corporate influence on public policy that subverts the public interest in favor of corporate profit (Hall and Lobina 2012). This argument is based largely on the following interrelated notions:

- Companies do not have an economic incentive to promote sustainable conditions beyond their fence lines (apart from public relations gains).
- Companies have an interest in weak water-related governance, which allows them to continue socially and environmentally harmful, yet profitable, practices and as such strive to undermine democratic processes.
- Water risks are not "shared"; many issues that companies consider risks, namely regulation, are actually a boon to communities and ecosystems.

This chapter sets out to explore, and in some instances challenge, these notions. It acknowledges the real threat of corporate policy capture and greenwashing as well as the inherent asymmetries in conceptions of water risk—yet raises the notion that current conditions offer a much greater incentive for companies to align their water-related policies and practices with the public interest than has been the case in the past. This position is nuanced and requires careful consideration of numerous aspects of water policy, among them inclusiveness, transparency, and collaboration. We argue, however, that there is a great opportunity for companies, governments, civil society groups, communities, and others to collaborate to achieve shared water-related objectives and that this is a worthy endeavor if done in a way that advances the public interest and aligns with global sustainable development goals. This potential lies at the heart of the "shared risk proposition" and the prospect of better aligning water challenges as shared objectives with mutually beneficial outcomes, as opposed to each stakeholder "fighting its corner" to the detriment of the others.

It is important to note that the debate on shared risk is separate and distinct from critiques of the privatization of water service provision. The privatization debate focuses on the legitimacy of private sector companies (as opposed to public agencies) being tasked with the delivery of water services (Gleick et al. 2002). The debate being explored here, in contrast, focuses on the role of private companies that use large volumes of water or cause water pollution, or both, and as a result can either exacerbate or potentially solve local water-related challenges.

The Business Case for Investing in Sustainable Water Management

Is there an economic and strategic business case for developing sustainable water management strategies? There is growing evidence supporting such a case, as we discuss here. This evidence includes both direct economic advantages that can accrue to forward-looking companies and social and community advantages that come from proactively identifying and reducing water-related risks.

Questioning the Business Incentive for Action

One critique against a private sector role in advancing sustainable water management suggests that companies do not have an economic incentive to do so and therefore cannot be relied upon to fulfill this role (Hall and Lobina 2012). This argument suggests that company efforts to mitigate adverse impacts caused by their operations or to advance sustainable water management more generally—by, for example, investing in improving the water-use efficiency of other actors or facilitating aquifer recharge schemes—are simply greenwash; that is, they are pursued to create the appearance of a responsible business for public relations gains without a good faith effort to advance the public interest. The PSIRU article cited earlier (Hall and Lobina 2012), referring to a Coca-Cola project aimed at recharging aquifers in India, states:

> Although increasing recharge of aquifers is a genuine way of reducing local water stress, these initiatives are not sustainable ways of delivering it. The companies do not have any direct economic incentive to fund such recharges—the economic return is a public relations gain from being seen to act responsibly. In effect, the incentive for water efficiency is created

entirely by public campaigns against the abstractions, and by general public and political pressure for greater environmental responsibility.

This argument rightly implies that corporate initiatives driven purely by altruistic motives cannot be relied upon as a long-term solution to water sustainability challenges. There is no evidence to suggest that companies will behave in this manner consistently and reliably over the long term. However, this argument also suggests that corporate water sustainability efforts are driven solely by a company's desire to be perceived as responsible despite growing evidence to the contrary (see CDP 2012). It overlooks the wide range of water-related business risks that create strong economic incentives for companies to invest in sustainable water management throughout their supply chains and in the watersheds in which they operate as a means of promoting long-term business viability (Pegram et al. 2009; Hepworth 2012; Ceres 2009; Larson et al. 2012). Moreover, it has also been shown that there can be real economic advantages to improving water-use efficiency or reducing the volume of wastewater produced that must then be treated (Ceres 2009).

Unsustainable Water Conditions and Business Risk

Water-related business risks are driven as much, if not more, by unsustainable watershed conditions beyond a company's fence lines, such as water scarcity or pollution, as they are by the company's water-related performance (Pegram et al. 2009; Larson et al. 2012; Morrison et al. 2010). As a case in point, an ultra-water-efficient factory located in a region of severe water stress or facing other water-related challenges can still face significant water risk. In 2012, 53 percent of Fortune Global 500 companies responding to a CDP questionnaire[1] reported that they had experienced detrimental water-related impacts during the previous five years, while 68 percent identified water as a risk to their business (CDP 2012). This reflects trends and real incidents that are moving the goalposts for the way companies view their internal efforts to drive operational efficiencies—no longer as an endgame of sustainability performance. The external basin conditions and contexts, where water risk ultimately resides, necessitate a broader endgame. This new awareness, and the reality that a business's water-related challenges can be fully addressed only through external engagement beyond the factory fence line, are being captured under the emerging paradigm of "water stewardship" (Hepworth and Orr 2013).

A lack of water in the most basic sense limits the amount of water a company can use and therefore the amount of goods it can produce. For example, in 2011, Gap Inc. cut its annual profit forecast by 22 percent as a result of production limitations driven by water shortages in Texas, India, Pakistan, and Brazil (Larson et al. 2012). Water scarcity also has secondary adverse impacts on business, with significant implications, for example, on the production of energy on which industry relies. In 2001, energy production in São Paulo, Brazil, was highly constrained as a result of both severe drought and governmental energy tariff policies (CLSA 2006). In order to prevent blackouts, the government imposed quotas aimed at reducing energy consumption by 10–35 percent. Many industries

---

1. CDP (formerly called the Carbon Disclosure Project) is an NGO based in the United Kingdom that leverages market forces such as investors, customers, and governments to incentivize companies and cities across the world to measure and disclose information about their environmental and sustainability performance, including that related to water management.

based in Brazil's southeast were plagued by reductions in operational capacity, production delays, or increased production costs (CLSA 2006). Additionally, for many industries, degraded ambient water quality increases the level of treatment, and therefore cost, required to purify water to appropriate levels for industrial production (JPMorgan 2008).

Societal expectations for corporate sustainability, including efficient and responsible water-related policies and practices, are also on the rise. Companies perceived to mismanage scarce water resources are likely to suffer damaged reputations, especially when company operations negatively affect basic human and environmental needs or contravene legal requirements. Effective advocacy campaigns in India have already forced the Coca-Cola Company to close its plant in the village of Plachimada, Kerala, and pay $48 million in damages because of the belief that Coca-Cola's groundwater pumping hindered communities' ability to extract the water needed to maintain their livelihoods (India Resource Center 2010). The loss of such goodwill can be a real economic loss.

Furthermore, consumers increasingly choose products on the basis of the perceived sustainability and social responsibility of companies' practices and policies. In the United Kingdom, expenditure on ethical goods and services has tripled in the past decade (The Co-Operative Bank 2009). Eighty-one percent of Koreans, 70 percent of Singaporeans, and nearly half of British consumers are willing to pay a premium for environmentally friendly products (Czarnowski 2009). Companies perceived to have better water-related practices may thus see improved brand value and a subsequent expansion in sales. The Alliance for Water Stewardship's March 2013 release of its beta standard for water stewardship suggests that stakeholders' ability to assess companies with respect to the adequacy of their water-related activities is arriving quickly (Alliance for Water Stewardship 2013). Furthermore, the financial community is increasingly seeking to invest in companies that are managing short- and long-term water-related risks and that are striving to meet stakeholders' expectations regarding water. Shareholder resolutions on water—mostly focused on the food, beverage, oil, and chemical industries—more than quadrupled over the past ten years (Ceres 2009).

Ensuing water-related development challenges, along with the changing landscape of stakeholder expectations, are creating very real economic incentives for companies to behave in a manner that is deemed responsible. It would be shortsighted to dismiss these potentially paradigm-shifting levers toward corporate sustainability because they diverge from previous corporate actions. In the past, the pressures and the companies' responses to water differed from those being pursued in today's highly branded, globalized, and increasingly water-stressed world (Hepworth and Orr 2013). Instead, we should ask how these new levers can drive changes that benefit businesses, communities, and ecosystems alike and how we can avoid one-sided outcomes and greenwashing that undermine such opportunities.

Do Businesses and Other Watershed Actors Have a Shared Interest in Robust Governance and Sustainable Water Management?

Businesses, communities, ecosystems, governments, and others may face common water-related challenges, but they may not share an interest in the same solutions to such challenges. Indeed, this is another core argument of those skeptical of a meaningful private sector role in driving more sustainable water management. Considering the notion of regulatory water-related business risks, for instance, the PSIRU (Hall and Lobina 2012, 15) contends that

> for communities and ecosystems, regulation is not a risk but a positive opportunity for democratic and peaceful limitation of competing (including corporate) behavior. By contrast, the [work of the United Nations' CEO Water Mandate] has the anti-democratic implication that companies would be subject to less risks if there was no democratic government and no civil society.

This is, at least in part, a valid argument. Companies can be negatively affected by strict regulations that limit their activities, and thus they often consider them a business risk (as would, for that matter, a local smallholder farmer facing limited water allocations). The 2012 CDP Global Water Report states that 51 percent of companies consider "tightening withdrawal limits" a risk, while 30 percent consider "restricted water operational permits" a risk (CDP 2012). However, water-intensive companies are also negatively affected by a lack of regulation on other actors in their watershed that eventually leads to the depletion or degradation of a shared resource. This creates a not uncommon scenario in which a wide range of actors using a common pool resource are threatened both by the possibility of their own resource use being regulated and by the unregulated resource use of other actors.

Regulation that is coherent and consistently applied is positive for society and reduces risk to business. We live in a world where ecosystems and communities do face regulatory risks when government fails to regulate polluters, allocations, permits, fines, fees, and use, or to oversee and manage the socioeconomic trade-offs inherent in particularly stressed basins. Water-using companies are similarly negatively affected by such poor public management and oversight. The same CDP report found that 54 percent of companies responding consider "regulatory uncertainty" a risk (CDP 2012). Nonetheless, for detractors to suggest that risks would be lower for companies if government and civil society were silent or absent truly misses the point. We recognize that some companies have a long history of polluting waterways, have failed to meet minimum legal requirements, or have operated where the rules of the game are poorly enforced—and that companies in turn have benefited from these situations. Such companies should be exposed and duly penalized. Yet in even those parts of the world where the rules are most lax or where the issue of water pollution, for example, is most acute, things are changing. Companies increasingly see higher risk where water challenges are great and effective water governance is not in place. Water is the one resource for which some companies actually seek clear and consistent regulatory signals and policies and in which coherence allows them to plan, protect, and avoid negative impacts (Pegram et al. 2009). As expressed in the 2012 CDP report, companies may indeed complain when they do not have enough water to operate their facilities, yet the alternative is to be exposed to wide criticism and fear of regulatory response, which can be draconian and reactive.

In Beilun District in the city of Ningbo, Zhejiang Province, China, the local government is trying to balance the need for both industrial development and ecological protection. Local authorities have literally shut down more than seventy-two factories, large to small, that were failing to adhere to local water regulations (*Shanghai Daily* 2013). A similar picture is emerging all over the developing world, where, because of the scale of the issue and lack of capacity, governments are closing down facilities that breach pollution controls. On the one hand, this solves an immediate problem, but it also creates the space for more engaged dialogue with companies seeking to adhere to rules and comply with local standards. This is especially true as corporate awareness of water-related risks is strengthened by information from emerging risks tools (see WRI 2013 and WWF 2013). It also appears that a company's ability to ensure a license to operate—social and legal—is increasingly predicated on its playing by formal rules and, where these are not in place, on adhering to the best practices defined internationally.

A Coca-Cola plant in the Indian village of Kaladera, Rajasthan, is located in a watershed where groundwater levels dropped from 9 to 38 meters below ground level between 1986 and 2006, reportedly leading to increased costs for irrigation for local farmers, reduced availability of drinking water, and reduced milk yield (Karnani 2012). The Coca-Cola plant, established in 1999, undoubtedly contributed to this phenomenon during its tenure in the watershed. However, between 1984 and 1996 the rate of decline already averaged 0.5 meter per year, and in 1998 the Central Ground Water Board had already deemed the area "overexploited," before the Coca-Cola plant was operational (Karnani 2012). Thus, groundwater depletion in Kaladera has coincided not only with Coca-Cola's presence but also with rising water demand due to population growth, urbanization, and more water-intensive lifestyles (Karnani 2012). The plant employs up to 250 people and accounts for less than 2.7 percent of local water extraction for 70 percent of the time and less than 0.9 percent for 40 percent of the time (Karnani 2012). In May 2004, at its highest levels, the plant accounted for 8 percent of local water extraction (Karnani 2012). As a well-resourced and well-known actor that clearly contributes to groundwater depletion, Coca-Cola is deemed largely responsible for this problem by local stakeholders, who are calling for closure of the plant.

Arguably, the core problem to be resolved in Kaladera, and in many other places around the world, is not just the exploitation of water resources by one industrial water user but widespread overexploitation enabled by the lack of a functioning management regime to regulate the utilization of a limited shared resource. A robust water governance system (with direct regulatory oversight) would incentivize widespread water-use efficiency while also denying permits to those unable to demonstrate responsible use. Coca-Cola has the financial and technical resources to facilitate and support such water governance. By contrast, shutting down the Coca-Cola plant may or may not slow the rate of groundwater depletion. The company has also been rolling out implementation of its "water neutrality" policy, which includes investing in watershed projects (e.g., rainwater harvesting) geared toward offsetting the amount of water use in the company's direct operation. It can be argued that closing the plant by itself would do little to solve the region's systemic long-term groundwater overdraft problem as long as governmental authorities do not have the resources, capacity, and will to effectively manage the watershed.

Do Companies, Communities, Ecosystems, and Others Have a Shared Risk Driven by Water Challenges?

The belief that companies have an incentive to invest in sustainable water management beyond their fence lines is grounded in the concept of shared risk. Though water risks are often distributed unevenly, the idea that companies share a need with the public for reliable water services and sustainable water management is getting traction (Hepworth and Orr 2013; Newborne and Mason 2012). In Colombia, for example, the indigenous vegetation of mountainous upland ecosystems is being cleared to make room for agriculture and cattle grazing. Without this vegetation, which traps water and protects soil, there is greater sedimentation in local waterways, on which the downstream city of Bogotá relies for its water supply. In recent years, this sedimentation has severely degraded water quality in the area, raising treatment costs for the local water utility, residential users, and businesses in the area, such as SABMiller's Bavaria. In order to address this shared risk, Bavaria has teamed up with The Nature Conservancy, other local NGOs, the Bogotá water company, and governmental agencies to remove sedimentation (Water Futures Partnership 2013).

Today in Kenya, integrated approaches to water, energy, and land are being undertaken by a range of actors living and working in the region, including small landholders, major horticultural businesses, the tourism sector, and local, national, and international authorities. Lake Naivasha, Kenya, has experienced declines in water supply and quality attributed to a variety of causes. The multiple water challenges have created risks for all water users who rely on healthy ecosystems for their livelihoods and future. Even though the export farmers have improved agricultural water-use efficiency to reach record levels, the cumulative effect of multiple water users had driven the lake to record low inflows. The consumer markets and retail companies purchasing Lake Naivasha's produce faced high reputational risks because of their poor understanding of the water demands of their suppliers and the perceptions of their customers (ERD 2012). The concept of shared risk was used to bring users together to work closely on joint action plans, funding local water-user associations, promoting and setting up reallocation plans, and creating incentives for government to better protect and manage the lake. These outcomes, supported and largely driven by the private sector, have been integral in advancing more sustainable management of the region. The underlying shared need for effective governance of the lake has united the stakeholders (WWF 2012). The Lake Naivasha example illustrates that while the criticism of shared risk has focused mainly on multinational companies with visible brands, the private sector in Kenya was largely represented by small and medium enterprises (SMEs). These companies, Kenyan and international, saw a clear business case based on economic incentives and opportunities to engage proactively with other water users. The focus on stewardship and shared risk should not negate the fact that most implementation of projects on the ground will be negotiated through SMEs—either through supply chain pressures around buyer protocol for stewardship or because these companies are embedded in their local communities and understand the water situation and business case from multiple perspectives.

Shared risk does not imply that water challenges create an equal and similar burden or sense of urgency for all stakeholders. There are specific challenges in facilitating this shift, resulting from the generally differing languages and expectations of various stakeholders regarding needs, time frames, and modes of communication. It is necessary to ask: *Risk of what and risk to whom?* The risk to an individual differs from societal or business risks, and certain groups will be more vulnerable than others. Water scarcity and pollution can be subjective in this sense. For a farmer, the danger may be back-to-back years of below-average rainfall. For the owner of a processing plant, the risk might be a sudden, temporary cessation of stream flow during peak operation time. For a government, risks might include the increasing costs of accessing water for utilities and the implications of higher energy costs, or failure to deliver on economic growth and development pathways because of poor water management (Orr and Cartwright 2010).

Rather, the concept of shared risk elevates local water challenges as a shared problem and in doing so suggests that we address these challenges in proactive and collaborative ways. Consider the alternative to this, which is the proposition that stakeholders facing similar challenges need not (or should not) speak or engage with one another over shared resources and should be ruled only by governmental policy, regulation, and fines, a premise that leaves those operating in geographies absent of good water governance to instead fight over resources. Critiques against shared risk, knowingly or not, indirectly (or directly) appear to advocate this position, yet collaboration and inclusion of all water users, including corporate and industrial actors, lies at the heart of sustainable water resource management and particularly integrated water resource management (IWRM)—the chief paradigm of reconciling water management challenges among shared users. Collaboration and integrated management also lie at the heart of the "soft path for water" approach (Gleick 2002; Wolff and Gleick 2002). IWRM is a process that involves identifying the multiple adverse impacts and trade-offs associated with current or proposed sectoral policies so that they can be subjected to a political process of informed decision making (Rama Mohan Rao et al. 2003). IWRM evolved as a response to the perception that this identification of trade-offs had historically been bypassed—that water management had been "unintegrated," with various governmental ministries managing and using water independently and the voices of multiple users not being heard.

How Do Other Water Users Benefit from Private Sector Involvement?

Regardless of whether companies benefit from robust water governance, for the "shared risk" concept to function, water managers and those advocating for the public interest must themselves benefit from companies operating in a stressed watershed. Companies can provide vital employment and tax revenues for local communities, contribute to gross domestic product and foreign exchange, and help governments deliver on poverty, growth, and development challenges. The example of Lake Naivasha showed that 10 percent of foreign exchange to Kenya could be linked to the output of the lake's produce and horticulture trade (WWF 2012). Government for the first time saw how water-related risk not only could drive businesses to relocate elsewhere but also could feed the perception that Kenya was an unstable place to do business, which could undermine much-needed foreign investment over the long term.

Companies can begin to contribute to sustainable water management in simple yet beneficial ways. For example, they can provide sorely needed financial resources to augment water supply, especially in places where water governance is weak and capital is limited. Sasol, a global integrated energy and chemical company with its main production facilities in South Africa, has recognized water security as a material challenge to its operations, some of which are highly reliant on the Vaal River system. Sasol uses about 4 percent of the catchment yield; municipalities use approximately 30 percent, of which water losses can be as high as 45 percent of urban use because of the aging infrastructure (Greenwood et al. 2012). Sasol has approached a number of municipalities to implement water conservation initiatives (Greenwood et al. 2012). One such project used Sasol funds to adjust the pressure management system within a township, reducing water leakage in off-peak hours and boosting water supply. This project saves 28 megaliters per day at a modest capital cost of $500,000, funded by Sasol with an operation and maintenance (O&M) cost of US$0.02 per cubic meter (Sasol 2010). By comparison, a project to improve internal water-use efficiency at a Sasol plant, which was also being undertaken at the time, required $50 million in capital expenditures and saved only 18 megaliters a day, with an O&M cost of US$2.00 per cubic meter (Sasol 2010). Sasol's implementation of this "beyond the fence line" water strategy has created systemwide water savings and supported local water security to the benefit of all while reducing water management costs for both local government and the business itself. This is a salient example of a win-win scenario, where the company was able to invest in the most cost-effective savings and where private and public interests were fully aligned.

Financial resources are just one of the tools industry can bring to the table. Companies can also help introduce new technology to improve water-use efficiency and water quality, collection of water-related data that supports informed water management decisions, and privileged access to national decision-making processes that NGOs and small communities may lack. For example, Intel teamed up with the City of Chandler, Arizona, to devise a collaborative approach to water management that includes building an advanced reverse osmosis facility to treat rinse water from Intel's manufacturing facility to drinking water standards before being returned to the municipal groundwater source and nearby farmlands (Morrison et al. 2010).

The lack of financial, technological, and informational resources in many watersheds underscores the reality that current governmental expenditures and investments in sustainable water resource management are low in many places across the globe (Ginneken et al. 2011; WaterAid 2011). In many cases, this may be due to corruption or a lack of awareness of the importance of sustainable water management. In other cases, it is because many governments face many competing demands for resources and policy attention and may lack the resources needed to manage water in a sustainable and equitable manner (WaterAid 2011). The private sector can play a critical role in bridging this resource and capacity deficit, especially in the context of partnerships and collaborative projects that harness the expertise, knowledge, and legitimacy of NGOs, academia, and governmental agencies. Playing this role, however, also requires that serious and careful attention be paid to protecting the public interest, addressing equity concerns, and preventing "policy capture."

Utilizing Corporate Resources While Ensuring Public Interest Outcomes and Preventing Policy Capture

Critical to this debate is the question of how to use corporate resources toward mutually beneficial outcomes while ensuring that companies act in a manner that aligns with the public interest and that avoids undue corporate influence. Though we see emerging levers that incentivize companies to act in a responsible manner, there are also important and powerful conflicting interests and competing incentives that may lead to perverse and unbalanced outcomes (Morrison et al. 2010; Hepworth et al. 2010). Below is a list of some of the most salient tensions and barriers to beneficial private sector involvement in water policy and management that must be navigated.

> *Many companies will not actively promote stringent regulatory frameworks that increase operational costs, limit production, or significantly undermine*

company influence in water decision making. There is a tension between a company's desire for a governance system that effectively manages others' water use (and therefore prevents challenges related to water scarcity and pollution) and its desire to prevent limitations to its own water use or stringent guidelines on water quality that drive up its own operational costs. These conflicting interests may not disincentivize company investment in or promotion of water-use efficiency among other water users in a particular basin, but they do call into question a company's incentive to meaningfully facilitate development of a governance scheme that could potentially limit its own production or increase operational costs. Similarly, while many companies may seek to implement projects that have public interest outcomes (e.g., addressing municipal nonrevenue water losses or riparian restoration or source water protection) and build their reputation among local stakeholders in doing so, only the rare company will choose to promote water governance processes such that their own influence on water decision making is significantly lessened.

Companies are unlikely to be the leading advocates for the prioritization of domestic water uses (i.e., human right to water and sanitation) over their own water use. In light of the recently affirmed human right to water and sanitation (United Nations 2010), water governance systems and legal frameworks in the twenty-first century are increasingly likely to include provisions that prioritize domestic and other water uses above those for industry. For example, the state water policy in Rajasthan (where Kaladera is located) stipulates that water allocations should be prioritized to drinking water, irrigation, power generation, and industry, in that order (Department of Water Resources 2000). While water uses to meet basic human needs are a clear societal priority (and in fact typically represent only a small proportion of water use in a basin), companies may be not be inclined to proactively advocate for water management regimes that deprioritize their own water use, and thus champions of such causes will typically need to come from other segments of society, including explicit governmental commitment.

There exists the potential for greenwashing of company claims and for industry initiatives that create the perception of responsible practice without tangible water sustainability and public interest outcomes. As demonstrated in this chapter, there are economic drivers for companies to invest in and promote sustainable water management inside and outside of their fence lines such that watershed challenges and associated business risks are better managed. However, history also shows that some of these drivers, especially those regarding license to operate and brand value, can hinge on the *perception* of action rather than genuine action itself. This leaves open the potential for greenwashing initiatives that foster a perception of good practice without any tangible benefits to the public interest. As the issue of water is tackled by a growing number of companies, the likelihood increases that claims regarding stewardship and industry-led initiatives, which perhaps serve the corporate social responsibility agenda far more than a strategic one, will proliferate. The onus will be on many of the practitioners and NGOs that support the corporate water stewardship paradigm, as well as academics and peer businesses, to differentiate between spurious claims and substantive efforts. While many critics will use these examples as straw men to criticize all corporate action on water, the case for validating claims and auditing performance should create stronger accountability and monitoring toward genuine shared outcomes.

Some investors and corporate executives may be prone to react to short-term pressures that do not enhance sustainable water management rather than to tackle long-term sustainability drivers. Much of the rationale regarding shared risk and collective action implicitly hinges on companies' assessing and managing risk on a long-term time horizon. For instance, companies considering long-term business viability will be much more likely to invest in water-use efficiency measures and sustainable water supply than will companies seeking to maximize profit in the near term. Likewise, many investors who have a short-term perspective and continue to orient around quarterly earnings statements may find it difficult to fathom the justification for long-term engagement in water governance in a particular region that will garner positive results only over a five- to ten-year time horizon. Even within a company, corporate social responsibility and sustainability departments may seek approval for programs and corollary expenditures based on long-term considerations and planning only to be confronted by unsupportive senior executives and legal departments who require shorter-term, more guaranteed returns on such company investments. This creates another tension. Though water-intensive companies certainly have incentives to support sustainable water management over the long term, such investments may be derailed or undermined by those seeking more immediate returns.

Some business models and operational contexts may not offer strong drivers for action. A significant number of companies have strong incentives to ensure sustainable access to water in a single location over the long term. These businesses, such as geographically bound mining or oil and gas companies, typically must operate in particular locations to be economically viable. Furthermore, SMEs may not have the economic means to simply move away from water-stressed settings. Others, however, have greater flexibility in the location of their facilities and supply chain and may not intend to remain in one watershed for an extended period of time. Such enterprises might be inclined to shift operations and supply chains away from water-challenged areas, thus leading to less dependence on those specific regions and perhaps even improving their perception among some key stakeholders when doing so. This "move away from the problem" alternative undermines incentives for companies to make substantial long-term investments in improving water governance in water-stressed areas.

Moving Forward: Unlocking Mutually Beneficial Corporate Action on Water

The existence of competing economic incentives for long-term sustainable thinking and short-term profit motives reveals a number of complex and difficult questions, such as the following:

- Multinational companies in many cases are a great boon to local economies, in both developed and emerging economic contexts. What is the appropriate balance between "high economic value" water uses and public interest water uses?

- How can companies already driving action on water be encouraged to offer a louder voice within their value chains and the wider business community?

- Should industry avoid water-stressed regions or instead focus its investment in these areas?

- In many parts of the world, businesses and governments alike are complicit in a system that often results in unbalanced corporate representation in water policy. What role can and should companies play in facilitating more democratic processes (especially in areas of weak governance)?

- What can companies that are genuinely interested in facilitating sustainable water management do to ease the skepticism of potential partners?

- How can companies be encouraged to consider more long-term profit drivers in the face of more immediate short-term profit motives and shareholder pressure that may incentivize unsustainable practices?

- How can companies and others encourage increased investment in areas experiencing water stress and weak governance when these areas are identified as high-risk locations?

These are difficult questions, and the conflicts inherent within them will certainly invite some to argue that companies should not play a meaningful role in public water management. It must be recognized, however, that companies have always engaged in water policy, and given the very significant challenges facing global water security, dogmatic stances for or against corporate engagement not only fail to reflect reality but also are likely to be counterproductive (Hepworth and Orr 2013). Instead, these tensions and questions should underscore the need to strengthen positive incentives for responsible and sustainable private sector action and to begin a meaningful discussion on what that can and should look like. Such a discussion can shed light on unsustainable practices where they exist, highlight examples of beneficial corporate action, and ultimately help determine the most appropriate role for the private sector in addressing our shared twenty-first-century water challenges. There is compelling evidence to suggest that businesses' role should not be limited to simply implementing operational efficiencies and being accountable for direct water-related adverse impacts on ecosystems and communities; it should be expanded such that companies are encouraged, if not expected, to make a positive contribution to broader watershed challenges that affect a wide range of stakeholders and actors.

References

Alliance for Water Stewardship. 2013. The AWS International Water Stewardship Standard: Beta Version for Stakeholder Input and Field Testing. http://www.allianceforwaterstewardship.org/Beta%20AWS%20Standard%2004_03_2013.pdf?utm_content=nicole.tanner@wwfus.org&utm_source=VerticalResponse&utm_medium=Email&utm_term=Beta%20Standard%26nbsp%3Bhere&utm_campaign=Update%20from%20the%20Alliance%20for%20Water%20Stewardshipcontent (accessed March 26, 2013).

CDP. 2012. Collective Responses to Rising Water Challenges. CDP Global Water Report 2012. Authored by Deloitte. https://www.cdproject.net/CDPResults/CDP-Water-Disclosure-Global-Report-2012.pdf (accessed March 26, 2013).

Ceres. 2009. Water Scarcity and Climate Change: Growing Risks for Businesses and Investors. Authored by Pacific Institute. http://www.pacinst.org/reports/business_water_climate/full_report.pdf (accessed October 14, 2010).

CLSA. 2006. *Remaining Drops: Freshwater Resources, A Global Issue.* Hong Kong: CLSA Blue Books.

Co-Operative Bank. 2009. Ten Years of Ethical Consumerism: 1999–2008. http://www.goodwithmoney.co.uk/assets/Ethical-Consumerism-Report-2009.pdf (accessed October 14, 2010).

Czarnowski, A. 2009. Ethical purchasing: Ethics still strong in a cold climate. *Brand Strategy*, 52–53. http://www.allbusiness.com/trends-events/investigations/11914169-1.html (accessed October 14, 2010).

Department of Water Resources. 2000. Water Allocation Priorities. Government of Rajasthan, India. http://waterresources.rajasthan.gov.in/5allocation.htm (accessed March 26, 2013).

European Report on Development (ERD). 2012. Confronting Scarcity: Managing Water, Energy, and Land for Inclusive and Sustainable Growth. Brussels: European Report on Development. http://erd-report.eu/erd/report_2011/documents/erd_report%202011_en_lowdef.pdf (accessed March 27, 2012).

Ginneken, M., U. Netterstrom, and A. Bennett. 2011. More, Better, or Different Spending? Trends in Public Expenditure on Water and Sanitation in Sub-Saharan Africa. Water Papers 67321-AFR. Washington, DC: World Bank. http://water.worldbank.org/sites/water.worldbank.org/files/publication/Water-Report-Dec-11.pdf (accessed March 26, 2013).

Gleick, P. H. 2002. Soft water paths. *Nature* 418:373. http://www.pacinst.org/topics/water_and_sustainability/soft_path/nature_07252002.pdf (accessed April 8, 2013).

Gleick, P. H., G. Wolff, E. L. Chalecki, and R. Reyes. 2002. The New Economy of Water: The Risks and Benefits of Globalization and Privatization of Fresh Water. Pacific Institute. http://www.pacinst.org/reports/new_economy_of_water/new_economy_of_water.pdf (accessed April 8, 2013).

Greenwood, R., R. Willis, M. Hoenig, G. Pegram, H. Baleta, J. Morrison, P. Schulte, and R. Farrington. 2012. The CEO Water Mandate Guide to Water-Related Collective Action. Authored by Ross Strategic. http://ceowatermandate.org/files/guide_to_collective_action.pdf (accessed March 26, 2013).

Hall, D., and E. Lobina. 2012. Conflicts, Companies, Human Rights, and Water—A Critical Review of Local Corporate Practices and Global Corporate Initiatives. London: University of Greenwich, Public Services International Research Unit. http://www.psiru.org/reports/conflicts-companies-human-rights-and-water-critical-review-local-corporate-practices-and-glo (accessed March 26, 2013).

Hepworth, N. 2012. Open for business or opening Pandora's box? A constructive critique of corporate engagement in water policy: An introduction. *Water Alternatives* 5 (3): 543–562. http://waterwitness.org/wwwp/wp-content/uploads/2012/03/An-introduction-to-corporate-engagement-on-water-policy.pdf (accessed August 5, 2013).

Hepworth, N., and S. Orr. 2013. Corporate water stewardship: New paradigms in private sector water engagement. In *Water Security: Principles, Perspectives, and Practices*, edited by B. A. Lankford, K. Bakker, M. Zeitoun, and D. Conway. London: Earthscan.

Hepworth, N., J. C. Postigo, and B. Güemes Delgado. 2010. Drop by Drop: Understanding the Impacts of the UK's Water Footprint through a Case Study of Peruvian Asparagus. London: Progressio, CEPES, and Water Witness International. http://www.progressio.org.uk/sites/default/files/Drop-by-drop_Progressio_Sept-2010.pdf

India Resource Center. 2010. Coca-Cola Liable for US$ 48 Million for Damages—Government Committee. http://www.indiaresource.org/news/2010/1003.html (accessed March 26, 2013).

JPMorgan. 2008. Watching Water: A Guide to Evaluating Corporate Risks in a Thirsty World. Global Equity Research. http://pdf.wri.org/jpmorgan_watching_water.pdf (accessed October 15, 2010).

Karnani, A. G. 2012. Corporate Social Responsibility Does Not Avert the Tragedy of the Commons—Case Study: Coca-Cola India. University of Michigan Ross School of Business Paper No. 1173. http://papers.ssrn.com/sol3/papers.cfm?abstract_id=2030268 (accessed March 28, 2013).

Larson, W. M., P. L. Freedman, V. Passinsky, E. Grubb, and P. Adriaens. 2012. Mitigating corporate water risk: Financial market tools and supply management. *Water Alternatives* 5 (3): 582–603. http://www.water-alternatives.org/index.php?option=com_content&task=view&id=224&Itemid=1 (accessed March 26, 2013).

Morrison, J., P. Schulte, J. Christian-Smith, S. Orr, N. Hepworth, and G. Pegram. 2010. The CEO Water Mandate Guide to Responsible Business Engagement with Water Policy. United Nations Global Compact and Pacific Institute. http://www.unglobalcompact.org/docs/issues_doc/Environment/ceo_water_mandate/Guide_Responsible_Business_Engagement_Water_Policy.pdf (accessed February 16, 2011).

Newborne, P., and N. Mason. 2012. The private sector's contribution to water management: Re-examining corporate purposes and company roles. *Water Alternatives* 5 (3): 604–619. http://www.water-alternatives.org/index.php?option=com_content&task=view&id=224&Itemid=1 (accessed March 26, 2013).

Orr, S., and A. Cartwright. 2010. Water scarcity risks: Experience of the private sector. In *Re-thinking Water and Food Security: Fourth Botín Foundation Water Workshop*, edited by L. Martinez-Cortina, A. Garrido, and E. Lopez-Gunn. London: CRC Press.

Pegram, G., S. Orr, and C. Williams. 2009. Investigating Shared Risk in Water: Corporate Engagement with the Public Policy Process. Report prepared by Pegasys Consulting for World Wide Fund for Nature. http://assets.wwf.org.uk/downloads/investigating_shared_risk.pdf (accessed March 26, 2013).

Rama Mohan Rao, M. S., C. H. Batchelor, A. J. James, R. Nagaraja, J. Seeley, and J. A. Butterworth, eds. 2003. Andhra Pradesh Rural Livelihoods Programme Water Audit Report. Rajendranagar, Hyderabad 500030, India. http://www.nri.org/projects/wss-iwrm/Reports/APRLPwra/APRLPwra_fullA4.pdf (accessed March 27, 2013).

Sasol. 2010. Managing the Challenge of Water Scarcity. Sasol Sustainability Report 2009. http://www.sasolsdr.investoreports.com/sasol_sdr_2009/material-challenges/managing-the-challenge-of-water-scarcity/ (accessed October 15, 2010).

Shanghai Daily. 2013. After 7-year battle, a village enjoys its fruits. March 27. http://www.shanghaidaily.com/nsp/National/2013/03/27/After%2B7year%2Bbattle%2Ba%2Bvillage%2Benjoys%2Bits%2Bfruits/ (accessed March 27, 2013).

United Nations. 2010. General Assembly, Human Rights Council Texts Declaring Water, Sanitation Human Right. UN GA/SHC/3987. http://www.un.org/News/Press/docs/2010/gashc3987.doc.htm (accessed April 8, 2013).

WaterAid. 2011. Off-Track, Off-Target: Why Investment in Water, Sanitation, and Hygiene Is Not Reaching Those Who Need It Most. http://www.wateraid.org/~/media/Publications/water-sanitation-hygiene-investment.pdf (accessed March 26, 2013).

Water Futures Partnership. 2013. Colombia. http://www.water-futures.org/countries/colombia.html (accessed March 26, 2013).

Wolff, G., and P. H. Gleick. 2002. The soft path for water. In *The World's Water 2002–2003: The Biennial Report on Freshwater Resources*, edited by P. H. Gleick, 1–32. Washington, DC: Island Press. http://www.pacinst.org/publications/worlds_water/worlds_water_2002_chapter1.pdf (accessed April 8, 2013).

World Resources Institute (WRI). 2013. Aqueduct: Measuring, Mapping, and Understanding Water Risks around the Globe. http://aqueduct.wri.org/ (accessed March 28, 2013).

World Wide Fund for Nature (WWF). 2012. Shared Risk and Opportunity in Water Resources: Seeking a Sustainable Future for Lake Naivasha. Report prepared by Pegasys—Strategy and Development. http://awsassets.panda.org/downloads/navaisha_final_08_12_lr.pdf (accessed March 26, 2013).

———. 2013. The Water Risk Filter. http://waterriskfilter.panda.org/ (accessed March 28, 2013).

CHAPTER 3

Sustainable Water Jobs

Eli Moore, Heather Cooley, Juliet Christian-Smith, and Kristina Donnelly

Many jobs, in different sectors of the economy, are required to manage, deliver, and treat the freshwater required to satisfy our commercial, institutional, industrial, agricultural, and domestic needs for goods and services. A range of sustainable water strategies to address twenty-first-century water challenges have emerged that reach far beyond the conventional water sector. Growing attention is being given to sustainable measures such as low-impact development, water reuse, watershed restoration, water conservation and efficiency, and many other proven and promising practices. As the country shifts to more sustainable water management, new services, occupations, and markets are emerging. In addition, over the next decade there will be a need for major investments in the nation's aging infrastructure for wastewater, stormwater, and drinking water along with ongoing operation and maintenance jobs to sustain that infrastructure. Many of these jobs are eligible for public funding and have been the focus of increased funding through the American Recovery and Reinvestment Act of 2009 (ARRA) and other federal and state policies. However, information is needed to help local, state, and federal agencies; utilities; companies; unions; and nonprofit entities adopt strategies that maximize job creation and other potential benefits of these practices.[1]

The need to take advantage of any opportunities for increased quantity and quality of employment opportunities is acute: 12 million workers in the United States who are interested in working and available to work are unemployed (US BLS 2012a). Millions more are working but remain in financial distress, with basic needs, such as health care, unmet. Some 10.5 million US workers were employed for more than half of the past year yet still have an income below the official poverty level (US BLS 2012b). For working parents in poverty, constraints on their ability to provide stable and appropriate child care can pass their hardship onto their children. Research now shows that these forms of inequality are harmful to economic growth and that regions and countries with greater equality experienced more vigorous economic growth (Treuhaft et al. 2011).

The jobs generated by sustainable water strategies have rarely been analyzed or tracked, yet a growing body of research points to significant numbers of jobs in certain sectors and in certain places. The types of jobs involved in implementing sustainable water strategies cover a broad range of occupations. In a recent study conducted by the Pacific Institute (Moore et al. 2013), 136 water-related occupations were identified in agriculture, urban residential and commercial settings, restoration and remediation, alter-

---

1. This chapter is modified and updated from a recent study by the Pacific Institute titled "Sustainable Water Jobs: A National Assessment of Water-Related Green Job Opportunities," available at http://www.pacinst.org/reports/sustainable_water_jobs/.

native water sources, and stormwater management. The number of employment opportunities created by sustainable water practices can be very high. The data available point to the creation of 10–15 jobs for every $1 million invested in alternative water supplies; 5–20 in stormwater management; 12–22 in urban conservation and efficiency; around 15 in agricultural efficiency and quality; and 10–72 jobs per $1 million invested in restoration and remediation.

This research suggests that water is a worthy arena for exploring green jobs and that if we continue to separate our efforts in water sustainability and economic opportunity, we may miss opportunities to strengthen both through integration. There is now greater awareness of the promise of green jobs—the potential for crafting economic strategies that reduce environmental impacts while creating high-quality work opportunities. But these efforts have largely focused on energy efficiency and renewable energy. This chapter summarizes the conclusion that the growing efforts to develop green jobs by investing in sustainable water systems are merited.

At the conclusion of this chapter, we also evaluate some of the challenges of developing these water jobs, including the need for better information for various stakeholders, such as workforce development and community-based organizations, labor unions, utilities and public water agencies, and businesses. The comprehensive assessment (Moore et al. 2013) looks in more detail at the full spectrum of sustainable water strategies and provides foundational information regarding the following questions:

- What are the best practices across sectors for sustainable water management and use?
- What activities and occupations are involved when these practices are adopted?
- What data are available to quantify the jobs generated by these practices?
- What is the quality of these sustainable water occupations, what is their growth in the overall economy, and what are the demographics of the workforce in these professions?
- How can disadvantaged communities be linked to job opportunities in the water sector?

Water Challenges in Today's Economy

As we enter the second decade of the twenty-first century, the United States faces a complex and evolving set of freshwater challenges (Christian-Smith and Gleick 2012). Despite the fact that the nation is, on average, a comparatively water-rich country, we are reaching absolute limits on our ability to take more water from many river systems, such as the Colorado, Sacramento–San Joaquin, and Chattahoochee River systems. We are also overpumping groundwater aquifers, including those in the Great Plains and California's Central Valley. Wetlands and aquatic ecosystems and fisheries are in decline. Continued population and economic growth are adding new demands for water, in competition with other uses. Many of the nation's water bodies are contaminated, and 42 percent of the nation's total stream length is considered to be in poor condition (US EPA 2006).

Much of our water infrastructure has not been adequately maintained, and confidence in our tap water system is falling. The US Environmental Protection Agency (EPA) projects that if we maintain current levels of investment in water systems through 2020, the percentage of the nation's water pipes considered "poor," "very poor," or "life elapsed" will reach 44 percent (US EPA 2002). Unresolved public health threats from contaminated drinking water exist in a growing number of communities, and the affordability of water for low-income users is a growing concern. Climate changes compound many of these challenges by altering water availability and quality and increasing the risk of both floods and droughts. A nationwide study analyzing water demand and supply under future climate change scenarios found that 70 percent of US counties may be at moderate to extreme risk of their water demand surpassing water supply by 2050 (NRDC 2010).

In the past, the traditional approach to meeting these challenges has relied on building massive, centralized, capital-intensive infrastructure, such as large dams and reservoirs. This approach has brought many benefits, but it has also come at great social, economic, and environmental costs, many of which were either ignored, undervalued, or unknown at the time (Gleick 2002). Currently, nearly 40 percent of North American freshwater and diadromous fish species are imperiled because of physical modifications to rivers and lakes (Jelks et al. 2008). Other consequences of traditional water development include loss of valuable ecological services that aquatic ecosystems provide, such as water filtration and retention, as well as massive energy demands to move and treat water and substantial and heavily subsidized costs.

The water challenges our nation faces must be addressed with strategies that avoid the negative outcomes of the past. To develop these strategies, water managers and others are rethinking approaches to ensure that sufficient water resources are available to meet anticipated needs in ways that improve, rather than ignore, social equity, ecological conditions, and long-term sustainability of human-ecological systems. Many of these approaches reflect "soft-path" principles, which include taking advantage of the potential for decentralized facilities, efficient technologies, flexible public and private institutions, innovative economic instruments, and human capital (Gleick 2002). We define sustainable water strategies as those that reduce or eliminate water contamination, restore watershed systems, and increase efficient use of natural, societal, and financial resources. We group the diverse range of techniques that meet these criteria into five overarching strategies:

1. Urban water conservation and efficiency
2. Stormwater management
3. Restoration and remediation
4. Alternative water sources
5. Agricultural water efficiency and quality

For each of these strategies, we review the technical and economic processes involved, identify the occupations of workers carrying out these activities, and assess the numbers of jobs created for each $1 million invested.

Improving the Efficiency of Urban Water Use

Techniques to improve urban water conservation and efficiency allow for the increased production of goods and services while maintaining or even reducing overall water use in urban areas. These efforts are also referred to as demand management or water productivity improvements. Urban water conservation and efficiency measures include installing efficient appliances and fixtures, improving water-use metering, improving landscape efficiency, xeriscaping, and replacing or repairing pipes to reduce water loss within the water conveyance and distribution system (see table 3.1) (Gleick et al. 2003). Water treated and reused on-site at a smaller scale is often referred to as grey water, which includes water from clothes washers, showers, and faucets for use on outdoor landscapes, for flushing toilets, and for other nonpotable uses.

Despite efficiency gains over the past twenty-five years, current urban water use in the United States remains wasteful. Inefficient fixtures and appliances are still commonplace, particularly in homes built prior to 1994,[2] in a range of commercial, institutional, and industrial settings. Even in a relatively dry and densely populated state such as California, where many water agencies have taken the concerns about water supply constraints seriously, far more can be done. A 2003 analysis by the Pacific Institute found that existing, cost-effective technologies and policies can reduce California's urban water demand by more than 30 percent (Gleick et al. 2003), and a follow-up study done in 2005 found continued room for improvement (Gleick et al. 2005). These findings have been echoed by studies and programs in other regions. For example, a Seattle study found that installing new, water-efficient fixtures and appliances reduced single-family indoor use by nearly 40 percent (Mayer et al. 2000). Experiences from other countries, such as Australia, offer insights into the levels of urban efficiency that can be achieved under difficult circumstances. For example, in South East Queensland, Australia, during a severe, long-term drought, total urban demand dropped to only 67 gallons per capita per day, and residential demand, including both indoor and outdoor uses, reached 43 gallons per capita per day (Queensland Water Commission 2010; Heberger 2012).

In addition to saving water, water conservation and efficiency measures provide a number of other benefits. Conservation and efficiency techniques can reduce costs for drinking water and wastewater treatment systems by reducing overall water delivery requirements. Reductions in demand can also reduce energy use and associated greenhouse gas emissions for water systems. Landscape efficiency improvements reduce chemical and fertilizer needs and subsequent runoff into local streams and waterways. High-efficiency appliances can also reduce detergent requirements for dishwashers and washing machines.

TABLE 3.1 Urban Water Conservation and Efficiency Techniques

Appliances and fixtures	Installing water-efficient appliances and fixtures in residences and in commercial, industrial, and institutional (CII) settings.
Landscaping	Utilizing landscape efficiency improvements, including advanced irrigation technologies and controllers, mulching, and low-water-use plants.
Reduction of water loss	Replacing or repairing pipes to reduce water loss within the water conveyance and distribution system.
Grey water	Collecting, treating, and using water from clothes washers, showers, and faucets on-site for outdoor irrigation, toilet flushing, or other nonpotable uses.

Source: Gleick et al. (2003).

---

2. National water efficiency standards for some fixtures were signed into law in 1992; implementation began in 1994.

The economic process of implementing urban conservation and efficiency measures involves a range of research, manufacturing, installation, and maintenance activities. Research and development, for example, generates new water-efficient technologies and practices, including mobile applications, soil moisture sensors, weather-based irrigation controllers, appliances and fixtures, and sensors (e.g., for cooling towers). Manufacturing activity produces these new tools and products, which are then distributed through wholesale and retail sales to businesses, utilities, and home owners. Design and planning programs at water agencies can help provide information to consumers or develop pricing strategies that encourage investment or more careful water use.

Figure 3.1 provides a summary of the occupations associated with urban water efficiency projects. Workers in these categories range from scientists developing new technologies to surveyors and landscape architects measuring water use and planning for improvements. Plumbers, construction laborers, and landscaping workers install and often maintain water-efficient technologies and vegetation, while public relations specialists raise awareness among community members and consumers.

Urban water efficiency projects create jobs, although limited data are available to develop accurate estimates (table 3.2). Early efforts to quantify the number of jobs created by urban water efficiency projects include the 1992 Madres del Este de Los Angeles Santa Isabel partnership, which estimated that a community-based program to install low-flush toilets created twenty-five full-time and three part-time jobs (Lerner 1997). In a more comprehensive analysis, the Alliance for Water Efficiency modeled the jobs created by investments in seven types of water conservation and efficiency programs, finding that 14.6 to 21.6 jobs in total employment were generated per $1 million invested (Alliance for Water Efficiency 2008). A similar study in Los Angeles by the Economic Roundtable gathered information on eleven urban water efficiency projects, ranging from bathroom retrofits to a green garden program, and found that the projects generated 11.8 jobs per $1 million invested, including 6 direct, 2.6 indirect, and 3.3 induced jobs.

Managing Stormwater

Stormwater runoff is a major cause of water pollution in the United States. Ten millimeters (a little more than three-eighths of an inch) of rain over a one-hectare (approximately two-acre) parking lot produces 100 cubic meters (over 26,000 gallons) of stormwater runoff and can carry a noxious mixture of soils, sediments, oils, chemicals, and other substances into local waterways. In some areas, combined sewers transport both stormwater and wastewater to treatment systems, which can be overwhelmed during high flows and release untreated wastewater into local waterways. For many years, water agencies have been trying to improve stormwater management, redesign and rebuild sewers, and reduce these untreated releases.

Upgrading existing conventional infrastructure to manage stormwater comes at considerable financial cost, which has encouraged urban areas to look for alternative ways to address stormwater runoff. According to the EPA's 2008 Clean Watersheds Needs Survey, approximately $189 billion is needed for pipe repair, new pipes, combined sewer overflow corrections, and stormwater management programs (US EPA 2008). Low-impact development (LID), or "green infrastructure," consists of alternative approaches that include systems and techniques intended to infiltrate, evaporate, and reuse stormwater on

RESEARCH AND DEVELOPMENT
e.g. development of water efficient technologies/fixtures for residential and commercial

- Environmental Engineers
- Soil and Plant Scientists
- Geologists
- Environmental Scientists and Specialists, Including Health Industrial Designers
- Conservation Scientists

MANUFACTURING AND DISTRIBUTION
e.g. production of water efficient fixtures

- Truck Drivers–Heavy and Tractor-Trailer
- Laborers and Freight, Stock, and Material Movers, Hand Milling and Planing Machine Setters, Operators, and Tenders, Metal and Plastic
- Shipping, Receiving, and Traffic Clerks
- Truck Drivers, Light or Delivery Services
- Stock Clerks and Order Fillers
- Industrial Designers

WHOLESALE AND RETAIL SALES
e.g. sales of low-flow toilets or drip irrigation equipment

- Sales Representatives, Wholesale and Manufacturing, Technical and Scientific Products
- Stock Clerks and Order Fillers
- Cost Estimators

DESIGN AND PLANNING
e.g. conducting an audit of residential or commercial water use

- Architectural and Civil Drafters
- Cartographers and Photogrammetrists
- Landscape Architects
- Surveyors
- Surveying and Mapping Technicians
- Environmental Engineers
- Commercial and Industrial Designers
- Environmental Restoration Planners
- Social and Community Service Managers
- Forest and Conservation Technicians
- Graphic Designers

INSTALLATION
e.g. install and upgrade efficient fixtures and appliances

- First-Line Supervisors/Managers of Construction Trades and Extraction Workers
- Construction Managers
- Landscaping & Groundskeeping Workers
- Carpenters
- Construction Laborers
- Operating Engineers and other Construction Equipment Operators
- Electricians
- Pipelayers
- Plumbers, Pipefitters, and Steamfitters
- Helpers–Carpenters
- Helper–Construction Trades, All Other Helpers—Pipelayers, Plumbers, Pipefitters, and Steamfitters
- First-Line Supervisors of Landscaping, Lawn Service, and Groundskeeping Workers
- Control and Valve Installers and Repairers, Except Mechanical Door
- Farmworkers and Laborers, Crop, Nursery, and Greenhouse
- Recreation Workers
- Tree Trimmers and Pruners
- Pesticide Handlers, Sprayers, and Applicators, Vegetation
- First-Line Supervisors/Managers of Farming, Fishing, and Forestry Workers
- First-Line Supervisors of Agricultural Crop and Horticultural Workers
- Agricultural Workers, All Others
- Forest and Conservation Workers

OPERATIONS AND MAINTENANCE
e.g. repair, maintain, and upgrade efficient fixtures and appliances

- Landscaping & Groundskeeping Workers
- Operating Engineers and Other Construction Equipment Operators
- Maintenance and Repair Workers, General
- Helpers--Installation, Maintenance, and Repair Workers
- Septic Tank Servicers and Sewer Pipe Cleaners
- General and Operations Managers
- Farmworkers and Laborers Crop, Nursery, and Greenhouse
- Control and Valve Installers and Repairers, Except Mechanical Door
- Grounds Maintenance Workers, All Other
- Home Appliance Repairers
- Janitors and Cleaners, Except Maids and Housekeeping Cleaners
- Tree Trimmers and Pruners
- Pesticide Handlers, Sprayers, and Applicators, Vegetation

FIGURE 3.1 OCCUPATIONS IN URBAN WATER CONSERVATION AND EFFICIENCY

TABLE 3.2 Jobs per $1 Million Invested in Urban Conservation and Efficiency Projects

Project Type	Total Employment (Jobs per $1 Million Invested)	Units and Methods	Source
Eleven water conservation projects in Los Angeles County	11.8	Total employment, including direct employment from expense reports and modeled indirect and induced employment	Burns and Flaming (2011)
Evapotranspiration irrigation controller rebate/direct install programs	20.4	Total employment modeled using categorized program expenses as inputs	Alliance for Water Efficiency (2008)
Water system loss control	21.6	(same as previous)	Alliance for Water Efficiency (2008)
High-efficiency toilet rebate program	18	(same as previous)	Alliance for Water Efficiency (2008)
High-efficiency toilet direct install program	17.2	(same as previous)	Alliance for Water Efficiency (2008)
Industrial water/energy survey and retrofit program	15.6	(same as previous)	Alliance for Water Efficiency (2008)
Retrofit cooling towers with conductivity and pH controllers	15.4	(same as previous)	Alliance for Water Efficiency (2008)
Restaurant surveys and direct install equipment retrofits	14.6	(same as previous)	Alliance for Water Efficiency (2008)
Installation of 5,439 grey water systems in new residential properties in Los Angeles County	5.3	Modeled total jobs (direct/indirect/induced) using IMPLAN system 2009 data and 2011 software	Burns and Flaming (2011)

the site where it is generated.[3] Of the $189 billion funding gap identified by the survey, $17.4 billion is related to green infrastructure projects (US EPA 2008).

Sustainable stormwater management practices and techniques include rainwater harvesting, bioswales and other forms of stormwater retention, permeable pavement, green roofs, and other efforts to reduce impervious areas (see table 3.3). Portland, Oregon, and other cities have programs to disconnect downspouts, rerouting rooftop drainage pipes to rain barrels, cisterns, or permeable areas instead of the storm sewer.

---

3. According to the EPA, low-impact development is "an approach to land development (or re-development) that works with nature to manage stormwater as close to its source as possible" (US EPA 2012b). In this chapter we use the terms "low-impact development" and "green infrastructure" interchangeably.

TABLE 3.3 Stormwater Management Techniques

Downspout disconnection	Downspouts are disconnected and rooftop drainage pipes are rerouted to drain rainwater into rain barrels, cisterns, or permeable areas instead of the storm sewer.
Rainwater harvesting	Rainwater is harvested and stored for later use.
Rain gardens	Rain gardens (also known as bioretention or bioinfiltration cells) collect runoff from rooftops, sidewalks, and streets into shallow, vegetated basins.
Planter boxes	Planter boxes are urban rain gardens with vertical walls and open or closed bottoms that collect and absorb runoff from sidewalks, parking lots, and streets.
Bioswales	Bioswales retain and filter stormwater as it moves through vegetated, mulched, or xeriscaped channels.
Permeable pavements	Permeable pavements infiltrate, treat, or store rainwater where it falls.
Green roofs	Green roofs are covered with growing media and vegetation and enable rainfall infiltration and evapotranspiration of stored water.
Urban forestry	Trees in urban areas reduce and slow stormwater runoff by intercepting precipitation in their leaves and branches.
Land conservation	Land conservation protects open spaces and sensitive natural areas within and adjacent to cities.

Source: US EPA (2012a).

The environmental benefits of these approaches include minimizing the impacts of urban runoff on local streams and the marine environment, reducing local flooding, recharging local groundwater supplies, and improving the reliability and flexibility of water supplies.

These approaches are typically less costly than conventional stormwater management techniques, with capital savings ranging from 15 percent to 80 percent (US EPA 2007). A recent report showed that 75 percent of 479 green infrastructure projects either kept costs the same or reduced costs, compared with the same projects that used more traditional infrastructure (American Rivers et al. 2012). Additionally, research has consistently shown a positive correlation between proximity to green spaces and home and property values (Crompton 2005).

It is difficult to determine how much green infrastructure has already been installed. According to the Greenroof & Greenwall Projects Database, more than 970 projects have resulted in the installation of more than 14 million square feet of green roofs in the United States (Greenroofs.com 2012), though these figures likely underestimate the true extent of this practice. Other practices, such as changes in landscaping and drainage systems, construction of rain gardens, and urban tree planting, have not yet been comprehensively quantified.

Figure 3.2 shows occupations involved in stormwater management. The range of occupations includes broad construction professions, septic tank servicers, sewer pipe cleaners, welders, cutters, and fitters for manufacturing, and more. Most of the occupations listed are traditionally within the landscaping and engineering sector and utilize similar techniques to address stormwater issues, though more sustainable water strate-

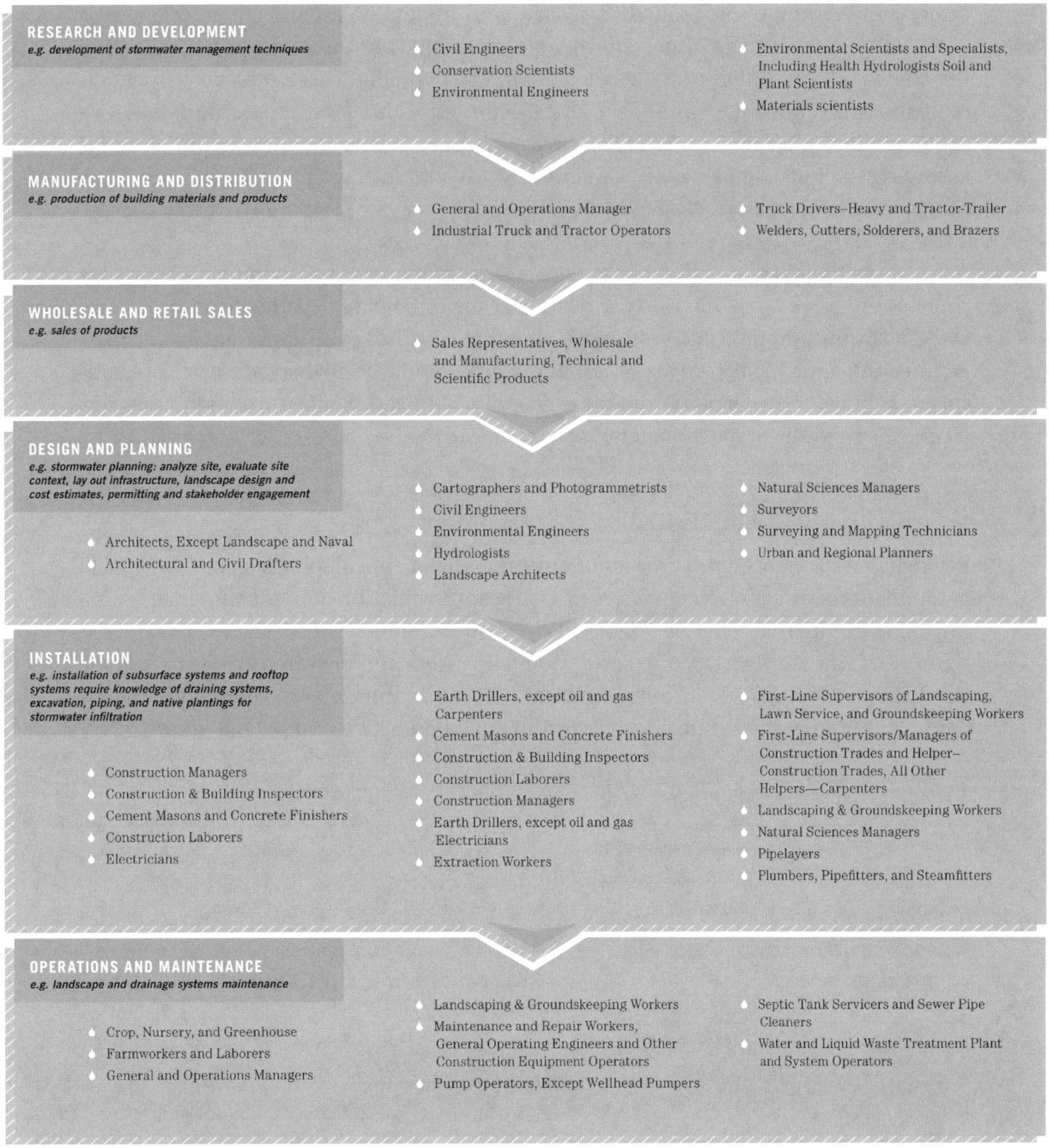

FIGURE 3.2 OCCUPATIONS IN STORMWATER MANAGEMENT

RESEARCH AND DEVELOPMENT
e.g. development of stormwater management techniques

- Civil Engineers
- Conservation Scientists
- Environmental Engineers
- Environmental Scientists and Specialists, Including Health Hydrologists Soil and Plant Scientists
- Materials scientists

MANUFACTURING AND DISTRIBUTION
e.g. production of building materials and products

- General and Operations Manager
- Industrial Truck and Tractor Operators
- Truck Drivers–Heavy and Tractor-Trailer
- Welders, Cutters, Solderers, and Brazers

WHOLESALE AND RETAIL SALES
e.g. sales of products

- Sales Representatives, Wholesale and Manufacturing, Technical and Scientific Products

DESIGN AND PLANNING
e.g. stormwater planning: analyze site, evaluate site context, lay out infrastructure, landscape design and cost estimates, permitting and stakeholder engagement

- Architects, Except Landscape and Naval
- Architectural and Civil Drafters
- Cartographers and Photogrammetrists
- Civil Engineers
- Environmental Engineers
- Hydrologists
- Landscape Architects
- Natural Sciences Managers
- Surveyors
- Surveying and Mapping Technicians
- Urban and Regional Planners

INSTALLATION
e.g. installation of subsurface systems and rooftop systems require knowledge of draining systems, excavation, piping, and native plantings for stormwater infiltration

- Construction Managers
- Construction & Building Inspectors
- Cement Masons and Concrete Finishers
- Construction Laborers
- Electricians
- Earth Drillers, except oil and gas Carpenters
- Cement Masons and Concrete Finishers
- Construction & Building Inspectors
- Construction Laborers
- Construction Managers
- Earth Drillers, except oil and gas Electricians
- Extraction Workers
- First-Line Supervisors of Landscaping, Lawn Service, and Groundskeeping Workers
- First-Line Supervisors/Managers of Construction Trades and Helper–Construction Trades, All Other Helpers—Carpenters
- Landscaping & Groundskeeping Workers
- Natural Sciences Managers
- Pipelayers
- Plumbers, Pipefitters, and Steamfitters

OPERATIONS AND MAINTENANCE
e.g. landscape and drainage systems maintenance

- Crop, Nursery, and Greenhouse
- Farmworkers and Laborers
- General and Operations Managers
- Landscaping & Groundskeeping Workers
- Maintenance and Repair Workers, General Operating Engineers and Other Construction Equipment Operators
- Pump Operators, Except Wellhead Pumpers
- Septic Tank Servicers and Sewer Pipe Cleaners
- Water and Liquid Waste Treatment Plant and System Operators

---

gies are designed to prevent and control stormwater runoff rather than simply diverting it to stormwater drains. New skills for these traditional occupations include knowledge of green roof design and installation, experience in working with new building materials (e.g., permeable pavement), and familiarity with land-use planning, site design, slope, and drainage.

Several studies have sought to quantify the jobs created through sustainable stormwater management projects (see table 3.4). A recent analysis found that four stormwater projects in Los Angeles generated 13.8 jobs per $1 million invested, with average wages of $52,800. An estimated 73 percent of workers involved in these projects lived within the county, an indication of the local economic activity that these projects can generate. Likewise, stormwater projects in the District of Columbia's green roofs initiative produced 19.7 jobs for every $1 million invested. A prospective analysis of proposed investments in Philadelphia, Pennsylvania, finds that committing $1.6 billion over the next twenty years will create over 8,600 jobs, or about 5.3 jobs per $1 million invested (BUCIP 2010).

However, there is one caveat to mention in these economic and job impacts. Estimates show promising overall workforce growth, but the anticipated demand for entry-level work in stormwater projects is lower than in other sectors. Professional services such as architectural and engineering services, financial institutions, and scientific and technical consulting tend to benefit from these economic impacts, while the construction services experience disproportionately low growth (Burns and Flaming 2011; BUCIP 2010).

Restoring and Remediating Ecosystems

Ecological restoration is the process of returning the chemical, physical, and biological components of a degraded ecosystem to a close approximation of predisturbance conditions. Environmental remediation—a subset of restoration—is the process of targeted removal of specific toxic substances or pollutants from soil or water. Whereas restoration generally focuses on returning an ecosystem to its natural state, remediation targets lands contaminated with substances harmful to human health and the environment.

TABLE 3.4 Jobs per $1 Million Invested in Stormwater Management Projects

Project Type	Total Employment (Jobs per $1 Million Invested)	Units and Methods	Source
Washington, DC, green roofs initiative	19.65	Total jobs based on total green roof area covered, in square feet	American Rivers and Alliance for Water Efficiency (2008)
National investment in green stormwater infrastructure	10	Modeled total jobs (direct/indirect/induced) using macroeconomic and spending multiplier	Green for All et al. (2011)
Operation and maintenance budget of four stormwater projects in Los Angeles	13.8	Modeled total jobs (direct/indirect/induced) using IMPLAN system 2009 data and 2011 software	Burns and Flaming (2011)
Montgomery County, Maryland, stormwater system construction	10.8	Montgomery County expects to employ 3,300 for a three-year project	Chesapeake Bay Foundation (2011)
Philadelphia, Pennsylvania, stormwater investment	5.3	Modeled total jobs (direct/indirect/induced) using IMPLAN system over next twenty years	BUCIP (2010)

Restoration can improve water quality and aquatic ecosystem health and can be achieved using both natural and artificial techniques. The EPA has identified three restoration techniques that deal with water quality (US EPA 1995):

1. In-stream techniques applied within the water system. For example, streams that have been channeled can be restored to achieve more natural geometry, meander, sinuosity, substrate composition, structural complexity, aeration, and stream bank stability.

2. Riparian techniques applied to the land area surrounding or bordering the water system. This can include planting of native riparian vegetation or construction of fencing to protect delicate riparian plants.

3. Upland or surrounding watershed techniques designed to reduce non–point source pollution from upstream lands within the watershed.

Although they are integral aspects of environmental restoration, techniques used for upland or surrounding watershed zones are cataloged here under stormwater management and agricultural efficiency. Table 3.5 lists some examples of in-stream and riparian restoration techniques.

Removal of contaminants in remediation projects can also be important, especially in aquatic systems, where pollutant transport is a risk. As shown in table 3.6, a wide range of techniques can be used to remediate groundwater and surface water, including phytoremediation, in which plants are used to absorb harmful contaminants, and installation of permeable reactive barriers, in which membranes are used to capture or treat groundwater contaminant plumes. Another important aspect of environmental remediation is the use of new technologies to characterize and monitor contamination (US EPA 2012c).

The work of restoration and remediation planning and design involves various technical assessments, coordinated planning, permitting, ecological research, design, and cost estimation. Planning and assessment activities include the review of aerial photographs, topographic maps, and results from soil tests and borings and engagement with regulatory

TABLE 3.5 Restoration Techniques

In-stream	• Streambed reconfiguration
	• Restoration of natural meander patterns
	• Daylighting of urban creeks from stormwater sewers
	• Root wad/tree revetments
	• Live stakes, live fascines, brush mattresses, branch packings, brush layering, vegetated geogrids, and live cribwalls
	• Channel deflectors and channel constrictors
	• Boulder clusters
	• Log drop structures
	• Removal of artificial structures
Riparian	• Wetland restoration
	• Restoration of riparian corridor vegetation
	• Levee setbacks or removal
	• Control of timing, location, and extent of water diversions
	• Construction of fences and gates in riparian corridor to control access

Source: US EPA (1995).

agencies and stakeholders (NOAA, n.d.). Design of restoration projects typically entails the application of engineering, landscape architecture, river ecology, and expertise in geomorphology. Cost estimates are developed using standardized unit material and labor costs and regional or municipal multipliers. The construction phase of restoration and remediation projects is typically labor-intensive and involves some use of heavy machinery. In-stream and riparian restoration is typically implemented with a mix of hand labor and heavy machinery for digging and bank alteration.

A large number of occupations are involved in restoration and remediation activities (figure 3.3). Workers in these occupations range from the environmental and civil engineers who develop designs to sales representatives and truck drivers who sell and deliver products, to scientists with expertise in river systems and aquatic ecology, to masons and laborers who construct the projects. Workers in newer occupations, such as brownfield redevelopment specialists and conservation technicians, also have skills required for these projects.

The number of jobs created by restoration and remediation projects is highly variable, ranging from about 15 to 72 total jobs per $1 million invested (table 3.7). For example, $10.6 million spent on restoring the Cache River basin in Illinois generated approximately 36 jobs for each $1 million invested (Caudill 2008). This includes 22 workers directly employed in the project and an additional 14 indirect and induced jobs, according to a study commissioned by the US Fish and Wildlife Service. About half of the direct jobs and 85 percent of the indirect and induced jobs were created locally in the four counties where the project took place. Likewise, dam removal on Ohio's Euclid Creek cost more than $500,000 and generated 38 jobs, or 72 jobs per $1 million invested.

Studies have found that the number of jobs created can vary depending on how much labor-intensive or equipment-intensive work must be done. For instance, a study by the University of Oregon's Institute for a Sustainable Environment estimates that direct employment in equipment-intensive watershed contracting is only 4.8 jobs per $1 million invested because labor is a relatively small proportion of total spending (36 percent) and worker payroll costs are relatively high (about $55,000 per job). By contrast, direct employment in labor-intensive projects is 13.8 jobs per $1 million, with labor making up 67 percent of spending and payroll costs averaging $31,000 per job, resulting in more and lower-wage jobs (Nielsen-Pincus and Moseley 2010).

TABLE 3.6 Remediation Techniques

Extraction or removal techniques	• Multiphase extraction • Natural attenuation
Containment techniques	• Evapotranspiration covers
In situ treatment techniques	• Thermal treatment • Flushing • Oxidation • Air sparging • Permeable reactive barriers • Phytotechnologies • Groundwater-circulating wells • Nanotechnology • Natural attenuation

RESEARCH AND DEVELOPMENT
e.g. development of new restoration techniques, and conducting watershed assessments

- Chemists
- Civil Engineers
- Environmental Engineers
- Environmental Scientists and Specialists, Including Health Hydrologists
- Landscape architects

MANUFACTURING AND DISTRIBUTION
e.g. production of heavy equipment and tools

- Chemists
- Farmworkers and Laborers, Crop, Nursery, and Greenhouse
- Mechanical Engineers
- Production Workers
- Truck Drivers
- Welders, Cutters, Solderers, and Brazers

WHOLESALE AND RETAIL SALES
e.g. sales of plants and equipment

- Sales Representatives, Wholesale and Manufacturing, Technical and Scientific Products

DESIGN AND PLANNING
e.g. soil and water testing, project design and cost estimates, permitting and stakeholder engagement

- Architects
- Architectural and Civil Drafters
- Brownfield Redevelopment Specialists and Site Managers
- Civil Engineers
- Commercial and Industrial Designers
- Environmental Engineers
- Environmental Restoration Planners
- Environmental Scientists and Specialists, Including Health
- Executive Secretaries and Administrative Assistants
- Hydrologists
- Industrial Ecologists
- Landscape Architects
- Natural Sciences Managers
- Soil and Plant Scientists
- Soil Conservationists
- Training and Development Specialists
- Urban and Regional Planners
- Zoologists and Wildlife Biologists

INSTALLATION
e.g. removal of artificial structures, contaminated soil and invasive species; alteration to river-bed; planting new vegetation

- Agricultural Workers
- Brownfield Redevelopment Specialists and Site Managers
- Cement Masons and Concrete Finishers
- Construction Laborers
- Construction Managers
- Electrical Engineers
- Electricians
- Engineering Managers
- Environmental Restoration Planners
- Environmental Science and Protection Technicians, Including Health
- Executive Secretaries and Administrative Assistants
- Farmworkers and Laborers, Crop, Nursery, and Greenhouse
- First-Line Supervisors/Managers of Construction Trades and Extraction Workers
- First-Line Supervisors/Managers of Farming, Fishing, and Forestry Workers
- First-Line Supervisors/Managers of Landscaping, Lawn Service, and Groundskeeping Workers
- Forest and Conservation Technicians
- Forest and Conservation Workers
- General and Operations Managers
- Hazardous Materials Removal Workers
- Industrial Truck and Tractor Operators
- Laborers and Freight, Stock, and Material Movers
- Landscaping and Groundskeeping Workers
- Occupational Health and Safety Technicians
- Operating Engineers and Other Construction Equipment Operators
- Pipelayers
- Plumbers, Pipefitters, and Steamfitters
- Refuse and Recyclable Material Collectors
- Training and Development Specialists
- Truck Drivers, Heavy and Tractor-Trailer

OPERATIONS AND MAINTENANCE
e.g. landscape maintenance and soil and water testing

- Farmworkers and Laborers, Crop, Nursery, and Greenhouse
- First-Line Supervisors/Managers of Landscaping, Lawn Service, and Groundskeeping Workers
- Industrial Truck and Tractor Operators
- Laborers and Freight, Stock, and Material Movers
- Tree Trimmers and Pruners

FIGURE 3.3 OCCUPATIONS IN RESTORATION AND REMEDIATION

TABLE 3.7 Jobs per $1 Million Invested in Restoration and Remediation Projects

Project Type	Total Employment (Jobs per $1 Million Invested)	Units and Methods	Source
Cache River restoration in Illinois	36	Includes direct jobs recorded in project reports and modeled indirect and induced jobs	Caudill (2008)
Everglades restoration	20	Modeled total jobs (direct/indirect/induced) using generic Sector 36 national multipliers	Everglades Foundation (2010), 125
In-stream restoration projects in Oregon	14.7	Modeled total jobs (direct/indirect/induced) using data from survey of grants and contractors	Nielsen-Pincus and Moseley (2010)
Riparian restoration projects in Oregon	23.1	Modeled total jobs (direct/indirect/induced) using data from survey of grants and contractors	Nielsen-Pincus and Moseley (2010)
Wetlands restoration projects in Oregon	17.6	Modeled total jobs (direct/indirect/induced) using data from survey of grants and contractors	Nielsen-Pincus and Moseley (2010)
Restoration of four rivers in Massachusetts	10–13	Modeled total jobs (direct/indirect/induced) using generic state-level multipliers	Massachusetts Department of Fish and Game (2012)
Removal of two dams near Watervliet, Michigan	19	Total jobs noted in project summary	Great Lakes Coalition (2012)
Removal of dam on Euclid Creek in Ohio	72.2	Jobs noted in project summary; may not be full-time equivalents	Gershman and Alexander (2012)

Developing Alternative Water Sources

Traditional water sources include rivers, lakes, and groundwater, as well as artificial reservoirs created by dams. In the past few years, however, alternative water sources have begun to play an increasingly important role in supplementing water systems where traditional sources have become harder or more expensive to develop. Such alternative sources include a range of unconventional supplies, such as rainwater, stormwater, grey water, and reclaimed water. Projects to capture these alternative water sources can be implemented by a water utility or at the facility level by households or businesses. For the purposes of this report, this strategy focuses on activities at the utility scale; other actions at the site level, such as rainwater harvesting and grey water reuse, are captured in other strategies. We also describe conjunctive use, or aquifer storage and recovery (ASR), as a means of storing these alternative sources for later use.

Broadly, water reuse refers to the process of treating and reusing wastewater for a beneficial purpose. Potential uses include agricultural and landscape irrigation, industrial processing and cooling, domestic uses, dust control, construction activities, concrete mixing, artificial lakes, replenishment of groundwater basins (referred to as groundwater recharge, discussed in more detail below), and even direct potable reuse in some circumstances. Treatment levels can be tailored for the intended purpose and the level of human contact, which can help save money and energy when the quality does not have to meet drinking water standards. Water can be distributed from a wastewater treatment facility or treated and reused directly on-site (such as at a home or an industrial facility). Techniques to treat and reuse grey water on-site are discussed in the urban water conservation and efficiency strategy.

A growing number of communities across the United States are already beginning to move in the direction of encouraging and expanding the use of reclaimed water. According to the EPA's Clean Watersheds Needs Survey, an investment of approximately $4.4 billion is needed to build, rehabilitate, or replace infrastructure for distribution of reclaimed water (American Rivers et al. 2012). In 2004, the EPA estimated total wastewater reuse in the United States at 1.7 billion gallons per day and growing at a rate of 15 percent per year. The same report estimates that 32 billion gallons of wastewater are produced each day, of which 12 billion gallons per day are discharged directly into an ocean or estuary. Portions of these flows could be made available for reuse (US EPA 2004).

Alternative sources can be captured and stored underground for later use. The term "conjunctive use" describes the coordinated use of groundwater and surface water to optimize supply and storage. When surface water supplies are plentiful, they can be used instead of groundwater or to recharge groundwater basins. Also known as aquifer storage and recovery (ASR), this technique can recharge groundwater through surface spreading, infiltration pits and basins, or subsurface injection. This practice not only increases the availability of water when it is most needed but also can add an additional treatment step; reduce evaporation and evapotranspiration, as compared with surface storage; and enhance the sustainability of groundwater storage by preventing subsidence. The EPA is aware of 1,203 aquifer recharge and ASR injection wells that are capable of operating; however, it is unclear how much water is injected into these wells each year (US EPA 2009).

Multiple activities are involved throughout the process of developing alternative water supply sources.[4] Research and development includes the development of new membranes with higher recovery rates and sensors that can provide real-time water quality readings. Manufacturing and distribution involves production and delivery of all of the products used throughout the process, including membranes, chemicals, and pumps. The planning phase can be extensive and can include a technical analysis of existing water supply conditions and future needs. Construction of treatment and storage facilities includes grading and paving the site and constructing the buildings. Ongoing operation and maintenance work is needed to ensure compliance with treatment requirements and to read meters, replace membranes, and repair the system.

---

4. The focus of this section is utility-scale alternative water supplies. Grey water systems are included as an urban water efficiency strategy.

Figure 3.4 summarizes the many occupations engaged in developing alternative water supplies throughout the development process. The planning phase, for example, employs environmental engineers, hydrologists, landscape architects, urban planners, and other highly skilled professionals. Facilities and water and sewage operations staff are responsible for ongoing operations and maintenance.

Studies of the number of jobs generated by alternative water supply projects are limited. Table 3.8 is based on several case studies in Los Angeles County, where major investments have been made in recycled water and groundwater facilities (Burns and Flaming 2011). These case studies suggest the creation of roughly 10 to 14 jobs per $1 million invested. For example, the design and installation of recycled water projects produced more than 12 jobs per $1 million invested, including 6.6 direct jobs, 2.3 indirect jobs, and 3.7 induced jobs. Most of the direct work was performed by companies located in the area. Likewise, groundwater remediation projects in Los Angeles generated 12.8 jobs per $1 million invested.

Improving Agricultural Water Efficiency and Quality

Agriculture is the largest consumer of water in the nation, and while great advances in water efficiency in this sector have been made, additional gains are available through implementation of improved water management techniques (Christian-Smith and Gleick 2012). These practices are meant to reduce water waste and, in some cases, improve crop quality and yield. They include improving irrigation technologies and scheduling, reducing erosion, lining canals, increasing pump efficiency, restoring riparian areas, recycling tailwater on-farm, and constructing spill reservoirs at the water supplier scale.

Practices such as these can help reduce on-farm water use as well as improve water quality. Farmers are constantly implementing new technologies and management practices to improve their products. In many cases, water savings are a cobenefit of practices that are meant to decrease input costs or improve crop quality. Precision irrigation, in particular, has been shown to do both. And, according to the Census of Agriculture, flood irrigation declined by 5 percent nationwide between 2003 and 2008, replaced by more precise sprinkler and drip irrigation. Nevertheless, 39 percent of irrigated land nationwide is still flood irrigated, and there is significant room for improvement in all areas of agricultural water management. The 2007 Census of Agriculture found that at least 23 percent of farms and ranches surveyed utilized conservation methods such as no-till or conservation tillage, runoff collection or filtration, fencing of animals from streams, and other practices (NASS 2007). Agricultural runoff is also a major source of pollution of surface water and groundwater, and thus conservation practices designed to reduce runoff can also contribute to improved water quality and ecosystem health.

Agricultural efficiency and runoff management is designed to improve the productivity of agricultural water use and water quality through more efficient irrigation management techniques, such as irrigation scheduling, tailwater recycling, drip irrigation, and conservation tillage (see table 3.9). Figure 3.5 shows some of the occupations associated with this strategy. We note, however, that thus far little attention has been given to the number and types of jobs related to improved agricultural water management. Additional research and analysis are needed to develop a more comprehensive list.

Installation and operation and maintenance are the two most labor-intensive phases of the process. Installation can entail trenching, piping, earthmoving, and installing

Sustainable Water Jobs

RESEARCH AND DEVELOPMENT
e.g. identify water supply and demand in area, and develop alternative water sources to meet future need

- Agricultural Engineers
- Chemists
- Environmental Engineers
- Environmental Scientists and Specialists, Including Health Geologists
- Hydrologists
- Soil and Plant Scientists

MANUFACTURING AND DISTRIBUTION
e.g. manufacturing of pipes

- Reinforcing Iron and Rebar Workers
- Laborers and Freight, Stock, and Material Movers, Hand Milling and Planing Machine Setters, Operators, and Tenders, Metal
- and Plastic
- Shipping, Receiving, and Traffic Clerks
- Stock Clerks and Order Fillers
- Tank Car, Truck, and Ship Loaders
- Truck Drivers–Heavy and Tractor-Trailer
- Truck Drivers, Light or Delivery Services
- Welders, Cutters, Solderers, and Brazers

WHOLESALE AND RETAIL SALES
e.g. estimate cost for ratepayers, purchase materials for installation

- Cost Estimators
- Purchasing Agents and Buyers,
- Farm Products
- Sales Representatives, Wholesale and Manufacturing, Technical and Scientific Products
- Shipping, Receiving, and Traffic Clerks
- Stock Clerks and Order Fillers

DESIGN AND PLANNING
e.g. site plan and design new alternative water supply (conjunctive use, recycled water) for area needs this includes hydrologic and water quality data collection, monitoring, and analysis. coordinate demand projections among utilities and water suppliers

- Architects, Except Landscape and Naval
- Architectural and Civil Drafters
- Cartographers and Photogrammetrists
- Civil Engineers
- Engineering Technicians, Except Drafters, All OtherEnvironmental Engineers
- Graphic Designers
- Landscape Architects
- Landscape Architecture & Engineering Occupations
- Surveying and Mapping Technicians
- Surveyors
- Urban and Regional Planners

INSTALLATION
e.g. Installation of piping to distribute recycled water

- Carpenters
- Cement Masons and Concrete Finishers
- Construction and Building Inspectors
- Construction Laborers
- Construction Managers
- Control and Valve Installers and Repairers, Except Mechanical Door
- Earth Drillers, Except Oil and Gas
- Electricians
- Excavating and Loading Machine and Dragline Operators
- First-Line Supervisors of Landscaping, Lawn Service, and Groundskeeping Workers
- Helpers--Extraction Workers
- Helpers--Pipelayers, Plumbers, Pipefitters, and Steamfitters
- Landscaping & Groundskeeping Workers
- Managers of Construction Trades Workers
- Occupational Health and Safety Technicians
- Operating Engineers and other Construction Equipment Operators
- Paving, Surfacing, and Tamping Equipment Operators
- Pesticide Handlers, Sprayers, and Applicators, Vegetation
- Pipelayers
- Plumbers, Pipefitters, and Steamfitters
- Tree Trimmers and Pruners

OPERATIONS AND MAINTENANCE
e.g. facilities operations and maintenance for distributing water to end users

- Construction and Building Inspectors
- Control and Valve Installers and Repairers, Except Mechanical Door
- Farm Equipment Mechanics
- General and Operations Managers
- Grounds Maintenance Workers, All Other
- Helpers—Installation, Maintenance, and Repair Workers
- Janitors and Cleaners, Except Maids and Housekeeping Cleaners
- Landscaping & Groundskeeping Workers
- Maintenance and Repair Workers, General
- Operating Engineers and Other Construction Equipment Operators
- Pesticide Handlers, Sprayers, and Applicators, Vegetation
- Tree Trimmers and Pruners
- Water and Wastewater Treatment Plant and System Operators

FIGURE 3.4 OCCUPATIONS IN THE DEVELOPMENT OF ALTERNATIVE WATER SOURCES

pumps, water filters, water meters, water control devices, and irrigation equipment. A majority of this work is performed by construction laborers and agricultural technicians. In order for these technologies to work effectively, operation and maintenance must be performed regularly by maintenance and repair workers, agricultural inspectors, and service repair workers. Irrigation systems are often installed by construction crews, typi-

TABLE 3.8 Jobs per $1 Million Invested in Alternative Water Supply Projects

Project Type	Total Employment (Jobs per $1 Million Invested)	Units and Methods	Source
Eighteen recycled water design and installation projects in Los Angeles and surrounding region	12.5	Modeled total jobs (direct/indirect/induced) using IMPLAN system 2009 data and 2011 software	Burns and Flaming (2011)
Operation and maintenance of two recycled water projects	9.8	Modeled total jobs (direct/indirect/induced) using IMPLAN system 2009 data and 2011 software	Burns and Flaming (2011)
Two groundwater remediation design and installation projects	12.8	Modeled total jobs (direct/indirect/induced) using IMPLAN system 2009 data and 2011 software	Burns and Flaming (2011)
Operation and maintenance of Tujunga Wellfield Liquid Phase Granular Activated Carbon Project Two	13.9	Modeled total jobs (direct/indirect/induced) using IMPLAN system 2009 data and 2011 software	Burns and Flaming (2011)

TABLE 3.9 Agricultural Water Management Techniques

Agricultural water efficiency techniques	• Improving irrigation scheduling • Improving irrigation technology (e.g., sprinkler and drip irrigation systems) • Lining canals and employing other seepage control options • Recycling tailwater on-farm • Increasing pump efficiency • Constructing spill reservoirs and conducting district reoperation to reduce water waste • Utilizing mulching and other techniques to increase soil water-holding capacity • Capturing stormwater flows for later use (e.g., on-farm ponds for frost and heat control and irrigation)
Agricultural water quality improvement techniques	• Planting cover crops • Constructing fencing around water bodies and streams • Utilizing conservation tillage or no-till • Restoring riparian zones or constructing buffer zones • Improving irrigation scheduling and using technology that reduces runoff

Sustainable Water Jobs   53

cally made up of a foreman and skilled laborers. The crews are often trained by the dealer or contractor or by the water supplier and have very specific skill sets that are difficult to find. The labor-intensive components occur in implementing the installation and operations stage, while the irrigation technologies dominate the business aspect of the process. Other occupations essential to installation include precision agriculture technician, agricultural technician, purchasing agent, and buyer of farm products.

A recent analysis found that implementation of a range of agricultural best management practices would generate 14.6 jobs per $1 million invested (Rephann 2010). Re-

RESEARCH AND DEVELOPMENT
e.g. development of new irrigation techniques, and conducting soil and crop analyses

- Agricultural Engineers
- Agricultural Science Teachers,
- Postsecondary
- Environmental Engineers
- Hydrologists
- Materials Engineer
- Natural Sciences Managers
- Precision Agriculture Technicians
- Soil and Plant Technicians
- Soil Scientists

MANUFACTURING AND DISTRIBUTION
e.g. production of irrigation technologies and products

- Electricians
- General and Operations Manager
- Machinist
- Materials Engineer
- Milling and Planing Machine Setters, Operators, and Tenders, Metal and Plastic
- Team Assembler
- Truck Drivers, Heavy and Tractor-Trailer
- Welders, Cutters, Solderers, and Brazers

WHOLESALE AND RETAIL SALES
e.g. sales of irrigation products

- Buyers and Purchasing Agents, Farm Products
- Sales Representatives, Agricultural

DESIGN AND PLANNING
e.g. irrigation and landscape design and cost estimates, permitting and stakeholder engagement

- Conservation of Resources Commissioners
- Conservation Policy Analysts and Advocates
- Engineering Managers
- Engineering Technicians, Except Drafters, All Other Irrigation Designer
- Permaculture Designers/Contractors
- Precision Agriculture Technicians
- Surveying and Mapping Technicians
- Surveyors
- Sustainable Agriculture Specialists

INSTALLATION
e.g. trenching, connecting water source, installing valves, assembling pipes and pump stations, connecting wires, installing sprays, landscape drip

- Agricultural Technician
- Agricultural Workers, All Other Agriculture Irrigation Specialist
- Carpenters
- Cement Masons and Concrete Finishers
- Construction Laborers
- Construction Managers
- Electricians
- Farm Labor Contractors
- Farmworkers and Laborers, Crop, Nursery, and Greenhouse Helpers—Carpenters
- Operating Engineers and Other Construction Equipment Operators
- Paving, Surfacing, and Tamping Equipment Operators
- Pipelayers
- Plumbers, Pipefitters, and Steamfitters
- Precision Agriculture Technicians
- Pump Operators, Except Wellhead Pumpers
- Soil and Plant Technicians
- Truck Drivers, Heavy and Tractor-Trailer
- Welders, Cutters, Solderers, and Brazers

OPERATIONS AND MAINTENANCE
e.g. landscape maintenance, soil and crop testing

- Agricultural Inspectors
- Agricultural Workers, All Other
- Agriculture Irrigation Specialist
- Construction Laborers
- Construction Managers
- Farm Equipment Mechanics and Service Technicians
- Farm Labor Contractors
- Farmers, Ranchers, and Other Agricultural Managers
- Farmworkers and Laborers, Crop, Nursery, and Greenhouse
- Farmworkers, Farm, Ranch, and Aquacultural Animals
- Maintenance and Repair Workers, General
- Precision Agriculture Technicians

FIGURE 3.5 OCCUPATIONS IN AGRICULTURAL WATER EFFICIENCY AND QUALITY

searchers found that if full implementation of agricultural practices outlined in the 2005 Commonwealth of Virginia Chesapeake Bay Nutrient and Sediment Reduction Tributary Strategy were achieved, 11,751 person-years of employment would be created (Chesapeake Bay Foundation 2011). The majority of direct jobs generated would be in construction, construction-related services, and agricultural and forestry support industries. Additional indirect and induced jobs would be created in other industries.

An additional set of occupations that appear in multiple sustainable water strategies are general roles that we did not include in the descriptions of different strategies. These include the following:

- Accountants and auditors
- Bookkeeping, accounting, and auditing clerks
- Building cleaning workers, all other
- Business operations specialists, all other
- Chief executives
- Executive secretaries and administrative assistants
- First-line supervisors and managers of office and administrative support workers
- General and operations managers
- Human resources managers
- Management analysts
- Managers of office and administrative support workers
- Marketing managers
- Office and administrative support workers, all other
- Office clerks—general
- Public relations specialists
- Receptionists and information clerks
- Secretaries, except legal, medical, and executive
- Vocational education teachers, postsecondary

Job Quality and Growth in Sustainable Water Occupations

Who is currently working in sustainable water occupations, and in what conditions? Which of these occupations would be appropriate for focused efforts to increase opportunities for disadvantaged communities? These questions are key to identifying strategies that advance ecological preservation and address economic inequality. It is clear from the previous section and other analyses that significant jobs are created when many types of sustainable water strategies are implemented. For these strategies to achieve the goal of advancing economic equality, there must be sufficient opportunities and appropriate pathways for disadvantaged communities to access these opportunities (Moore et al. 2013).

Our approach to identifying green job opportunities is to focus on jobs that are already growing in the overall economy, recognizing that we may be able to take action to make them grow greener and faster. This is rooted in some important lessons from past green job efforts. Much of the previous investment in green jobs went into preparing workers for occupations that were green and entirely new, such as solar installers and energy auditors, yet the demand for graduates of these programs fell short. Because these workers were not prepared with a broader set of skills that qualified them for conventional occupations, they were left with limited options for employment. We follow the Center on Wisconsin Strategy (White et al. 2012) and others who recognize that all occupations will likely change as sustainable strategies reshape the tasks they carry out, but the priority for training new workers should be on green skills embedded in a broader set of skills for growing occupations. Our analysis to identify sustainable water occupations that are more opportune is guided by the following questions:

- Which sustainable water occupations are expected to have significant future growth beyond any growth in the water-related industries?
- What level of education and experience is required to gain employment in these occupations?
- What are the wages and the level of union representation in these occupations?
- What are the demographics of the current workforce in sustainable water occupations?
- In which occupations are workers required to have unique skills in order to be able to work on sustainable water projects?
- What educational pathways and barriers may exist to workers from disadvantaged communities seeking employment in these occupations?

We focus on occupations for which at least 100,000 job openings are projected for the year 2020. This is the threshold used by the Occupational Information Network (O*NET) for its Bright Outlook status.

The level of education and experience required for workers to access employment can be a barrier or a pathway, depending on the accessibility of educational opportunities that match time availability, language, cost, and prerequisites to the worker. The appropriate level of education and experience also depends on aspirations of the prospective workers. An occupation that requires higher education may be completely appropriate for a community college student setting a ten-year goal, whereas an adult parent with little time to go back to school may be looking for a shorter pathway into an occupation with family-supporting wages.

Twenty-seven sustainable water-related occupations have more than 100,000 job openings projected for 2020 in the overall economy and have relatively accessible education and experience requirements consistent with job zones 1 through 3 as defined by O*NET. Nine of these occupations require little or no previous work experience or education (job zone 1); eight require some experience and education (job zone 2); and ten require experience and some formal education (job zone 3).

Of the twenty-seven sustainable water-related occupations with a high number of projected openings, half have median wages above the national median wage of $16.57 per hour. Of those in job zone 1, the median hourly wage ranges from $10.50 to $14.50. In job zone 2, median hourly wage ranges from the $11.50 earned by laborers and freight movers to the $20 of operating engineers. Median wages of occupations in job zone 3 vary widely, with recreation workers and maintenance and repair workers earning $10.50 and $17, respectively, and business operations specialists and agricultural managers earning more than $30 per hour.

Unionization in these occupations varies from the low 4–7 percent of farmworkers and recreation workers to 20 percent of construction workers and plumbers. For the management-level occupations, union representation is not applicable, but for occupations such as brownfield site manager, business operations specialist, and first-line office supervisor, workers may be eligible for union representation. These latter three occupations are notable because they have low union representation yet have among the highest median hourly wages. The occupations with higher levels of union representation are also among those with extremely low percentages of female workers.

The gender makeup of the workforce in these sustainable water occupations suggests a dividing line between clerical occupations and those that involve managerial or manual labor roles. For all but four of the fifteen occupations for which worker demographic data are available, the percentage of female workers is well below the national average of 47 percent. The four occupations with an above-average number of female workers are office clerk; receptionist and information clerk; bookkeeping, accounting, and auditing clerk; and business specialist, all of which are at least two-thirds women.

The racial makeup of workers in sustainable water occupations ranges from the predominantly white agricultural managers to the disproportionately Latino and African American laborers and truck drivers. Latino, African American, and Asian American workers constitute 30 percent of US workers and make up a similar proportion of workers in clerical occupations and carpenters. These workers are underrepresented in positions as farmers, ranchers, and other agricultural managers (6 percent); general and operations managers (17 percent); bookkeeping, accounting, and auditing clerks (20 percent); electricians (22.5 percent); operating engineers and other construction equipment operators (22 percent); and business operations specialists (25 percent).

The underrepresentation of people of color working as agricultural managers and general managers is concerning, given that these occupations are the two of the highest-paid sustainable water occupations that do not require advanced degrees. Agricultural managers and general managers are projected to have 235,000 and 410,000 openings in 2020 and currently have median hourly wages of $31 and $45.75, respectively.

Ten sustainable water occupations are projected to have more than 100,000 job openings by 2020 and require higher education and experience consistent with job zones 4 and 5. These occupations tend to have higher median wages and lower percentages of women, workers of color, and union representation than those requiring less education and training. Median hourly wages range from the $21 and $22 earned by executive secretaries and graphic designers to above $35 for sales representatives of wholesale products and manufacturers of technical products, civil engineers, management analysts, and construction managers. Women represent less than half the national average percentage of the workforce in positions as construction managers, cost estimators, and civil engineers, and women are also underrepresented as chief executives and management analysts. Union representation among these occupations is expectably low, considering that they are commonly management-level positions. Chief executives are the only occupation categorized in job zone 5, requiring the highest level of education and experience, and receive a median wage of $80 hourly.

Conclusions

The United States faces a complex set of water-related challenges that can be addressed only with a greater commitment to invest in sustainable solutions. Failure to rise to this challenge will leave crumbling infrastructure, contaminated waterways, and water shortages that threaten public health, economic growth, and ecosystems in myriad ways. Sustainable approaches take advantage of the potential for decentralized facilities; efficient technologies; flexible public, private, and community-based institutions; innovative economics; and human capital. These solutions can be grouped into five major strategies: urban water conservation and efficiency, stormwater management, restoration and remediation, alternative water sources, and agricultural water efficiency and quality.

The types of jobs involved in implementing sustainable water strategies cover a broad range of occupations. Our full analysis (Moore et al. 2013) identified 136 occupations involved in the work of achieving more sustainable water outcomes in agriculture, urban residential and commercial settings, restoration and remediation, alternative water sources, and stormwater management.

The extent of employment created by sustainable water practices is substantial. The data available point to 10–15 jobs per $1 million invested in alternative water supplies; 5–20 in stormwater management; 12–22 in urban conservation and efficiency; 14.6 in agricultural efficiency and quality; and 10–72 jobs per $1 million invested in restoration and remediation. The view that environmental conservation produces net job losses is not supported by any of the data on sustainable water projects.

The potential for job creation through implementation of sustainable water strategies has been largely ignored by policy makers, scholars, and practitioners, with the efforts on green jobs mostly limited to energy efficiency and renewable energy activities. This has left a gap in understanding and action on these opportunities.

Many of the occupations involved in sustainable water projects are also projected to have high demand in the overall economy. Thirty-seven sustainable water occupations are projected in the overall economy to have more than 100,000 job openings by 2020. Workers trained to advance water sustainability in these occupations will likely experience high demand and have the competitive edge of holding green skills. The high number of workers who will move into these occupations also means that there may be a growing demand for occupational training, creating an opportunity to train a new generation with the green skills that will make their sector more sustainable.

Numerous sustainable water occupations are accessible to workers without advanced degrees. Twenty-seven of the thirty-seven occupations projected to have 100,000 job openings by 2020 generally require on-the-job training, with some requiring previous experience and associate's degrees or technical training but not graduate degrees. This translates to a more feasible pathway to employment for adults without formal education beyond high school.

While data are limited, the existing training and education programs preparing workers with the skills needed in sustainable water fields appear nascent and small in scale. Conventional educational and job training resources such as community colleges and union apprenticeships are only beginning to integrate green skills into their water-related occupational training. And the link to disadvantaged communities is even more tenuous. Our analysis identified fewer than a dozen independent nonprofit organizations in the United States that are providing disadvantaged communities with job training and certification for sustainable water-related careers.

Existing programs linking disadvantaged communities to sustainable water opportunities face multiple challenges in designing and implementing programs, meeting the scale of need, and placing program graduates. Programs have difficulty matching training to actual labor demand for particular occupations and skills. Certifications and licensing related to sustainable water occupations too often are accepted in too small a geographic area or too narrow an industry to provide workers and employers with the needed level of confidence in their value.

Existing training programs have also developed promising strategies for connecting disadvantaged communities to sustainable water jobs. Hybrid models that both train and hire workers through coordinated business and nonprofit branches have found greater success in placing graduates and maintaining stable funding. Organizations with contracts to provide operation and maintenance services for public and private entities have also found a more stable source of funding and practical work experience for participants. However, training programs are unlikely to achieve substantial economic improvements for disadvantaged communities unless they are coupled with policy that increases and targets demand for new workers.

Recommendations

The research points to several action areas in research, policy, and community organizing that would facilitate greater opportunities in sustainable water jobs. These opportunities can best be understood, planned for, and taken advantage of with future efforts that consider the following recommendations.

Improve and Expand Data Collection and Research

Agencies managing state revolving funds and other state and federal funding programs should require grantees and loan recipients to submit information on job types and numbers using a template aligned with standard occupational codes.

Models for financing the ongoing operation and maintenance of sustainable water projects must be developed, piloted, and refined to ensure that desired environmental and economic outcomes are sustained. This research should start with the lessons of leading efforts at the local scale, such as those related to stormwater in Portland and Philadelphia.

Workforce development and training organizations must increase their capacity to track and evaluate job placement and other program outcomes to strengthen the feedback loop that will improve programming and broader understanding of best practices.

Economic research projecting job generation of sustainable water projects should incorporate data on the occupations and types of firms specific to these projects, as in some cases they differ significantly from those involved in conventional water project industries.

Integrate Sustainable Job Priorities into Policy and Planning

Water utilities, state water agencies, planning departments, and other public entities funding and managing sustainable water projects should implement "high-road" strategies that consider job quality, training, and targeted hiring as an integral component of project design and implementation. This should include local hiring and minority hiring requirements and incentives that increase contracting with and hiring of individuals from local and disadvantaged communities.

Considering the strong interdependence between water and energy use, policy and workforce efforts to take advantage of the win-win solutions at this nexus should be strengthened.

Better planning for and investment in financing of the ongoing operation and maintenance of sustainable water projects is needed to ensure that the maximum environmental and employment benefits are realized.

Support Community Programs and Partnerships

Unions, community-based organizations, and environmental advocates should join together in envisioning and promoting policies and funding programs that incorporate high-road work opportunities into sustainable water projects.

Workforce development and training organizations must build stronger partnerships with unions and employers to ensure that training is well aligned with emerging occupations and skill sets, and they must increase placement rates for program participants.

Training programs seeking to improve access of disadvantaged communities to sustainable water jobs should focus on occupations that are projected to have high labor demand in the overall economy. Data on the labor demand generated by sustainable water projects alone is not robust enough to justify training workers for these jobs unless their training will also prepare them to qualify for conventional occupations in demand.

Partnerships between industry associations, labor unions, and training programs are needed to develop, deliver, and evaluate and update standardized training and certification that reflects actual skills needed and that is accessible to disadvantaged communities. This should include the creation of a centralized, up-to-date clearinghouse of training curriculum and certification standards.

References

Alliance for Water Efficiency. 2008. Transforming Water: Water Efficiency as Stimulus and Long-Term Investment. Position Paper. December 4. Chicago: Alliance for Water Efficiency.

American Rivers and Alliance for Water Efficiency. 2008. Creating Jobs and Stimulating the Economy through Investment in Green Water Infrastructure. http://www.americanrivers.org/assets/pdfs/green-infrastructure-docs/green_infrastructure_stimulus_white_paper_final.pdf (accessed August 16, 2013).

American Rivers, Water Environment Federation, American Society of Landscape Architects, and ECONorthwest. 2012. Banking on Green: A Look at How Green Infrastructure Can Save Municipalities Money and Provide Economic Benefits Community-wide. http://www.asla.org/uploadedFiles/CMS/Government_Affairs/Federal_Government_Affairs/BankingonGreen HighRes.pdf.

Burns, P., and D. Flaming. 2011. Water Use Efficiency and Jobs. Los Angeles: Economic Roundtable. http://www.lacountycleanwater.org/files/managed/Document/487/Water Use Efficiency %26 Jobs.pdf.

Business United for Conservation Industry Partnership (BUCIP). 2010. Capturing the Storm: Profits, Jobs, and Training in Philadelphia's Stormwater Industry. Philadelphia, PA: Sustainable Business Network of Greater Philadelphia. http://www.sbnphiladelphia.org/images/uploads/Capturing the Storm - BUC Needs Assessment.pdf.

Caudill, J. 2008. The Economic Impacts of Restoration and Conservation-Related Expenditures: The Cache River Watershed in Southern Illinois. Arlington, VA: US Fish and Wildlife Service, Division of Economics, Census of Agriculture.

Chesapeake Bay Foundation. 2011. Debunking the "Job Killer" Myth: How Pollution Limits Encourage Jobs in the Chesapeake Bay Region. Annapolis, MD: Chesapeake Bay Foundation. http://www.cbf.org/document.doc?id=1023.

Christian-Smith, J., and P. H. Gleick, eds. 2012. *A Twenty-First Century U.S. Water Policy*. New York: Oxford University Press.

Crompton, J. L. 2005. The impact of parks on property values: Empirical evidence from the past two decades in the United States. *Managing Leisure* 10 (4): 203–218.

Everglades Foundation. 2010. Measuring the Economic Benefits of America's Everglades Restoration: An Economic Evaluation of Ecosystem Services Affiliated with the World's Largest Ecosystem Restoration Project. Report prepared by Mather Economics. http://everglades.3cdn.net/8edd03d0943ae993fe_e0m6i4gx2.pdf (accessed August 16, 2013).

Gershman, D., and J. Alexander. 2012. Cleveland Great Lakes Restoration Projects Producing Results for People, Communities. Ann Arbor, MI: Great Lakes Coalition. http://healthylakes.org/wp-content/uploads/2012/09/Cleveland-Success-Stories-20121.pdf (accessed December 2012).

Gleick, P. H. 2002. Soft water paths. *Nature* 418:373.

Gleick, P. H., H. Cooley, and D. Groves. 2005. California Water 2030: An Efficient Future. Oakland, CA: Pacific Institute.

Gleick, P. H., D. Haasz, C. Henges-Jeck, V. Srinivasan, G. Wolff, K. Cushing, and A. Mann. 2003. *Waste Not, Want Not: The Potential for Urban Water Conservation in California*. Oakland, CA: Pacific Institute.

Great Lakes Coalition. 2012. Great Lakes Restoration Projects: Producing Results for People, Communities. Douglas, MI: Great Lakes Coalition. http://healthylakes.org/wp-content/uploads/2012/02/2012-Success-Stories.pdf (accessed December 2012).

Green for All, American Rivers, Economic Policy Institute, and Pacific Institute. 2011. Water Works: Rebuilding Infrastructure, Creating Jobs, Greening the Environment. Oakland, CA: Green for All. http://greenforall.org/wordpress/wp-content/uploads/2012/07/Green-for-All-Water-Works.pdf.

Greenroofs.com. 2012. International Greenroof and Greenwall Projects Database. http://www.greenroofs.com/projects/plist.php (accessed November 2012).

Heberger, M. 2012. Australia's millennium drought: Impacts and responses. In *The World's Water: The Biennial Report on Freshwater Resources*, edited by P. H. Gleick, 7:97–126. Washington, DC: Island Press.

Jelks, H. L., S. J. Walsh, N. M. Burkhead, S. Contreras-Balderas, E. Díaz-Pardo, D. A. Hendrickson, J. Lyons, N. E. Mandrak, F. McCormick, J. S. Nelson, S. P. Platania, B. A. Porter, C. B. Renaud, J. J. Schmitter-Soto, E. B. Taylor, and M. L. Warren Jr. 2008. Conservation status of imperiled North American freshwater and diadromous fishes. *Fisheries* 33 (8): 372–406.

Lerner, S. 1997. *Eco-Pioneers: Practical Visionaries Solving Today's Environmental Problems*. Cambridge, MA: MIT Press.

Massachusetts Department of Fish and Game. 2012. The Economic Impacts of Ecological Restoration in Massachusetts. Boston: Massachusetts Department of Fish and Game, Division of Ecological Restoration. http://www.mass.gov/eea/docs/dfg/der/pdf/economic-impacts-ma-der.pdf.

Mayer, P. W., W. B. DeOreo, and D. M. Lewis. 2000. Seattle Home Water Conservation Study: The Impacts of High Efficiency Plumbing Fixture Retrofits in Single-Family Homes. Report prepared for Seattle Public Utilities and US Environmental Protection Agency. Boulder, CO: Aquacraft.

Moore, E., H. Cooley, J. Christian-Smith, K. Donnelly, K. Ongoco, and D. Ford. 2013. Sustainable Water Jobs: A National Assessment of Water-Related Green Job Opportunities. Oakland, CA: Pacific Institute. http://www.pacinst.org/reports/sustainable_water_jobs/.

National Agricultural Statistics Service (NASS). 2007. 2007 Census of Agriculture. United States Department of Agriculture, National Agricultural Statistics Service. http://www.agcensus.usda.gov/Publications/2007/Full_Report/usv1.pdf.

National Oceanic and Atmospheric Administration (NOAA). n.d. Streams and Rivers Restoration. NOAA Habitat Conservation Restoration Center. http://www.habitat.noaa.gov/restoration/techniques/srrestoration.html (accessed September 7, 2012).

Natural Resources Defense Council (NRDC). 2010. Climate Change, Water, and Risk: Current Water Demands Are Not Sustainable. Report prepared by Tetra Tech. New York: Natural Resources Defense Council. http://www.nrdc.org/globalwarming/watersustainability/files/WaterRisk.pdf.

Nielsen-Pincus, M., and C. Moseley. 2010. Economic and Employment Impacts of Forest and Watershed Restoration in Oregon. Eugene: University of Oregon, Institute for a Sustainable Environment. http://ewp.uoregon.edu/sites/ewp.uoregon.edu/files/downloads/WP24.pdf (accessed May 15, 2012).

Queensland Water Commission. 2010. Annual Report 2010–11. City East, Queensland, Australia: Queensland Water Commission. http://www.dews.qld.gov.au/__data/assets/pdf_file/0016/31471/qwc-annual-report-1011.pdf (accessed May 2, 2013).

Rephann, T. J. 2010. Economic Impacts of Implementing Agricultural Best Management Practices to Achieve Goals Outlined in Virginia's Tributary Strategy. Charlottesville: University of Virginia, Weldon Cooper Center for Public Service, Center for Economic and Policy Studies. http://www.cbf.org/document.doc?id=467.

Treuhaft, S., A. G. Blackwell, and M. Pastor. 2011. America's Tomorrow: Equity Is the Superior Growth Model. Oakland, CA: PolicyLink. http://www.policylink.org/atf/cf/%7B97c6d565-bb43-406d-a6d5-eca3bbf35af0%7D/SUMMIT_FRAMING_WEB.PDF.

United States Department of Labor, Bureau of Labor Statistics (US BLS). 2012a. Employment Projections: Table 1.10, Replacement Needs, Projected 2010–2012. Washington, DC: US Department of Labor, Bureau of Labor Statistics. http://www.bls.gov/emp/ep_table_110.htm.

———. 2012b. A Profile of the Working Poor, 2010. Report 1035. Washington, DC: US Department of Labor, Bureau of Labor Statistics. http://www.bls.gov/cps/cpswp2010.pdf.

United States Environmental Protection Agency (US EPA). 1995. Ecological Restoration: A Tool to Manage Stream Quality. EPA-841-F-95-007. Washington, DC: US Environmental Protection Agency, Office of Water. http://water.epa.gov/type/watersheds/archives/11.cfm.

———. 2002. The Clean Water and Drinking Water Infrastructure Gap Analysis. EPA-816-R-02-020. Washington, DC: US Environmental Protection Agency, Office of Water. http://water.epa.gov/aboutow/ogwdw/upload/2005_02_03_gapreport.pdf (accessed December 18, 2012).

———. 2004. 2004 Guidelines for Water Reuse. EPA-625-R-04-108. http://www.epa.gov/nrmrl/pubs/625r04108.html (accessed December 18, 2012).

———. 2006. The Wadeable Streams Assessment: A Collaborative Survey of the Nation's Streams. EPA-841-B-06-002. Washington, DC: US Environmental Protection Agency, Office of Research and Development, Office of Water. http://water.epa.gov/type/rsl/monitoring/upload/2007_10_25_monitoring_wsa_factsheet_10_25_06.pdf (accessed February 4, 2011).

———. 2007. Watershed Assessment, Tracking, and Environmental Results System (WATERS). http://water.epa.gov/scitech/datait/tools/waters/ (accessed August 9, 2013).

———. 2008. Clean Watersheds Needs Survey 2008: Report to Congress. EPA-832-R-10-002. http://water.epa.gov/scitech/datait/databases/cwns/upload/cwns2008rtc.pdf (accessed December 18, 2012).

———. 2009. Class V Fact Sheet: Aquifer Recharge and Aquifer Storage and Recovery Wells. http://water.epa.gov/type/groundwater/uic/upload/fs_uic_class5_fact_sheet_on_aquifer_storage_and_recovery_wells.pdf (accessed December 18, 2012).

———. 2012a. Federal Regulatory Programs: Water; Green Infrastructure. June. http://water.epa.gov/infrastructure/greeninfrastructure/gi_regulatory.cfm (accessed December 18, 2012).

———. 2012b. Low Impact Development (LID). http://water.epa.gov/polwaste/green/.

———. 2012c. Remediation Technologies: Tools and Resources to Assist in Contaminated Site Remediation. Superfund. http://www.epa.gov/superfund/remedytech/remed.htm (accessed December 18, 2012).

White, S., L. Dresser, and J. Rogers. 2012. Greener Reality: Jobs, Skills, and Equity in a Cleaner U.S. Economy. Madison: Center on Wisconsin Strategy. http://www.cows.org/_data/documents/1306.pdf.

CHAPTER 4

Hydraulic Fracturing and Water Resources

What Do We Know and Need to Know?

Heather Cooley and Kristina Donnelly

According to some energy analysts, natural gas is "poised to enter a golden age" as a result of the availability and development of large volumes of new sources of unconventional natural gas, including coal bed methane, tight gas, and shale gas. Historically, natural gas production from unconventional reserves has been limited. In 2010, unconventional natural gas accounted for about 14 percent of total global natural gas production. The International Energy Agency (IEA) projects that by 2013 annual production from unconventional sources will triple and will represent about one-third of all natural gas production (IEA 2012). While North America, especially the United States and Canada, dominated unconventional gas production in 2010, growth in unconventional gas production is expected widely around the world (IEA 2012). China, in particular, is projected to experience major increases in production, becoming the second-largest producer after the United States. While shale gas accounts for the vast majority of growth in natural gas production, some growth is also projected for tight gas.

Natural gas is typically classified as conventional or unconventional. Conventional natural gas is generally held as a pocket of gas *beneath* a rock layer with low permeability and flows freely to the surface once the well is drilled. By contrast, unconventional natural gas is more difficult to extract because it is trapped *in* rock with very low permeability. Extracting natural gas from unconventional sources is more complex and costly than conventional natural gas recovery. Technological improvements, however, have made extraction from unconventional sources more economically viable in recent years. In particular, the combination of horizontal drilling and hydraulic fracturing has greatly increased the productivity of natural gas wells. These new techniques have also raised concerns about the adverse environmental and social consequences of these practices, especially effects on water resources.

To date, much of the debate about hydraulic fracturing has centered on the use of chemicals and concerns that these chemicals could contaminate drinking water. In response, numerous states have passed or are considering regulations requiring natural gas operators to disclose the chemicals used during well injection. Additionally, the Groundwater Protection Council and the Interstate Oil and Gas Compact Commission have established a public website that allows companies to voluntarily disclose water and chemical usage for wells since January 2011 that have been hydraulically fractured, although it is of note that these data are not subject to third-party verification and are not in a format that can be searched or aggregated.

The debate has been particularly controversial in the United States, where the majority of unconventional natural gas development has been concentrated. To better identify and understand the key issues, the Pacific Institute conducted extensive interviews with a diverse group of stakeholders, including representatives from state and federal agencies, academia, industry, environmental groups, and community-based organizations from across the United States. This chapter provides a short summary of the key issues identified in the interviews and in an initial assessment and synthesis of existing research. It especially examines the impacts of hydraulic fracturing and unconventional natural gas extraction on water resources and identifies areas in which more information is needed. More detail is available in the full report (Cooley and Donnelly 2012).

Overview of Hydraulic Fracturing

Hydraulic fracturing, or fracking, refers to the process by which fluid is injected into wells under high pressure to create cracks and fissures in rock formations that improve the production of these wells. These fissures can extend more than 300 meters (1,000 feet) from the well (Veil 2010). The fracturing fluid consists of water, chemical additives, and a propping agent. The propping agent—typically sand, ceramic beads, or another incompressible material—holds open the newly created fissures to allow the natural gas to flow more freely. In the first few days to weeks after completion of the fracturing process, the well pressure is released and some of the fracturing fluid (referred to as flowback) flows back to the surface through the well bore. Some unknown volume of fracturing fluid, along with its chemical additives, remains underground. Over longer time periods, any water naturally present in the ground (referred to as produced water) continues to flow through the well to the surface. The flowback and produced water, which can be considerably saltier than seawater and contain a variety of other contaminants (IOGCC and ALL Consulting 2006), are typically stored on-site in tanks or pits before reuse, treatment, or disposal. There are varying and conflicting reports on whether and to what extent wells will be fracked multiple times over their productive life (Nicot et al. 2011), although this will likely depend on local geology, spacing of wells, and natural gas prices.

Hydraulic fracturing was first developed in the early twentieth century but was not commercially applied until the mid- to late 1940s. Although initially developed to improve the production of oil and gas wells, hydraulic fracturing has been used in other applications, including development of drinking water wells (NHDES 2010), disposal of wastes, and enhancement of electricity production from geothermal energy sources. Hydraulic fracturing is standard practice for extracting natural gas from unconventional sources, including coal beds, shale, and tight sands, and is increasingly being applied to conventional sources to improve their productivity. While the process is the same, the various applications of hydraulic fracturing differ in their water requirements, the amount and types of chemicals employed, and the quantity and quality of wastewater generated.

We note that there is no single definition of "hydraulic fracturing." Some, including industry representatives, define hydraulic fracturing narrowly, referring only to the process by which fluids are injected into a well bore. They argue that some of the challenges, such as wastewater disposal, spills, and leaks, are common to all oil and gas operations and therefore are not specifically associated with hydraulic fracturing. Others, however, define the issue more broadly to include impacts associated with well construction and completion, the hydraulic fracturing process itself, and well production and closure (US EPA 2011c; ProPublica 2012). For these groups, hydraulic fracturing and unconventional natural gas production are synonymous because hydraulic fracturing has allowed for the development of these unconventional natural gas resources. Without hydraulic fracturing, shale gas production would be far more limited. For the purposes of this analysis, we use a broader definition of hydraulic fracturing to include impacts associated with well construction and completion, the hydraulic fracturing process itself, and well production and closure (Cooley and Donnelly 2012).

Concerns Associated with Hydraulic Fracturing Operations

Hydraulic fracturing has generated a tremendous amount of controversy. There are daily media reports from outlets across the United States, Canada, South Africa, Australia, France, England, and elsewhere about environmental, social, economic, and community impacts. In an effort to identify the key issues, the Pacific Institute interviewed sixteen representatives of state and federal agencies, academia, industry, environmental groups, and community-based organizations in the United States. Their responses are summarized in figure 4.1. Although the sample size was relatively small, the interviews were extensive, and the detailed responses from these diverse stakeholders are similar across the spectrum and indicative of the broad range of concerns associated with hydraulic fracturing raised in other forums.

All of the interviewees indicated that impacts on the availability and quality of water resources were among the primary concerns associated with hydraulic fracturing operations. Water-related findings of the interviews include the following:

- Spills and leaks were the most commonly cited concern, with fourteen of the sixteen people interviewed expressing concern.

- Thirteen of the interviewees considered wastewater treatment and disposal to be key challenges. One industry representative noted that wastewater management was perhaps a larger issue than chemical usage.

- Three-quarters of the interviewees were concerned about the water requirements of hydraulic fracturing. This concern was not limited to interviewees in the most arid regions; rather, it was expressed by people working in various regions across the United States. In some cases, the concern was directly related to the effects of large water withdrawals on the availability of water for other uses. In other cases, concern was related to how large withdrawals would affect water quality.

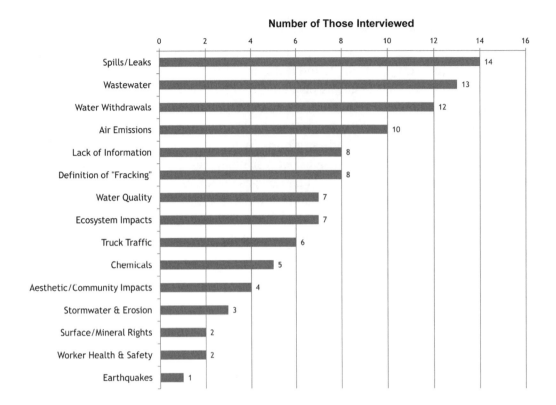

FIGURE 4.1 KEY CONCERNS IDENTIFIED BY INTERVIEWEES

Note: Results are based on interviews with sixteen representatives of state and federal agencies, academia, industry, environmental groups, and community-based organizations.

- Nearly half of the interviewees explicitly identified water quality as a key issue. Many of the other concerns mentioned, such as spills, leaks, and wastewater management, also imply concern about water quality. One interviewee expressed concern about surface water contamination associated with air emissions.

- Less than one-third of those interviewed specifically identified chemical usage and the associated risk of groundwater contamination as key issues, although many more expressed concern about groundwater contamination more broadly. Some of the interviewees thought that with so much attention given to chemical usage, inadequate attention is given to some of the other issues, such as wastewater disposal and methane migration, which may ultimately pose more serious risks.

- One issue identified in our interviews that was not directly related to environmental impacts was the overall lack of information, with half of those interviewed describing it as a key problem. Several commented on the complexity of the issues and the difficulty of explaining the technology to the general public.

Water Challenges

In this section, we summarize the available information on the following key water-related concerns identified by those interviewed: (1) water withdrawals; (2) groundwater contamination associated with well drilling and production; (3) wastewater management; (4) truck traffic and its impacts on water quality; (5) surface spills and leaks; and (6) stormwater management. This information is largely drawn from a review of the academic and gray literature and from media reports. Our focus throughout this chapter is on shale gas, although we discuss other unconventional natural gas sources for which information is readily available. Here, we evaluate the impacts associated with well construction and completion, the hydraulic fracturing process itself, and well production and closure.

Water Withdrawals

The drilling and fracking of a horizontal shale gas well uses large volumes of water, although the amount of water required is both variable and uncertain. The US Environmental Protection Agency (EPA) reports that fracturing of shale gas wells requires between 2.3 million and 3.8 million gallons of water per well (US EPA 2011c).[1] An additional 40,000–1,000,000 gallons is required to drill the well (GWPC and ALL Consulting 2009). This is considerably more water than is required for conventional gas wells and even for coal bed methane because the wells to access shale gas are deeper. Water requirements for hydraulic fracturing of coal bed methane, for example, range from 50,000 to 350,000 gallons per well (US EPA 2011c), although we note that these estimates may be outdated and may not include the application of more recent water-intensive processes.

New data, however, suggest that the water requirements for fracking of shale gas wells might be both much larger and more variable than is reported by the EPA (table 4.1). For example, Thomas Beauduy (2011) of the Susquehanna River Basin Commission finds that fracking in the Marcellus Shale region requires, on average, about 4.5 million gallons per well. Water requirements can be even greater within Texas's Eagle Ford Shale area, where fracking can use up to 13 million gallons of water per well (Nicot et al. 2011), with additional water required to drill the wells. These data highlight the significant variation among shale formations, driven in part by differences in the depth to the target formation, even among wells within close proximity of one another (Nicot et al. 2011). Estimation of water requirements is further complicated by uncertainty about how many times a single well will be fracked over the course of its productive life and by limited publicly available data.

Water for hydraulic fracturing is typically withdrawn from one location or watershed over several days (Veil 2010). Additionally, in some cases, the water is taken from "remote, often environmentally sensitive headwater areas" (Beauduy 2011, 34), where even small withdrawals can significantly affect the flow regime. As a result, while fracking may account for a small fraction of a state's or even a basin's water supply, there can be more severe local impacts. Additionally, much of the water injected underground either is not

---

1. While *The World's Water* volumes prefer to consistently use metric units, much of the research on water and fracking discussed here is based on work done in the United States, so we have opted to report the original units. For reference, one cubic meter contains 264.2 gallons. Additional conversion factors can be found in the Water Units, Data Conversions, and Constants section near the end of this book.

TABLE 4.1 Water Requirements for Hydraulic Fracturing by Shale Plays in Texas

Shale Play	Water Requirements (Gallons per Well)		
	Low Value	Median Value	High Value
Barnett Shale	< 1 million	2.6 million	> 8 million
Haynesville and Bossier Shales	< 1 million	5.5–6 million	> 10 million
Eagle Ford Shale	1 million	6–6.5 million	13 million
Woodford, Pearsall, and Barnett-PB Shales	< 1 million	0.75–1 million	< 5 million

Source: Estimated on the basis of data in Nicot et al. (2011).

recovered or is unfit for further use once it is returned to the surface, usually requiring disposal in an underground injection well. This water use represents a consumptive use if it is not available for subsequent use within the basin from which it was extracted. In some cases, water is treated and reused for subsequent fracking jobs, although this is still fairly uncommon, and no national estimate on the prevalence of this practice is available (US GAO 2012).

There is some evidence that the water requirements for hydraulic fracturing are already creating conflicts with other uses and could constrain future natural gas production in some areas. For example, in Texas, a major drought in 2011 prompted water agencies in the region to impose mandatory reductions in water use. Water agencies, some of which sold water to natural gas companies, indicated they might have to reconsider these sales if the drought persisted. Natural gas companies also tried to purchase water from local farmers, offering $9,500 to nearly $17,000 per million gallons of water (Carroll 2011). Likewise, at an auction of unallocated water in Colorado during the spring of 2012, natural gas companies successfully bid for water that had previously been largely claimed by farmers, raising concerns among some about the impacts on agriculture in the region and on ecosystems dependent on return flows (Finley 2012).

Concerns over water availability are not limited to drier climates. Pennsylvania is generally considered a relatively water-rich state. However, in August 2011, thirteen previously approved water withdrawal permits in Pennsylvania's Susquehanna River basin were temporarily suspended because of low stream levels; eleven of these permits were for natural gas projects (Susquehanna River Basin Commission 2011). While parts of the state were abnormally dry, the basin was not experiencing a drought at the time, suggesting that natural gas operations are already creating conflict with other water uses under normal conditions. In many basins, the application of fracking is still in its infancy and continued development could dramatically increase future water requirements and further intensify conflicts with other uses.

While water withdrawals directly affect the availability of water for other uses, water withdrawals can also affect water quality. For example, withdrawals of large volumes of water can adversely affect groundwater quality through a variety of means, such as mobilizing naturally occurring substances, promoting bacterial growth, causing land subsidence, and mobilizing lower-quality water from surrounding areas. Similarly, withdrawals from surface water can affect the hydrology and hydrodynamics of the source water (US EPA 2011c), and reductions in the volume of water in a surface water body can reduce the ability to dilute municipal or industrial wastewater discharges.

Given the proposed expansion of drilling in many regions, conflicts between natural gas companies and other users are likely to intensify. More and better data are needed on the volume of water required for hydraulic fracturing and the major factors that determine the volume, such as well depth and the nature of the geologic formation. Additional analysis is needed on the cumulative impacts of water withdrawals on local water availability, especially given that water for hydraulic fracturing can be a consumptive use of water. Finally, more research is needed to identify and address the impacts of these large water withdrawals on local water quality. This work must be done on a basin-by-basin level.

Groundwater Contamination Associated with Well Drilling and Production

Groundwater contamination from shale gas operations can occur through a variety of mechanisms. Natural gas is located at various depths, often (but not always) far below underground sources of drinking water. A well bore, however, must sometimes be drilled through these drinking water sources in order to access the gas. Chemicals and natural gas can escape the well bore if it is not properly sealed and cased. While there are state requirements for well casing and integrity, accidents and failures still occur, as was demonstrated by an explosion in Dimock Township, Pennsylvania (see box 4.1 for more information). Old, abandoned wells can also potentially serve as migration pathways (US EPA 2011b) for contaminants to enter groundwater systems. States have estimated that there are roughly 150,000 undocumented and abandoned oil and gas wells in the United States (IOGCC 2008). Natural underground fractures, as well as those potentially created during the fracturing process, could also serve as conduits for groundwater contamination (Myers 2012). Finally, coal bed methane is generally found at shallower depths and in closer proximity to underground sources of drinking water, and therefore accessing natural gas from this source might pose a greater risk of contamination.

Much of the debate about groundwater contamination—and some of the most striking visual images showing water and burning natural gas coming out of home faucets—is related to reports of methane contamination in drinking water. Nearly 90 percent of shale gas is composed of methane. A study in New York and Pennsylvania found that methane levels in drinking water wells in active gas production areas (less than 1 kilometer, or about five-eighths of a mile, from wells) were seventeen times higher than in those outside of active gas production areas. An isotopic analysis of the methane suggests that the methane in the active gas production areas originated from deep underground (Osborn et al. 2011).

Methane is not currently regulated in drinking water, although it can pose a public health risk. Robert B. Jackson of Duke University and his colleagues (2011) note that methane is not regulated in drinking water because it is not known to affect water's potability and does not affect its color, taste, or odor. Methane, however, is released from water into the atmosphere, where it can cause explosions, fires, asphyxiation, and other health or safety problems. The 2009 New Year's Day drinking water well explosion in Dimock, Pennsylvania, for example, was due to methane buildup associated with natural gas production. The US Department of the Interior recommends taking mitigative action when methane is present in water at concentrations exceeding 10 milligrams per liter (mg/l) (Eltschlager et al. 2001). A recent study, however, notes that research on the health effects is limited and recommends that "an independent medical review be initiated to evaluate the health effects of methane in drinking water and households" (Jackson et al. 2011, 5).

BOX 4.1 Dimock Township, Pennsylvania

Dimock Township is located in northeastern Pennsylvania's Susquehanna County, the heart of some of the most productive drilling areas in the Marcellus Shale play. On New Year's Day in 2009, a residential water well in Dimock exploded as a result of methane buildup in the well. Further investigation found methane gas in drinking water wells and in the headspaces of drinking water wells that provide water to local residents. These water wells were located near drilling wells owned and operated by Cabot Oil & Gas Corporation, and in February 2009 the Pennsylvania Department of Environmental Protection (DEP) issued a notice of violation against the company, which stated that Cabot had discharged natural gas, failed to properly cement casings, and failed to prevent natural gas from entering fresh groundwater (PA DEP 2009). Pennsylvania has what is called a "rebuttable presumption" for drinking water pollution, whereby the oil and gas operator is assumed to be responsible for drinking water pollution that occurs within 1,000 feet and within six months of a drilling operation, unless the company can provide baseline data to refute the claim. In the absence of baseline data, the company is required to replace the water that has reportedly been lost or degraded (025 Pa. Code §78.51). Cabot was ordered to install methane detectors in nine homes and provide drinking water to four homes in the affected area (Lobins 2009).

The DEP conducted an investigation into the methane contamination and determined that Cabot was responsible for polluting thirteen drinking water wells, which was later revised to include an additional five wells (PA DEP 2010). Other violations were found, including several cases of improper or insufficient casings and excessive borehole pressure. In November 2009, the DEP entered into a consent order and settlement agreement with Cabot that required the company to permanently restore or replace water supplies for the affected homes and fix any wells identified to have improper or insufficient casing (PA DEP 2009). Cabot was also ordered to cease drilling in the area, and the company was later completely banned from fracking new or existing wells until authorized by the DEP.

Six well owners signed agreements with Cabot and had water treatment systems installed, including methane venting systems, although most were still using bottled water because they lacked confidence in the treatment systems. Twelve well owners refused to sign agreements with Cabot and took part in a civil suit. Cabot continued to provide temporary water service to these twelve homes. In October 2011, however, the DEP formally stated that Cabot had fully complied with the consent order and was no longer required to provide drinking water to Dimock residents (Legere 2011). The DEP allowed Cabot to stop providing water to the twelve homes that had not installed the water treatment systems because Cabot had provided a solution and the well owners had been given sufficient time to sign the agreement (US EPA 2011a).

> Despite a subsequent announcement in December 2011 from the US Environmental Protection Agency (EPA) that Dimock water was safe to drink, local residents submitted results from their own testing, which indicated that the water was polluted (McAllister and Gardner 2012). In January 2012, the EPA began sampling water at sixty-four homes in the area and supplying drinking water to four households that had shown elevated levels of contaminants that pose a health concern (US EPA 2012a). Results of the testing indicated that although five homes showed elevated levels of arsenic, barium, and manganese—all naturally occurring substances not necessarily linked to fracking—the private wells did not have contaminant levels that posed a health concern or exceeded the safe range for drinking water (US EPA 2012b). The EPA's testing also concluded that elevated levels of methane were present in some of the wells, although Cabot disputes whether the methane resulted directly from the drilling.

There is also significant concern about groundwater contamination from hydraulic fracturing fluids, although limited data are available. According to draft reports released in December 2011 and September 2012, however, EPA testing detected the presence of chemicals commonly associated with hydraulic fracturing in drinking water wells in Pavillion, Wyoming (US EPA 2011b, 2012c). Encana Oil and Gas Inc., the company responsible for the natural gas wells, disputed the findings of the study, criticizing the EPA's testing methods and assumptions as well as the processes used to construct the monitoring wells and analyze the results (see box 4.2 for additional information).

Real analysis of the likelihood and extent of groundwater contamination is hindered by a lack of baseline data and confusion about definitions. Without baseline data, it is difficult to confirm or deny reports of groundwater contamination. In 2009, regulatory officials submitted signed statements to the United States Congress asserting that there were no confirmed cases of groundwater contamination associated with the hydraulic fracturing process (NYSDEC 2011). Likewise, an American Petroleum Institute report states that "there are zero confirmed cases of groundwater contamination connected to the fracturing operation in one million wells hydraulically fractured over the last 60 years" (American Petroleum Institute 2010). Yet documented cases in Dimock, Pennsylvania; possibly in Pavillion, Wyoming; and elsewhere provide evidence of groundwater contamination. In these cases, however, the contamination was associated with well casing integrity and wastewater disposal, not the process of injecting fluids underground per se—and so the issue is clouded by definitions.

Wastewater Management

Natural gas drilling also produces liquid waste. After completion of the fracturing process, well pressure is released and some of the fracturing fluid, along with naturally occurring substances, returns to the surface through the well bore. This mixture, commonly referred to as flowback, returns to the surface over the course of several hours to weeks after the fracturing process is completed (GWPC and ALL Consulting 2009). The amount of fracturing fluid that is actually recovered has not been well quantified and is likely to be highly variable, depending on local formation characteristics. While various sources quote estimates for the fracture fluid recovery rate (Beauduy 2011; Hoffman 2010; US EPA 2011c), a report by the Groundwater Protection Council (GWPC) and ALL Consulting (2009, 67) notes that "it is not possible . . . to differentiate flow back water from natural formation water." Thus, these estimates are likely based on assumptions rather than on actual data.

In addition to flowback, natural gas operations may generate produced water. Produced water "is any water that is present in a reservoir with the hydrocarbon resource and is produced to the surface with the crude oil or natural gas" (Veil et al. 2004, 1). Produced water can consist of natural formation water, that is, groundwater; naturally occurring substances, such as radioactive materials, metals, and salts; and even some residual fracturing fluid. The physical and chemical properties of produced water depend on the local geology (Veil et al. 2004). Flowback and produced water often have very high levels of total dissolved solids (TDS), in some cases exceeding 200,000 mg/l (Kargbo et al. 2010), nearly three times higher than seawater. In a recent report, the US Government Accountability Office (US GAO 2012) found that the volume of produced water generated by a given well varies depending on the type of hydrocarbon produced, the geographic location of the well, and the method of production.

Wastewater resulting from natural gas production is temporarily stored in pits, embankments, or tanks at the well site and then transported, usually via pipeline or truck, to a disposal site. Pits can lead to groundwater contamination, particularly if the pits are unlined or if the integrity of the lining is compromised. In Pavillion, Wyoming, for example, high concentrations of benzene, xylenes, and other organic compounds associated with gasoline and diesel were found in groundwater samples from shallow monitoring wells near pits (US EPA 2011b) (see box 4.2 for additional information on Pavillion, Wyoming).

Wastewater from natural gas operations can be disposed of using a variety of methods. In most areas, the primary way to dispose of wastewater from natural gas operations is by injection into a Class II well.[2] In 1988, the EPA made a determination that oil and gas waste is exempt from hazardous waste regulations under the Resource Conservation and Recovery Act of 1976. As a result, oil and gas wastes can be disposed of in Class II wells rather than in Class I hazardous waste wells.[3] Class II wells are subject to less stringent requirements than are Class I wells, and therefore disposal in Class II wells presents a greater risk of contaminating groundwater and triggering earthquakes than in Class I wells (Hammer and VanBriesen 2012).

The EPA estimates that there are about 144,000 Class II wells in operation in the United States, about 20 percent of which are disposal wells for brine and other fluids from oil and natural gas production. A Class II well might be an on-site well operated by the natural gas company or, more commonly, an off-site well operated by a commercial third

---

2. An injection well is a site where fluids, such as water, wastewater, brine, or water mixed with chemicals, are injected deep underground into porous rock formations, such as sandstone or limestone, or into or below the shallow soil layer. Injection wells are used for long-term storage, waste disposal, enhancement of oil production, mining, and prevention of saltwater intrusion.

3. States can adopt more stringent regulations if desired.

BOX 4.2 Pavillion, Wyoming

The Pavillion gas field is located in central Wyoming in the Wind River basin, the upper portion of which serves as the primary source of drinking water for the area. Oil and gas exploration began in the area in the 1950s and increased dramatically between 1997 and 2006. The Pavillion gas field is composed of a mix of sandstone and shale; at the time of this writing, the field had 169 vertical gas production wells. Encana Oil and Gas Inc. owns the rights to the Pavillion field and began drilling in the area in 2004 after acquiring another drilling company. Encana has not drilled any new wells since 2007 (US EPA 2011b).

In 2008, domestic well owners began complaining about taste and odor problems, and residents believed these issues to be linked to nearby natural gas activities. In response to complaints from local residents, the EPA initiated an investigation, collecting four rounds of water samples from thirty-five domestic wells and two municipal wells between 2009 and 2011. The EPA also installed two deep monitoring wells in 2010 and took two rounds of samples from each of these wells. According to a draft report released in December 2011, EPA testing found chemicals commonly associated with hydraulic fracturing in drinking water wells in the area (US EPA 2011b). The EPA also found that concentrations of dissolved methane in the domestic wells were higher near the gas production wells. The report concluded that nearby drilling activities had "likely enhanced gas migration" (US EPA 2011b).

Encana is disputing the EPA's preliminary findings. According to Encana, methane is "commonly known" to occur in the shallow groundwater aquifers in the area (Encana Oil and Gas Inc. 2011b) and is expected, given that the Pavillion gas field is also quite shallow (Encana Oil and Gas Inc. 2011a). Furthermore, Encana argues that Pavillion has always had poor water quality, referencing historical reports that levels of sulfate, total dissolved solids, and pH "commonly exceed state and federal drinking water standards" (Encana Oil and Gas Inc. 2011b). A 2011 report from the Wyoming Water Development Commission confirms that Pavillion's water is generally of poor quality and has often had taste and odor problems. However, the Commission states that nearly all of the private wells meet federal and state drinking water standards (James Gores & Associates 2011). One of the challenges associated with the EPA's analysis is that baseline data are not available to support claims about impacts on groundwater quality.

Encana continues to dispute the findings of the study, criticizing the EPA's testing methods and assumptions as well as the processes used to construct the monitoring wells and analyze the results (Gardner 2012). Although the EPA has indicated its intention to submit the report to scientific review, the comment period on the draft report has been extended several times, most recently until September 2013, while the EPA collects and distributes new information and meets with stakeholders (*Federal Register* 2013).

party (Veil 2010). In some cases, wastewater receives partial treatment prior to disposal to avoid clogging the well (Hammer and VanBriesen 2012).

With the proper safeguards, disposing of wastewater by underground injection reduces the risk of releasing wastewater contaminants into the environment; however, it increases the risk of earthquakes and can require the transport of wastewater over long distances (Hammer and VanBriesen 2012; Keranen et al. 2013). Some states do not have sufficient injection well capacity to handle the volume of wastewater generated from expanding hydraulic fracturing operations, so wastewater is transported to neighboring states for disposal (Veil 2010). For example, as of late 2010, Pennsylvania had only seven active disposal wells, and some wastewater had been hauled to Ohio, West Virginia, and other states for disposal (STRONGER 2010; Veil 2010).[4]

Flowback and produced water have been treated at municipal wastewater treatment plants (GWPC and ALL Consulting 2009), although this practice is both uncommon and controversial. Municipal systems are not typically designed to handle this type of wastewater, which can potentially disrupt the treatment process and discharge salts and other contaminants into the environment. In 2008 and 2009, TDS levels exceeded drinking water standards along Pennsylvania's Monongahela River, a major source of drinking water that receives discharges from facilities handing wastewater from natural gas production (STRONGER 2010). In 2009, excess TDS, primarily from mining discharges, "wiped out 26 miles of stream" in Greene County, Pennsylvania (STRONGER 2010, 22). In response, regulations for new or expanded facilities that accept oil and gas wastewater, including municipal wastewater treatment plants and centralized treatment plants, were passed in 2010 that set strict monthly discharge limits for TDS, chlorides, barium, and strontium (STRONGER 2010). Municipal wastewater treatment plants in Pennsylvania can still receive wastewater from "grandfathered" natural gas operations, although this has now been virtually eliminated (Hammer and VanBriesen 2012).

Wastewater reuse is becoming more common, driven in large part by the challenges associated with wastewater disposal and in part by the growing difficulty of finding new sources of water for fracking operations. Reusing wastewater for new fracking activities reduces the total volume of water required, helping to minimize impacts associated with water withdrawals. Wastewater can also be reused for irrigation, dust control on unpaved roads, and deicing of roads (US EPA 2011c; Hammer and VanBriesen 2012). In most cases, the wastewater must be treated prior to reuse, but in others it is simply blended with freshwater to bring the levels of TDS and other constituents down to an acceptable range (Veil 2010). Treatment for reuse can occur at the well site using a mobile plant or at a centralized industrial facility. Some downsides of reuse include the need for more on-site storage, energy requirements for the treatment processes, and additional transportation needed to haul wastewater to the treatment plant and among sites. Additionally, concentrated treatment residuals, including brine, must be disposed of in some manner and may require dilution (NYSDEC 2011).

Wastewater treatment and disposal associated with hydraulic fracturing may prove to be a larger issue than some of the other water-related risks. Yet to date there has been little discussion about the risks that wastewater treatment and disposal pose. In some areas, they may physically or economically constrain natural gas operations. Additional work is needed

---

4. Applications for at least twenty additional disposal wells are presently before the EPA (STRONGER 2010).

to understand the nature of the risk of wastewater treatment and disposal to human health and the environment and to identify where it may constrain natural gas operations.

Truck Traffic

Hydraulic fracturing operations generate a large amount of truck traffic. All of the materials and equipment needed for activities associated with hydraulic fracturing, including water and chemicals, are typically transported to the site by trucks (US EPA 2011c). Additionally, wastewater from natural gas operations is usually removed by tanker truck to the disposal site or to another well for reuse. Using information from the natural gas industry, the New York State Department of Environmental Conservation estimates that high-pressure hydraulic fracturing in a horizontal well would require 3,950 truck trips per well during early development of the well field (NYSDEC 2011), two to three times more than is required for conventional vertical wells (see table 4.2). Much of the truck traffic is concentrated over the first fifty days following well development. Truck traffic could be reduced by nearly 30 percent if pipelines were used to move water between sites, although pipelines can create other concerns, such as leaks, spills, and right-of-way controversies.

Truck traffic raises a variety of other water-related social and environmental concerns. Trucks increase wear and erosion on local roads and increase the risk of spills, both of which can pollute local surface water and groundwater. In addition, because so much of new drilling is occurring in rural locations, new roads must be built to accommodate the truck traffic, increasing habitat fragmentation and ecological disturbances.

TABLE 4.2 Truck Traffic Estimates for Vertical and Horizontal Wells

	Horizontal Well		Vertical Well	
Well Pad Activity	Heavy Truck	Light Truck	Heavy Truck	Light Truck
Drill pad construction	45	90	32	90
Rig mobilization	95	140	50	140
Drilling fluids	45		15	
Non-rig drilling equipment	45		10	
Drilling (rig crew, etc.)	50	140	30	70
Completion chemicals	20	326	10	72
Completion equipment	5		5	
Hydraulic fracturing equipment	175		75	
Hydraulic fracturing water hauling	500		90	
Hydraulic fracturing sand	23		5	
Produced water disposal	100		42	
Final pad preparation	45	50	34	50
Miscellaneous	—	85	—	85
Total one-way, loaded trips per well	1,148	831	398	507
Total vehicle round trips per well	**3,950**		**1,810**	

Source: NYSDEC (2011).

Note: Light trucks have a gross vehicle weight rating that ranges from 0 to 14,000 pounds. Heavy trucks have a gross vehicle weight rating in excess of 26,000 pounds. The gross vehicle weight is the maximum operating weight of the vehicle, including the vehicle's chassis, body, engine, engine fluids, fuel, accessories, driver, passengers, and cargo but excluding any trailers.

Surface Spills and Leaks

All fossil-fuel extraction activities come with some risk of surface water or groundwater contamination from the accidental or intentional release of waste. In the case of hydraulic fracturing, common wastes of concern include fracking fluid, additives, flowback, and produced water. Fluids released onto the ground from spills or leaks can run off into surface water and seep into groundwater.

Spills can occur at any stage during the drilling life cycle. Chemicals are hauled to the site, where they are mixed to form the fracturing fluid. Accidents and equipment failure during on-site mixing of the fracturing fluid can release chemicals into the environment. Above ground storage pits, tanks, or embankments can fail. Vandalism and other illegal activities can also result in spills and improper wastewater disposal. For example, in Canton Township, Pennsylvania, a January 2012 spill of 20,000 gallons of hydraulic fracturing wastewater is being investigated as "criminal mischief" (Clarke 2012). In a larger incident in March 2012, criminal charges were filed against a waste-hauling company and its owner for illegally dumping millions of gallons of produced water into streams and mine shafts and on properties across southwestern Pennsylvania (Pennsylvania Attorney General 2012). Given the large volume of truck traffic associated with hydraulic fracturing, truck accidents can also lead to chemical and wastewater spills. In December 2011, a truck accident in Mifflin Township, Pennsylvania, released fracking wastewater into a nearby creek (Reppert 2011).

While there are reports of spills and leaks associated with hydraulic fracturing operations, the national extent of the problem is not yet well understood. A recent report from Pennsylvania documented a string of violations in the Marcellus Shale region, many of which could result in surface spills and leaks, including 155 industrial waste discharges, 162 violations of wastewater impoundment construction regulations, and 212 faulty pollution prevention practices (Pennsylvania Land Trust Association 2010), during the thirty-two-month period from January 2008 to August 2010. New research provides documentation of twenty-four cases in six states of adverse health effects on humans, companion animals, livestock, horses, and wildlife associated with natural gas operations, including spills and leaks (Bamberger and Oswald 2012). Additional research is needed on the frequency, severity, cause, and impact of spills associated with hydraulic fracturing.

Stormwater Management

Stormwater runoff carries substances from the land surface that can be detrimental to water quality and ecosystem health and deposits them into local waterways. While runoff is a natural occurrence, human disturbances to the land surface have increased the timing, volume, and composition of runoff. According to the EPA, a 0.4-hectare (one-acre) construction site with no runoff controls can contribute thirty to forty metric tons of sediment each year, comparable to the runoff from six and one-half hectares (sixteen acres) of natural vegetated meadow (US EPA 2007a; Schueler 1994). Drilling for natural gas contributes to this problem, as the process requires disturbances to the land surface. Modern natural gas drilling requires the clearing of three or more hectares (typically seven to eight acres) per well pad, which includes area for the pad itself plus additional land for access roads, waste pits, truck parking, equipment, and more (Johnson 2010). Runoff can also contain pollutants from contact with drilling and construction equipment as well as with storage facilities for fracking fluid and produced water.

Stormwater discharges are regulated by state and local governments. The National Pollutant Discharge Elimination System (NPDES) program regulates stormwater runoff at the federal level, although states can receive primacy to administer their own permitting program. At the federal level, oil and gas operations have been afforded special protections and are exempt from provisions in the Clean Water Act. Consequently, oil and gas operators are not required to obtain a stormwater permit unless, over the course of operation, the facility generates stormwater discharge containing a reportable quantity of oil or hazardous substances or the facility violates a water quality standard (40 CFR 122.26(c)(1)(iii)).[5] In 2005, the definition of oil and gas exploration and production was broadened to include construction and related activities, although regulations still require well pads larger than one acre to apply for an NPDES stormwater permit (Wiseman 2012).[6] A 2005 study of the surface water impacts of natural gas drilling noted the difficulty of monitoring and suggested that few facilities were monitoring in a way that would allow them to determine whether they even required an NPDES permit (US EPA 2007b).

Conclusions

Energy analysts project massive increases in domestic natural gas production over the next twenty-five years. This increase is expected to be largely supplied by unconventional sources, especially shale gas. Although previously too expensive to develop, unconventional natural gas resources have become more economically viable in recent years as a result of the application of horizontal drilling and hydraulic fracturing. These technological advances have allowed for a rapid expansion of natural gas development both in areas accustomed to natural gas operations and in new areas.

Hydraulic fracturing has generated a tremendous amount of controversy in recent years. Hydraulic fracturing is hailed by some as a game changer that promises increased energy independence, job creation, and lower energy prices. Others have called for a temporary moratorium or a complete ban on hydraulic fracturing because of concern over environmental, social, and public health concerns. There are daily media reports on this topic from outlets across the United States and in a host of other countries, including Canada, South Africa, Australia, France, and England.

In an effort to identify the key issues, the Pacific Institute interviewed a diverse set of representatives of state and federal agencies, academia, industry, environmental groups, and community-based organizations in the United States. Despite the diversity of viewpoints, there was surprising agreement about the range of concerns and issues associated with hydraulic fracturing. Interviewees identified a broad set of social, economic, and environmental concerns, foremost among which are impacts of hydraulic fracturing on the availability and quality of water resources. In particular, key water-related concerns identified by the interviewees included (1) water withdrawals; (2) groundwater contamination associated with well drilling and production; (3) wastewater management; (4) truck traffic and its impacts on water quality; (5) surface spills and leaks; and (6) stormwater management (Cooley and Donnelly 2012).

Much of the media attention on hydraulic fracturing and its risk to water resources has centered on the use of chemicals in the fracturing fluids and the risk of groundwater

---

5. This requirement will not be met by sediment discharges alone.
6. States can implement stronger requirements if desired.

contamination. The mitigation strategies identified to address this concern have centered on disclosure and, to some extent, the use of less toxic chemicals. Risks associated with fracking chemicals, however, are not the only issues that must be addressed. Indeed, interviewees more frequently identified as key issues the overall water requirements of hydraulic fracturing and the quantity and quality of wastewater generated.

Most significantly, a lack of credible and comprehensive data and information is a major impediment to identifying or clearly assessing the key water-related risks associated with hydraulic fracturing and to developing sound policies to minimize those risks. Given the nature of the business, industry has an incentive to keep the specifics of its operations secret in order to gain a competitive advantage, avoid litigation, and so forth. Additionally, there are few peer-reviewed scientific studies on the process and its environmental impacts. While much has been written about the interaction of hydraulic fracturing and water resources, the majority of this writing is either industry or advocacy reports that have not been peer-reviewed. As a result, the discourse around the issue is largely driven by opinion. This hinders a comprehensive analysis of the potential environmental and public health risks and identification of strategies to minimize these risks.

Finally, the dialogue about hydraulic fracturing has been marked by confusion and obfuscation because of a lack of clarity about the terms used to characterize the process. For example, the American Petroleum Institute and other industry groups, using a narrow definition of fracking, argue that there is no link between their activities and groundwater contamination (American Petroleum Institute 2010), despite observational evidence of groundwater contamination in Dimock, Pennsylvania, and Pavillion, Wyoming, that appears to be linked to the integrity of the well casings and of wastewater storage. Additional work is needed to clarify terms and definitions associated with hydraulic fracturing to support more fruitful and informed dialogue and to develop appropriate energy, water, and environmental policy.

References

American Petroleum Institute. 2010. Freeing Up Energy—Hydraulic Fracturing: Unlocking America's Natural Gas Resources. http://www.api.org/~/media/Files/Policy/Exploration/HYDRAULIC_FRACTURING_PRIMER.ashx.

Bamberger, M., and R. E. Oswald. 2012. Impacts of gas drilling on human and animal health. *New Solutions: A Journal of Environmental and Occupational Health Policy* 22 (1): 57–77.

Beauduy, T. W. 2011. Hearing on Shale Gas Production and Water Resources in the Eastern United States, October 20, 2011. US Senate Committee on Energy and Natural Resources, Subcommittee on Water and Power. http://energy.senate.gov/public/index.cfm/files/serve?File_id=0da002e7-87d9-41a1-8e4f-5ab8dd42d7cf.

Carroll, J. 2011. Worst drought in more than a century strikes Texas oil boom. Bloomberg, June 13. http://www.bloomberg.com/news/2011-06-13/worst-drought-in-more-than-a-century-threatens-texas-oil-natural-gas-boom.html.

Clarke, C. R. 2012. Police: Spill at gas well site may have been vandalism. *Williamsport Sun-Gazette*, January 13. http://www.sungazette.com/page/content.detail/id/573278/Police--Spill-at-gas-well-site-may-have-been-vandalism.html?nav=5011.

Cooley, H., and K. Donnelly. 2012. Hydraulic Fracturing and Water Resources: Separating the Frack from the Fiction. Oakland, CA: Pacific Institute.

Eltschlager, K. K., J. W. Hawkins, W. C. Ehler, and F. Baldassare. 2001. Technical Measures for the Investigation and Mitigation of Fugitive Methane Hazards in Areas of Coal Mining. Pittsburgh, PA: US Department of the Interior, Office of Surface Mining Reclamation and Enforcement,

Appalachian Regional Coordinating Center. http://www.osmre.gov/resources/newsroom/News/Archive/2001/090601.pdf.

Encana Oil and Gas Inc. 2011a. Encana Pavillion news media conference call—Technical briefing, December 20. http://www.encana.com/pdf/news-stories/encana-pavillion-technical-briefing.pdf.

Encana Oil and Gas Inc. 2011b. Why Encana Refutes U.S. EPA Pavillion Groundwater Report. Press release. *Business Wire*, December 12. http://www.encana.com/news-stories/news-release/details.html?release=632327.

Federal Register. 2013. Draft research report: Investigation of ground water contamination near Pavillion, WY. EPA Notice of Extension of Public Comment Period. *Federal Register* 78 (8): 2396–2397.

Finley, B. 2012. Colorado farms planning for dry spell losing auction bids for water to fracking projects. *Denver Post*, April 1. http://www.denverpost.com/environment/ci_20299962/colorado-farms-planning-dry-spell-losing-auction-bids.

Gardner, T. 2012. Update 1—EPA to retest Wyoming water said tainted by fracking. Reuters, March 9. http://uk.reuters.com/article/2012/03/09/usa-epa-fracking-idUKL2E8E9C4220120309.

Ground Water Protection Council (GWPC) and ALL Consulting. 2009. Modern Shale Gas Development in the United States: A Primer. Report prepared for US Department of Energy, Office of Fossil Energy; and National Energy Technology Laboratory. http://www.all-llc.com/publicdownloads/ShaleGasPrimer2009.pdf.

Hammer, R., and J. VanBriesen. 2012. In Fracking's Wake: New Rules Are Needed to Protect Our Health and Environment from Contaminated Wastewater. Natural Resources Defense Council. http://www.nrdc.org/energy/files/Fracking-Wastewater-FullReport.pdf.

Hoffman, J. 2010. Susquehanna River Basin Commission Natural Gas Development. Presentation at Science of the Marcellus Shale Symposium, January 29, Lycoming College, Williamsport, PA. http://www.srbc.net/programs/docs/SRBC%20Science%20of%20the%20marcellus%20012910.pdf.

International Energy Agency (IEA). 2012. Golden Rules for a Golden Age of Gas: *World Energy Outlook* Special Report on Unconventional Gas. Paris: International Energy Agency. http://www.worldenergyoutlook.org/media/weowebsite/2012/goldenrules/WEO2012_GoldenRulesReport.pdf.

Interstate Oil and Gas Compact Commission (IOGCC). 2008. Protecting Our Country's Resources: The States' Case; Orphaned Well Plugging Initiative. Report prepared for US Department of Energy, National Energy Technology Laboratory. http://groundwork.iogcc.org/sites/default/files/2008-Protecting-Our-Country%27s-Resources-The-States%27-Case.pdf.

Interstate Oil and Gas Compact Commission (IOGCC) and ALL Consulting. 2006. A Guide to Practical Management of Produced Water from Onshore Oil and Gas Operations in the United States. Report prepared for US Department of Energy, National Petroleum Technology Office. http://fracfocus.org/sites/default/files/publications/a_guide_to_practical_management_of_produced_water_from_onshore_oil_and_gas_operations_in_the_united_states.pdf.

Jackson, R. B., B. R. Pearson, S. G. Osborn, N. R. Warner, and A. Vengosh. 2011. Research and Policy Recommendations for Hydraulic Fracturing and Shale-Gas Extraction. Durham, NC: Duke University, Center on Global Change. http://www.nicholas.duke.edu/cgc/HydraulicFracturingWhitepaper2011.pdf.

James Gores & Associates. 2011. Pavillion Area Water Supply Level I Study. Report prepared for State of Wyoming Water Development Commission. http://wwdc.state.wy.us/agency_publications/PavillionWaterSupplyLl_exesum-2011.pdf.

Johnson, N. 2010. Pennsylvania Energy Impacts Assessment Report 1: Marcellus Shale Natural Gas and Wind. The Nature Conservancy. http://www.nature.org/media/pa/tnc_energy_analysis.pdf.

Kargbo, D. M., R. G. Wilhelm, and D. J. Campbell. 2010. Natural gas plays in the Marcellus Shale: Challenges and potential opportunities. *Environmental Science and Technology* 44 (15): 5679–5684.

Keranen, K. M., H. M. Savage, G. A. Abers, and E. S. Cochran. 2013. Potentially induced earthquakes in Oklahoma, USA: Links between wastewater injection and the 2011 M_w 5.7 earthquake sequence. *Geology*, published online March 26. doi:10.1130/G34045.1.

Legere, L. 2011. DEP: Cabot OK to stop Dimock water deliveries. *Scranton Times-Tribune*, October 20. http://thetimes-tribune.com/news/dep-cabot-ok-to-stop-dimock-water-deliveries-1.1220855#axzz1bLXITzCt.

Lobins, S. C. 2009. Notice of Violation: Letter from Pennsylvania Department of Environmental Protection to Cabot Oil & Gas Corporation. February 27. http://s3.amazonaws.com/propublica/assets/methane/pdep_nov_cabot_090227.pdf.

McAllister, E., and Gardner, T. 2012. EPA may truck water to residents near fracking site. Reuters, January 6. http://www.reuters.com/article/2012/01/06/us-usa-fracking-epa-idUSTRE8041YE20120106.

Myers, T. 2012. Potential contaminant pathways from hydraulically fractured shale to aquifers. *Groundwater* 50 (6): 872–882. doi:10.1111/j.1745-6584.2012.00933.x.

New Hampshire Department of Environmental Services (NHDES). 2010. Well Development by Hydro-fracturing. Environmental Fact Sheet WD-DWGB-1-3. http://des.nh.gov/organization/commissioner/pip/factsheets/dwgb/documents/dwgb-1-3.pdf.

New York State Department of Environmental Conservation (NYSDEC). 2011. Revised Draft, Supplemental Generic Environmental Impact Statement on the Oil, Gas, and Solution Mining Regulatory Program—Well Permit Issuance for Horizontal Drilling and High-Volume Hydraulic Fracturing to Develop the Marcellus Shale and Other Low-Permeability Gas Reservoirs. Albany: New York State Department of Environmental Conservation, Division of Mineral Resources, Bureau of Oil and Gas Regulation. http://www.dec.ny.gov/data/dmn/rdsgeisfull0911.pdf.

Nicot, J.-P., A. K. Hebel, S. M. Ritter, S. Walden, R. Baier, P. Galusky, J. Beach, R. Kyle, L. Symank, and C. Breton. 2011. Current and Projected Water Use in the Texas Mining and Oil and Gas Industry. Report prepared for Texas Water Development Board. https://www.twdb.texas.gov/publications/reports/contracted_reports/doc/0904830939_MiningWaterUse.pdf.

Osborn, S. G., A. Vengosh, N. R. Warner, and R. B. Jackson. 2011. Methane contamination of drinking water accompanying gas-well drilling and hydraulic fracturing. *Proceedings of the National Academy of Sciences* 108 (20): 8172–8176. doi:10.1073/pnas.1100682108.

Pennsylvania Attorney General. 2012. Greene County Business Owner Charged with Illegally Dumping Millions of Gallons of Gas Drilling Waste Water and Sewage Sludge. Press release. Accessed May 24, 2012. http://www.attorneygeneral.gov/press.aspx?id=6030.

Pennsylvania Department of Environmental Protection (PA DEP). 2009. Consent Order and Agreement. http://s3.amazonaws.com/propublica/assets/natural_gas/final_cabot_co-a.pdf.

———. 2010. Modification to Consent Order and Agreement Dated November 4, 2009. http://www.marcellus-shale.us/pdf/Cabot_Consent-Mod_4-15-2010.pdf.

Pennsylvania Land Trust Association. 2010. Marcellus Shale Drillers in Pennsylvania Amass 1614 Violations Since 2008—1056 Identified as Most Likely to Harm the Environment. http://conserveland.org/violationsrpt.

ProPublica. 2012. What Is Hydraulic Fracturing? Fracking—Gas Drilling's Environmental Threat. http://www.propublica.org/special/hydraulic-fracturing-national.

Reppert, J. 2011. Collision spills fracking fluid on state route. *Williamsport Sun-Gazette*, December 27. http://www.sungazette.com/page/content.detail/id/572658/Collision-spills-fracking-fluid-on-state-route.html?nav=5011.

Schueler, T. R. 1994. The importance of imperviousness. *Watershed Protection Techniques* 1 (3): 100–111.

State Review of Oil and Natural Gas Environmental Regulations (STRONGER). 2010. Pennsylvania Hydraulic Fracturing State Review. http://www.shalegas.energy.gov/resources/071311_stronger_pa_hf_review.pdf.

Susquehanna River Basin Commission. 2011. What's New: 13 Water Withdrawals Remain on Hold to Protect Streams in the Susquehanna Basin. http://www.srbc.net/whatsnew/Newsletters/article_58.asp.

United States Environmental Protection Agency (US EPA). 2007a. Developing Your Stormwater Pollution Prevention Plan: A Guide for Construction Sites. EPA-833-R-06-004. http://www.epa.gov/npdes/pubs/sw_swppp_guide.pdf.

———. 2007b. Summary of the Results of the Investigation Regarding Gas Well Site Surface Water Impacts. http://www.epa.gov/npdes/pubs/oilandgas_gaswellsummary.pdf.

———. 2011a. ATSDR Record of Activity/Technical Assist. http://www.epa.gov/aboutepa/states/dimock-atsdr.pdf (accessed May 24, 2012).

———. 2011b. Investigation of Ground Water Contamination near Pavillion, Wyoming: Draft. Washington, DC: US Environmental Protection Agency, Office of Research and Development. http://www.epa.gov/region8/superfund/wy/pavillion/EPA_ReportOnPavillion_Dec-8-2011.pdf.

———. 2011c. Plan to Study the Potential Impacts of Hydraulic Fracturing on Drinking Water Resources. EPA/600/R-11/122. Washington, DC: US Environmental Protection Agency, Office of Research and Development. http://water.epa.gov/type/groundwater/uic/class2/hydraulicfracturing/upload/hf_study_plan_110211_final_508.pdf.

———. 2012a. Action Memorandum—Request for Funding for a Removal Action at the Dimock Residential Groundwater Site, Intersection of PA Routes 29 & 2024, Dimock Township, Susquehanna County, Pennsylvania. Philadelphia, PA: US Environmental Protection Agency, Region 3. http://www.fossil.energy.gov/programs/gasregulation/authorizations/Orders_Issued_2012/58._EPA_III.pdf.

———. 2012b. EPA Completes Drinking Water Sampling in Dimock, PA. Press release, July 25. http://yosemite.epa.gov/opa/admpress.nsf/d0cf6618525a9efb85257359003fb69d/1a6e49d193e1007585257a46005b61ad!opendocument.

———. 2012c. Investigation of Ground Water Contamination near Pavillion, Wyoming, Phase V Sampling Event: Summary of Methods and Results. Denver, CO: US Environmental Protection Agency, Region 8. http://www.epa.gov/region8/superfund/wy/pavillion/phase5/PavillionSeptember2012Narrative.pdf.

United States Government Accountability Office (US GAO). 2012. Energy-Water Nexus: Information on the Quantity, Quality, and Management of Water Produced during Oil and Gas Production. GAO-12-256. Washington, DC: US Government Accountability Office. http://www.gao.gov/assets/590/587522.pdf.

Veil, J. A. 2010. Water Management Technologies Used by Marcellus Shale Gas Producers. Report prepared by Argonne National Laboratory, Environmental Science Division, for the US Department of Energy, Office of Fossil Energy, National Energy Technology Laboratory. http://fracfocus.org/sites/default/files/publications/water_management_in_the_marcellus.pdf.

Veil, J. A., M. G. Puder, D. Elcock, and R. J. Redweik Jr. 2004. A White Paper Describing Produced Water from Production of Crude Oil, Natural Gas, and Coal Bed Methane. Report prepared by Argonne National Laboratory for US Department of Energy, National Energy Technology Laboratory. http://www.ipd.anl.gov/anlpubs/2004/02/49109.pdf.

Wiseman, H., and F. Gradijan. 2012. Regulation of Shale Gas Development, Including Hydraulic Fracturing. University of Tulsa Legal Studies Research Paper No. 2011-11.

CHAPTER 5

Water Footprint

Julian Fulton, Heather Cooley, and Peter H. Gleick

We use water for a variety of purposes in our daily lives: for drinking, bathing, washing our clothes, watering our gardens, and more. These direct uses of water, however, are only part of the story. Water is also required to produce nearly everything we use and consume, from the food we eat and the clothes we wear to the technological devices that are integral to our modern society. Because this indirect water isn't visible in the product, it is also often referred to as "embedded," "embodied," or "virtual" water. A full measure, therefore, of the water footprint of an individual, industrial sector, or society is the combination of direct water use and the water used indirectly to provide the goods and services consumed.

As pressures on water resources intensify globally, there is growing interest in evaluating the complex ways in which everyday human activities affect the world's water resources. Traditionally water has been thought of as a local or regional issue, but as globalization has forged increasing interconnectedness among people and economies, better understanding is needed of the ways in which observed impacts to water systems have important global dimensions. The "water footprint" has emerged as one tool for identifying and quantifying these impacts.

In recent years, there has been an explosion of interest in developing the methods and data necessary for evaluating and comparing water footprints. In this chapter we discuss the concept of a water footprint and how it is related to other footprint indicators, such as the carbon footprint and the ecological footprint; how the water footprint is calculated; some findings from recent research; and finally what the water footprint means in the context of California.

The Water Footprint Concept

The water footprint of a product (good or service) has been defined as the quantity of freshwater consumptively used, both directly and indirectly, throughout its supply chain (Hoekstra et al. 2011). Consumptive and nonconsumptive uses of water occur at almost every step along the supply chain of a product, but the water footprint counts only the consumptive portion because the nonconsumptive portion can theoretically be used for other purposes (see box 5.1). For example, an average of 140 liters (37 gallons) of water is consumptively used to produce a standard 125-milliliter (4.2-ounce) cup of coffee, including the water required to grow the coffee crop and process the beans. In fact, all food products require water in their production. Meat and dairy products tend to be very

> **BOX 5.1** A Note About Water Use
>
> The water literature is rife with confusing and often misleading terminology to describe water use—"water withdrawal," "consumptive use," "nonconsumptive use," "real water," "paper water," and more. It is important to clarify these terms, as different meanings can lead to different or conflicting conclusions. A water *withdrawal* is commonly defined as water taken from a source and used for some human need. Water withdrawals can be divided into two categories, consumptive and nonconsumptive use. *Consumptive use* typically refers to "water withdrawn from a source and made unavailable for reuse in the same basin, such as through conversion to steam, losses to evaporation, seepage to a saline sink, or contamination" (Gleick 2003, 278). Additionally, water that is incorporated into goods or plant and animal tissue is unavailable for reuse and thus is also considered a consumptive use. *Nonconsumptive use*, on the other hand, refers to water that can be made available for reuse within the basin from which it was extracted, such as through return flows or recoverable wastewater discharge into water bodies.

water intensive because water is required not only for bathing and hydrating the animals but also, and especially, for all the feed they eat over their life span.

It isn't just food that requires water to produce. The water footprint of cotton clothing, for example, includes the water for growing cotton, processing and dyeing fabric, and various finishing procedures—and the footprint of a typical T-shirt has been estimated at almost 2,500 liters (over 650 gallons) (Chapagain et al. 2006; Mekonnen and Hoekstra 2010).[1] Many industrial products, such as electronics and cars, can require vast amounts of water in their production, from the mining of various metals to the transport, processing, and disposal of a wide range of materials. Energy production is also a significant consumer of water. Generation of a megawatt-hour (MWh) of electricity, enough to supply the average US household for about a month, consumes an average of 1,100 liters (288 gallons) in the western United States (Cooley et al. 2011), but it can consume around 250,000 liters (65,000 gallons) for inefficient hydroelectricity produced in a hot, dry climate (Mekonnen and Hoekstra 2012). The production of goods is often much more water intensive than the provision of services, but even service industries such as hospitality (e.g., hotels), food (restaurants), health care (hospitals), and recreation (golf courses, reservoirs) can require significant levels of consumptive water use.

These are examples of just some of the products that are a part of our everyday lives, and they all require some amount of consumptive water use before we purchase and use or consume them. When we consume them, there is a connection between our action and the water that was consumptively used because our action has generated a demand for that use. The water footprint is therefore considered a consumption-based metric

---

1. This figure does not include the water used to wash the product over its life.

because it attributes the water use to the consumer rather than the producer. While this is the opposite of the way we have historically thought about water use, it is actually conceptually very close to other, more familiar ways of measuring people's interactions with the global environment.

Water, Carbon, and Ecological Footprints and Nexus Thinking

Just as the water footprint is a consumption-based metric, so too are the carbon footprint and the ecological footprint. A carbon footprint measures the greenhouse gas emissions associated with people's activities and consumption habits in order to create a better understanding of their contribution to global climate change. Transportation and other energy-intensive activities, for example, tend to generate large carbon footprints because so much of the global population's energy system is based on fossil fuels instead of renewable forms of energy. The ecological footprint measures the amount of biologically productive land and water area that is appropriated by a product or activity; it is measured as the amount of "global hectares" required to sustain an individual or population (see http://www.footprintnetwork.org).

Table 5.1 compares water, carbon, and ecological footprint indicators. There may be synergies as well as trade-offs in these indicators. For example, some renewable energy technologies that have a lower carbon footprint, such as photovoltaics, also have a lower water footprint. However, crop varieties that are less water intensive might be grown only in regions requiring more extensive land use or more transportation fuels to get them to market. It is important to engage in "nexus thinking" (e.g., the water-energy nexus) by considering the interactions of a range of environmental indicators when making decisions, whether personal, business, or governmental.

Furthermore, these consumption-based indicators should not be understood to replace or obviate the need for production-based indicators. Traditional regulation and other ways of working with producers are still directly relevant to reducing impacts to the local environment. Rather, consumption-based indicators provide a perspective that creates options for new leverage points in addressing their respective concerns. For the water footprint, this concern has to do with water availability in the places where products are produced. The next section discusses details of the water footprint and how it is measured.

Calculating a Water Footprint

A water footprint provides a metric and methodology for quantifying consumptive use, including additional qualitative features about the water used, such as where the water comes from, the kinds and quality of water used, and more, as discussed below. The basic approach in calculating a water footprint is to combine consumptive water-use factors (volume per unit of production) for individual goods and services with statistics on production, trade, and consumption of those goods and services. The Water Footprint Network (WFN) advanced the methodology described in the *Water Footprint Assessment Manual* (Hoekstra et al. 2011).

The water footprint is often reported as three components: green water, blue water, and grey water. Green water is the amount of precipitation and soil moisture that is di-

TABLE 5.1 Comparison of Footprint Indicators

Indicator	Measurement	Concern
Water footprint	Consumptive use	Water availability and quality
Carbon footprint	Greenhouse gases	Climate change
Ecological footprint	Global hectare	Land and bioproductivity degradation

rectly consumed in an activity, as in growing crops. Blue water is the amount of surface water or groundwater that is applied and consumed in an activity, as in growing crops or manufacturing an industrial good. Finally, grey water[2] is the water needed to assimilate pollutants from a production process back into water bodies at levels that meet governing standards, regardless of whether those standards are actually met (Hoekstra et al. 2011). The green, blue, and grey water metrics are calculated for individual processes and then aggregated on the basis of the consumption patterns of the unit of interest.

Because a water footprint is based on the goods and services consumed, it can be calculated at different levels of consumer activity, that is, for individuals, households, regions, states, nations, or even all of humanity. Trade is an important consideration when calculating water footprints, as the various processes required to produce a final good often take place in different countries. Using the example of a cotton T-shirt, the cotton may be grown in India, woven into fabric and dyed in China, and finished in Mexico before heading to market in the United States. Thus, the water footprint of that T-shirt will be geographically distributed according to the structure of its supply chain (Chapagain et al. 2006).

Green, blue, and grey water footprints are often combined and reported as a single value in the literature. Each, however, has a distinct ecological and social context. Green water pertains to rainwater and soil moisture occurring where crops are grown and thus may be significant only to the degree it makes water unavailable for other land uses, alternative crops, or native vegetation. Blue water, by contrast, represents a deliberate extraction and allocation of surface water or groundwater resources, often with pumping and conveyance systems used to extract and deliver water where it is needed. Grey water is an indicator of water quality rather than a measure of consumptive water use. Even though the contamination of surface waters is by definition a consumptive use, contaminated water can often still serve multiple uses, such as navigation or cooling. Thus, in order to eliminate double counting of upstream grey water footprints by downstream blue water uses, as well as to avoid confusing their distinct ecological contexts, we think it is appropriate to present each type of water separately.

The quantitative measure of a green, blue, or grey water footprint, by itself, might reveal little about the consequences or local impacts of that water use. Additional information (and often subjective evaluation) is required to estimate the impact of a water footprint. If the water consumed comes from areas where water is relatively abundant, the social and environmental impacts may be far less than for the same consumption from a region where water is scarce. Such a contextualized analysis of the relative impacts of a water footprint is generally called a water footprint sustainability assessment in the literature (Hoekstra et al. 2011). These assessments are rarer in the literature, so here we present only the quantitative dimensions of water footprint analyses in addition to our more contextualized report on California's water footprint.

---

2. Not to be confused with wastewater that is reused directly on a site, often also called grey water or greywater.

Water Footprint Findings

The water footprint concept has developed substantially in scientific literature over the past decade and resulted in numerous publications and extensive data sets, many of which have come about through the work of the Water Footprint Network. With some exceptions, this research has focused on the scale of national consumption and international trade. This is understandable, given that the national level is often where most production and trade statistics are gathered and reported, but ideally the science would advance to be able to identify water footprints of more hydrologically appropriate units, such as river basins. Another limitation of these findings is that they account only for bilateral trade flows, thus missing consumptive water uses that may have occurred further upstream in the supply chain of a product. Additionally, we note that while the scope of products covered is substantial, these figures do not include the water footprints of energy products, mining activities, or traded service industries such as tourism and call centers.

Beginning at the global level, Arjen Hoekstra and Mesfin Mekonnen (2012) of the University of Twente in the Netherlands found that the water footprint of global consumption, averaged between 1998 and 2005, was about 9.1 trillion cubic meters (around 7.4 billion acre-feet) per year.[3] By comparison, that is about 500 times the average annual flow of the Colorado River in its natural state (Cohen et al. 2001). Around three-quarters of the global water footprint is green water, 11 percent is blue water, and 15 percent is grey water. Ninety-two percent is associated with the consumption of agricultural products, while industrial products make up 5 percent and direct water consumption makes up just 4 percent (figure 5.1). Twenty-two percent of the global water footprint is "external," meaning that it pertains to internationally traded products.

Water footprints, however, vary tremendously from country to country. The United States, for example, has one of the highest per capita water footprints in the world, at

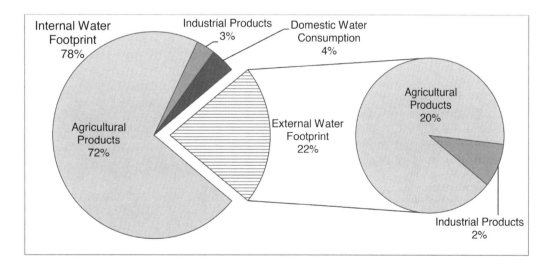

FIGURE 5.1 GLOBAL WATER FOOTPRINT BROKEN DOWN BY TYPES OF CONSUMPTIVE USE
Source: Mekonnen and Hoekstra (2011).

---

3. One acre-foot is the amount of water that would fill one acre to a depth of one foot and is approximately 325,851 gallons, or 1,233,482 liters.

nearly 8,000 liters per capita per day (LPCD), or 2,100 gallons per capita per day (GPCD) (Mekonnen and Hoekstra 2011). Other developed countries have an average water footprint of around 5,000 LPCD (1,300 GPCD), and the global average is just under 4,000 LPCD (1,000 GPCD). Several developing countries, such as China and India, have per capita water footprints that are well below the global average; however, their large populations and rapidly rising standards of living, if accompanied by rising water use, will have substantial impacts on the world's water resources. Figure 5.2 shows the per capita total water footprint for nearly every country in the world.

Another important dimension of a nation's water footprint is its relative dependence on internal versus external water resources (figure 5.3). As mentioned earlier, 22 percent of the entire global water footprint pertains to traded goods, and many nations rely heavily on international markets to make up for local water deficits. For the most part, these are countries with very little agricultural production capacity within their territory to begin with, including small, geographically isolated nations such as Malta and Brunei; nations with challenging topography, such as Japan and Switzerland; and nations with extreme climates, such as Iceland and Kuwait. One interesting standout case is the Netherlands, which is 95 percent externally dependent despite its large agricultural economy. Conversely, nations whose water footprint is mostly internal typically either are more isolated economically, as is Myanmar, or have abundant resources for agricultural production, as does the United States. In fact, the United States not only has one of the largest water footprints in the world but also is a massive exporter of water-intensive products to other countries.

The Question of Scale: California's Water Footprint

Delving beneath the level of national water footprint assessments is difficult in terms of data availability, but it is important to understand how water footprint patterns and outcomes might change with scale. California, as a case of a subnational unit, is interesting because it is the state with the largest population and gross domestic product in the nation (about one-eighth on both counts). California represents a substantial share of US

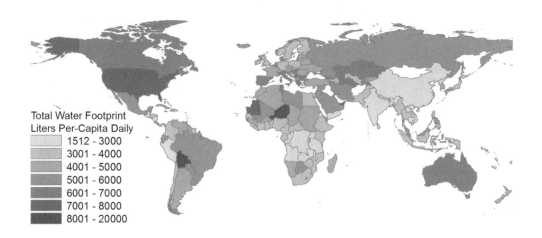

FIGURE 5.2 NATIONAL PER CAPITA WATER FOOTPRINTS
Source: Mekonnen and Hoekstra (2011).

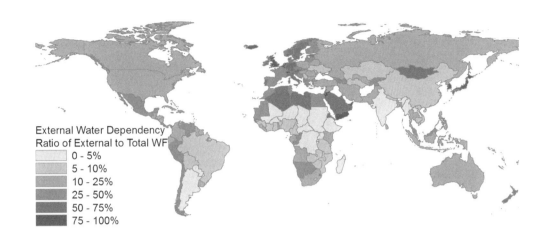

FIGURE 5.3 EXTERNAL WATER DEPENDENCE OF NATIONS
Source: Mekonnen and Hoekstra (2011).

economic activity, in terms of both consumption and production, so we might expect its water footprint to be similar to that of the United States as a whole. Yet the state is unique climatically and hydrologically, with minimal precipitation during the summer and fall and very little runoff flowing out of its borders. In fact, California derives about 10 percent of its annual surface water supplies from its neighbors (CDWR 2009).

The Pacific Institute recently completed an assessment of California's water footprint, and we briefly present the results here (Fulton et al. 2012). Our analysis, summarized in figure 5.4, found that the water footprint of the average Californian is about the same as that of the average American in terms of size and breakdown of products. However, it is very different in two important respects. First, whereas the United States as a whole is a net virtual water exporter, California is a net importer. This creates a unique situation wherein decision makers in California must consider the fact that the water footprint of in-state consumption pertains more to water resources in other states and other countries than to California's own water resources. On the one hand, this means that if California experiences a severe drought, the impact on consumers may not be so detrimental. This situation might be strategic in that California enjoys the benefits of being able to consume water-intensive products without having to consume scarce local water resources (though California already consumes a substantial part of its own water resources). Conversely, this may make California consumers more susceptible to droughts or water-related issues elsewhere, and Californians must consider the effect their consumption habits may have on exacerbating those impacts. Further research is needed to understand the specific water-related impacts in locations where the goods that Californians consume are produced.

The second important difference relates to the "type" of water. Despite the per capita water footprints being nearly equal, California's has a much larger share of blue water than does that of the United States as a whole. This means that the state relies more on applied surface water (primarily in irrigation), which has trade-offs with other uses, such as supporting ecosystems. Additionally, blue water irrigation systems may have specific vulnerabilities with respect to climate change that are different from the vulnerabilities of production systems that depend on green water. More research is also needed in this area. Nevertheless, these particular aspects of California's water footprint suggest a set of vulnerabilities and policy options that would likely be overlooked in a national assessment.

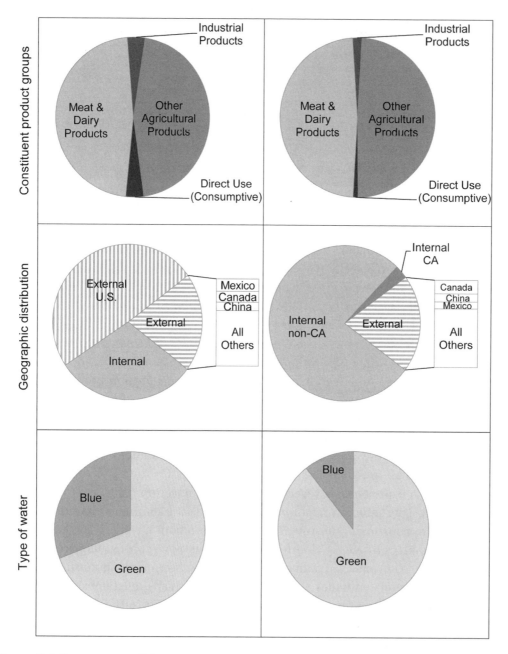

FIGURE 5.4 COMPARISON OF US AND CALIFORNIA WATER FOOTPRINTS BY PRODUCTS, GEOGRAPHIC DISTRIBUTION, AND TYPE OF WATER

California's per capita water footprint (left column) and that of the United States (right column) are compared along three dimensions: constituent product groups (top row), geographic distribution (middle row), and type of water (bottom row).

Conclusion

Continued population growth and improved living standards are creating new patterns of production and consumption that are likely to place increasing demands on the world's water resources. Water resources are already strained in many locations, and the complex ways in which these processes occur require further attention. The water footprint tool, still evolving in concept and method, has begun to illuminate some of the important dimensions of the ways in which people, economies, and water resources are interconnected. The concept is reasonably analogous to the carbon footprint and the ecological footprint, tools that have made substantial inroads with international policy and business strategy. Important insights have already resulted from footprint analyses, and the approach offers promise for assisting decision making at a number of levels, from individuals to businesses to governing bodies at various scales.

National-level water footprints have been calculated for nearly every country on Earth, and we now have a good sense of how affluence and increased consumption (particularly of some products, such as meat) affect demand for the world's water resources. Trade is an important way in which countries can overcome local water scarcity by importing virtual water. But trade also provides a way in which distant water resources might be overexploited or mismanaged without the knowledge of the consumer who is benefiting from that water use. The more we know about how these causes and effects occur, the better the world's water resources can be managed.

The case of California illustrates how a water footprint assessment at the state level can yield very different results from assessments at larger scales. This dynamic is very important to understand, since California consumers and state agencies will be able to make better-informed decisions than if they were to rely on a nationwide assessment. Specifically, our assessment shows that California is more dependent on water resources outside of its borders than is the United States as a whole and that the state is more reliant on blue water. These aspects of California's water footprint make decision making in relation to such challenges as climate change unique.

But the state level is not necessarily the ideal scale at which water footprint information best applies. Many economic and water management decisions are made at the state level, but there has also been a movement in the past ten to fifteen years to shift many important management decisions to the scale of river basins. This seems appropriate, given that populations living within river basins (e.g., the Colorado and Nile River basins) have a hydrologic commonality and should to some degree share in water-related decision making. For example, upstream populations can shift their water footprint away from local resources by importing products produced downstream, thereby allowing water to provide ecological benefits along more of the river's reach. These types of basin-level decisions are more difficult when they require the crossing of political boundaries, and the field of transboundary water management has a lot to offer on this subject. Water footprint research is just beginning to move in the direction of using the river basin as a more appropriate unit of analysis.

Other exciting applications and improvements of the water footprint method have to do with the study of how water footprints change over time. This information is especially important in examining how particular countries or regions develop and change their production and consumption habits. For example, while the US population and economy have continued to grow, domestic water withdrawals have been basically steady since about 1980 (Gleick 2003). In part, this is the result of improvements in efficiency of water use through various technologies in the agricultural, power production, manufacturing, commercial, and residential sectors. But the economy overall has also changed, with more of the products we consume being produced outside of our borders. The water footprint method, applied over a time series, can shed light on the extent to which the apparent leveling of overall water use in the United States has been in part an externalizing of water use.

Overall, the water footprint tool is useful in answering a number of questions pertaining to the relationship between people and the world's water resources. At its core, it provides a description of the quantity, type, and location of water resources required to produce products and to fulfill the consumption habits of individuals or populations. As populations grow and change, the connections between their actions and the global environment need to be better understood. The world's water resources are being pressured on numerous fronts, as evidenced in the other chapters in this book. The water footprint tool, though still evolving, is beginning to provide us with another way in which to better understand and reduce human impacts on those resources.

References

California Department of Water Resources (CDWR). 2009. California Water Plan Update 2009. Sacramento: California Department of Water Resources.

Chapagain, A. K., A. Y. Hoekstra, H. H. G. Savenije, and R. Gautam. 2006. The water footprint of cotton consumption: An assessment of the impact of worldwide consumption of cotton products on the water resources in the cotton producing countries. *Ecological Economics* 60 (1): 186–203. doi:10.1016/j.ecolecon.2005.11.027.

Cohen, M. J., C. Henges-Jeck, and G. Castillo-Moreno. 2001. A preliminary water balance for the Colorado River delta, 1992–1998. *Journal of Arid Environments* 49:35–48. doi:10.1006/jare.2001.0834.

Cooley, H., J. Fulton, and P. H. Gleick. 2011. Water for Energy: Future Water Needs for Electricity in the Intermountain West. Oakland, CA: Pacific Institute. http://www.pacinst.org/wp-content/uploads/2013/02/water_for_energy3.pdf.

Fulton, J., H. Cooley, and P. H. Gleick. 2012. California's Water Footprint. Oakland, CA: Pacific Institute. http://www.pacinst.org/reports/ca_water_footprint/ca_ftprint_full_report.pdf.

Gleick, P. H. 2003. Water use. *Annual Review of Environment and Resources* 28 (1): 275–314. doi:10.1146/annurev.energy.28.040202.122849.

Hoekstra, A. Y., A. K. Chapagain, M. M. Aldaya, and M. M. Mekonnen. 2011. *The Water Footprint Assessment Manual: Setting the Global Standard*. London: Earthscan.

Hoekstra, A. Y., and M. M. Mekonnen. 2012. The water footprint of humanity. *Proceedings of the National Academy of Sciences* 109 (9) (February 28): 3232–3237. doi:10.1073/pnas.1109936109.

Mekonnen, M. M., and A. Y. Hoekstra. 2010. The Green, Blue, and Grey Water Footprint of Crops and Derived Crop Products. Value of Water Research Report Series No. 47. Delft, Netherlands: UNESCO-IHE Institute for Water Education.

———. 2011. National Water Footprint Accounts: The Green, Blue, and Grey Water Footprint of Production and Consumption. Value of Water Research Report Series No. 50. Delft, Netherlands: UNESCO-IHE Institute for Water Education.

———. 2012. The blue water footprint of electricity from hydropower. *Hydrology and Earth System Sciences* 16 (1): 179–187. doi:10.5194/hess-16-179-2012.

CHAPTER 6

Key Issues for Seawater Desalination in California
Cost and Financing

Heather Cooley and Newsha Ajami

In June 2006, the Pacific Institute released *Desalination, with a Grain of Salt*, an assessment of the advantages and disadvantages of seawater desalination for California (Cooley et al. 2006). At that time, there were twenty-one active seawater desalination proposals along the California coast. Since then, only one project, a small plant in Sand City, has been permitted and built. A second project, in Carlsbad, has all of the necessary permits, finally secured financing in December 2012, and is now under construction. Interest in seawater desalination, however, remains high in California, and many agencies are conducting technical and environmental studies and pilot projects to determine whether to develop full-scale facilities.

In 2011, the Pacific Institute initiated a new research project on seawater desalination. As part of that effort, we conducted some twenty-five one-on-one interviews with industry experts, water agencies, environmental and community groups, and regulatory agencies to identify some of the key outstanding issues for seawater desalination projects in California. Among the key issues raised were the cost and financing of seawater desalination projects, the marine impacts, and the energy requirements and associated greenhouse gas emissions. In this chapter, we provide a short summary of the costs of various projects and the spatial and temporal variability associated with these costs. We also describe some of the project-related risks associated with desalination plants. Finally, we provide two case studies of seawater desalination projects in the United States to highlight some of the issues involving cost and financing of these projects.

How Much Does Seawater Desalination Cost?

The biggest barrier to widespread use of desalination technology remains its relatively high cost. Most desalination capacity worldwide has been built where natural renewable water resources are extremely limited and where energy resources critical for operating desalination systems are abundant and inexpensive, such as in the Arabian Gulf region.

In other regions, however, alternative sources of water supply and improvements in water efficiency are still largely less costly than the water produced at desalination facilities. Costs have declined considerably over the past forty years, although they are unlikely to decline much further, barring a major technological breakthrough.

There are many components and definitions of the cost of a desalination project, and uncertainty about these terms can lead to confusion about cost comparisons, especially among those without a finance background. In this chapter, we explain the cost terminologies. We also compare costs among proposed and recently constructed plants and describe some of the factors that contribute to the large variability among projects. We conclude by comparing the cost of seawater desalination with that of other water alternatives.

It is important to note that this chapter describes *financial* costs. There are other costs that may not be captured in a financial analysis because they are not internalized in the project developer's cost stream or are not subject to market-like transactions (NRC 2008). For example, seawater intake screens kill marine organisms. These impacts have a cost associated with them that some other group or individual bears, for example, reductions in fish populations that affect local fishermen. These costs may be captured in an economic analysis. But unless they are internalized in the project through, for example, a requirement to restore or enhance wetlands, they will not be factored into the financial feasibility of a project and are not included in this report.

Cost Terminology

In this section, we provide a short introduction to some of the terms used to describe the financial costs of a desalination project.

Capital Costs and Operation and Maintenance Costs

The primary cost components of a seawater desalination plant are commonly divided into two major categories: capital costs and operation and maintenance (O&M) costs. Capital costs are those costs incurred during construction of the project and include expenses associated with planning, permitting, designing, and constructing the project. Other capital costs include expenses associated with buying or leasing equipment, purchasing the site, and mitigating environmental impacts.

Capital costs may also include the cost to finance the project—referred to as the cost of capital. The capital required to develop desalination projects is not free, and the project developer must pay to access it. The cost of capital includes the cost of debt (for those projects financed with debt) and the cost of equity (for those projects financed with private equity). The cost of debt is based on the interest rate of the debt incurred for the project, that is, the bond or loan, and reflects the compensation a lender demands for the risks involved in the project (Pemberton 2003). The cost of equity is the return paid to private equity investors to compensate for the risk they undertake by investing their capital. Generally, the greater the perceived risks of the project, the harder and costlier it is to secure capital.

In addition to capital costs, there are O&M costs associated with desalination plants. O&M costs represent the ongoing costs of operating the plant, including expenses associated with replacement membranes for reverse osmosis systems, pre- and post-treatment chemicals, energy to run the plant, environmental monitoring, and labor for plant operators. Some O&M costs, such as labor, are fixed; they do not vary with respect to the amount of water produced. Other O&M costs, such as the costs of energy and chemicals, are variable.

Capital and O&M costs are often reported in different ways, which can complicate project cost comparisons. Capital costs are reported as onetime, fixed costs. O&M costs, by contrast, are typically reported as annual costs. In order to determine the total annual cost of a desalination plant, the capital and O&M costs must be put on similar footing. The most common approach is to spread the capital costs, which include the cost to finance the project, over the life of the project, for example, for twenty or thirty years, to produce an annualized capital cost. The annualized capital cost and the annual O&M cost are added together to produce a total annual cost.

Figure 6.1 shows an annual cost breakdown of a typical seawater desalination plant. Although variable from project to project, the annualized capital cost typically accounts for about one-third of the annual cost of a seawater desalination plant. O&M costs account for the remaining two-thirds. Generally, energy costs—at 36 percent of the annual cost—represent the single largest O&M cost and are of similar magnitude as the annualized capital cost. Depending on the desalination process and local site conditions, chemical costs can also be significant, accounting for about 12 percent of the total annual cost.

Unit Cost

In addition to reporting capital and O&M costs, water managers often report the unit cost of water, that is, the cost per cubic meter (or acre-foot or thousand gallons) of water produced. The unit cost of water is derived from the annual cost (the annualized capital cost plus the O&M cost) and the annual production of water from the plant. Specifically, it is determined by the following formula:

$$\text{unit cost} = \frac{(\text{annualized capital cost} + \text{annual O\&M cost})}{\text{annual amount of water produced}}$$

Given that the project cost is driven, in part, by the size of the plant, the unit cost effectively normalizes the cost of the project by the volume of water produced. It can also be useful for comparing projects of different sizes and for comparing various water supply and demand management alternatives.

Desalination Cost Estimates

Discussions about the actual cost of desalination plants have been muddled and muddied because estimates have been provided in a variety of units, years, and ways that are not directly comparable. Some report the cost of the desalination plant alone. Others include the costs for additional infrastructure needed to integrate the desalination plant into the rest of the water system, which can be significant. Some estimates include the cost to finance the project, while others do not. In some regions, governmental subsidies hide some of the real costs. And local conditions, such as water salinity, options for disposing of brine, level of environmental regulation, and labor or land costs, vary widely. For example, the capital cost of the proposed Carlsbad Desalination Project is currently estimated at $537 million. An additional $239 million is needed to build a pipeline to transport the water to customers and make other water system improvements

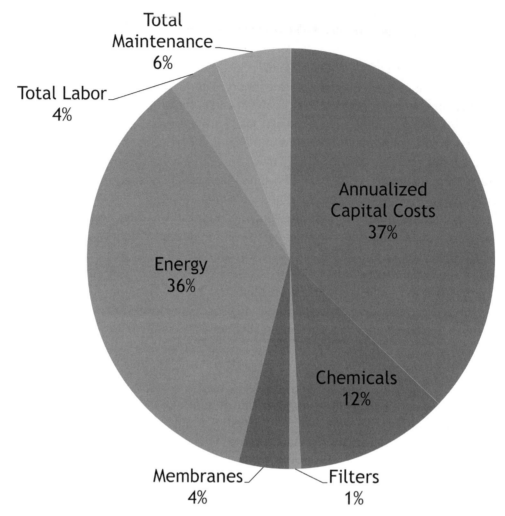

FIGURE 6.1 ANNUAL COST BREAKDOWN OF A TYPICAL SEAWATER DESALINATION PLANT
Source: NRC (2008).

Note: Estimate is for a reverse osmosis plant and is based on the following assumptions: capacity of 190,000 cubic meters (m^3) per day (50 million gallons per day); constant energy costs at $0.07 per kilowatt-hour (kWh); membrane life of five years; nominal interest rate of 5 percent; and a depreciation period of twenty-five years.

(Lee 2013).[1] Given the inconsistent nature of estimates, the public, media, and analysts must exercise caution when comparing claims of costs for different projects.

Table 6.1 shows capital, O&M, and unit cost estimates for proposed plants in California and several recently constructed plants in Florida and Australia. O&M costs shown here are based on estimates made the year the project was completed; these costs change over time and may not reflect current O&M costs. Costs for desalination plants in Australia were converted into US dollars using the conversion rates that were in effect in the year that the plant was completed. We have not adjusted the apparent year of each reported cost for inflation because inflation varies from country to country. Even without

---

1. Note that these estimates do not include financing costs, which substantially increase the lifetime cost of the project.

adjustment to current-year dollars, however, it is apparent from the table that costs vary far more widely than can be explained solely by inflation or currency exchange rates.

As shown in table 6.1, large plants tend to have higher capital and O&M costs than small plants, but this is not always the case. For example, the Kwinana Desalination Plant in Perth, Western Australia, has a capital cost of less than $330 million and an O&M cost of $17 million per year. By contrast, the Gold Coast Desalination Plant in Tugun, Queensland, is slightly smaller but has much higher capital and O&M costs, at $890 million and $30 million per year, respectively. The Gold Coast plant was considerably more expensive than the Perth plant because steel and labor costs were dramatically higher when the plant was constructed. Additionally, the Gold Coast plant used a more complicated intake design and required the construction of a longer pipeline to distribute the water to customers (Crisp 2012).

Unit costs are also highly variable among projects. Variability among desalination projects is driven by a range of site- and project-specific factors. Land prices and labor costs, for example, can vary considerably among projects. Likewise, energy prices can have a major influence on the project cost. For example, the average electricity price of the Gold Coast Desalination Plant was $0.15 per kilowatt-hour (kWh) in 2009–2010 (Robertson 2010). Currently, electricity prices for industrial customers in California are about $0.10 per kilowatt-hour (US EIA 2012).[2] Other project- and site-specific factors that affect the cost include source water temperature and salinity, intake and outfall design, environmental mitigation requirements, interest rates and loan period, project team experience, permitting requirements, and the presence of hidden and visible subsidies (Cooley et al. 2006; WateReuse Association 2012).

It is important to note that desalination costs are not static but can vary over time. Many of the estimates shown in table 6.1 are engineering estimates based on the initial project design. Actual production costs can be higher because of increases in the capital and O&M costs or a reduction in the amount of water produced. Capital costs may change if the price of materials, such as cement or steel, or the cost of borrowing money changes. Major project delays are also likely to increase capital costs.

O&M costs can also change over the life of the desalination plant as a result of changes in input costs or in the lifetime of major equipment. Energy, for example, represents a major cost component (figure 6.1), and short- and long-term increases in energy prices can increase production costs. If energy accounts for 36 percent of the annual production cost, a 25 percent increase in energy costs would increase the cost of produced water by 9 percent. Similarly, membranes typically last for five to seven years, although in some cases ten years is possible. If the actual lifetime of the membranes exceeds five to seven years, the production cost may be lower than the initial estimate. Conversely, if the membranes foul more quickly, O&M costs increase.

The cost of produced water from a desalination plant is especially sensitive to changes in the amount of water produced. As described previously, the unit cost of water is determined by dividing the annual costs by the amount of water of produced. Typically, unit cost estimates during the design and planning of a project are based on the assumption that the desalination plant will operate at or near full capacity. If the plant produces less water, however, the unit cost increases (figure 6.2). Take, for example, a plant that

---

2. This was the price in 2012 for industrial customers.

TABLE 6.1 Desalination Costs for Proposed and Recently Constructed Plants

Project	Capacity (m³/d)	Capital Cost[a] ($ Millions)	O&M Cost[b] ($ Millions)	Unit Cost[c] ($/m³)	Dollars[d] are in...	Status	Data Source
Santa Cruz/ Soquel Creek Water District	9,480	$114	$3–$4	n/a	2012 dollars	Proposed	Luckenbach (2012)
California American Water[e]	18,600–30,300	$175–$207	$7.77–$11	$2.06–$2.63	2012 dollars	Proposed	Separation Processes, Inc. (2012)
Deep Water Desal[e]	18,600–30,300	$134–$160	$9.38–$12.3	$1.94–$2.53	2012 dollars	Proposed	Separation Processes, Inc. (2012)
The People's Moss Landing Water Desal Project[e]	18,600–30,300	$161–$190	$7.06–$10.1	$1.90–$2.42	2012 dollars	Proposed	Separation Processes, Inc. (2012)
Tampa Bay, Florida	94,800	$158	n/a	$0.89	2007 dollars	Complete	See chapter text
Gold Coast Desalination Plant[f]	125,000	$888	$30	$1.61	2009 dollars	Complete	Pankratz (2012)
Kwinana Desalination Plant, Perth[f]	144,000	$330	$17	$0.99	2008 dollars	Complete	NWC (2008)
Carlsbad Desalination Plant	190,000	$776	$49–$54	$1.54–$1.78	2012 dollars	Proposed	Lee (2013); SDCWA (2012a)
Camp Pendleton	190,000–379,000	$1,300–$1,900	$45–$105	$1.54–$1.90	2009 dollars	Proposed	SDCWA (2012d)
Kurnell Desalination Plant, Sydney[f]	208,000	$1,565	$47	$1.95	2008 dollars	Complete	NWC (2008)
Southern Seawater Desalination Plant[f]	273,000	$1,466	n/a	$2.29	2012 dollars	Complete	Pankratz (2012)
Port Stanvac, Adelaide[f]	273,000	$1,878	$136	$1.94	2012 dollars	In progress	Pankratz (2012); Kemp (2012)
Wonthaggi Desalination Plant, Melbourne[f]	413,000	$3,651	$63	$5.31	2012 dollars	Complete	Pankratz (2012)

(continued)

TABLE 6.1 *continued*

Note: n/a = not available.

[a]Capital costs here are based on capital expenditures and include the cost of the desalination plant plus any other infrastructure required to integrate the desalination plant into the rest of the water system. The cost to finance the project is not included.

[b]O&M costs include the operation and maintenance costs of the desalination plant and any other infrastructure required to integrate the desalination plant into the rest of the water system.

[c]The unit cost captures capital and O&M costs as well as the cost to finance the project.

[d]Costs have not been adjusted for inflation. Unless otherwise indicated, we assume that cost estimates are provided in the year in which the project was completed.

[e]The actual capacity of these projects may differ from what is stated here. The study these estimates were drawn from (Separation Processes, Inc. 2012) adjusted the plant capacities to make a more accurate comparison among projects in the region.

[f]Costs for plants in Australia were converted into US dollars using conversion rates for the year in which the cost estimates were made.

produces 190,000 cubic meters (m^3) of water per day, or 50 million gallons per day (MGD), at a cost of $2.17 per m^3 ($2,700 per acre-foot) at full capacity; an estimated one-third of the cost is fixed, and the remaining two-thirds of the cost is variable. If the operator reduces water production to 150,000 m^3 per day (40 MGD), on average, then the unit cost of water increases to $2.53 per m^3 ($3,100 per acre-foot). If the operator further reduces water production to 38,000 m^3 per day (10 MGD), then the unit cost increases to more than $7.96 per m^3 ($9,800 per acre-foot).

Cost Comparisons

Seawater desalination has traditionally been considerably more expensive than most other water supply and demand management options, but cost comparisons made on the basis of the cost of existing water supplies can be misleading. The costs of existing supplies are often far below the investment needed to construct and maintain these systems because of heavy subsidies from state and federal investments. Additionally, current water prices often fail to include costs for adequately maintaining and improving water systems. As a result, the public often has a somewhat distorted perception of the cost of a reliable supply of high-quality water, making it difficult for water utilities to justify their investment in more expensive water supply options, such as desalination.

Consequently, a comparison of the cost of new seawater desalination systems should be based on the *marginal cost* of water: the cost of providing, or saving, the next increment of water. The marginal cost of water varies considerably among communities and changes over time. As a result, these costs should be evaluated periodically on a case-by-case basis. A recent analysis of water alternatives for San Diego County, California—a region where construction of multiple desalination plants is under consideration—finds that seawater desalination has the highest marginal cost of all plausible alternatives (table 6.2). The marginal cost of recycled water is also relatively high, although treating the water to potable levels and using the existing distribution system can reduce this cost considerably. New surface water and groundwater are less expensive than recycled water or seawater desalination, although these supplies are limited. In many cases, conservation and efficiency improvements have the lowest marginal cost.

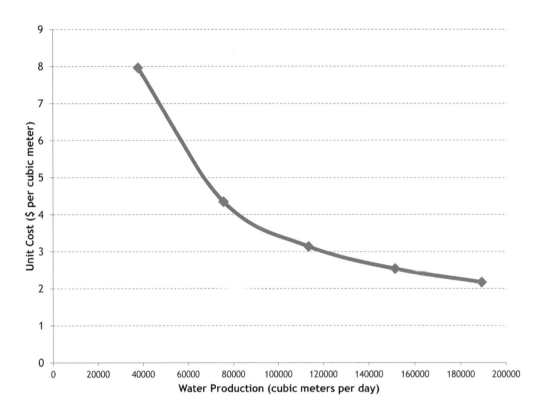

Figure 6.2 An Example of the Relationship between Desalination Plant Production and Unit Cost
Note: Based on a desalination plant with a capacity of 190,000 m³ per day (50 MGD).

Table 6.2 Example of the Marginal Cost of Water in San Diego County

Water Alternative	Cost per Cubic Meter	Cost per Acre-Foot
Imported water	$0.71–$0.79	$875–$975
Surface water	$0.32–$0.65	$400–$800
Groundwater	$0.30–$0.89	$375–$1,100
Seawater desalination	$1.46–$2.27	$1,800–$2,800
Recycled water (nonpotable)	$1.30–$2.11	$1,600–$2,600
Recycled water (potable)	$0.97–$1.46	$1,200–$1,800
Water conservation and efficiency	$0.12–$0.81	$150–$1,000

Source: Equinox Center (2010).

Note: All costs are in 2010 dollars. Cost estimates for water conservation and efficiency are based on costs to the water agency, for example, estimated expenditures on educational initiatives or incentives for conservation measures divided by the cumulative water savings, and do not include costs or savings to the customer. One acre-foot equals 1,233 cubic meters.

The cost associated with wastewater conveyance and treatment is another factor that is often excluded from cost comparisons. The introduction of a new source of water increases the amount of wastewater that must be collected, treated, and disposed of. Some communities may have adequate wastewater treatment capacity, and the additional costs would simply be the variable O&M costs associated with that treatment. In other communities, however, wastewater treatment capacity may need to be expanded, which represents an additional, and in some cases significant, capital cost to the community. While these costs would apply to all of the water supply projects under consideration to meet demand, the costs would not be incurred if water demand were met through water conservation and efficiency improvements.

Desalination Projects and Risk

Large-scale projects, such as desalination plants, include a variety of risks that affect the ability to attract financing, the rate of return, and the overall viability of a project. Each project will have a unique set of risks that vary in their relative importance. Some of the risks that have been identified for seawater desalination projects include the following (Pemberton 2003; Wilcox and Whitney 2004; Wolfs and Woodroofe 2002; Schiller 2012):

> *Permitting risks* are risks that governmental permits and approvals required to construct or operate the project will not be issued or will be issued only subject to onerous conditions.
>
> *Design and technology risks* are risks that can be attributed to the use of a design or technology that fails to deliver at expected service levels.
>
> *Construction risks* arise during the construction period. Construction risks can be attributed to poor project design, ineffective project management, inexperienced contractors, or lack of effective communication among the various contractors working on a project.
>
> *Operational and performance risks* arise during the operation and maintenance period and can be attributed to inexperienced operators, higher than anticipated operating costs, product water that fails to meet required standards, or underperformance of equipment.
>
> *Financial risks* are influenced by the financial status of the borrower, changes in interest rates, fluctuation in foreign exchange rates (if the project relies on foreign investment), and changes to pricing of construction or operational phase insurance.
>
> *Market risks* can be caused by unrealistic demand assessments and water production assumptions, reluctance of customers to buy the produced water, and reluctance of those with rate-making authority to approve rate increases to cover new costs.
>
> *Political risks* are associated with political decisions that affect the cost or viability of the project and can relate to laws, tax rates, governmental permissions, and other factors. These risks may be associated with decisions made by elected or appointed officials or by the public through, for example, a referendum.[3]
>
> *Force majeure risks* are the risks of events that render the construction or operation of the project impossible, either temporarily or permanently. A

---

3. For example, in November 2012, Santa Cruz residents passed Measure P, which guarantees residents the right to vote on whether to pursue a seawater desalination project in the future.

force majeure event includes an extraordinary event or circumstance that prevents one or both parties from fulfilling their obligations under the contract and is beyond the control of the parties, such as an earthquake, flood, hurricane, war, strike, or riot.

Recent experience suggests that a key risk associated with seawater desalination plants is overly optimistic assumptions about future demand for the produced water and underappreciation of the availability of less expensive alternatives. This is a type of market risk that is also often referred to as demand risk. Water managers often have multiple sources of water to choose from to meet water demand. Desalination is among the most expensive options, and water managers may reduce the output of a desalination plant when demand drops or when less expensive options are available. As noted above, reducing the output can increase the overall unit cost of the produced water considerably, which can further reduce demand or make other supply options even more economically attractive. In response, water managers may temporarily or permanently shut down the desalination plant. This can reduce the variable operating costs associated with the plant, especially energy and chemical costs, but ultimately leave ratepayers to pay off a plant while receiving little to no benefit from it. Additionally, as desalination expert Tom Pankratz notes, "frequent stopping and starting, or long-term mothballing, may not only affect the short-term performance of a plant, it may actually be detrimental to the system itself" (Pankratz 2011).

Recent experience in Australia highlights demand risk (table 6.3). The Australian government has made a massive investment in seawater desalination plants in response to a severe and persistent drought, developing six major plants since 2006. The first plant, with a capacity of 37 MGD, was built in Perth, became operational in late 2006, and continues to operate at or near full capacity. A second plant, the Southern Seawater Desalination Plant, is being developed in the same region and should be completed in 2013. Other plants were built in New South Wales, Victoria, and South Australia. While many consider the first Perth plant to be a success, the four other plants built in recent years have been or will be put into standby mode as a result of reductions in demand and the availability of less expensive alternatives.

For example, the Kurnell Desalination Plant in Sydney, New South Wales, was completed in January 2010 but was mothballed in July 2012 because less expensive surface water sources are available.[4] Likewise, the Wonthaggi Desalination Plant in Victoria was completed in December 2012. Although the plant performed as designed, the government announced it would not be ordering water from the plant in 2013. Similarly, the Port Stanvac Desalination Plant, south of Adelaide, South Australia, is expected to be completed in 2013, but in an October 2012 statement, SA Water's chief executive, John Ringham, announced that "to keep costs down for our customers, SA Water is planning to use our lower-cost water sources first, which will mean placing the desalination plant in stand-by mode when these cheaper sources are available" (Kemp 2012). The desalination plant, which cost nearly $1.9 billion, is slated to go on standby mode in 2015.

Experience in Santa Barbara, California, and in Tampa Bay, Florida, provides further evidence of demand risk. During a severe multiyear drought in California (1987–1992), the City of Santa Barbara partnered with several local water agencies to build a seawater desalination plant. The plant was completed in March 1992, and shortly thereafter the

---
4. If reservoir levels fall below 80 percent capacity, the desalination plant will be restarted.

TABLE 6.3 Operational Status of Recently Constructed Desalination Plants

Facility Name	Location	Date Operational	Plant Capacity (MGD)	Plant Capacity (m³/day)	Status
Tampa Bay Desalination Plant	Tampa Bay, Florida	2007	25	95,000	In intermittent use
Gold Coast Desalination Plant[a]	Tugun, Queensland, Australia	February 2009	33	125,000	In standby mode since December 2010 because of high operating cost and operational issues
Kurnell Desalination Plant[b]	Sydney, New South Wales, Australia	January 2010	66	250,000	In standby mode since July 2012 because reservoirs are full
Kwinana Desalination Plant	Perth, Western Australia	November 2006	38	144,000	Operational
Southern Desalination Plant	Binningup, Western Australia	September 2011 (phase 1); March 2013 (phase 2)	72	270,000	Phase 1 is operational; phase 2 is under construction
Wonthaggi Desalination Plant[c]	Victoria, Australia	December 2012	109	413,000	In standby mode since January 2013
Port Stanvac Desalination Plant[d]	Adelaide, South Australia	2013	72	270,000	Will be put in standby mode in 2015

Sources:
[a] Marschke (2012).
[b] AAP Newswire (2012).
[c] Hosking (2012).
[d] Kemp (2012).

drought ended. The plant was eventually decommissioned, as the cost to produce the water was too high to warrant its use during non-drought periods. In addition, the high cost of building the plant and connecting to the California State Water Project raised local water prices enough to encourage substantial additional conservation, further decreasing need for the plant. Similarly, Florida's Tampa Bay Seawater Desalination Project was completed in 2008 in an effort to reduce groundwater pumping. Since it was built, that plant has operated far below its full capacity as a result of periodic operational issues, reductions in regional water demand, and availability of less expensive alternatives.

Project Delivery Methods

Planning, designing, financing, constructing, and operating major water projects, such as desalination plants, can be challenging. Such projects are often accomplished by public agencies or through various types of public-private partnerships. The delivery of large-scale water projects may involve different parties and stakeholders, including the owner, designers and engineers, construction contractors, operation contractors, external financiers, and consumers. The project delivery method determines how these multiple groups relate to one another. There is a range of project delivery methods; among the most common are the following:

Design-bid-build (DBB). This is a traditional project delivery approach for municipal infrastructure projects in which the owner, that is, the public water provider, determines that a project is needed and secures financing for the project. The owner works with an engineer to define the project and develop plans with detailed specifications; the owner then solicits bids from multiple contractors to construct the project; the contractor that provides the lowest bid that meets all of the criteria is selected; and operations are performed by the owner or by a contract operator.

Design-build (DB). The owner develops the project concept and secures financing; the owner then hires a design-build team to construct the project for a lump-sum fixed price or a guaranteed maximum price; operations are performed by the owner or by a contract operator.

Design-build-operate (DBO). The owner develops the project concept and secures financing; the owner hires a single team responsible for the development, design, construction, and long-term operation of the project.

Design-build-own-operate-transfer (DBOOT). The contractor is responsible for the design, construction, and long-term operation of the project. In contrast with the more traditional project delivery methods, the contractor must secure financing for the project and initially owns the facility; the water agency agrees to purchase the water at an agreed-upon price over a certain period (these purchase agreements allow the contractor to secure the necessary financing); the contract contains provisions for the transfer of ownership of the facility to the water provider on an agreed-upon date.

Design-build-own-operate (DBOO). In this variation of the DBOOT delivery method the contractor is responsible for the design, financing, construction, and long-term operation of the project, and the water agency agrees to purchase the water at an agreed-upon price over a certain period. Unlike the situation with a DBOOT, however, there is no asset transfer at the end of the project.

Each project delivery method has advantages and disadvantages with respect to how risk is shared among parties, the length of the process, and other factors. Some of these advantages and disadvantages are described in table 6.4. A key difference among the project delivery methods is the degree to which innovation in technology and design is integrated into the project. The water sector in the United States has traditionally employed the DBB delivery model. With DBB, the public utility or agency (owner of the project) plays a central role, selecting different groups to design, construct, and possibly operate the project. DBB contracts are generally awarded on the basis of lowest cost (NRC 2008). Thus, there is often little incentive to implement innovative designs and technologies that may have a higher capital cost but that reduce operating costs and result in a less expensive project overall. With a DBO delivery method, by contrast, the DBO contractor is responsible for operation of the project. Thus, the contractor has an incentive to promote innovative designs and technologies that can optimize and improve the operability of the project.

Another key difference among the various project delivery methods is the party required to secure financing. Under the DBB, DB, and DBO delivery methods, the water provider secures project financing, largely by issuing debt (bonds or loans). Under the DBOO and DBOOT delivery methods, by contrast, the contractor secures financing by issuing debt or obtaining private equity or both. One possible drawback in using this approach, however, is the high cost of private capital, which is reflected in the final price of desalinated water. Proponents of this delivery method, however, argue that the higher costs (which are borne by the consumer) are offset by lower risk for the water provider, higher efficiency of the contractor, and technology performance guarantees (Texas Water Development Board 2002).

Risk Allocation among Project Partners

A key element of any public-private partnership is the allocation of risk among the project partners. With DBB contracts, the owner—typically the public—carries most of the risks associated with the project. By contrast, with DBO and DBOOT contracts, the contractor bears more of the risk (figure 6.3). By assuming a specific risk, the project partner is responsible for any additional cost (or savings) associated with that risk and cannot pass that cost along to the other project partner. For example, if Partner A assumes the construction risk for the project, and the cost of construction is higher than anticipated, Partner A must pay the additional cost of construction and cannot pass those costs along to Partner B. Likewise, if construction costs are lower than anticipated, Partner A captures that savings and does not pass the savings on to Partner B.

The project sponsor is expected to identify the risks associated with the project and determine the appropriate risk and responsibility allocation prior to the procurement process and contract development. This allows the interested parties to understand the

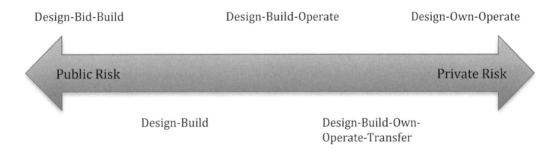

FIGURE 6.3 RISK ALLOCATION UNDER VARIOUS PROJECT DELIVERY METHODS

underlying risks and responsibilities and price their proposal accordingly, since the contractual risk structure is directly related to pricing (Wilcox and Whitney 2004). In theory, a particular risk should be allocated among parties such that the party with the greatest ability to control the outcome takes on that risk. There are some risks, however, that are outside the control of both parties.

As described above, demand risk is one of the key risks associated with desalination projects. The quantity of water that will be purchased is set forth in a long-term water purchase agreement—also referred to as an off-take agreement. This agreement is negotiated prior to construction of the project and is often needed to secure financing. Among the most common forms of off-take agreement is the "take-or-pay" contract. Under a take-or-pay contract, the buyer agrees to pay for a minimum amount of water from the seller on a certain date even if the buyer does not need the water. This type of contract provides guaranteed revenue for the seller but commits the buyer to a purchase even if actual demand drops. A minimum commitment to a large volume of water, however, provides a disincentive for water agencies to pursue more cost-effective water supply and water conservation and efficiency programs. Thus, take-or-pay contracts, if structured poorly, can result in significant exposure to demand risk and higher costs for the buyer—a problem noticed long ago in the context of energy take-or-pay contracts.[5]

Case Studies

In the following section, we provide two case studies that emphasize several elements of seawater desalination projects, including their cost, how they are financed, the structure of the public-private partnership, and the risks associated with the projects. The first case study looks in detail at the Tampa Bay Seawater Desalination Project, which is the largest plant built in North America to date and has been the subject of several in-depth assessments. The second case study focuses on the proposed desalination project in Carlsbad, California, which has been in development for more than a decade, has received all of the necessary permits, and finally secured financing in December 2012.

Tampa Bay Seawater Desalination Project

Like many communities around the United States, the Tampa Bay, Florida, region has been experiencing significant growth combined with increasing constraints on water availability. Overpumping of local groundwater resources in the region has forced communities to think about alternative water solutions. In the late 1990s, some water managers and planners thought seawater desalination might be a key element of the region's long-term water future. The Tampa Bay Seawater Desalination Project faced several major delays but was completed in 2007 (Cooley 2009). It is the first large-scale seawater desalination project constructed in the United States.

---

5. A summary of problematic energy take-or-pay contracts can be found here: http://www.jbsenergy.com/energy/papers/blueprint/blueprint.html#_Toc2576605. Similarly, take-or-pay contracts for natural gas caused serious problems for both suppliers and consumers and required federal intervention in the 1980s and 1990s (http://www.naturalgas.org/regulation/history.asp). See also the Carlsbad case study in this chapter for some issues that can arise with a take-or-pay contract.

Since its completion in 2007, the plant has operated far below its design capacity of 95,000 m³ per day (25 MGD). Figure 6.4 shows the plant's average monthly water production between 2003 and October 2012. During some short periods, the plant operated at or near capacity. In others, the plant operated at reduced capacity or was in standby mode. In 2013, Tampa Bay Water plans to operate the plant at a higher capacity (averaging 42,000 m³ per day, or 11 MGD) because a local water reservoir is undergoing repair. Once this is complete, however, production from the desalination plant will likely be reduced once again because cheaper water supply alternatives are available. A recent analysis by Tampa Bay Water found that implementing cost-effective conservation and efficiency programs in the region would reduce water demand by 2035 by 150,000 m³ per day (40 MGD) (Tampa Bay Water 2012), significantly more water than is produced by the seawater desalination plant and at a lower cost.

Project Delivery Method

In response to ongoing water supply concerns, the West Coast Regional Water Supply Authority—later to become Tampa Bay Water—issued a request for proposals (RFP) to design, build, operate, and own a desalination plant in 1996. The RFP allowed the project developer to select the size and location of the plant as well as the operations and method of seawater desalination. In July 1999, after a two-and-a-half-year solicitation process, Tampa Bay Water entered into a water purchase agreement with S&W Water, LLC—a partnership between Stone & Webster and Poseidon Resources Corporation.

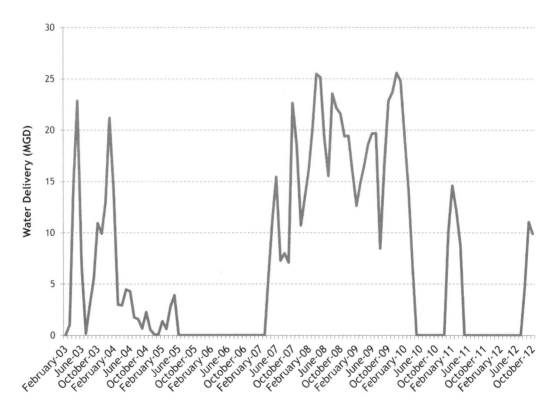

FIGURE 6.4 AVERAGE MONTHLY WATER PRODUCTION FROM TAMPA BAY SEAWATER DESALINATION PLANT, 2003–2012

Note: The design capacity of the plant is 95,000 cubic meters per day (25 MGD).

TABLE 6.4 Advantages and Disadvantages of Various Project Delivery Methods

	Advantages	Disadvantages	Works best when...
Design-bid-build (DBB)	• Well understood by all involved parties • Potential for high degree of owner control and involvement • Independent oversight of construction contractor	• Segments design, construction, and operation and reduces collaboration • Linear process increases schedule duration • Prone to disputes and creates opportunities for risk avoidance by the designer and construction contractor • Low-bid contractor selection reduces creativity and increases risks of performance problems • Risks are mostly borne by the owner • May not allow for economies of scale in operations • For new technologies, operability may not be the primary design concern	• Operation of facility is minimal or well understood by owner • Project requires high degree of public oversight • Owner wants to be extensively involved in the design • Schedule is not a priority
Design-build (DB)	• Collaboration between designer and contractor • Parallel processes reduce duration • Reduces design costs • Reduces potential for disputes between designer and construction contractor • Single point of accountability • Can promote design innovation • Provides more certainty about costs at an earlier stage • Allows owner to assign certain risks to DB team	• Owner may not be as familiar with DB process or contract terms • Reduces owner control and oversight • Design and "as-built" drawings not as detailed • Eliminates "independent oversight" role of the designer • Does not inherently include incentives for operability and construction quality, as does DBO or DBOOT approach • Higher cost to compete	• Time is critical, but existing conditions and desired outcomes are well defined • Project uses conventional, well-understood technology • Owner is willing to relinquish control over design details • Operational and aesthetic issues are easily defined • Early contractor input will likely save time or money

Design-build-operate (DBO)	• Allows collaboration among designer, construction, contractor, and operator • Parallel processes reduce duration • Operator input on new technologies and design • DBO contractor has built-in incentive to ensure quality, since contractor will be long-term operator • Single point of accountability • Allows owner to assign certain risks to DBO contractor • Economies of scale for operations • Collaboration, long-term contract, and appropriate risk allocation can cut costs • Defines long-term expenses for rate setting	• Reduces owner involvement • Owner may not be familiar with DBO contracting • High cost to compete may limit competition • Depending on contract terms, may give operator incentives to overcharge for ongoing renewals and replacements or to neglect maintenance near the end of the contract term • Operations contract may limit long-term flexibility • Requires multiphase contract	• Owner's staff does not have experience in operating the type of facility • Input conditions to the facility can be well defined and the number of external influences affecting plant operations is limited • Owner is comfortable with less direct control during design, construction, and operation
Design-build-own-operate-transfer (DBOOT)	• Same as with DBO, plus the following: • Can be used where project expenditures exceed public borrowing capacity • Beneficial when preserving public credit for other projects is important • Can isolate owner from project risk	• Same as with DBO, but lack of public financing increases the cost of money	• Public financing cannot be obtained • Transfer of technology risk is important

Source: Adapted from Texas Water Development Board (2002) and NRC (2008).

Under the initial agreement, S&W Water would develop a desalination plant capable of supplying 95,000 m^3 per day (25 MGD) to Tampa Bay Water and would sell that water at a fixed price. The agreement was structured as a BOOT arrangement that would allow Tampa Bay Water to buy the plant after thirty years or earlier, if desired.

Soon after the water purchase agreement was signed, the project began encountering problems. In June 2000, Stone & Webster declared bankruptcy, leaving Poseidon without an engineering and construction partner. Poseidon then hired Covanta Energy to construct the plant, and a new partnership emerged: Tampa Bay Desal. Obtaining financing for the project, however, proved to be difficult. Covanta Energy had a poor bond rating and was unable to secure financing for the construction bonds. Tampa Bay Water, however, had a more favorable AA bond rating, which would allow it to obtain financing at a lower interest rate and thereby reduce the financing cost. In early 2002, Tampa Bay Water bought Poseidon's interest in the project while retaining the engineering and construction company (Covanta Construction) to complete the project. At the time, the plant's design and permitting were complete, while construction was only 30 percent complete (Smith 2009). Tampa Bay Water signed a thirty-year O&M contract with Covanta Tampa Bay, Inc. (a subsidiary of Covanta Energy) to operate the facility.[6] This move effectively restructured the DBOOT procurement into a DBO contract, whereby Tampa Bay Water assumed ownership and many of the risks associated with the plant.

Before Covanta Tampa Bay could commence the separate thirty-year O&M contract, the plant was required to meet certain performance standards. Covanta Construction was unable to meet the test standards after repeated attempts and finally declared bankruptcy in 2003. In 2004, the parties agreed to a settlement whereby Tampa Bay Water paid Covanta nearly $5 million to end its contract. In return, Tampa Bay Water retained full control of the facilities and the operating contract. In late 2004, Tampa Bay Water signed a DBO contract with American Water-Pridesa, a joint venture between American Water and Acciona Agua, to repair deficiencies and operate the plant on a long-term basis.

Under the current operating contract, Tampa Bay Water must pay American Water-Pridesa about $7 million per year regardless of whether water is produced by the plant. This base contract covers the fixed O&M costs for the plant plus some return to the operator. Tampa Bay Water pays variable operating costs (chemicals and electricity) separately. This contract allows Tampa Bay Water to reduce the plant's production if cheaper alternatives are available.

Project Cost

The initial capital cost for the desalination plant and the fifteen-mile pipeline to connect it to the water system was estimated to be $110 million. Initial performance tests in 2003, however, revealed thirty-one deficiencies in the plant. To fix the plant, Tampa Bay Water signed a $29 million, two-year contract with American Water-Pridesa. Major improvements were made to the pre-treatment, reverse osmosis, and post-treatment system, and the final capital cost of the project was $158 million, more than 40 percent above the original cost.

Initially, desalination proponents were excited by the Tampa Bay project and by the apparent price breakthrough. In the initial water purchase agreement, S&W Water made a binding commitment to *deliver* desalinated water in the first year of operation at an unprecedented wholesale cost of $0.45 per m^3 ($557 per acre-foot), with a thirty-year

---

6. The O&M contract was worth $300–$360 million and was Covanta Tampa Bay's only asset (Wright 2003).

average cost of $0.55 per m^3 ($678 per acre-foot) (Heller 1999). This was well below the cost of water from other recent desalination plants. By comparison, in that same year the Singapore Public Utilities Board announced plans to build a large desalination plant to *produce* water at an estimated cost of between $1.98 and $2.31 per m^3 ($2,440 and $2,850 per acre-foot, respectively), with an additional cost to deliver water to customers (*U.S. Water News* 1999).

The project developers were unable to deliver on the initial cost estimates, and the cost estimates have increased. Additionally, as noted above, the production of water from the plant has varied dramatically since the project was completed in 2007, further affecting the unit cost (figure 6.4). According to recent estimates from Tampa Bay Water, the unit cost of water from the desalination plant is $1.05 per m^3 ($1,300 per acre-foot) if the plant is operating at full capacity (Vatter 2012).[7] Officials at Tampa Bay Water, however, indicate that the plant will operate at an average production of 40,000 m^3 per day (11 MGD) from October 2012 through October 2013, which is considerably less than full capacity. At this reduced capacity, the unit cost of water is estimated at $1.30 per m^3 ($1,600 per acre-foot), more than double the original cost estimate (Vatter 2012).

Project Financing

The Tampa Bay Seawater Desalination Project was originally structured as a BOOT agreement, which usually employs some combination of debt and equity financing. Tax-exempt private-activity bonds provided 90 percent of the project's capital cost (Lokiec and Kronenberg 2001), and the remaining 10 percent was provided as an equity stake by the project developer. In 2007, however, Tampa Bay Water bought out Poseidon's interest in the project using short-term, variable interest rate bonds and became the sole owner of the plant. Tampa Bay Water selected this financing strategy because interest rates were low at the time. Furthermore, the bonds would be repaid shortly after the project was completed because the Southwest Florida Water Management District (SWFWMD), a regional agency that regulates and permits water resources for west-central Florida, agreed to provide $85 million of the project's eligible capital costs over an eighteen-month period upon completion. The SWFWMD obtained these funds through voter-approved, locally collected ad valorem property taxes designated for water supply creation.

Summary

When the Tampa Bay desalination plant was announced, it was hailed as a breakthrough for desalination in the United States and sparked enormous interest and excitement. But the project encountered several technical and managerial challenges. The experiences of Tampa Bay Water can help to inform the development of projects under consideration elsewhere. Some of the lessons that can be learned about cost, financing, and risk include the following:

- Anticipate the transfer of ownership at any point of the project and consider what actions will be taken if that transfer occurs.

- Exercise caution when transferring ownership under a DBOOT contract. The new owners must recognize that they are assuming many of the risks associated with the project.

---

7. This estimate includes O&M plus the annual debt service cost, assuming that the full capital cost of the project was financed at Tampa Bay Water's average bond rate over thirty years.

- Ensure that those submitting proposals have relevant experience with desalination projects.
- Minimize demand risk by implementing cost-effective water supply alternatives prior to the development of the project.
- Consider establishing a minimum operating contract, rather than a take-or-pay contract, to allow the water supplier to adjust production of the plant if cheaper alternatives are available.

Carlsbad Desalination Project

Poseidon Resources Corporation is working to build a 190,000 m^3 per day (50 MGD) desalination facility at the Encina Power Station in the city of Carlsbad. The project would utilize the Encina Power Station's existing surface intake and outfall structures. Poseidon began developing the project in 1998, and in 2003 it constructed a demonstration facility at the site. The project has been highly contested, with local and environmental groups filing more than a dozen legal challenges, including lawsuits and permit appeals, against the project. By November 2009, Poseidon had received all of the permits required to build and operate the plant and announced it would begin to secure financing. Financing was finally secured in December 2012. Here we provide a detailed case study of the cost, financing, and partnerships associated with the project.

Project Cost

Numerous and contradictory cost estimates for the Carlsbad Desalination Project have been put forth since the project was initiated in the late 1990s. In May 2002, initial capital cost estimates for the desalination plant were $260 million (Green 2002), dramatically less than other plants of similar size at the time. Estimates for the desalination plant and distribution pipeline were quoted at $270 million in 2002 (Jimenez 2002) and $320 million in 2009 (Schwartz 2009). By October 2011, however, projected costs had increased to $950 million, more than three and one-half times the original cost. It is difficult to determine why these costs grew so dramatically over the ten-year period because there has been a lack of transparency regarding assumptions about financing, inflation rates, energy costs, interest repayment, related facilities, and other factors.

The water purchase agreement provided detailed cost estimates for the project. According to these figures, the capital cost of the project is $767 million (Lee 2013). Of that amount,

- $537 million is needed for the planning, permitting, design, construction, mitigation, and legal costs associated with the desalination plant;
- $159 million is needed to build a ten-mile pipeline to transport the desalinated water from the coast to the regional water distribution system; and
- $80 million is needed for indirect costs associated with water system improvements, including relining and rehabilitating a five-and-one-half-mile section of the distribution pipeline and modifying the Twin Oaks Valley Water Treatment Plant to accommodate desalinated water flows.

In addition to the capital costs, operating and maintaining these facilities would cost an additional $49–$54 million each year (Lee 2013).[8] Financing costs for the project, including bond repayments and the return to the equity investors, exceed $1 billion (Lee 2013).

As described previously, the unit cost of water is determined by dividing the annual capital and O&M costs by the amount of water produced. In 2002, Poseidon estimated that the unit cost of water would be $0.65 per m^3 ($800 per acre-foot) (Green 2002). In the initial contracts with local water agencies, Poseidon agreed to sell desalinated water at the same price that the agencies paid for imported water. Poseidon was willing to take a loss on the project in the short term, with the belief that imported water supplies would eventually be more expensive than the desalinated water and that they would recover these costs over the long term. Today, water agencies pay less than $0.81 per m^3 ($1,000 per acre-foot) for treated water from the Metropolitan Water District of Southern California (MWD), which is more expensive than Poseidon's original cost estimate but considerably less than the most recent estimates for the project.

Like the total cost, the unit price of water has also grown tremendously as the project has developed. By July 2010, the unit cost for water had increased to $0.89 per m^3 ($1,100 per acre-foot) plus an additional $0.41 per m^3 ($500 per acre-foot) (in 2010 dollars) for delivery charges (SDCWA 2010). Since that time, the unit cost has continued to rise in response to changes in the plant design, inflation rates, improved financing estimates, and more. The most recent figures from the San Diego County Water Authority (SDCWA, or Water Authority) estimate that the cost to produce desalinated water and to integrate it into the existing distribution system is about $1.55–$1.76 per m^3 ($1,900–$2,200 per acre-foot) (in 2012 dollars) (SDCWA 2012a).[9] Note that the cost estimates—both the total cost and the unit cost—will likely change as the project advances.

A key question is how the cost of desalination compares with the cost of other water supply alternatives. A recent analysis by the Water Authority finds that the marginal cost of new supply within the San Diego area ranges from $1.20 to $1.93 per m^3 ($1,475–$2,375 per acre-foot) (in 2011 dollars) (table 6.5). The cost of water from the proposed seawater desalination plants—Carlsbad and Camp Pendleton—are at the high end of the range. Desalinated brackish groundwater and recycled water are considerably less expensive. This analysis, however, does not include an evaluation of the cost of water conservation and efficiency measures, many of which provide water for considerably less than $1.22 per m^3 ($1,500 per acre-foot) (Gleick et al. 2003).

Project Delivery Method

The Carlsbad desalination project has been in development since 1998 as a public-private partnership. The project partners—and the relationships between those partners—have changed considerably over the past fourteen years. In December 2001, Poseidon completed an initial study for the City of Carlsbad and proposed to build a desalination plant at the Encina Power Station. The original proposal was structured as a DBOO agreement, whereby Poseidon would retain ownership of the facility and would ask the regional water agencies, including the City of Carlsbad and the Water Authority, to sign a thirty-year contract to purchase water directly from Poseidon (La Rue 2001).

---

8. This includes O&M costs for the entire project, that is, the desalination plant, the conveyance pipeline, and the efficiency losses at the Twin Oaks Valley Water Treatment Plant.

9. The range in the unit price reflects a purchase of 48,000 to 54,000 acre-feet per year.

TABLE 6.5 Marginal Unit Cost of Water

Water Alternative	Cost per Cubic Meter	Cost per Acre-Foot
Brackish groundwater (Mission Basin Narrows)	$1.39	$1,717
Brackish groundwater (Otay River)	$1.20–$1.69	$1,475–$2,086
Indirect potable reuse (City of San Diego)	$1.60–$1.93	$1,975–$2,375
Indirect potable reuse (North San Diego County Regional Reuse)	$1.32–$1.40	$1,628–$1,730
Seawater desalination (Camp Pendleton)	$1.54–$1.90	$1,900–$2,340

Source: SDCWA (2012d).

Note: These costs refer to estimates for specific projects and are directly from the SDCWA. The costs shown in this table refer to average estimates across the regions. All costs are in 2011 US dollars and exclude any grants or incentives. The cost listed for the City of San Diego's indirect potable reuse project includes wastewater-related costs that could reduce the unit cost by up to $600 per acre-foot ($0.49 per cubic meter). This figure also excludes the costs of conservation and efficiency, water transfers from agriculture, and additional water from regional providers.

By 2002, the partnership had shifted to a DBOOT arrangement. In November 2002, the Water Authority approved the outline of a partnership with Poseidon to build a desalination plant in Carlsbad. Under the proposed agreement, Poseidon would obtain the permits, secure the financing, and supervise the construction. The power plant's owner, Cabrillo Power, would operate and maintain the facility. The Water Authority would be the sole purchaser of water and would be responsible for building the pipelines that distribute the water. Poseidon would initially own the plant but would sell it to the Water Authority after five years of operation (Jimenez 2002).

Negotiations between Poseidon and the Water Authority were on-again, off-again between 2003 and 2005. Unable to reach an agreement, the two sides resolved to pursue separate, competing projects at the Encina Power Station: one by the Water Authority and the other by Poseidon. In mid-2006, however, the Water Authority announced it would no longer pursue a desalination plant in Carlsbad (Rodgers 2006). Poseidon continued to pursue water purchase agreements with local agencies. By late 2009, Poseidon had received all of the permits it needed to build the plant and had signed water purchase agreements with a number of local agencies (Burge 2009). With these agreements in place and with the withdrawal of the Water Authority, the Carlsbad Desalination Project in effect shifted to a DBOO arrangement.

In mid-2010, the structure of the project abruptly shifted back to a DBOOT arrangement. In July 2010, the Water Authority approved key terms to be the sole purchaser of water from Poseidon. Water Authority staff began negotiating specific elements of the agreement, including the water purchase price, allocation of risk, and options to eventually purchase the project's pipeline and the entire desalination plant (SDCWA 2010). In September 2012, after more than two years of negotiation, the Water Authority released the water purchase agreement for a public review period prior to a vote by the county board of directors. Under this agreement, Poseidon would be responsible for designing, permitting, financing, constructing, and operating the desalination plant and would be the owner of the plant during the contract term.[10] The Water Authority would have the

---

10. The contractor for the project would be a joint venture of Kiewit Infrastructure West and J.F. Shea Construction. IDE Technologies would provide the process technology (e.g., the pumps and membranes) and would also operate and maintain the desalination plant.

right to purchase the plant after ten years of commercial operation under previously specified terms and conditions or to take over ownership of the plant after thirty years for $1. The Water Authority may also purchase the plant if Poseidon is unable to secure financing or if the venture goes bankrupt (SDCWA 2012c). It is of note that in Tampa Bay, Florida, the project contractors were unable to secure financing, prompting Tampa Bay Water to purchase Poseidon's interest in the project and assume many of the project-related risks.

The pipeline needed to convey the desalinated water from the coast to the distribution system would be developed under a separate DB agreement between the Water Authority and Poseidon. The Water Authority would own and maintain the pipeline and would finance the project using municipal bonds. Poseidon, however, would assume responsibility for payment of the bonds in the event that Poseidon is unable to complete the desalination project. The Water Authority would also assume responsibility for completing and financing other water system improvements, specifically, relining the conveyance pipeline and upgrading the Twin Oaks Valley Water Treatment Plant.

The process for selecting Poseidon as the project partner was different from the typical process in a DBOOT arrangement. Ordinarily, the public agency solicits proposals for a project. Interested parties then submit their qualifications and, if accepted, submit a proposal that includes some cost estimate. The public agency then selects a proposal on the basis of a variety of factors, including cost. The Water Authority, however, did not allow other parties to submit proposals and determine which one offered the best alternative. Without this process, it is difficult to determine whether the best proposal was selected.

Risk Allocation

One key element of the water purchase agreement between Poseidon and the Water Authority is the allocation of risk among the project partners. As shown in table 6.6, project-related risks are split between Poseidon and the Water Authority. Poseidon, for example, assumes the risks associated with construction of the plant; this includes ensuring that the project is completed on time, on cost, and according to design standards. Poseidon also assumes certain operational risks, including those associated with the amount of energy consumed, the costs of membrane performance and replacement, and chemical costs. Additionally, if Poseidon fails to produce water, then Poseidon must cover the debt service cost for the desalination plant and the Water Authority's conveyance pipeline.

The Water Authority also assumes several project-related risks that could further increase the price of water. For example, the Water Authority assumes the risk associated with changes in law or regulations that generally apply to water treatment facilities and wastewater dischargers. Thus, any additional costs associated with complying with these laws or regulations would allow Poseidon to increase the price of water. The State Water Resources Control Board, for example, is considering amendments to the California Ocean Plan that could increase the costs associated with brine discharge. Under the purchase agreement, capital costs of up to $20 million (in 2010 dollars) could be passed along to the Water Authority. The Water Authority also assumes the risk associated with inflation and rising electricity prices, and thus Poseidon can increase the price of water because of costs associated these factors.

The Water Authority also assumes the risk associated with sustaining demand for the water produced by the desalination plant. The water purchase agreement is structured as a take-or-pay contract, and under these terms the Water Authority must pay for the water

TABLE 6.6 Project Risk Allocation

Risk Description	Poseidon	San Diego County Water Authority
Construction risk—facility is not completed on time, on cost, and according to design standards	X	
Permitting risk—current permit and environmental mitigation requirements increase	X	
Change in law risk—future unanticipated laws or regulations increase operating costs	X	X
Technology risk—plant technology does not perform as expected	X	
Output risk—plant is not able to produce the projected volume of water	X	
Operating margin risk—the price of water is not adequate to generate enough revenue to pay expenditures or may increase more than projected	X (budget cap)	X (subject to CPI)
Pipeline operating risk—the pipeline connecting the plant to the regional aqueduct system and associated facilities is unable to transport acceptable water to the SDCWA's wholesale customers	X	X
Electricity—changes in the cost or amount of electricity used will affect the water price	X (electricity consumption)	X (electricity price)
Force majeure events	X (insurable)	X (uninsurable)

Source: SDCWA (2012b).

Note: The San Diego County Water Authority (SDCWA) assumes those risks, and costs, that affect all water treatment plant operators and wastewater dischargers and the cost of power plant improvements (with a cap of up to $20 million in capital and $2.5 million in annual operating costs; these estimates are in 2010 dollars).

regardless of whether that water is needed. In a July 2012 meeting, SDCWA staff indicated that current regional water demands do not justify a fixed commitment to purchase the entire output of the plant, which is equivalent to 69,000 megaliters (Ml) per year. Rather, staff suggested that the agreement be based on a lower contract minimum of 59,000 Ml per year and that fixed costs be allocated on the basis of that purchase amount.

Based on the staff's recommendation, the 2012 water purchase agreement establishes a take-or-pay contract with a minimum demand commitment of 59,000 Ml per year. Thus, the Water Authority must purchase at least 59,000 Ml per year of the most expensive water available, even if demand declines or if cheaper alternative supplies are available. This exposes the Water Authority to significant demand risk and could have major cost implications for the Water Authority and its ratepayers. One option is for the Water Authority to consider a contract that establishes an even lower minimum commitment. While this may increase the unit cost of water, it would allow the Water Authority to reduce some operating costs, especially energy and chemical costs.

Project Financing

Efforts to finance the project began in late 2009. At that time, the cost estimate for the project was $300 million. Stephen Howard, head of the infrastructure group at Barclays Capital, which was seriously considering the plant at the time, was quoted as saying that a $300 million project was feasible, but "if this project was north of $800 million to $1 billion, it would probably be very challenging" (Schwartz 2009). Today, costs are squarely within this range.

Financing for the project was finally secured in December 2012, shortly after the water purchase agreement was approved. Nearly 80 percent of the project's cost was financed through tax-exempt private-activity bonds. The remaining 20 percent was provided by cash equity from Stonepeak Infrastructure Partners (Brennan 2013).[11] Standard & Poor's assigned a "BBB–" credit rating to the project, which is the lowest investment rating and was expected to increase the cost of financing the project. Despite the low rating, the project received a 4.78 percent yield on the bonds for the desalination plant and 4–4.5 percent for the pipeline. Higher rates of return are expected for the equity providers. The Water Authority estimates that the cost to finance the project would be $227 million.

Summary

It is not yet clear whether the Carlsbad Desalination Project will succeed. However, the project raises several issues that should be considered by other developers seeking to pursue desalination project, particularly the following:

- The project was negotiated in a noncompetitive environment, which raises concerns about whether the best project was selected, especially in relation to the cost, design, and operation of the plant.

- Demand risk is a major concern for the project. Alternative supplies and conservation and efficiency improvements are available at lower cost than the desalination project.

- The take-or-pay contract establishes a high minimum commitment to purchase water. This would prevent the Water Authority from pursuing cheaper water alternatives as they become available.

- The Water Authority should consider a lower minimum commitment, which would allow it the flexibility to reduce plant output and capture potentially significant financial savings from lower O&M expenses.

Conclusions

Economics—including both the cost of the water produced and the complex financial arrangements needed to develop a project—are important factors that will determine the ultimate success and extent of desalination in California. In this chapter we provided detailed information about the cost of seawater desalination projects, how they are financed, and some of the risks associated with these projects.

---

11. Stonepeak Infrastructure Partners, formerly the infrastructure investment division of Blackstone, is an equity investment firm that focuses on traditional infrastructure assets, especially in North America.

How Much Does Desalination Really Cost?

Our analysis finds that the cost to produce water from a desalination plant is highly variable. Recent estimates for plants proposed in California range from $1.54 to $2.43 per m^3 ($1,900 to more than $3,000 per acre-foot). Although the cost of seawater desalination has declined considerably over the past twenty years, desalination costs remain high, and major cost breakthroughs are unlikely to occur in the near to mid term. Indeed, desalination costs may increase in response to rising energy prices.

The public and decision makers must exercise caution when comparing cost estimates for different seawater desalination projects. In many cases, costs are reported in ways that are not directly comparable. For example, some report the cost of the desalination plant alone, whereas others include the additional infrastructure, such as conveyance pipelines, needed to integrate the desalination plant into the rest of the water system. Some estimates include the cost to finance the project, and others do not. Even when there is an apples-to-apples comparison, a number of site- and project-specific factors, such as energy, land, and labor costs, make cost comparisons difficult.

Furthermore, costs associated with wastewater conveyance and treatment (and many other indirect costs to ecosystems) are often excluded from desalination cost comparisons. The introduction of a new source of water increases the amount of wastewater that must be collected, treated, and disposed of. Some communities may have adequate wastewater treatment capacity, and the additional costs would simply be the variable O&M costs associated with that treatment. In other communities, however, wastewater treatment capacity may need to be expanded, which represents an additional, and in some cases significant, capital cost to the community. While these costs would apply to all of the water supply projects under consideration to meet demand, the costs would not be incurred if water demand were met through water conservation and efficiency improvements.

What Are Some of the Risks Associated with Seawater Desalination Projects?

Several risks can affect a seawater desalination project's cost, ability to attract financing, and overall viability. Many of these risks are not unique to seawater desalination projects; rather, they apply broadly to all major infrastructure projects. These include risks associated with permitting, construction, operations, and changes in law.

But as recent experience in the United States and Australia has shown, desalination projects entail risks specific to large water supply projects, including demand risk. Demand risk is the risk that water demand will be insufficient to justify continued full-time operation of the desalination plant as a result of the availability of less expensive water supply and demand management alternatives. In Australia, for example, four of the six desalination plants that have been developed since 2006 are being placed in standby mode. Likewise, the Tampa Bay Seawater Desalination Project operates considerably below full capacity because demand is lower than expected and less expensive water supply options are available. Demand risk raises serious concerns about the size and timing of desalination projects.

In some regions, seawater desalination can make an important contribution to the availability and reliability of water resources. However, it remains among the most expensive options available to meet water demands. Additionally, project developers may pursue large plants in an effort to capture economies of scale and reduce the unit cost of water. This can, however, lead to oversized projects that ultimately increase demand risk and threaten the long-term viability of a project.

How Are Desalination Projects Structured?

Issues regarding financing and the allocation of project costs and risks are tied to the way the project is structured. Many project developers in California are using some form of public-private partnership. The private sector's involvement in seawater desalination projects is not new. The private sector has developed several small plants to supply high-quality water for specific industrial purposes, such as for use on oil and gas platforms. Likewise, a desalination project in Santa Barbara, completed in 1992, was operated and partially owned by a private company. In some cases, the private sector's involvement is limited to conducting feasibility studies and preparing environmental documents as requested by the public project developer. In other cases, however, a private entity owns and operates the desalination plant and sells water directly to a public agency. Public-private partnerships provide a mechanism to access private capital and allocate risks among the project partners. They can, however, be highly contentious because of concerns about openness and transparency of data and financial information and the allocation of the risk among the project partners.

Additionally, utilities that are developing seawater desalination plants must be sure that there is a demand for that water, especially when establishing minimum commitments under take-or-pay contracts. A take-or-pay contract provides guaranteed revenue for the seller but commits the buyer to a purchase even if actual demand drops. This exposes the buyer to demand risk and provides a disincentive for water agencies to pursue more cost-effective water supply and water conservation and efficiency programs.

References

AAP Newswire. 2012. Desal plant closure not waste: NSW govt. Australian Associated Press, June 26. http://news.smh.com.au/breaking-news-national/desal-plant-closure-not-waste-nsw-govt-20120626-20z7d.html.

Brennan, D. S. 2013. Low bond rates cut desalination costs. *San Diego Union-Tribune*, January 24. http://www.utsandiego.com/news/2013/jan/24/poseidon-desalination-carlsbad-water-authority/.

Burge, M. 2009. Carlsbad desalination plant gets final OK: Coastal panel gives permit to Poseidon. *San Diego Union-Tribune*, November 4. http://www.utsandiego.com/news/2009/Nov/04/carlsbad-desalination-plant-gets-final-ok/.

Cooley, H. 2009. Tampa Bay desalination plant: An update. In *The World's Water 2008–2009: The Biennial Report on Freshwater Resources*, edited by P. H. Gleick, 123–125. Washington, DC: Island Press.

Cooley, H., P. H. Gleick, and G. Wolff. 2006. *Desalination, with a Grain of Salt: A California Perspective*. Oakland, CA: Pacific Institute. http://www.pacinst.org/publication/desalination-with-a-grain-of-salt-a-california-perspective-2/.

Crisp, G. 2012. Global business leader, desalination, at GHD. Personal communication, October 25.

Equinox Center. 2010. San Diego's Water Sources: Assessing the Options. Report prepared by Fermanian Business and Economic Institute. Encinitas, CA: Equinox Center. http://www.equinoxcenter.org/assets/files/pdf/AssessingtheOptionsfinal.pdf.

Gleick, P. H., D. Haasz, C. Henges-Jeck, V. Srinivasan, G. Wolff, K. K. Cushing, and A. Mann. 2003. Waste Not, Want Not: The Potential for Urban Water Conservation in California. Oakland, CA:

Pacific Institute. http://www.pacinst.org/wp-content/uploads/2013/02/waste_not_want_not_full_report3.pdf.

Green, K. 2002. Water panel OKs talks on desalination plant; rival companies jockey for position. *San Diego Union-Tribune*, May 24.

Heller, J. 1999. Water board green-lights desalination plant on bay. *Tampa Bay Business News*, March 16.

Hosking, W. 2012. Water from Wonthaggi desalination plant flows into Cardinia Reservoir. *Herald Sun*, Melbourne, September 26.

Jimenez, J. L. 2002. Desalination partnership gets water authority nod; directors agree to work with Poseidon. *San Diego Union-Tribune*, November 15.

Kemp, M. 2012. Silver lining to $1.8bn Port Stanvac desalination plant white elephant. *The Advertiser*, October 4.

La Rue, S. 2001. Desalting plant for some county water proposed for Carlsbad. *San Diego Union-Tribune*, December 5.

Lee, M. 2013. Public affairs representative, San Diego County Water Authority. Personal e-mail communication, February 6.

Lokiec, F., and G. Kronenberg. 2001. Emerging role of BOOT desalination projects. *Desalination* 136 (1–3): 109–114.

Luckenbach, H. 2012. Desalination program coordinator. Personal e-mail communication, October 9.

Marschke, T. 2012. Call to keep desal plant running. *Gold Coast Sun*, July 12.

National Research Council (NRC). 2008. *Desalination: A National Perspective*. Washington, DC: National Academies Press.

National Water Commission (NWC). 2008. Emerging Trends in Desalination: A Review. Waterlines Report Series No. 9. Canberra, ACT, Australia: National Water Commission.

Pankratz, T. 2011. Plant mothballing: To be or not to be. *Water Desalination Report* 47 (3): 2–3.

———. 2012. Editor of Global Water Intelligence's *Water Desalination Report*. Personal communication, November 9.

Pemberton, C. 2003. Financing Water and Sewerage Systems: A Caribbean Perspective. St. Lucia: Caribbean Basin Water Management Programme Inc. http://www.bvsde.paho.org/bvsacd/cwwa/cecil.pdf.

Robertson, T. 2010. Pre-Hearing Non-Government Question on Notice, Estimates Committee C 2010. Government of Queensland, Australia. www.jeffseeney.com.au/getdata.do?source=3&id=191 (accessed August 14, 2013).

Rodgers, T. 2006. County drops Carlsbad desalination plan; private company is free to pursue own proposal. *San Diego Union-Tribune*, July 28.

San Diego County Water Authority (SDCWA). 2010. Consideration of Key Terms and Conditions of a Potential Water Purchase Agreement with Poseidon Resources (Channelside) for Desalinated Seawater from the Carlsbad Desalination Project. Water Planning Committee. July 16. http://www.sdcwa.org/sites/default/files/files/desal-board-memo-2010-07-16.pdf.

———. 2012a. Favorable Desalination Project Bond Sale Saves Ratepayers $200 Million. Press release, December 14. http://www.sdcwa.org/favorable-desalination-project-bond-sale-saves-ratepayers-200-million (accessed January 22, 2013).

———. 2012b. Member Agency Purchases of Water Authority–Owned Supplies from the Carlsbad Desalination Project: Special Board of Directors' Meeting, July 12, 2012. http://es.slideshare.net/waterauthority/desalmemberagency (accessed August 7, 2013).

———. 2012c. Overview of Key Terms for a Water Purchase Agreement between the San Diego County Water Authority and Poseidon Resources: Water Plan Committee Meeting, September 27, 2012.

———. 2012d. Public Workshop on Issues Related to the Carlsbad Desalination Project: San Diego County Water Authority, Special Water Planning Committee Meeting, October 2, 2012. http://www.sdcwa.org/sites/default/files/files/board/2012_presentations/2012_10_02presentations.pdf (accessed August 7, 2013).

Schiller, E. A. 2012. Successful Procurement and Finance Methods: The Southwest Florida Experience. http://texaswater.tamu.edu/readings/desal/tampaflaproject.pdf (accessed November 9, 2012).

Schwartz, N. 2009. San Diego Water Board OKs Huge Desalination Plant. Associated Press, Los Angeles Metro Area, May 13.

Separation Processes, Inc. 2012. Evaluation of Seawater Desalination Projects. Report prepared for Monterey Peninsula Regional Water Authority. Carlsbad, CA: Separation Processes, Inc.

Smith, A. L. 2009. Water Supply Case Study: Tampa Bay Seawater Desalination Plant. KDI/ADB/ADBI/WBI Conference, Knowledge Sharing of Infrastructure Public-Private Partnerships in Asia, May 19–21, 2009, Seoul, Korea. http://www.adbi.org/files/2009.05.21.cpp.day3.sess7.smith.usa.water.supply.pdf.

Tampa Bay Water. 2012. Tampa Bay Water Demand Management Plan Avoided Supply Cost Technical Memorandum. Clearwater, FL: Tampa Bay Water.

Texas Water Development Board. 2002. Alternative Project Delivery. Technical report prepared by RW Beck Inc. Austin: Texas Water Development Board. http://www.twdb.texas.gov/publications/reports/contracted_reports/doc/2000483347.pdf.

United States Energy Information Administration (US EIA). 2012. Electric Sales, Revenue, and Average Price: With Data for 2011. http://www.eia.gov/electricity/sales_revenue_price/.

U.S. Water News. 1999. Tampa Bay desalinated water will be the cheapest in the world. March. http://www.uswaternews.com/archives/arcsupply/9tambay3.html.

Vatter, L. 2012. Budget analyst, Tampa Bay Water. E-mail communication, November 5.

WateReuse Association. 2012. Seawater Desalination Costs. WateReuse Association Desalination Committee White Paper. Alexandria, VA: WateReuse Association.

Wilcox, J., and S. Whitney. 2004. Considerations of desalination projects. In *Desalination and Water Reuse*, edited by S. Nicklin, 22–26. Leicester, England: Tudor Rose.

Wolfs, M., and S. Woodroffe. 2002. Structuring and financing international BOO/BOT desalination projects. *Desalination* 142:101–106.

Wright, A. G. 2003. Desalination dispute leaves a bitter taste: Judge orders Tampa Bay Water and Covanta to name a mediator to work out plant dispute. McGraw-Hill Construction Industry Headlines. December. http://enr.construction.com/news/environment/archives/031208.asp.

CHAPTER 7

Zombie Water Projects

Peter H. Gleick, Matthew Heberger, and Kristina Donnelly

Terraforming is not science fiction but reality: humans are remaking the surface of our planet. We are altering the composition of the planet's atmosphere and fundamentally changing the entire global climate. We remove mountains and dig miles beneath the surface for minerals. We cut down and destroy entire forests. We wipe out species and communities of species. We divert and consume entire rivers. We've stored so much water behind artificial reservoirs that we've actually (very modestly) changed the planet's orbital dynamics. But we've lost sight of a fundamental principle: "can" does not mean "should." Modern civilization must learn that the ability to do something doesn't mean that we actually should, especially in the field of large-scale geoengineering. And this is especially true in the field of large water projects.

A key characteristic of the world's hydrologic cycle is the uneven distribution of water in both space and time. While the planet, as a whole, has plenty of water, problems of availability arise in regions and seasons in which water is limited. These problems with distribution have led to a wide variety of solutions to store and move water from regions or periods of abundance to regions or periods of scarcity. In ancient times, moving water meant carrying it on one's back, or digging short irrigation ditches from rivers to agricultural fields, or, by the time of the ancient Romans, constructing elaborate gravity-fed aqueducts tens or even hundreds of kilometers long. But it wasn't until the industrial age, with the energy of fossil fuels and the invention of machines that could pump water uphill, that humans began to conceive of and implement truly long-distance transfers of water. Today, there are no technical constraints on how far water can be moved; our limitations are political, economic, and environmental. But even these constraints do not stop people from dreaming of ever more grandiose schemes to try to help solve global water problems.

As any consumer of popular culture today knows, a zombie is an undead creature, something that was killed but came back to "life," in some form, to terrify, entertain, and gross people out. Zombies are big business. Zombie books, movies, costumes, makeup, computer games, and more are probably worth billions to our economy (not to mention the value of extra sales of axes, chain saws, and shotguns to people who never hunt or cut down trees).

But not all zombies are fictional, and some are potentially really dangerous—at least to our pocketbooks and environment. These include zombie water projects, which we define as large, costly water projects that are proposed, killed off for one reason or another, and brought back to life, even if the project itself is socially, politically, economically, or environmentally unjustified. We offer here—in an only partly tongue-in-cheek effort— a list of some major zombie water projects from the past century: serious proposals for massive projects to use, divert, or transfer water that died only to come back to life in one form or another.

We do not believe that any of the following zombie water projects will ever be built or implemented. But, as in movie zombies, we can never really be sure if they are truly dead.

The North American Water and Power Alliance— NAWAPA

One of the largest water projects ever proposed was the North American Water and Power Alliance (NAWAPA). Stretching over the entire North American continent, it would have altered the hydraulic landscape in massive, dramatic ways. While never built, it remains a paper monument to the engineering mentality that in the past century dominated, and to some extent still dominates, water planning and thinking.

History

The North American Water and Power Alliance, or NAWAPA, was a plan first proposed in the 1950s to divert the flows of large rivers in western Canada and Alaska to the United States and Mexico. The project, named for the intergovernmental alliance that would run the project, would modify watersheds and water flows for much of the North American continent, and backers claimed it would "completely solve the water shortage problem" for generations to come (Special Subcommittee on Western Water Development 1964).

The grandiose plan has been called the "most outsize engineering project ever conceived" (Rothfeder 2004, 111) and "the water scheme to beat all water schemes," grander in scale than any river manipulation project ever imagined (Worster 1992, 315). Its total cost was estimated by backers in 1964 at $100–$200 billion over thirty years, a large sum considering that total federal government spending was about $110 billion in 1964. Accounting for inflation alone, and not considering increases in land value or construction costs, the project would cost $0.7–1.4 trillion in 2013, nearly 10 percent of the United States' annual economic output, or 40 percent of annual federal spending.

The project to divert rivers in Alaska and the Pacific Northwest to the lower forty-eight states was first imagined in the late 1950s by Donald McCord Baker, an engineer with the Los Angeles Department of Water and Power. He approached Ralph M. Parsons, president of the Ralph M. Parsons Company, one of the country's largest engineering and construction companies,[1] headquartered in nearby Pasadena. Parsons was so taken with the idea that he vigorously promoted it his entire life, including endowing the NAWAPA Foundation to advance the cause. In addition, in the 1960s, Parsons assembled "several former Bureau of Reclamation engineers . . . to make money consulting and designing resource projects for countries around the world" (Worster 1992, 315).

The project also gained the attention of US senator Frank E. Moss, a Democrat from Utah, for whom the project became a lifelong interest. At Moss's instigation, the Sen-

---

1. It remains so to this day, having built large projects, such as a Washington, DC, airport and the Goddard Space Flight Center in Maryland, and securing a $243 million contract in 2004 to build 150 hospitals in Iraq. The latter project erupted in controversy when it was later found that only twenty hospitals were completed (Mandel 2006).

ate Public Works Committee studied the proposal, and the Senate held a series of hearings on the subject. The committee report concluded that the project "warrants a serious analysis" while acknowledging that merely to conduct a "thorough engineering study" to find out whether the project was actually feasible and cost-effective would take years and cost millions. Such a study was never authorized by Congress, but Moss continued to push for the project. He wrote a book promoting NAWAPA called *The Water Crisis* in 1967, and, after losing his bid for reelection in 1976, he was hired by Parsons as a lobbyist. The proposal was also treated seriously by the executive branch of the federal government: the US secretary of state traveled to Canada to discuss the project, and the NAWAPA Foundation sent groups of American dignitaries north to tour project sites.

Promoters of NAWAPA emphasized impending water shortages and the need for traditional technological and infrastructural solutions. Concern was especially great for the arid and semi-arid Southwest, where groundwater reserves were dwindling in Texas's Upper Rio Grande region and in Colorado, the Great Basin, and Southern California. Because NAWAPA would take thirty years to construct, its backers argued, it was important to begin before it was too late: "the water crisis is a problem of serious and far-reaching implications" that will "grow steadily worse until it reaches alarming proportions in the years 1980 and 2000" (Special Subcommittee on Western Water Development 1964, 11). Frank Moss, the Democratic senator from Utah, noted that current water withdrawals in the United States stood at 300 billion gallons per day, or 27 percent of stream flow. By 1980, he argued, population growth and expanded industry and irrigation would increase demand to 900 billion gallons per day, or 80 percent of available stream flow. The forecasted increases in water use never occurred. According to the US Geological Survey, total water withdrawals peaked at 350 billion gallons per day in 1980 and have declined slightly since then, even while the country's population and agricultural output have continued to rise steadily (Gleick 2003; Kenny et al. 2009).

Project Design

It was perhaps inevitable that some ambitious planner or engineer, confronted with the aridity and water scarcity of the American Southwest, would eventually look north. Some of the continent's largest rivers, in terms of flow, are found in Southeast Alaska, the Canadian province of British Columbia, and the Yukon Territory and "pour wasted and unused" into the Pacific and Arctic Oceans (*NAWAPA* 1964). NAWAPA's goal was nothing less than "total water management" on a continental scale, requiring the construction of 3,151 individual projects, including dams, canals, pipelines, tunnels, and pumping stations to divert freshwater from this region of plenty to cities and farms on the Canadian prairie, in the United States, and as far away as Mexico. The catchment area of the project covers an area of 3.4 million square kilometers (1.3 million square miles) and by some estimates accounts for a quarter of all runoff in North America (see figure 7.1).

The plan called for diverting up to 20 percent of the flow of the Tanana, Susitna, and Yukon Rivers in Alaska. In British Columbia, it would tap the Churchill, Blackstone, Slave, Coppermine, Peace, and Mackenzie Rivers. Diversions from these rivers would be redirected to fill a natural valley that runs the length of British Columbia, the Rocky Mountain Trench, second in size only to the Great Rift Valley in East Africa. When filled, the reservoir would extend 800 kilometers (500 miles), average 16 kilometers (10 miles) in width, and be capable of storing 500 cubic kilometers (400 million acre-feet) of water.

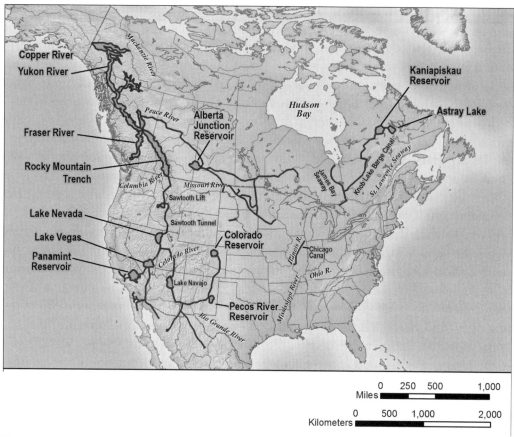

FIGURE 7.1 MAP OF NORTH AMERICA SHOWING THE MAJOR WATER PROJECTS FOR NAWAPA
Source: Map by Matthew Heberger.
NAWAPA's water projects extend from northwestern British Columbia, the Yukon Territories, and Alaska to Mexico and the St. Lawrence Seaway.

This would be the largest reservoir on the continent, sixteen times bigger than Lake Mead, behind Hoover Dam on the Colorado River.

From the southern end of this new reservoir, a canal would extend southeast through the prairies in the provinces of Alberta, Saskatchewan, Manitoba, and Ontario and into Lake Superior, delivering irrigation water along the way and sending flow into the Great Lakes (figure 7.1). Among its purported benefits would be to "alleviate falling levels and degraded quality of the Great Lakes" (*NAWAPA* 1964). The "Canadian–Great Lakes Waterway" would also open the long-dreamed-of waterway across the continent, allowing barge traffic from Alaska all the way to New Orleans or Montreal.

At the southern end of the Rocky Mountain Trench, another set of canals, tunnels, and lift pumps would send water to the Columbia River basin, the high plains of Idaho and Montana, and farther south. In northern Nevada, the main canal would split. A western fork would direct water southwest toward Las Vegas, Los Angeles, San Diego, the Colorado River delta, and the Baja California Peninsula. The eastern fork would send irrigation water to Colorado, Utah, New Mexico, Texas, and the northern Mexican states of Sonora and Chihuahua.

Among the other astounding facts about the project are that it would require the movement of 25 million cubic meters (32 million cubic yards) of earth and the use of

27 million metric tons (30 million tons) of steel (*NAWAPA* 1964). These figures were announced in a promotional film produced by Parsons and were apparently meant to stoke the imagination of Americans, who, it has been argued, "have a predilection for technological solutions to problems, especially if the solution involves the construction of an impressive structure" (Sewell et al. 1967, 11).

Purported Benefits

Parsons's spokesmen admitted that "at first or casual glance, the size of the concept may seem unrealistic or fanciful." The company insisted, however, that more careful analysis revealed it to be "feasible from an engineering standpoint and economically realistic" (*NAWAPA* 1964). Much of the contemporary criticism of NAWAPA cited here comes from a twenty-page article, "NAWAPA: A Continental Water System," printed in the *Bulletin of the Atomic Scientists* and containing an analysis of the project by prominent experts in the fields of economics, political science, engineering, and fisheries (Sewell et al. 1967).

While the plan relied on contemporary technologies such as concrete arch and earthen dams, there was legitimate concern about whether it was even technically realistic. The plan called for a dam 570 meters (1,700 feet) high, taller than a 170-story building and taller by far than the highest dam that had ever been built, Hoover Dam, at 220 meters (725 feet). (In 2013, the world's tallest dam is the Nurek Dam in Tajikistan, at 300 meters, or 980 feet.) Critics also noted that much of the construction was to take place in the Rocky Mountains and would involve the blasting and moving of solid rock.

Among the main benefits of NAWAPA would be increased water supply for agriculture and cities. The planners foresaw a West with a rapidly growing population and increasing water use. Overall, the system was designed to deliver 150 cubic kilometers (120 million acre-feet) per year of water annually to thirty-three American states and seven Canadian provinces. To put this in perspective, this is about one-third of the average outflow of the Mississippi River and about eight times the natural flow of the Colorado River. This flow would allow Mexico to triple its irrigated acreage, would nearly double irrigated area in the United States by adding 16 million hectares (40 million acres), and would add 3 million hectares (7 million acres) in Canada. Further, it would allow the United States to recover thousands of acres of formerly productive farmland that had been "degraded by the Colorado River's excessive mineral content."

Project proponents played into growing unease about population growth by saying that America had a duty to produce more food and calling NAWAPA the "only hope of averting worldwide famine." Furthermore, "additional reclamation in the West," it was claimed, would support the "development of 230,000 family-size farms, . . . 40,000 local retail enterprises, and $4 billion in retail trade annually." Even with promised water deliveries, this would have been unlikely, as the irrigated western small family farm had long been proven an elusive fantasy. The mythos of Reclamation—converting dry and barren land into irrigated cropland, divided into 160-acre parcels to support family farms—was, by the 1960s, no longer a viable dream. These relatively small holdings, by western standards, and the high cost of farm inputs made it tough to break even. While irrigation water is used to grow valuable citrus fruits and vegetables in Southern California and Arizona, irrigation in the High Plains has never been very productive because of the short growing season, with most irrigation water today used to grow pasture for grazing cattle. Imported water would also potentially reduce pressure on the overpumped

Ogallala Aquifer, which was being drained at alarming rates by farmers in the High Plains states from Nebraska to Texas.

Another benefit would be an increase in freshwater inflows to the Great Lakes, which in the 1960s were in deplorable condition as a result of pollution from industry and pesticides and fertilizer in agricultural runoff. Further, residents became alarmed when water levels in Lakes Huron and Michigan dropped to their lowest levels in one hundred years. But to economist James Crutchfield, attempting to dilute pollution with additional inflow would have been foolishly expensive and unlikely to succeed. A cheaper and more effective solution would have been to "reduce pollution load" through the "establishment and enforcement of adequate receiving water standards" (Sewell et al. 1967, 20). Indeed, this strategy was later pursued through federal clean water legislation.

Electrical generation was another promised benefit; hydroelectric dams would generate 100 million kilowatts of electrical power, 30 percent of which would be needed to lift water over mountain ranges. The remainder would be sold for competitive prices on the open market, which, proponents stated, would stimulate industry and new economic activities. It would have increased North America's electrical output by one-sixth with "clean hydroelectric power—no pollution, no CO_2, no acid rain" (Reisner 1986, 505). University of Washington professor of civil engineering E. Roy Tinney called the allocation of water and power facilities suboptimal, declaring the project little more than "a map of potential hydropower sites" (Sewell et al. 1967, 23). The project could be accused of planning unnecessary hydroelectric facilities to make up for revenue shortfalls from water projects that could never pay for themselves. Journalist Marc Reisner would later criticize the US Bureau of Reclamation for bundling bad water projects with "cash-register dams" to make projects look economically feasible when considered from a "river basin perspective" (Reisner 1986, 42).

Abundant electricity and inexpensive canal transportation would benefit Canada, NAWAPA advocates claimed, by opening access to its mineral-rich northern regions. However, enthusiasm for these new opportunities was tempered by the reality that land would have been flooded: "The reservoirs not only flood good farming land and destroy natural habitat, they also eliminate large forest resources and inundate as yet unknown mineral deposits" (Sewell et al. 1967, 23). Reservoirs would also have flooded the lands of some First Nations people and cities such as Prince George, British Columbia, whose population today is 70,000.

Among the more dubious claims by Parsons were that "fish and game would increase" and the creation of "new scenic waterways" would enhance tourism. Even contemporary observers were skeptical of these claims, which in hindsight seem even more preposterous. A critic wrote, "I boggle at the thought that western Canadians, or American visitors, would regard a giant man-made lake as an acceptable substitute for the magnificent recreational area that it would despoil" (Sewell et al. 1967, 19). The project would have been devastating to salmon fisheries in every major river in British Columbia. Dams and diversions would have either cut off or submerged breeding habitat for migrating salmon on the Tanana, Bella Coola, Dean, Chilcotin, and Fraser Rivers, which were "among the most important salmon rivers on earth" since the destruction of the fisheries of the Columbia and Sacramento Rivers.

Luna Leopold, professor of hydrology at the University of California, Berkeley, wrote: "The environmental damage that would be caused by that damned thing can't even be described. It could cause as much harm as all the dam-building we have done in a hundred years" (quoted in Reisner 1986, 510). Today, several Pacific salmon are listed as federally protected endangered species in the United States, making the construction of new dams on America's northwestern rivers increasingly unlikely. In fact, in order to help recovering salmon populations, the Bonneville Power Administration, Oregon's state-owned hydropower utility, spends $170 million per year on a fish and wildlife recovery program in the Snake River and Columbia River watersheds (Garrick et al. 2009, 378), and some dams are even being removed. The United States Congress passed the Wild and Scenic Rivers Act in 1968 with the intention of protecting rivers from damming and development for the enjoyment of future generations. The National Environmental Policy Act followed in 1969, requiring federally funded projects to undergo a thorough environmental review, which by law must consider the "no project" alternative.

Public Acceptance

At the time, economist James Crutchfield called the NAWAPA proposal "symptomatic of a growing public dissatisfaction with the planning of water development" in the United States. Despite huge investments, the problems of pollution and overuse were evident everywhere. Cold war rivalry may have also played a role. Historian Donald Worster wrote that the project "had about it the irresistible logic of an imperial history," comparing it to rival plans in the Soviet Union to reverse the flow of Siberia's north-flowing rivers to water a new breadbasket in the country's arid central plains (Worster 1992, 316). Project backers certainly played into the public's anxieties, proclaiming that "ultimately, the decision to build NAWAPA—or a project similar to it—will determine, in some part, the future economic well-being in North America" (*NAWAPA* 1964).

Despite its appeal, Worster wrote, the scheme arrived both too early and too late to win broad acceptance: too early because at the time there were other, more readily exhaustible water supplies that could be tapped. Planned or ongoing projects such as the California State Water Project or the Central Arizona Project were already controversial and massive drains on the public purse. Critics in the 1960s pointed out that even if there were a need for massive new water supplies, the NAWAPA scheme would not be the most economical way to provide them. Possible alternatives included desalination, recycling, "depollution," reduction of leaks in cities, and decrease of waste through better technology and pricing. And NAWAPA arrived too late because in the mid-1960s the environmental movement was growing and the Reclamation era was coming to an end. Rachel Carson had published the groundbreaking *Silent Spring* in 1962, awakening Americans to the perils of pollution from pesticides and herbicides, and many people were already acquainted with Aldo Leopold's "land ethic" (Lee 1978, 557).

Reception by Canada

At the time of its introduction, it was noted that the proposal "aroused relatively little enthusiasm in Canada," but as the idea gained momentum, more Canadians became "openly hostile to the whole idea, viewing it as an attempt to plunder Canadian resources" (Sewell et al. 1967, 9). Canada is endowed with more freshwater per person than any other nation: it ranks first in renewable freshwater but, with 34 million people, only thirty-sixth in terms of population. A prominent American political scientist noted that, at the time, some Americans held the view that Canadians had "an obligation to share their water resource with the United States" and suggested in 1967 that the United States could conceivably "apply pressure on the Canadian government to change its policy, backed up perhaps by sanctions of various kinds" (Sewell et al. 1967, 9)—a strategy unlikely to further endear the project to Canadians.

Canada has historically opposed most water export schemes (Rothfeder 2004). Federal water policy promulgated by the environment ministry in 1987, during the administration of conservative Brian Mulroney, stated, "The Government of Canada emphatically opposes large-scale exports of our water. We have another reason for our opposition; the inter-basin diversions necessary for such exports would inflict enormous harm on both the environment and society, especially in the North, where the ecology is delicate and where the effects on Native cultures would be devastating" (Environment Canada 1987). However, the lack of a federal law regarding water transfers left the issue to the provinces, and there is no formal federal ban on such transfers.

Few Canadians believed the project would ever be built, but any mention of NAWAPA by an American politician continues to draw a vigorous negative response from the Canadian press. Despite widespread public opposition, the plan also receives occasional support from British Columbia politicians. Currently, one of the province's largest sources of income is its logging industry, which harms the environment and is also subject to boom-and-bust cycles, depending on the economy and the housing market in the United States. By contrast, agriculture is more dependable: "water could be sold through forty-year contracts . . . ensuring a steady, predictable income every year" (Reisner 1986, 511).

Statements by the United Nations and international trade agreements may further limit the Canadian government's ability to impose a ban on all water transfers. At the United Nations' 1992 International Conference on Water and the Environment in Dublin, the delegates issued a statement recognizing water as an "economic good" in response to the increasing scarcity, conflict, and overuse. To some, the declaration made it seem "inconceivable that Canada could continue to oppose water transfers and reject NAWAPA" (Rothfeder 2004, 111).

Trade liberalization may also play a role in facilitating international water projects. In 1994, the United States, Canada, and Mexico signed the North American Free Trade Agreement (NAFTA) with a goal of reducing barriers to trade and investment by eliminating tariffs on many imports among the three countries. A provision of the agreement (chapter 11) "backs the right of companies to sue if they believe that their ability to trade freely is hampered in any way" (Rothfeder 2004, 112). In fact, this occurred in 1999, when a California company, Sun Belt Water, sued the Canadian government over its prohibition on shipping freshwater by marine tanker from British Columbia (Boyd 2003).

Renewed Interest and Current Status

As with many zombie water projects, interest in NAWAPA seems to fade away only to be revived decades later. In the 1980s, the American economy was recovering from recession and inflation at the same time a severe drought occurred across the western states, sparking widespread concern over food prices. At a 1980 meeting in California titled "A High-Technology Policy for U.S. Reindustrialization," sponsored by the fringe Fusion Energy Foundation, Dr. Nathan Snyder of the Parsons Company "reintroduced NAWAPA to a large and enthusiastic audience" (Reisner 1986, 508).

As recently as 2008, Michael Campana, professor of hydrogeology at Oregon State University, wrote that he had "heard talk about 'bringing NAWAPA back,' just as I have heard people suggest reviving a plan to study the diversion of Columbia River water to the Southwest USA" (Campana 2008). In recent years, the project has again been promoted by controversial political figure Lyndon LaRouche through his advocacy organization LaRouchePAC (LaRouche was behind the Fusion Energy Foundation, mentioned above). LaRouche has run for the US presidency eight times, and although he began his political career in the radical leftist student politics of the 1960s, his views are now considered ultraconservative. His reputation has been tarnished by a tax fraud conviction, for which he served fifteen years in prison; apparent anti-Semitism; climate change denial; and the promotion of 9/11 conspiracy theories (*Wikipedia* 2013).

Although it seems doubtful that NAWAPA, as originally planned in the 1950s and 1960s, will ever be built, some of its individual components have already come to fruition. The overall scheme of NAWAPA included diversions and hydropower on rivers flowing into Hudson Bay in eastern Canada. Beginning in 1971, the government of Quebec has built hydroelectric dams on rivers in the James Bay watershed, including the La Grande and Eastmain Rivers. The James Bay Project currently generates 16,000 megawatts (equivalent to the output of twenty-four average coal power plants), providing half of the province's electricity and making electricity one of the province's major exports. The James Bay Project continues to expand, with construction of dams on the Rupert River having begun in 2010.

Lest we consider NAWAPA finally laid to rest, consider the scene described by Marc Reisner in the epilogue to his classic book *Cadillac Desert*. In 1981, British Columbia premier Bill Bennett visited San Francisco to give a talk at the Commonwealth Club. When an audience member asked whether British Columbia would sell water to the United States, he replied with a firm "No." He then added, "But come and see me in twenty years" (Reisner 1986, 513). And consider the words of Tom McCall, governor of Oregon from 1967 to 1975: "This is a plan that will not roll over and die. It may be fifty years, or it may be a hundred years, but something like it will be built."

The Reber Plan

Water policy and planning in the twentieth century was dominated in California, as it was in many other places, by an engineering mentality. A massive imbalance between when and where precipitation fell, where runoff flowed, and where demand for water occurred required major replumbing, and this led to the construction of some of the world's largest water transfer systems in California, including the Los Angeles Aqueduct, the California State Water Project, and the federal Central Valley Project. Today, these and other systems transfer billions of cubic meters of water every year from the wetter, mountainous northern regions of the state to the drier agricultural lands of the Central Valley and the cities along California's coast.

California also served as a critical region during World War II, providing food, major seaports and airports, railroad terminuses, munitions plants, shipyards, troop-staging areas, and workers for the massive Pacific war effort. It may be no surprise, therefore, to learn that the combination of the war effort and the engineering approach to water produced one of the most significant water zombies ever designed—the Reber Plan for San Francisco Bay.

History

Freshwater runoff from California's two largest rivers, the Sacramento and San Joaquin, merges in the Central Valley Delta region, once one of the biggest inland aquatic ecosystems in the world (figure 7.2). That water then flows into San Francisco Bay and out the Golden Gate to the Pacific Ocean. So much freshwater once flowed out of the bay during peak runoff season that early explorers reported finding plumes of freshwater outside the Golden Gate in the Pacific Ocean (Presidio Trust, 2012). This system also supported some of the world's largest anadromous fisheries—primarily salmon—with millions of fish being born in the rivers, living part of their lives in the oceans, and then returning to their upstream habitats to spawn.

The Reber Plan—also called the San Francisco Bay Project—was developed by John Reber, a schoolteacher and theatrical producer. Reber's idea was to transform San Francisco Bay from the largest natural estuary and coastal ecosystem on the West Coast into a completely artificial engineered system. Two major earth and rock-fill dams would be constructed, one between the city of Richmond and Marin County and one between Oakland and San Francisco (where the current Richmond–San Rafael Bridge and San Francisco–Oakland Bay Bridge now stand). These dams would turn the upper and lower parts of the bay into freshwater reservoirs, permitting the destruction of 20,000 acres (80 square kilometers) of marshes and wetlands with landfill on most of the eastern side of the bay. Highways and railroads would cross the dams, and a lock system the size of the Panama Canal would be built to permit shipping to reach the ports of Oakland, Stockton, and Sacramento (figures 7.2 and 7.3).

Gray Brechin, a historical geographer from the University of California, Berkeley, wrote that part of the motivation for the plan was fear that Japan would attack the West Coast. "Political cartoons of the day show that people were terrified," Brechin says. "The Reber Plan seemed to provide the answer; people could be moved in and out of the area relatively quickly" (Sinclaire 2011).

Supporters of the Reber Plan also argued that the freshwater reservoirs created by the dams would solve California's water problems by capturing water that was "lost" out the Golden Gate. That water could then be sent throughout the state to meet growing demands. Many newspapers supported parts of the plan when it was proposed, the Joint Army and Navy Board requested a report (Nishkian 1946), and Congress even held some initial hearings in December 1949.

Project Review and Demise

The US Army Corps of Engineers was ordered to build a working hydraulic model of San Francisco Bay to see whether the Reber Plan would work. In 1957 the corps built the model, more than one and one-half acres in size, in Marin County. It created a physical representation of an area from the Pacific Ocean to Sacramento and Stockton, including San Francisco Bay, San Pablo Bay, and Suisun Bay and a portion of the Sacramento–San Joaquin Delta. (KQED, n.d.) Testing began in 1959, and corps engineers determined that the Reber Plan would be catastrophic for the ecosystem, flows, and chemistry of the estuary and for San Francisco Bay. The reservoirs, instead of storing freshwater, would become enormous evaporation ponds, incapable of providing any discernible amount of freshwater. Contaminants would build up in the bay, and commercial and natural fish-

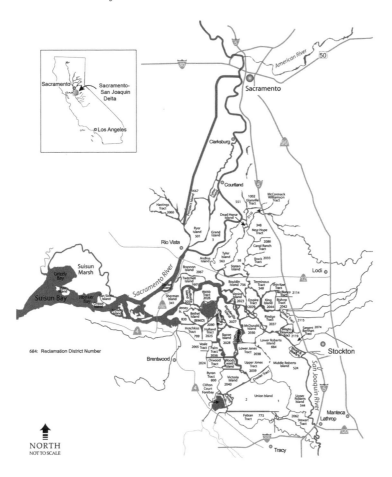

FIGURE 7.2 THE SACRAMENTO–SAN JOAQUIN DELTA IN NORTH-CENTRAL CALIFORNIA
Source: US Geological Survey.
The Sacramento–San Joaquin Delta, where California's two largest rivers meet, was a historically rich ecosystem but now provides much of the water transferred south for urban and agricultural developments.

eries would be completely eliminated. The idea for the project was also overwhelmed by changing perceptions of the value of San Francisco Bay, a growing national environmental movement, and a decrease in the perceived need for the military benefits the plan ostensibly offered. While the Reber Plan attracted considerable attention, it was ultimately opposed by the State of California, the US Bureau of Reclamation, and the US Army Corps of Engineers, and it never proceeded beyond the initial hearings and physical testing.

Alaskan Water Shipments

Most large-scale water transfers have been limited to water that can be moved overland through pipes and aqueducts. Transfers across bodies of water, especially salt water, have been constrained by distance and technology and limited by needs that could be satisfied with cheaper and more practical local alternatives. Despite this, proposals periodically arise for large-scale bulk water transfers by sea from regions of water abundance

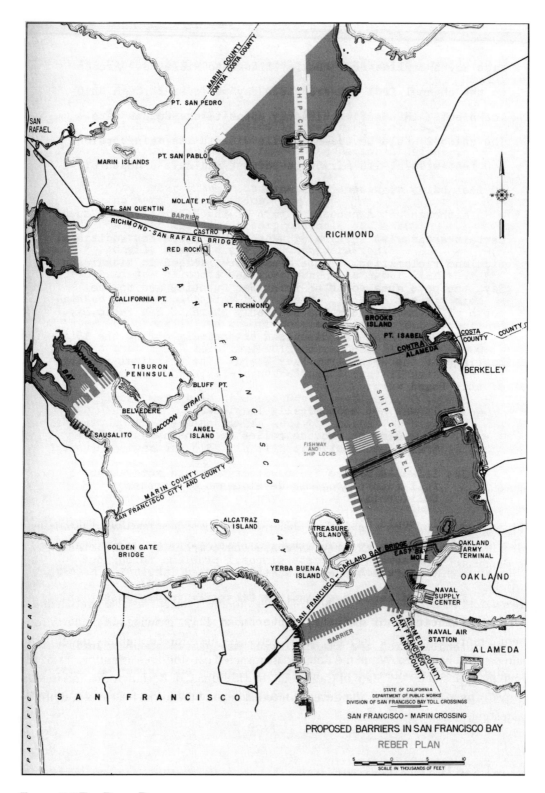

FIGURE 7.3 THE REBER PLAN

Source: University of California, Water Resources Center Archives.

Conceived in the 1940s, the Reber Plan would have turned San Francisco Bay into a completely artificial, militarized system. The proposed water benefits turned out to be unrealistic.

to regions of severe scarcity. Until the late twentieth century, such transfers that actually occurred were done with bulk tankers, like the kind used to ship oil, typically called very large crude carriers (VLCCs) or ultra large crude carriers (ULCCs).

Occasionally tankers have been constructed solely for the purpose of carrying potable water, which requires very strict water quality protections. Over the past two decades, however, some entrepreneurs have sought to develop and commercialize giant reusable, towable plastic bladders dedicated to the business of water transfers. We wrote about this technology in volume 1 of *The World's Water* (Gleick 1998), describing some short-term commercial experiments with bulk water bags in the Mediterranean and the long-term efforts of entrepreneur Terry Spragg to develop "Spragg Bags" for both water transfers and emergency response.

In the late 1990s, a company called Aquarius Water Trading and Transportation developed and built water bags that were briefly used for commercial deliveries of water from Piraeus, Greece, to Aegina—a distance of around twenty kilometers (about twelve and a half miles). A Norwegian company, Nordic Water Supply, also developed a bag system designed to deliver water from Turkey to northern Cyprus. This project also failed to be economically and politically viable (Gleick 1998). More recent proposals include the following example of an effort to move water from Alaska to commercial markets overseas.

History

In the past decade, a proposal to ship water from Sitka, Alaska, to Asia or the Middle East has been proposed and promoted several times. Each time, the idea has run into the wall of economics, collapsed, lain fallow for a while, and then, like most zombie water projects, gotten up and staggered around again (Gleick 2011).

Here are the basics of the idea: In 2006, the town of Sitka in southeastern Alaska built a forty-two-inch-diameter pipeline connecting a local reservoir (Blue Lake Reservoir in Tongass National Forest) and the local port at Sawmill Cove. The waters of the lake are replenished by nearly one hundred inches (2.5 meters) of annual precipitation and some additional glacier melt. The pipeline is capable of delivering 22,500 gallons of water per minute, or around 32 million gallons per day (Alaska Water Project 2013). That same year, a local company, True Alaska Bottling Company (TAB), and a financial partner—S2C Global Systems from San Antonio, Texas—secured a contract from the town to sell just around 3 billion gallons of water a year (11 million cubic meters) in bulk from this reservoir at a rate of around one cent per gallon. The company proposed that water be fed through an underground pipe system leading 2,000 feet offshore to permit loading of deepwater ships. In an early proposal, S2C Global Systems said it was developing a hub with an Indian port on the Arabian Sea (Barclay 2010).

Project Limitations

The company has never been able to capitalize on the contract, and it missed a series of deadlines imposed by the town to complete financial deals for the water. In late 2010, for the third time in four years, TAB and S2C Global Systems missed a deadline, for a minimum annual shipment of 50 million gallons by early December 2010. No water has been shipped in the years since the contract was first signed, in 2006 (Walton 2010b). The city amended and extended the contract each time but increased export minimums and imposed fees on the company.[2]

Despite the project's limitations, the Sitka Economic Development Association has been a strong supporter. In 2010, Garry White, executive director of the association, commented on the project. "This water is falling into the ocean less than a mile from the lake," he noted. "The community sees an opportunity to take this resource that flows out into the ocean and make it a driver for us."

The biggest problem with the project, as with many such large-scale water transfer ideas, is the high cost of transportation. Moving water from Sitka, Alaska, to Asia or the Middle East involves massive energy costs for operating large tankers. Typical medium-sized water tankers can carry as much as 100 million gallons per trip. If the water is worth even a penny a gallon (more than typical desalinated water costs), a tanker with 100 million gallons is worth only around a million dollars. Capital and operating costs for tankers far exceed the value of water as a cargo. The costs are increased by the fact that even high-quality water, as is available in Sitka, will deteriorate in quality in a tanker over time, requiring additional treatment and expense. Professor James McNiven at Dalhousie University in Nova Scotia observed (Barclay 2010):

> If that water sits in the hold of a tanker for weeks traveling across the ocean, when it arrives it's not spring water anymore; you're going to have to clean it up. As a business proposition this gets to be very expensive and chances are the economics don't work.

In an interview with Circle of Blue in August 2010, S2C's president, Rod Bartlett, optimistically suggested that the cost of purchasing and shipping water from Sitka to India would be on the order of $0.07 per gallon, ten times the cost of even the most expensive desalinated water. Why there would ever be a buyer for this water is hard to explain (Walton 2010a) when the alternative is a cheaper (albeit still costly) desalination plant built and owned locally.

Nevertheless, the project proponents are still pursuing buyers for the water. As of early January 2013, the City and Borough of Sitka's website included an ad for bulk water exports that linked to a commercial Sawmill Cove solicitation for "proposals from companies or individuals who are interested in bulk fresh water export. We currently have 19,047 acre feet available annually for sale" (City and Borough of Sitka, Alaska 2013).

Las Vegas Valley Pipeline Project

Nevada is a water-short state, given the paucity of natural rainfall, high average temperatures, and large and rapidly growing population centers of Reno and, especially, Las Vegas. Water management around Las Vegas was the responsibility of individual water districts until 1991, when seven neighboring water agencies joined together to form the Southern Nevada Water Authority (SNWA). The seven member agencies are the cities of Las Vegas, North Las Vegas, Boulder City, and Henderson and the water and reclamation districts of Las Vegas Valley, Big Bend, and Clark County.

---

2. In early 2013, the TAB website said the company was owned by Cove Partners LLC, an investment and hotel company located in Lebanon, Oregon.

Although the individual water districts maintain control over their own daily operations, the SNWA is in overall charge of managing the region's water resources and ensuring future supply (SNWA 2007). As part of this responsibility, the SNWA considers and evaluates long-term plans, including costly and politically controversial proposals. One of these is the Las Vegas Valley Pipeline Project—a potential water zombie.

History

Las Vegas is a desert. The city itself was created as a consequence of the construction of Boulder (now Hoover) Dam on the Colorado River and the creation of Lake Mead, the only significant source of water in the region (SNWA 2006). In the late 1920s, when seven western states that share the Colorado River were negotiating how to apportion flows, the southern Nevada region had a very low population and little to no agricultural or industrial water demand as compared with the other basin states. Partly as a result, during the negotiations Nevada was allocated 300,000 acre-feet per year (AFY) of water, just 4 percent of the total amount estimated for the lower basin. Today, nearly all of it goes to the Las Vegas region (Clark County), where 2 million people—72 percent of Nevada's total population—reside.

As Nevada's population has grown through the twentieth century, there have been efforts to modify or renegotiate the Colorado River "Law of the River"—the set of compacts, agreements, contracts, regulations, and court decisions that govern management of the Colorado River (US BOR 2007). However, these efforts are highly controversial and are largely ignored by the other basin states, which also are experiencing increased demand for a limited resource. Part of the argument against reallocation of the Colorado River flow was based on the fact that Nevada had other in-state resources that could be developed, as the other lower basin states had done.

As a result, southern Nevada has considered several options for increasing its available water supplies, including development of five northern groundwater basins. In 1989, the Las Vegas Valley Water District (LVVWD) applied to the Nevada State Engineer (NSE) for permission to drill 146 groundwater wells in twenty-eight different sub-basins in White Pine, Lincoln, and northern Clark Counties. The permits requested up to 840,000 AFY (over a billion cubic meters per year) from wells as much as 500 kilometers (300 miles) from the city (Reinhold 1991) and drew substantial opposition from local residents, public interest groups, and federal agencies. More than 3,600 protests were filed with the State Engineer's Office (Reinhold 1991). When asked about the groundwater project in 1994, the SNWA's director, Patricia Mulroy, said that although others were calling it "the singularly most stupid idea anyone's ever had," concern over the applications helped draw attention to Nevada's water needs (Christensen 1994). The groundwater permits, along with other water development proposals, were used to argue for changes to the Law of the River (Greene and Hynes 1996). One article notes that Mulroy said that the applications for rural groundwater would be dropped if the agency's Colorado River apportionment were to increase (Egan 1994). Within a few years, Las Vegas would end up with new rights to the Virgin River, a second pipeline into Lake Mead, and a water-banking agreement with Arizona.

Recent Developments

The groundwater development project was revived in 2004 in the form of a proposed major pipeline to northeastern Nevada counties during a prolonged drought in the Colorado River basin (Casey 2004). Before the drought, the SNWA had been expecting future water deliveries to come from surplus and banked supplies on the Colorado; however, falling lake levels put new pressures on those resources, and the SNWA could no longer count on their supply. Projections suggested that if drought conditions continued, at least one of the SNWA's water intakes from Lake Mead could be unusable as early as 2012. Plans were made to build a third, deeper intake, and in February 2004 the SNWA's board approved a plan to move forward with new plans to tap northern Nevada groundwater resources. The SNWA submitted a right-of-way application for construction of the pipeline to the US Bureau of Land Management (BLM) in 2004.

State permitting agencies approved new SNWA groundwater applications in Spring Valley in April 2007 and applications in Delamar, Dry Lake, and Cave Valleys in July 2008. However, these rulings were followed by nearly four years of efforts to delay implementation, including two court cases, required resubmission of the appropriations applications, and public hearings, meetings, and a mandatory reopening of the protest period. In March 2012, the NSE again approved the applications for the four basins, with even greater pumping allowances as well as modified requirements for research, monitoring, and reporting. The rulings also required expanded hydrologic and biological monitoring, preparation of annual reports, completion of baseline studies, and development of efforts to mitigate adverse effects.

The BLM published a final environmental impact statement (EIS) for the pipeline system and associated infrastructure in August 2012 and approved the right-of-way in December. The EIS recommended approval of a modified plan that would support transfer of nearly 115,000 acre-feet per year through over 300 miles of pipeline, at an estimated cost of just under $4 billion for the "preferred alternative" (US BOR 2012) (see table 7.1). A coalition of ranchers, farmers, rural local governments, and environmentalists appealed the state engineer's 2012 ruling. Oral arguments were scheduled for June 2013, and if the decision is appealed (which is likely regardless of the ruling) the case could move into 2014 (Eggen 2012).

Concerns about the project have changed little since its inception. They focus on the effects that groundwater pumping and withdrawals would have on local ecosystems and existing water users in the northern valleys, equity concerns about massive water transfers from rural communities to urban centers, the long-term impact of climate change on water needs and availability, and the science behind the groundwater models and demand assumptions. Opponents are particularly concerned that groundwater in the extraction areas is already overallocated and that pumping at the proposed levels will draw down the water table by hundreds of feet. Although the SNWA has promised to monitor and to stop pumping should major drawdown occur, opponents argue that once negative effects are recognized, it will be too late to recover and too difficult politically to reduce extraction (Deacon 2009). Some opponents describe the project as a twenty-first-century version of the Los Angeles water grab from the Owens Valley, or the movie *Chinatown* with its political intrigue, corruption, and land speculation. In that instance, in the early 1900s Los Angeles obtained the rights to take water from a distant rural community and disrupted the farming culture and environmental conditions there.

TABLE 7.1 Withdrawal Volumes for the Groundwater Development Project (Acre-Feet per Year)

Location	SNWA Applications	NSE Ruling 2007/2008	NSE Ruling 2012	BLM EIS, Alternative F
Spring Valley	91,224	60,000	61,127	84,370
Delamar Valley	11,584	2,493	6,042	6,591
Dry Lake Valley	11,584	11,584	11,584	11,584
Cave Valley	11,584	4,678	5,235	11,584
Total	125,976	78,755	83,988	114,129
Snake Valley	50,679			

Source: US BLM (2012).

The Special Problems with Snake Valley

Although the proposals for groundwater extraction have always included plans to pump water from Snake Valley, obtaining approval for this piece of the effort has been considerably more complicated than for the other basins. The Snake Valley basin straddles the border of Nevada and Utah and is the only basin in the proposal that is shared with another state (Deacon 2009).[3] In 2004, Congress passed an act that would allow the groundwater development project to be constructed on federal lands; however, it also required the SNWA to come to an agreement with Utah over Snake Valley (P.L. 108-424). The two states reached a tentative agreement in 2009 (State of Utah 2009); however, as of early 2013, Utah still had not signed. And then, suddenly and unexpectedly, the governor of Utah, Gary Herbert, announced in April 2013 that he would not sign the agreement to share the groundwater in the joint basin. Governor Herbert said that he based his decision on feedback from local Snake Valley residents, including ranchers, landowners, Native American tribes, and governmental officials, most of whom opposed the agreement.

"A majority of local residents do not support the agreement with Nevada," the governor said. "Therefore, I cannot in good conscience sign the agreement because I won't impose a solution on those most impacted that they themselves cannot support" (*Salt Lake Tribune* 2013).

The SNWA considered suing Utah for not signing the agreement, but as of mid-2013 it had not yet taken action.

Las Vegas Pipeline Summary

Is this a zombie water project? We're not sure. It is unclear whether the project is politically, economically, or environmentally viable and will move forward, although the SNWA maintains it is indispensable to the future of the city. The economic recession that began in 2008 slowed growth in the Las Vegas area, and increased conservation efforts have improved the water situation, though not by much and perhaps only temporarily. The city's heavy dependence on the Colorado River, combined with very little industrial or agricultural use, means that the bulk of the efforts to modify demands must fall on

---

3. However, it is not the only transboundary basin potentially affected.

residential users. As some studies have suggested, far more could be done in this area (Cooley et al. 2007). Ensuring that southern Nevada has a sustainable supply of water will certainly require changes in both supply and demand management, but in the long run, there will have to be changes to regional development policies as well.

The project doesn't appear to be dead yet, and indeed, it may actually be built. But it is clear that it represents the kind of project that would have been built in the twentieth century, when society didn't care about, or didn't understand, the social and environmental implications of unrestrained long-distance water extraction and transfers.

Diverting the Missouri River to the West

Another proposal for a massive interbasin water diversion is the idea of moving Missouri or Mississippi River water west to the arid and semi-arid parts of the country. While far smaller in scope than the NAWAPA project described earlier, water diversion from the Missouri or Mississippi basin would still be an enormous undertaking. Water would be moved in massive pipes and aqueducts, requiring large amounts of energy to pump water over mountains and leading to ecological and social dislocations in downstream regions.

History

There is a long history of ideas, efforts, actual expenditures, and construction projects designed to move Missouri River water somewhere else. As early as 1889, the Constitutional Convention of North Dakota considered a proposal to build a canal to divert Missouri River water in Montana to the Red River for irrigation. The infamous Garrison Diversion Project and its related components (proposed and partially built in the 1930s, 1940s, and 1950s) is a case in which hundreds of millions of dollars have been spent to try to do things that ultimately have failed to satisfy political, environmental, and social needs (Springer 2013).

Recent Efforts at Revival

The concept has always been seen as far-fetched, yet like traditional zombies, it is never seen as really dead. In 2012, the idea was raised again when the US Bureau of Reclamation included it as one of many possibilities for rethinking management of the Colorado River. The study by the Bureau of Reclamation ("the Basin Study") was the culmination of a two-year effort to develop and analyze future supply and demand scenarios for the seven states that share the Colorado River (Arizona, California, Colorado, New Mexico, Nevada, Utah, and Wyoming) (US BOR 2012). In the bureau's analysis, the project was described as a 670-mile pipeline to divert water from one of the Mississippi River's major tributaries to help seven arid states in the West (Salter 2012). The core of the idea is to build diversion dams along the Missouri and its tributaries to divert water during periods of high flow and move water west. If the water goes to users in the upper part of the Colorado River basin, for example, downstream users hope that upstream users would forgo some of their formal Colorado River legal allotments, permitting increased use lower in the basin. They suggest that the project would cost an estimated $11.2 billion and take thirty years to complete, though such initial cost estimates have a long history of being largely guesswork.

The Basin Study is a multiyear, multi-participant assessment of the growing challenges facing the most important river basin in the arid western United States—the Colorado. It shows that water demands in the basin already exceed the river's available supply—a situation the Pacific Institute previously defined in a paper in the *Proceedings of the National Academy of Sciences* as having reached "peak water" (Gleick and Palaniappan 2010). The situation will worsen as water demands grow with population and if, as scientists now expect, water supply in the river diminishes with climate changes. Just as important as identifying water scarcity problems is identifying the options for addressing or preventing them. The Basin Study offers over a hundred different options and strategies for increasing supply or reducing demand. Many make sense; a few are far-fetched and seriously out of touch with reality, including the idea to divert the Missouri River. It did, however, receive considerable attention in the media when it was released, reflecting the fascination the public and the media have with massive engineering ideas.

This kind of proposal has been made especially forcefully by the city of Las Vegas, which (as described earlier in this chapter) faces serious water constraints as a result of uncontrolled development. As stated in 2011 by Patricia Mulroy, director of the Southern Nevada Water Authority (Velotta 2011),

> it makes no difference to the corn and the alfalfa whether it gets Colorado River water or Mississippi water or Missouri water. . . . You could improve the transportation and cargo transports on the Mississippi River, which have been severely impaired this year by flood conditions, and at the same time provide some security for those communities that have lost everything by pulling some of that water off and moving it. . . . It's more water than the system down there can handle. Let's use it. Let's recharge the Ogallala aquifer, let's replace some Colorado River users. Let them use some of this and leave the other water in the Colorado River for those states that are west of the Colorado.

From a narrow technological perspective, there is no doubt that such a pipeline can be built: a treatment plant, massive pumping stations, and a 600-mile-plus pipeline to move water. Technically, this is standard engineering, albeit on a large scale.

Project Limitations

But some observers note that it is a dangerous idea whose time has come and gone again, and in this era of fiscal limits, the federal government should not spend one penny even studying it, much less invest in such a thing. Why? The project is predicated on three assumptions—all false:

> First, it assumes that conservation and efficiency will not be sufficient to limit water demand and that we must find more "supply."

> Second, it assumes that money, energy, and the environment don't matter.

> Third, it assumes that there is excess water in the Missouri River to transfer.

Water demands should not be assumed to grow without limits. In the first case, water managers, planners, local officials, and others are all trained to assume that water demands must grow inexorably with population and the economy and that the only response is to find new sources of water. Recent experience in the western United States and many other regions—indeed, in the United States as a whole—shows that we can cut water use substantially by improving efficiency, cutting waste, and changing use patterns while populations and economies continue to grow. Figure 7.4 shows that this has been the case in the United States for over thirty years, during which time total water use has actually declined—and new studies show that vast untapped and relatively low-cost efficiency improvements remain. Moreover, it is long past time we had a real conversation about land use, urban planning, and population policies instead of assuming they have no implications for water policy.

In the second case, a pipeline taking water west from the Missouri River makes sense only if you deeply discount or ignore economic, energy, and environmental factors. Such a pipeline would cost—at a minimum—tens of billions of dollars to build and would require massive energy inputs to run. Here is just one piece of this puzzle: moving water from the Missouri River at Leavenworth, Kansas (often described as a logical diversion point), to Denver, Colorado, would require a pipeline at least 600 miles long pumping water uphill around 5,000 feet (see figure 7.5). Even doing the math on the back of an envelope shows that the energy costs for pumping alone, with conservative assumptions, would be at least $1,000 per acre-foot of water. Add in the additional costs of financing and building such a pipeline and the power plants to run it, of operating it, and of treating and distributing the water, and we're looking at some of the most expensive water

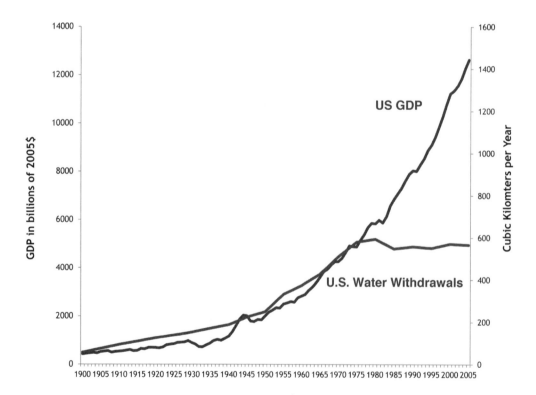

FIGURE 7.4 US WATER USE AND GROSS DOMESTIC PRODUCT
Source: Gleick and Palaniappan 2010.
The graph shows US water withdrawals from 1900 to 2005 in cubic kilometers per year plotted against US gross domestic product in 2005 adjusted dollars. Note that water withdrawals are no longer growing with population and the economy but have actually declined.

in the world (not including bottled water—another story). Even more ambitious, and costly, would be trying to move any of the water over the Rocky Mountains, where the additional pumping and energy costs would skyrocket.

Finally, the idea is also largely predicated on the assumption that the Missouri River has excess water. This, of course, depends on how you define "excess" and who is doing the defining. Some of the proposals try to argue that water would be diverted only during infrequent high flow, or flood, events, which would further dramatically increase the unit cost of water, since huge, costly infrastructure would be idle for long periods of time. It is unlikely that, if actually built, the system would be operated in this manner. But it is also ironic that the idea was presented at exactly the same time—the 2012–2013 seasons—that massive, severe drought cut the flows of the Missouri and Mississippi Rivers to record low levels, barges were being stranded, downstream users were demanding more upstream releases, and Midwest agriculture was suffering from water shortages. The US Army Corps of Engineers—the federal agency responsible for managing the

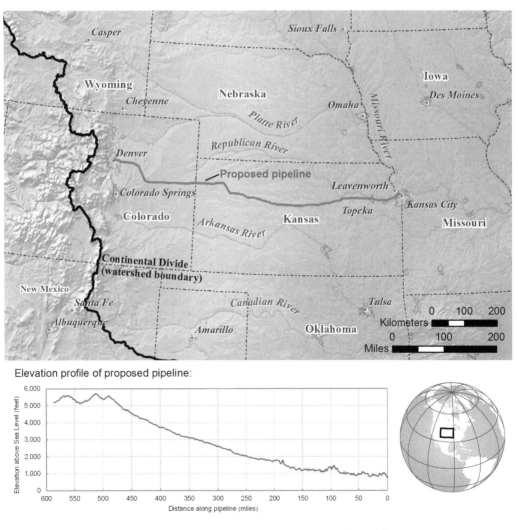

FIGURE 7.5 ONE POSSIBLE ROUTE FOR THE DIVERSION OF MISSOURI RIVER WATER TO DENVER AND THE COLORADO FRONT RANGE

This figure also presents the distance/elevation profile showing that water would have to be moved nearly 600 miles and be pumped nearly 5,000 feet up, at a huge energy cost.

river—declined requests for more water from the Missouri River to alleviate shortages on the Mississippi because such releases would have come at the expense of upstream users, drinking water supplies, environmental conditions, and hydropower generation.

Conclusions

We've reached the era of peak water, even on big rivers such as the Missouri or for untapped resources such as northern-flowing rivers in Alaska and Canada. There are many other potential candidates for the category of "zombie water projects." There is the proposal to divert water from the Gulf of California back north to the Salton Sea. There is the proposal to cut a canal or run a pipeline from the Mediterranean Sea or the Red Sea to the Dead Sea (see the In Brief report in this volume for a detailed update on this idea). There has been talk for decades about building a massive canal in the Jonglei region of Sudan on the Nile to dry up the ecologically valuable inland Sudd swamps there and cut evaporative losses on the Nile. The Chinese are building a massive south–north canal system that may be quite alive.

While we can continue to fantasize about developing new water supplies by building larger and larger engineering projects to move water longer and longer distances, destroying ever more distant ecosystems in the name of satisfying uncontrolled and unquestioned sprawl and development in arid regions, perhaps it is time to ask: Just because we can, should we? And once we ask that question, the answer typically comes back quickly: No, we shouldn't. There are often far more environmentally, politically, and economically sound solutions that should be considered first. And many of these zombie water projects should be left for dead.

References

Alaska Water Project. 2013. Bulk Water Pipeline. http://www.alaskabulkwater.com/Water-Pipeline.html (accessed January 2, 2013).

Barclay, E. 2010. Alaska town eyes shipping water abroad. *National Geographic Daily News*, June 25. http://news.nationalgeographic.com/news/2010/06/100625-freshwater-alaska-shipping-water/.

Boyd, D. R. 2003. *Unnatural Law: Rethinking Canadian Environmental Law and Policy*. Vancouver: University of British Columbia Press.

Campana, M. E. 2008. Canadian water exports: Will NAWAPA return? *WaterWired*, January 25. http://aquadoc.typepad.com/waterwired/2008/01/kennedy-to-cana.html.

Casey, J. V. 2004. Officials speed up plans to tap rural Nevada water. *Las Vegas Review-Journal*, February 27.

Christensen, J. 1994. Las Vegas wheels and deals for Colorado River water. *High Country News*, February 21. https://www.hcn.org/issues/4/118.

City and Borough of Sitka, Alaska. 2013. Bulk water export advertisement. http://www.sitka.net/sitka/resources.html (accessed January 2, 2013).

Cooley, H., T. Hutchins-Cabibi, M. Cohen, P. H. Gleick, and M. Heberger. 2007. Hidden Oasis: Water Conservation and Efficiency in Las Vegas. Oakland, CA: Pacific Institute. http://www.pacinst.org/publication/hidden-oasis-water-conservation-and-efficiency-in-las-vegas/.

Deacon, J. E. 2009. Potential environmental effects of the Southern Nevada Groundwater Project. *Mojave Applied Ecology Notes* 2 (3): 6–8.

Egan, T. 1994. Las Vegas stakes claim in 90's water war. *New York Times*, April 10. http://www.nytimes.com/1994/04/10/us/las-vegas-stakes-claim-in-90-s-water-war.html?pagewanted=all&src=pm.

Eggen, L. 2012. Pipeline resolution not likely in 2013. *Ely Times*, Nevada, December 21. http://www.elynews.com/news/article_4f0810be-4ace-11e2-af85-0019bb2963f4.html.

Environment Canada. 1987. Federal Water Policy. http://www.ec.gc.ca/eau-water/default.asp?lang=En&n=D11549FA-1.

Garrick, D., M. A. Siebentritt, B. Aylward, C. J. Bauer, and A. Purkey. 2009. Water markets and freshwater ecosystem services: Policy reform and implementation in the Columbia and Murray-Darling basins. *Ecological Economics* 69 (2): 366–379.

Gleick, P. H. 1998. Water bag technology. In *The World's Water 1998–1999: The Biennial Report on Freshwater Resources*, edited by P. H. Gleick, 200–205. Washington, DC: Island Press.

———. 2003. Water use. *Annual Review of Environment and Resources* 28:275–314. doi:10.1146/annurev.energy.28.040202.122849.

———. 2011. Zombie water projects (just when you thought they were really dead . . .). Forbes Blogs, December 7. http://www.forbes.com/sites/petergleick/2011/12/07/zombie-water-projects-just-when-you-thought-they-were-really-dead/.

Gleick, P. H., and M. Palaniappan. 2010. Peak water limits to freshwater withdrawal and use. *Proceedings of the National Academy of Sciences* 107 (25): 11155–11162. doi: 10.1073/pnas.1004812107.

Greene, S., and M. Hynes. 1996. The politics of water: Law of the river guides water talks. *Las Vegas Review-Journal*, October 13.

Kenny, J. F., N. L. Barber, S. S. Hutson, K. S. Linsey, J. K. Lovelace, and M. A. Maupin. 2009. Estimated use of water in the United States in 2005. Circular 1344. Reston, VA: US Geological Survey. http://pubs.usgs.gov/circ/1344/.

KQED. n.d. The Reber Plan: A Big Idea for San Francisco Bay. San Francisco: KQED. http://education.savingthebay.org/the-reber-plan-a-big-idea-for-san-francisco-bay/.

Lee, L. B. 1978. 100 years of reclamation historiography. *Pacific Historical Review* 47 (4): 507–564. doi:10.2307/3637371.

Mandel, J. 2006. Report details problems with contract for Iraq health centers. *Government Executive*, May 1. http://www.govexec.com/management/2006/05/report-details-problems-with-contract-for-iraq-health-centers/21710/.

NAWAPA. 1964. Ralph M. Parsons Company. http://www.youtube.com/watch?v=mTissXa48yg&feature=youtube_gdata_player.

Nishkian, L. H. 1946. Report on the Reber Plan and Bay Land Crossing. San Francisco, CA: Joint Army-Navy Board on an Additional Crossing of San Francisco Bay.

Presidio Trust. 2012. *Before the Bridge: Sight and Sound at the Golden Gate*. San Francisco: Presidio Trust.

Reinhold, R. 1991. Battle lines drawn in sand as Las Vegas covets water. *New York Times*, April 23.

Reisner, M. 1986. *Cadillac Desert: The American West and Its Disappearing Water*. Rev. ed. New York: Viking.

Rothfeder, J. 2004. *Every Drop for Sale: Our Desperate Battle over Water in a World About to Run Out*. New York: Jeremy P. Tarcher/Penguin.

Salter, J. 2012. Major pipeline among ideas for aiding arid West. Associated Press, December 11. http://bigstory.ap.org/article/major-pipeline-among-ideas-aiding-arid-west.

Salt Lake Tribune. 2013. The right decision: Herbert right to reject water deal. April 3. http://www.sltrib.com/sltrib/opinion/56100950-82/herbert-deal-governor-nevada.html.csp.

Sewell, W. R. D., V. Ostrom, J. A. Crutchfield, E. R. Tinney, and W. F. Royce. 1967. NAWAPA: A continental water system. *Bulletin of the Atomic Scientists* (September): 8–27.

Sinclaire, J. 2011. The fitness of physical models. *Pacific Standard*, December 5.

Southern Nevada Water Authority (SNWA). 2006. Water Resource Plan. Las Vegas, NV: Southern Nevada Water Authority.

———. 2007. About the Southern Nevada Water Authority. http://www.snwa.com/html/about_index.html (accessed November 3, 2007).

Special Subcommittee on Western Water Development. 1964. A Summary of Water Resources Projects, Plans, and Studies Relating to the Western and Midwestern United States. 755-567 O-64. Washington, DC: United States Senate, Committee on Public Works.

Springer, P. 2013. The quest for water. *The Forum*. http://legacy.inforum.com/specials/century/jan3/week6.htm.

State of Utah. 2009. Snake Valley Agreement. Salt Lake City, UT: State of Utah, Utah Division of Water Rights. http://waterrights.utah.gov/snakeValleyAgreement/snakeValley.asp.

United States Department of the Interior, Bureau of Land Management (US BLM). 2012. Clark, Lincoln, and White Pine Counties Groundwater Development Project Final Environmental

Impact Statement. FES 12-33. Reno, NV: United States Department of the Interior, Bureau of Land Management. http://www.blm.gov/nv/st/en/prog/planning/groundwater_projects/snwa_groundwater_project/final_eis.html (accessed April 25, 2013).

United States Department of the Interior, Bureau of Reclamation (US BOR). 2007. Law of the River. Boulder City, NV: United States Department of the Interior, Bureau of Reclamation, Lower Colorado Region. http://www.usbr.gov/lc/region/pao/lawofrvr.html (accessed November 24, 2007).

———. 2012. Colorado River Basin Water Supply and Demand Study. Boulder City, NV: United States Department of the Interior, Bureau of Reclamation, Lower Colorado Region. http://www.usbr.gov/lc/region/programs/crbstudy/finalreport/index.html.

Velotta, R. N. 2011. Troubled waters: Las Vegas' perpetual quest to quench itself. *Vegas Inc.*, August 1. http://www.vegasinc.com/news/2011/aug/01/troubled-waters/.

Walton, B. 2010a. Bulk water company plans to export to India, East Asia, and the Caribbean. *Circle of Blue Water News*, August 27. http://www.circleofblue.org/waternews/2010/world/bulk-water-company-plans-to-export-to-india-east-asia-and-the-caribbean/.

———. 2010b. Bulk water shipping company misses deadline to export from Alaska. *Circle of Blue Water News*, December 13. http://www.circleofblue.org/waternews/2010/world/bulk-water-shipping-company-misses-deadline-to-export-from-alaska/.

Wikipedia. 2013. Lyndon LaRouche. http://en.wikipedia.org/w/index.php?title=Lyndon_LaRouche&oldid=551550653 (accessed February 2013).

Worster, D. 1992. *Rivers of Empire: Water, Aridity, and the Growth of the American West*. New York: Oxford University Press.

WATER BRIEF 1

The Syrian Conflict and the Role of Water

Peter H. Gleick

There is a long history of conflicts over water—consistently documented in each previous volume of *The World's Water* biennial assessments, going back to the first, in 1998. The Pacific Institute also maintains a searchable online chronology of such conflicts going back five thousand years.[1] There were dozens of new examples in the past two years, in countries from Latin America to Africa to Asia. Access to water and the control of water systems have been causes of conflict, weapons used during conflicts, and targets of conflict, but there are also growing risks of violence over the role that water plays in development disputes and economic activities. One especially disturbing example of a major conflict, with complicated but direct connections to water, has developed since 2010: the unraveling of Syria and the escalation of massive civil war there.

Syria and Water

Syria's political dissolution is, like almost all conflicts, the result of complex, interrelated factors. In this case, the conflict was influenced by an especially repressive and unresponsive political regime, the erosion of the economic health of the country, and a wave of political reform sweeping over the entire Middle East and North Africa region. But in a detailed assessment, Francesco Femia and Caitlin Werrell (2012) of the Center for Climate and Security in Washington, DC, noted that factors related to drought, agricultural failure, water shortages, and water mismanagement also played an important role in nurturing Syria's "seeds of social unrest" and contributing to violence. In particular, they argue that a very severe drought led to persistent crop failures, which in turn led to very significant dislocation and migration of rural communities to the cities. These factors in turn contributed to urban unemployment and economic dislocations and social unrest.

Water has always been a scarce resource in the region—one of the driest in the world. Syria receives, on average, less than 250 millimeters (not quite 10 inches) of rainfall annually. All of its major rivers (the Tigris, Euphrates, and Orontes Rivers and the Yarmouk, a tributary of the Jordan) are shared with neighboring countries. And Syria, like the region as a whole, experiences periodic droughts. Over the past century (from 1900 to 2005) there were six significant droughts in Syria, in which average monthly winter precipitation dropped to only one-third of normal. Five of these droughts lasted only one season;

---

1. The chronology can be found at http://www.worldwater.org/conflict.html, and a version appears in this volume of *The World's Water*.

the sixth lasted two. Starting in 2006, however, and lasting into 2011, Syria experienced a multi-season extreme drought and agricultural failures, described by Shahrzad Mohtadi as the "worst long-term drought and most severe set of crop failures since agricultural civilizations began in the Fertile Crescent many millennia ago" (Mohtadi 2012).

Robert Worth of the *New York Times* noted that this drought contributed to a series of social and economic dislocations (Worth 2010). The United Nations estimated that by 2011, the drought was affecting 2–3 million people, with 1 million driven into food insecurity. More than 1.5 million people—mostly agricultural workers and family farmers—moved from rural regions to cities and temporary settlements near urban centers, especially on the outskirts of Aleppo, Hama, Homs, Damascus, and Dara'a.

In 2008, a diplomatic cable from the US embassy in Damascus to the US Department of State in Washington warned of the implications of the drought, with a review of local opinions and concerns as well as insights from the Food and Agriculture Organization of the United Nations (FAO).[2] During the drought, FAO Syrian representative Abdullah bin Yehia provided a briefing on the impacts, described as a "perfect storm" when combined with other economic and social pressures. At that time, concerns were expressed that the population displacements "could act as a multiplier on social and economic pressures already at play and undermine stability in Syria." In July 2008, Syria's minister of agriculture stated publicly to UN officials that the economic and social fallout from the drought was "beyond our capacity as a country to deal with." This warning was perceptive and prescient: some of the earliest political unrest began around the town of Dara'a, which saw a particularly large influx of farmers displaced from their lands by crop failures.

The political problem was worsened by water mismanagement, poor planning, and policy errors. The regime of Bashar al-Assad traditionally provided large subsidies for the production of water-intensive crops, such as wheat and cotton, and flood irrigation systems are largely inefficient. Groundwater supplies were reportedly overpumped, leading to dropping groundwater levels and rising production costs. Water withdrawals upstream by Turkey for its own agricultural production in the southern Anatolia region further reduced surface flows inside Syria. All of these factors contributed to growing economic and political uncertainty.

Suzanne Saleeby, writing for *Jadaliyya*, a magazine published by the Arab Studies Institute, analyzed the links between economic and environmental conditions and the subsequent political unrest. Saleeby (2012) argued, "The regime's failure to put in place economic measures to alleviate the effects of drought was a critical driver in propelling such massive mobilizations of dissent."

Saleeby noted that the Assad regime made great strides in improving access to water, building water infrastructure that benefited rural communities and expanded agricultural production. But she also noted that the vast expansion of irrigated agriculture in the region, especially of nonfood industrial crops, worsened water shortages in rural areas and led to salination of farmland.

These effects increased vulnerability to subsequent droughts. At the same time, governmental sales of land in recent years to private agricultural interests led to growing mistrust and concern about corruption. Extensive exploitation of groundwater led to substantial drops in groundwater levels and, in some regions, contamination by salts and

---

2. These observations were cited in the diplomatic cable, published at http://wikileaks.org/cable/2008/11/08DAMASCUS847.html. The accuracy of this cable's contents has not been verified, but these statements appear to be in the public record.

nitrates, making local wells (for example, in the ar-Raqqah region) unfit for human use. In all of these cases, trends toward privatization "corroded customary law over boundary rights" (Saleeby 2012), further worsening local tensions. These complex factors contributed to the extent and severity of the unrest (Saleeby 2012):

> It is logical to conclude that escalating pressures on urban areas due to internal migration, increasing food insecurity, and resultant high rates of unemployment have spurred many Syrians to make their political grievances publically known. One might look to the city of Deir ez-Zor, one of Syria's most dangerously dry areas, to locate deeply rooted seeds in the harvest of dissent. The northeastern city experienced one of the strongest sieges by the Syrian army at the beginning of Ramadan after popular uprisings spread across its parched expanse. As a local activist told *Syria Today*, the citizens of Deir ez-Zor "are suffering and complain that they have had no help from the authorities who tell them what type of crops they have to plant, and have a monopoly on buying up what they produce." Also among the cities whose residents' livelihoods were most crippled by recent drought was Daraa, historically a "bread basket" of Syria. Additionally, Hama remains a major destination for drought-displaced farmers despite suffering its own water scarcity woes. In all three centers of popular uprisings lie important narratives of livelihoods lost and families left wanting.

Other Links between Water and Conflicts in Syria

After serious conflict in Syria developed, violence worsened and spread. In 2012, incidental impacts on urban water distribution systems were reported, as were specific, intentional attacks on water systems because of their strategic value. During fighting around the city of Aleppo in 2012, the major pipeline delivering water to the city was badly damaged, and in September the city of about 3 million people was suffering shortages of drinking water (BBC 2012). In late November 2012, anti-Assad Syrian rebels overran governmental forces and captured the Tishrin Dam, a hydroelectric dam on the Euphrates River, after heavy clashes (Mroue 2012). The dam supplies several areas of Syria with electricity and is considered to be of major strategic importance to the Syrian regime. In February 2013, anti-Assad forces captured the Tabqa Dam, also called the al-Thawrah Dam, which is the largest hydroelectric dam in the country and provides much of the electricity for the city of Aleppo (BBC 2013).

The Role of Climate Change

Indications for the future are not promising: on top of political unrest and economic dislocations, the region faces challenges posed by growing populations, the lack of international agreements over shared water resources, poor water management, and the increasing risks of climate change. In 2008, the World Bank was already warning of future climate risks (World Bank 2008):

> According to the latest IPCC assessment, the climate is predicted to become even hotter and drier in most of the MENA [Middle East and North Africa] region. Higher temperatures and reduced precipitation will result in higher frequency and severity of droughts, an effect that is already materializing in the Maghreb.

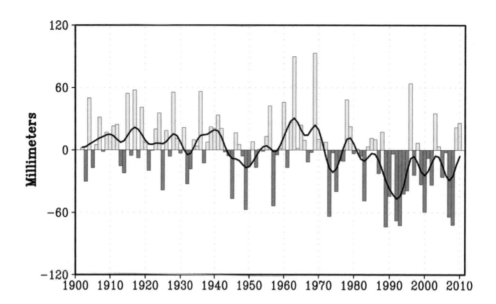

FIGURE WB 1.1 MILLIMETERS OF RAINFALL DURING WINTER PERIODS FROM 1902 TO 2010, SHOWING A DROP IN RAINFALL IN 1971–2010
Source: NOAA (2011).

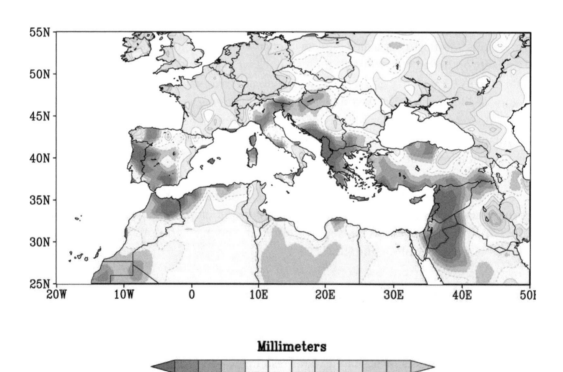

FIGURE WB 1.2 FOUR DARKEST SHADES INDICATE THE AREAS AROUND THE MEDITERRANEAN THAT EXPERIENCED SIGNIFICANTLY DRIER WINTERS DURING 1971–2010 THAN IN 1902–2010
Source: NOAA (2011).

A research paper published in 2012 suggested that climate change is already beginning to influence long-term droughts in the region including Syria by reducing winter rainfall (see figures IB 1.1 and IB 1.2) (Hoerling et al. 2012). That study suggests that winter droughts are increasingly common and that human-caused climate change is playing a role. Martin Hoerling of the National Oceanic and Atmospheric Administration's Earth System Research Laboratory, one of the study's authors, stated, "The magnitude and frequency of the drying that has occurred is too great to be explained by natural variability alone" (NOAA 2011).

If the international community wants to reduce the risks of local and international political conflicts and violence over water, more effort will have to be put into recognizing these risks and improving the tools needed to reduce them.

References

British Broadcasting Corporation (BBC). 2012. Aleppo water supply cut as Syria fighting rages. September 8. http://www.bbc.co.uk/news/world-middle-east-19533112.

———. 2013. Syria crisis: "Powerful" minibus explosion kills 13. February 11. http://www.bbc.co.uk/news/world-us-canada-21409661.

Femia, F., and C. Werrell. 2012. Syria: Climate Change, Drought, and Social Unrest. Center for Climate and Security, Washington, DC. http://climateandsecurity.org/2012/02/29/syria-climate-change-drought-and-social-unrest/.

Hoerling, M., J. Eischeid, J. Perlwitz, X. Quan, T. Zhang, and P. Pegion. 2012. On the increased frequency of Mediterranean drought. *Journal of Climate* 25:2146–2161. doi:10.1175/JCLI-D-11-00296.1.

Mohtadi, S. 2012. Climate change and the Syrian uprising. *Bulletin of the Atomic Scientists*, August 16. http://thebulletin.org/web-edition/features/climate-change-and-the-syrian-uprising.

Mroue, B. 2012. Activists: Syrian rebels seize major dam in north. *Daily Star*, Lebanon, November 26. http://www.dailystar.com.lb/News/Middle-East/2012/Nov-26/196180-activists-syrian-rebels-seize-major-dam-in-north.ashx#ixzz2KA8TTBbj.

National Oceanic and Atmospheric Administration (NOAA). 2011. NOAA Study: Human-Caused Climate Change a Major Factor in More Frequent Mediterranean Droughts. October 27. http://www.noaanews.noaa.gov/stories2011/20111027_drought.html.

Saleeby, S. 2012. Sowing the seeds of dissent: Economic grievances and the Syrian social contract's unraveling. *Jadaliyya*, February 16. Washington, DC, and Beirut: Arab Studies Institute. http://www.jadaliyya.com/pages/index/4383/sowing-the-seeds-of-dissent_economic-grievances-an.

World Bank. 2008. A Strategy to Address Climate Change in the MENA Region. Washington, DC: World Bank. http://go.worldbank.org/OIZZFRJZZ0.

Worth, R. F. 2010. Earth is parched where Syrian farms thrived. *New York Times*, October 13. http://www.nytimes.com/2010/10/14/world/middleeast/14syria.html?_r=0.

WATER BRIEF 2

The Red Sea–Dead Sea Project Update

Kristina Donnelly

The Dead Sea is a terminal lake with great historical, cultural, and economic significance. The sea sits at the confluence of East and West, on the borders of the Hashemite Kingdom of Jordan, Israel, and the Palestinian National Authority, an area that has transported people, goods, and ideas for generations. For at least the past 150 years, serious proposals have been made to engineer a connection between the Red Sea, the Mediterranean Sea, and the Dead Sea. The political, social, water, and environmental needs of the region have each, over time, contributed to the endurance of this idea; hammers, as they say, tend to manifest a great deal of nails, and humans have an affinity for large, complex hammers. Over time, the proposal has been offered as a means to facilitate transportation, generate electricity, supply freshwater through desalination, and advance broader regional peace and economic development. This Water Brief report summarizes both the history of and recent events in the proposals to build major infrastructure in this region.[1]

Projects to Improve Transportation

The first proposals date as far back as the 1850s, when the British suggested that a canal linking the Red Sea to the Mediterranean Sea could serve as an alternative to the Suez Canal as a route from Europe to India (Allen 1855). Political considerations regarding Egypt made finding alternatives to the Suez Canal attractive for the British and then for Israel (Ofner 1946; *New York Times* 1956). As early as 1883, a British member of Parliament noted the "immense importance of obtaining, if possible, a waterway to India independent of the Suez Canal, and all its untoward complications" (Martin 1883). The idea persisted through the mid-twentieth century (Wallenstein 1966). As proposed, creation of a waterway suitable for navigation would completely alter the local environment and would flood important cultural, religious, and historical sites (*San Francisco Chronicle* 1883). Ultimately, Israel chose to build a rail system over a canal to link the Mediterranean to the Red Sea, although even this project was never completed.

---

1. We chose not to include this section in the earlier chapter on zombie water projects because it isn't clear that the idea is fully dead—yet.

Projects to Expand Energy

As the lowest point on Earth, the Dead Sea has tremendous potential for hydroelectric power generation. In 1902, Theodor Herzl, the founder of modern Zionism, wrote about a canal that would take advantage of the 400-meter (1,310-foot) elevation difference between the Red Sea and the Mediterranean to generate hydroelectric power. As the cost of oil rose in the 1970s, Israel began studying the feasibility study of a canal project for energy generation, with navigation listed only as a potential "fringe" benefit (Torgerson 1979). Jordan objected to the plan, as one option would have passed through the Gaza Strip, and in response began studying the feasibility of a hydropower canal from the Red Sea to the Dead Sea (Khouri 1980). Over time, canal proposals have considered other kinds of energy generation besides hydropower, including nuclear power and solar energy fields (Wallenstein 1966; Torgerson 1979). Although energy generation would also become another "fringe" benefit of future canal proposals, it remains a key component to reducing overall project costs.

Projects to Produce Freshwater

The project began to take its modern shape during the Middle East peace talks in the early 1990s, when freshwater scarcity and drastically increasing populations were threatening the water security of the entire region (Wolf and Newton 2007). In order to address some of the region's critical water issues, the peace process brought about the formation of the Multilateral Working Group on Water Resources. At a meeting of the group in 1993, Israel proposed desalinating Red Sea water to provide 1 billion cubic meters of water to the region, particularly Jordan (Schmetzer 1993).

It was envisioned that this project would be part of a much larger regional economic development plan to facilitate peace, with the *London Observer* calling it "the development centerpiece of the Middle East peace accords" (Holmes 1994; *Wisconsin State Journal* 1994; Land 1994). Both sides began referring to the section of the Jordan Rift Valley between the Red Sea and the Dead Sea as "Peace Valley" (Peres 1994). However, by 1995 the project had been shelved as a result of onerous bureaucracy, deteriorating relations, lack of funding, and failure of the parties to agree on design and management. At that time, the project was estimated to cost $4.2 billion, and the World Bank and other international monetary officials were making it clear that the money would not come from the international community (Rodan 1995; Izenberg 1997). There would be little movement on the idea until changing circumstances offered a fresh application for the proposal.

Projects to Restore the Dead Sea

The Dead Sea has slowly been shrinking throughout the twentieth and twenty-first centuries. Israel, Jordan, and Syria began diverting the Jordan River in the 1960s, and everyone knew that the Dead Sea would slowly disappear as a result (Wallenstein 1966). Located in an area that receives less than 100 millimeters (under 4 inches) of rain per year and where temperatures often exceed 45°C, the Dead Sea is completely reliant on inflow for its continued existence. It is estimated that the total inflow to the Dead Sea has been reduced from around 1,250 million cubic meters (MCM) per year in 1950 to around 260 MCM per year in 2010. Furthermore, much of the water left in the lower Jordan River consists of raw sewage, agricultural runoff, fish pond water, and saline spring water (Gafny et al. 2010). Lucrative chemical production facilities on the southern shore of the sea have also contributed to the decline. In order to extract potash and other chemical products, the chemical industries pump water from the upper Dead Sea to the lower section, into solar evaporation ponds. Without this pumping, the southern Dead Sea would completely dry up. Paradoxically, the southern portion of the sea is now suffering from rising levels: waste materials left over from the mineral extraction process are causing the water levels to increase by 20 centimeters per year. This means that the majority of the tourism activity taking place on the Israeli side of the Dead Sea is at risk from flooding, and large-scale dredging plans are being discussed (Udasin and Hartman 2011).

The Dead Sea shoreline now sits 426 meters (m) below sea level, more than 30 m lower than in the 1960s. Without any intervention, the sea is expected to drop by another 150 m, stabilizing around 543 m below sea level, by the mid-twenty-second century (Coyne et Bellier et al. 2012).

Although one canal project noted the replenishment of the Dead Sea as a potential benefit as early as the 1980s (Khouri 1980), the idea would not take prominence until the late 1990s, when sinkholes began to appear on both the Israeli and Jordanian shores. Sinkholes form when declining sea levels move the interface of fresh and saline groundwater, bringing undersaturated water into contact with salt formations. This freshwater dissolves these formations, leaving behind underground caverns that can eventually collapse (Yechieli et al. 2002). These sinkholes have swallowed up roads, houses, and other infrastructure on the shorelines,[2] which has had, and will continue to have, negative consequences for the local and national economies.

A Combined Plan

In 2001, the Center for Middle East Peace and Economic Cooperation revived the concept, in a project dubbed the "Peace Conduit," this time with saving the Dead Sea as a major objective, with other benefits being the production of drinking water, generation of hydropower, and a reduction in regional conflict (Lazaroff 2001). Israel and Jordan announced their commitment to the project in 2002 during the World Summit on Sustainable Development in Johannesburg, South Africa (Gavrieli and Bein 2007).

In 2005, Israel, the Palestinian Authority, and Jordan signed an agreement for a $15.5 million feasibility study, funded through a multidonor trust fund administered by the World Bank (Al Bawaba 2005; Urquhart 2005). In 2013, the World Bank issued the final three draft reports: a feasibility study, an environmental and social assessment, and a study of alternatives.[3] The reports tentatively conclude that a $10 billion pipeline is feasible, though risky and not without environmental and social impacts.

The proposed plan would convey 2,000 MCM per year of Red Sea water to a desalina-

---

2. Sometimes even people: http://www.nbcnews.com/id/31475786/ns/world_news-environment/t/dead-sea-sinkholes-swallow-plans/#.UVgTQaLZ7jI.
3. The World Bank held a series of stakeholder meetings in February 2013 and planned to issue the final report documents in June (World Bank 2013).

tion plant in Jordan, south of the southern tip of the chemical industries' evaporation ponds. By 2060, the plant would produce more than 820 MCM per year, with 550 MCM per year for Jordan, 60 MCM per year for Israel, 60 MCM for Palestine, and 200 MCM left unallocated. Despite the inclusion of a hydropower facility in the preferred option, by 2060 the desalination plant would still require at least an additional 6,140 gigawatt-hours of energy per year.[4] Leftover brine waste would be used to replenish the Dead Sea, stabilizing the level around 416 m below sea level by 2054. Potential impacts from this brine include precipitation of gypsum, which could create whitening events, stimulate algae blooms, or otherwise alter the water's aesthetics.

Criticisms of the Project

It is unclear whether an effective multinational organization could form and carry out such a cumbersome and lengthy project. In order to access international support and financing, the project partners would need to set up a legal and institutional framework "founded on internationally accepted law and good practice" (Coyne et Bellier et al. 2012). That could mean Israel would need to recognize that the Palestinian Authority has rights to the Dead Sea, which it has not yet done. As noted in the feasibility study, "one particularly powerful lesson to be learnt from experience elsewhere is that good governance requires a strong and autonomous regulatory authority." It has not yet been made clear what kind of regulatory authority could be created, how it would operate, who would pay for it, and what its roles and responsibilities would be.

Moreover, anti-Israel sentiment is high in Jordan and the Palestinian Authority, and any collaborative projects are often widely perceived as "normalizing" the occupation in Palestine. The feasibility study and other reports have considered a Jordanian-led project, with bilateral agreements set up to sell water and electricity to Israel and Jordan. However, a unilateral project would still require multinational cooperation and agreement.

None of the individual countries has the financial means or political support to unilaterally move forward, and international financing is crucial. Although the feasibility study considered whether the project could be built in phases in order to alleviate some financial pressure, the study concluded that at least 75 percent of the project's total capacity would be needed in the first phase to make the project financially viable. The feasibility study concluded that a mix of funding sources would be needed in order to meet the $10 billion price tag, and it outlined a list of preconditions that international financiers would likely have prior to investing, including implementing multinational agreements, making improvements to existing water infrastructure, and increasing the institutional capacity of the water sector.

In addition, environmental organizations and the chemical industries remain critical, citing concerns about Dead Sea and Red Sea water quality and environmental impacts, as well as potential environmental and groundwater impacts in the Arava Valley. Both Israel and Jordan harvest chemicals from the rich, mineral-laden waters of the Dead Sea, and industries on both sides of the border are concerned that the impact of

---

4. Energy and cost estimates include the energy required to pump potable water to Amman but not to Israel or the Palestinian Authority.

mixing the desalination brine with water from the Dead Sea will threaten their lucrative operations.

The freshwater component of the project, while important, is less pressing now than it was at the beginning of the twenty-first century. Israel is already moving forward with its own coastal desalination program, which is expanding the freshwater available internally. The feasibility study concluded that it would be economically feasible to deliver water only to the low-elevation demand centers in the Dead Sea basin and the Arava Valley, where demand has been estimated to be 60 MCM per year. It is unclear whether the benefits of the project offer enough incentive for Israel's continued participation. In Jordan, the Disi Water Conveyance Project is expected to be completed in 2013 and will transport 100 MCM of water to Amman from the Disi Aquifer, which lies beneath southern Jordan and Saudi Arabia. Although the pressing need for freshwater in Jordan is expected to be temporarily assuaged by the project, water from this groundwater basin is not a sufficient or sustainable supply.

The idea to link the Mediterranean Sea or the Red Sea with the Dead Sea is not new, although the purpose and scope have changed considerably over time. The factors that have influenced these changes—population, politics, economics, and natural resource needs—are not static and will continue to shift. It is unclear whether the proposed alternative, or some other iteration, can meet all these needs simultaneously and for all stakeholders in the region.

References

Al Bawaba. 2005. Donor countries agree on providing funding to the study on Red Sea–Dead Sea Canal project. Al Bawaba News Service, July 7. http://www.albawaba.com/news/donor-countries-agree-providing-funding-study-red-sea-dead-sea-canal-project.

Allen, W. 1855. *The Dead Sea, a New Route to India: with Other Fragments and Gleanings in the East*. 2 vols. London: Longman, Brown, Green, and Longmans.

Coyne et Bellier, Tractebel Engineering, and Kema. 2012. Red Sea–Dead Sea Water Conveyance Study Program: Draft Feasibility Study Report. Report No. 12 147 RP 04. July. http://siteresources.worldbank.org/INTREDSEADEADSEA/Resources/Feasibility_Study_Report_Summary_EN.pdf.

Gafny, S., S. Talozi, B. Al Sheikh, and E. Ya'ari. 2010. Towards a Living Jordan River: An Environmental Flows Report on the Rehabilitation of the Lower Jordan River. Amman, Bethlehem, and Tel Aviv: Friends of the Earth Middle East. http://foeme.org/uploads/publications_publ117_1.pdf.

Gavrieli, I., and A. Bein. 2007. Formulating a regional policy for the future of the Dead Sea—The "Peace Conduit" alternative. *Water Resources in the Middle East* 2:109–116.

Holmes, C. W. 1994. At Israel-Jordan Summit, peace stirs hope for profits. *Austin American-Statesman*, July 24.

Izenberg, D. 1997. An insider's view of the Jordan rift. *Jerusalem Post*, May 9.

Khouri, R. J. 1980. Jordan reveals own plan for canal link to Dead Sea. *Washington Post*, August 30.

Land, T. 1994. Plan for desert canals links former enemies. *Washington Times*, January 23.

Lazaroff, T. 2001. A bitter end for the Dead Sea? *Jerusalem Post*, April 27.

Martin, H. J. 1883. The Palestine Canal. Letter to Mr. Jon Corbett, Member of Parliament. *New York Times*, August 6.

New York Times. 1956. Israel studies Suez substitute for Middle East link to Europe. August 11.

Ofner, F. 1946. British study Red Sea canal to rival Suez. *Chicago Daily Tribune*, May 15.

Peres, J. 1994. Israel, Jordan open border nations contemplate commercial possibilities of new "Peace Valley." *Lincoln (NE) Journal Star*, August 9.

Rodan, S. 1995. Getting down to business in the Middle East. *Jerusalem Post*, November 3.

San Francisco Chronicle. 1883. The Dead Sea canal scheme: One of the most important projects of the century. August 26.

Schmetzer, U. 1993. Canal could douse Mideast water wars. *Chicago Tribune*, October 29.

Torgerson, D. 1979. Israel sees Dead Sea as spring of life. *Los Angeles Times*, August 25.

Udasin, S., and B. Hartman. 2011. Tourism, environment ministers urge Dead Sea salt harvest. *Jerusalem Post*, May 24. http://www.jpost.com/Enviro-Tech/Tourism-environment-ministers-urge-Dead-Sea-salt-harvest.

Urquhart, C. 2005. Peace canal deal for thirsty Middle East. *The Guardian*, May 9.

Wallenstein, A. 1966. Ambitious proposal for Negev canal draws strong reactions. *Washington Post*, October 20.

Wisconsin State Journal. 1994. Peace good economic strategy. July 27.

Wolf, A. T., and J. T. Newton. 2007. Case Study Transboundary Dispute Resolution: Multilateral Working Group on Water Resources (Middle East). Transboundary Freshwater Dispute Database (TFDD), Oregon State University. http://www.transboundarywaters.orst.edu.

World Bank. 2013. Red Sea–Dead Sea Water Conveyance Study Program—Overview, Updated: January. Washington, DC: World Bank.

Yechieli, Y., D. Wachs, M. Abelson, O. Crouvi, V. Shtivelman, E. Raz, and G. Baer. 2002. Formation of sinkholes along the shore of the Dead Sea—Summary of the first stage of investigation. *GSI Current Research* 13:1–6. http://www.gsi.gov.il/_Uploads/178GSI-Curent-Research-vol13.pdf.

WATER BRIEF 3

Water and Conflict
Events, Trends, and Analysis (2011-2012)

Peter H. Gleick and Matthew Heberger

Violence over freshwater has a long and distressing history. For nearly two decades, the Pacific Institute has been tracking, analyzing, and cataloging instances of conflict over water resources (see box WB 3.1).

In recent years, there has been an increase in reported cases of water-related disputes and violence. Figure WB 3.1 shows the average number of events per year from 1931 to 2012, averaged over two-year reporting periods. Part of this increase is almost certainly due to improvements in reporting; new Internet tools that permit more comprehensive collection and dissemination of news, data, and information; and more widespread awareness of the issue. But it is also possible that part of the increase is due to growing tensions and disputes over limited freshwater resources and the unresolved political challenges associated with "peak water"—the limits imposed on the availability of both renewable and nonrenewable water resources (Gleick and Palaniappan 2010). Without more information and a more comprehensive analysis, we are not able at this point to make any definitive statement about how to attribute the observed trends in this field, but we put forward the hypothesis, subject to better data and analysis, that the risks of water-related disputes involving violence are increasing. Others have also expressed this hypothesis, based on estimates of growing absolute and per capita scarcity of water, water contamination, and the extensive reliance of agriculture and some urban uses on nonrenewable sources of water that are rapidly being depleted (see, e.g., CSIS and Sandia National Laboratories 2005; US DIA 2012; Leurig 2012).

This idea is also at odds with some academic and popular writing that argues that the fear of water wars is overblown (Kramer et al. 2013). Part of that argument is based on the observation that there are substantial numbers of historical agreements and political treaties over shared interstate water systems and a long history of cooperation and negotiation when disputes develop. But the argument against water "wars" is different from the broader issue of water-related conflicts. And it is somewhat of a "straw man" argument: the "water wars" discussions are almost always found in the popular media, not academic analyses, and while wars are almost never solely or primarily about water, water-related violence, at many different scales, does occur.

Furthermore, much of the discussion and analysis of water wars focuses on *transnational*, or *interstate*, disputes, while there appears to be a growing threat of *subnational*, or *intrastate*, water conflicts. This distinction is further discussed and analyzed below, but we note that the prevalence and availability of international mechanisms to reduce the risks of interstate conflicts may offer little or no help in the area of intrastate conflicts.

BOX WB 3.1 The Pacific Institute Water Conflict Chronology

The Pacific Institute maintains a comprehensive database, the Water Conflict Chronology, at http://www.worldwater.org. An update is also provided every two years in this biennial water report, *The World's Water*, published by Island Press. Using these data, the Pacific Institute produces theoretical research papers, historical reviews, and regional case studies on water conflicts. We have organized workshops on lessons from regional water disputes in the Middle East, Central Asia, and Latin America. We have brought together experts from the fields of traditional and nontraditional arms control and helped coordinate a workshop on the role of science and religion in reducing the risks of water-related violence, which was held at the Pontifical Academy of Sciences of the Vatican.

The full Water Conflict Chronology includes integrated Google Maps; time, location, and subject filters; and a separate searchable bibliography.* The nature of entries in the chronology can be described and categorized in different ways. The Institute has split the categories, or types, of conflicts as follows, though other groupings and distinctions can also be useful:

Military tool (state actors): water resources, or water systems themselves, are used by a nation or state as a weapon during a military action.

Military target (state actors): water resources or water systems are targets of military actions by nations or states.

Terrorism or domestic violence, including cyberterrorism (nonstate actors): water resources or water systems are the targets or tools of violence or coercion by nonstate actors. A distinction is drawn between environmental terrorism and ecoterrorism (see Gleick 2006).

Development dispute (state and nonstate actors): water resources or water systems are a major source of contention and dispute in the context of economic and social development.

The Water Conflict Chronology has appeared in every volume of *The World's Water* since 1998. It continues to be one of the most popular features of the Pacific Institute's work, and it is used regularly by the media and by academics interested in understanding more about both the history and character of disputes over water resources (Zakaria 2013).

*Water Conflict Chronology, http://worldwater.org/chronology.html.

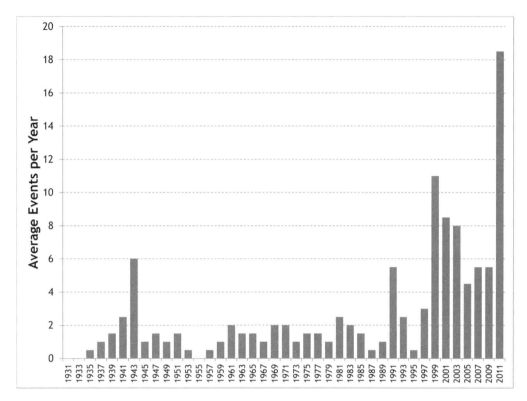

FIGURE WB 3.1 NUMBER OF REPORTED WATER CONFLICT EVENTS PER YEAR, 1931–2012 (AVERAGED OVER TWO-YEAR PERIODS TO SMOOTH FOR CROSS-YEAR EVENTS AND UNEVEN REPORTING)

As shown here, it is precisely these subnational conflicts that appear to be increasing in scope and severity. In that sense, it is not water wars that the international community must address but the far broader lethal causes of water conflicts overall, especially conflicts over equitable access to water, strategies for sharing during shortages, and water contamination.

There is also evidence of a shift in the nature of these conflicts, away from water disputes between nations and toward subnational and local violence over water access. Figure WB 3.2 shows the changes in the number and proportion of reported events at the transnational and subnational levels. The growing risk of subnational water conflicts was noted as far back as 1998 in the first volume of *The World's Water* (Gleick 1998, 105):

> Traditional political and ideological questions that have long dominated international discourse are now becoming more tightly woven with other variables that loomed less large in the past, including population growth, transnational pollution, resource scarcity and inequitable access to resources and their use.

Part of this shift is almost certainly due to improved local reporting of water conflicts, but part is also likely to reflect the greater availability of diplomatic and political tools at the international level that permit disputing parties to move toward cooperation rather than conflict. Such tools are notoriously weak at the local level, especially in countries with young or weak political institutions. And the greater prevalence of international

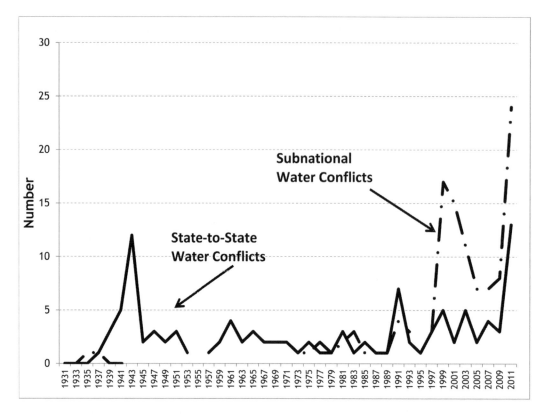

FIGURE WB 3.2 INCIDENCE OF REPORTED WATER-RELATED CONFLICTS FOR TRANSNATIONAL EVENTS (SOLID LINE) AND SUBNATIONAL EVENTS (DASHED LINE). IN THE PAST TWO DECADES, WATER CONFLICTS HAVE BEEN DOMINATED BY SUBNATIONAL DISPUTES.

mechanisms to reduce the risks of interstate conflicts over water does little or nothing to reduce the risks of intrastate, or subnational, violence.

The latest version of the Water Conflict Chronology includes events from 2011 and 2012 plus new events from other years added by our research, submissions from colleagues and researchers, and contributions from the public. We continue to welcome such contributions.

2011 and 2012 Update

In 2011 and 2012, violence over water was reported in every major developing region of the world, especially the Middle East, Africa, and Asia, with additional important examples in Latin America. Below, we provide an update with key examples from this two-year period. The full chronology and list are available online.[1]

---

1. Corrections, additions, and modifications are welcome. Send full citations and supporting information to pgleick@pacinst.org.

Southern and Western Asia (India, Afghanistan, Pakistan, Iran)

In Afghanistan, three separate events were reported, in April, July, and December 2012. In mid-April, 150 schoolgirls were reported to have been sickened by poison in their school water supply in an intentional attack thought to have been carried out by religious conservatives opposed to the education of women (Hamid 2012). In July, seven children were killed by a bomb thought to have been intended for Afghan police and planted at a freshwater spring in the Taywara District of western Ghor Province (Shah 2012). In December, in a sign of broader tensions over water between Afghanistan, Iran, and Pakistan, which share several rivers, Afghans defending the Machalgho Dam were killed. The dam is being developed for irrigation and local power supply. This dispute is just one of several surrounding the international waters of the region (Mashal 2012).

The year 2012 also saw a series of water-related conflicts within and between India, Pakistan, and Afghanistan. India has long experienced internal (state-to-state) disputes over water, especially between Karnataka and Tamil Nadu over the Cauvery River in a dispute that has been going on since at least 1991. At that time, court rulings allocating water from Karnataka to Tamil Nadu led to riots in Bangalore that killed twenty-three people, mostly Tamils. In 2002, a farmer in Karnataka jumped into a reservoir, killing himself, to protest water deliveries to Tamil Nadu (Circle of Blue 2012). Violence broke out again in 2012 when the state of Karnataka stopped the flow of water to Tamil Nadu, in violation of an order by the Supreme Court of India. In October 2012, thousands of farmers in Karnataka tried to prevent the release of water from two dams (Krishna Raja Sagar and Kabini Dams) on the Cauvery River. Injuries to protestors and police were reported (Circle of Blue 2012). The water releases were ordered by the Indian Supreme Court, which required that Karnataka meet deliveries to the downstream state of Tamil Nadu despite severe drought. The dispute continued later in the year when Karnataka again stopped releases downstream (*Indian Express* 2012).

Separately, scuffles and protests broke out around New Delhi during the summer of 2012 as residents surrounded water delivery trucks and fought over water (Reuters 2012a). That summer was the hottest in thirty-three years, leading to extensive energy and water shortages.

In an international dispute between Pakistan and India, Pakistani militants were reported to have attacked and sabotaged water systems, flood protection works, and dams in the Wullar Lake region of northern Kashmir. In August 2012, water engineers and workers were attacked and explosives were detonated at the unfinished Tulbul Navigation Lock/Wullar Dam, which Pakistan claims would violate the Indus Water Treaty with India by cutting water flows to Pakistan (Hassan 2012).

The Middle East and North Africa (MENA) Region

Serious examples of the use of water as a target and weapon of war were reported in Libya, during the civil war there, and in the long-standing dispute over borders, land, settlements, and water control in Israel and Palestine in 2011. Cross-border and subnational disputes over water were reported in northern Africa in 2012 involving Egypt, Ethiopia, and Sudan, and separately, within both Sudan/South Sudan and Egypt over water allocation and scarcity.

Libya

During the 2011 Libyan Civil War, forces loyal to dictator Muammar Gaddafi took control of a water operations center, cutting off the water supply to Tripoli. The system controls Libya's so-called Great Manmade River—a system of pumps, pipes, and canals that brings water from distant aquifers to Tripoli and other cities. Half the country was left without running water, prompting the United Nations and neighboring countries to mobilize tanker ships to deliver water to coastal cities (Circle of Blue 2011; UPI 2011).

Sudan, Egypt, and Ethiopia

Continuing violence in Sudan had displaced hundreds of thousands of refugees, leading to an increase in deaths from water shortages. Violence broke out in March 2012 in refugee camps, where large numbers of people faced serious water scarcity. Fighting was reported at the limited water points in the Jamam refugee camp (McNeish 2012). The international aid group Médicins Sans Frontières reported in June 2012 that as many as ten refugees were dying daily as water ran out in refugee camps in South Sudan (Ferrie 2012).

Within Egypt, farmers from the Abu Simbel region held over two hundred tourists hostage in June in a protest over the inadequate provision of irrigation water for local farms. The farmers captured the tourists after they had visited famous attractions in the region but released them after officials agreed to a temporary release of water (*Egypt Independent* 2012). Shortly after this, a series of public protests over shortages of drinking and irrigation water took place, extending across seven Egyptian governorates. Some of these protests were violent: in Beni Suef Governorate, one person was killed and many were injured during a conflict over irrigation water; in Minya, villagers clashed with officials over water shortages and water pollution; in Fayoum, hundreds of people protesting water shortages blocked a highway and set fires (Ooska News 2012).

In a transnational dispute involving Egypt, Ethiopia, and Sudan in September and October 2012, information was released about a possible secret agreement between Egypt and Sudan about an airfield that could be used to attack a major Ethiopian dam under construction. The dam, now known as the Grand Ethiopian Renaissance Dam (GERD), has also been called the Millennium Dam or the Hidase Dam. If built, GERD will be one of the largest dams in the world and the largest hydroelectric facility in Africa, located just upstream of the Ethiopian-Sudanese border on the Blue Nile. Egypt has expressed concerns about reductions in water flows reaching its citizens. The claim in 2012, strongly denied by Egypt, was that Egypt and Sudan had reached an agreement to permit Egypt to use Sudanese territories to launch attacks on GERD if diplomatic efforts failed to resolve water-sharing disputes between Egypt and Ethiopia. The allegations originated in an internal 2010 e-mail made available by WikiLeaks (*Sudan Tribune* 2012; Al Arabiya News 2012).

Yemen

In September 2011, political violence in Yemen's capital, Sana'a, led to secondary impacts on access to water. Damage to infrastructure contributed to "acute water and power shortages, forcing residents to rely on power generators and buy water extracted from wells and sold on a thriving black market" (AP 2011). The violence arose during the Yemeni uprising that followed the Tunisian Revolution in early 2011 and coincided with mass protests in Egypt and other parts of the Middle East. During the violence, governmental soldiers shelled neighborhoods and destroyed many traditional rooftop water tanks, worsening the problem of water access.

Israel and Palestine

Two separate incidents involving conflicts and violence over water were reported in 2011 in Israel and Palestine—a region with an especially long history of water disputes. In the first, Israel's military was reported to have destroyed nine water tanks in the Bedouin village of Amniyr in the South Hebron Hills in the West Bank, Palestine. Later, soldiers destroyed pumps and wells in the Jordan Valley villages of Al-Nasaryah, Al-Akrabanyah, and Beit Hassan (Abuwara 2011). In the second incident, Israeli settlers near Qasra, a West Bank village of 6,000, were reported to have destroyed crops—including olive trees—and a water well (Bsharat and Ramadan 2011).

Syria

In a period (2011–2012) in which there were a large number of water-related conflicts, one particularly disturbing event occurred with a complicated but direct connection to water: the unraveling of Syria and the escalation of massive civil war there (see In Brief 2). Syria's political dissolution, like almost all conflicts, is the result of complex and interrelated factors, including an especially repressive and unresponsive political regime, the erosion of the economic health of the country, and a wave of political reform sweeping over the entire Middle East and North Africa region. But in hindsight, factors related to drought, agricultural failure, water shortages, and water mismanagement appear to have played an important role in nurturing "seeds of social unrest" as well; see the separate In Brief on this topic in this volume.

Latin America

Violence over water was reported widely in Latin America in 2012. Incidents involved Brazil, Bolivia, Chile, Peru, and Guatemala. Three events were reported in Brazil in 2012: one associated with a controversy over control and management of indigenous lands, one during a dispute over dam construction, and violence associated with ongoing drought and water shortages. In the first instance, Brazil's federal police were investigating reports that water used by an indigenous tribe had been poisoned by nearby landowners who were attempting to gain control over disputed land. Since 2009, three men from this tribe had been reported killed in this dispute (AP 2012). In November 2012, work on the controversial $13 billion Belo Monte Dam was halted after protesters burned buildings at three dam sites. Several local groups and environmentalists fear that the dam across the Xingu River, a tributary of the Amazon, will damage local communities and their way of life, but the conflict also has economic roots (Phys.org 2012). Northeastern Brazil also saw growing conflicts after severe drought reduced water availability. News agencies report that on average, one person a day was being killed as a result of "water wars" that involved locals fighting over scarce supplies (Catholic Online 2012).

There is a growing dispute over water from the Silala/Siloli River, which originates in Bolivia in the highlands of the Andes and flows to Chile. Bolivia recently linked an ongoing argument about access to the Pacific coast with the flows of water in a shared river:

the Silala/Siloli (*The Economist* 2012a). In Guatemala, human rights activists protesting the impacts of mining on local water quality and water rights were attacked. In June 2012, a local activist was shot, mirroring another incident in 2010 when an activist had been shot while protesting the impacts of mining on water quality and local water rights (Amnesty International 2012).

In Peru, several incidents involving protests, injuries, and deaths were reported in regions with opposition to large-scale mining operations because of concerns over water quality and water rights. In July 2012, four people were reported to have been killed in clashes between protestors and police over the proposed multibillion-dollar Minas Conga gold mine (Yeager 2012). Protests focused on pollution, concern about water supplies, and economic issues have significantly slowed, and temporarily completely halted, all work at the mine (Reuters 2012b). A similar protest over the Xstrata Tintaya copper mine in May 2012 led to two deaths and fifty injuries (Reuters 2012c).

Central Asia

Central Asia, including the former Soviet republics, where almost all of the major rivers cross international borders, has been the scene of long-simmering tensions over water (Water Politics 2012). In this region, there are especially clear links between water and energy—a growing concern in other parts of the world as well. In 2012, a major dispute developed over water allocation and management among the Central Asian countries of Tajikistan, Kyrgyzstan, and Uzbekistan. Tensions among countries in the region have been growing for some time because of water scarcity, rivers that cross political borders, dam construction and management, and a lack of cooperative institutional agreements or institutions. In April 2012, Tajikistan protested that Uzbekistan had stopped deliveries of natural gas because of a disagreement over a hydropower dam. Uzbekistan cut gas deliveries needed by Tajikistan because of concerns over efforts to build a hydroelectric power station that Uzbekistan said would disrupt water supplies for its citizens. Gas flow was resumed after a new contract was signed (Kozhevnikov 2012).

This event was perceived to be part of a larger long-standing dispute over two major proposed hydropower dams in Kyrgyzstan and Tajikistan.[2] These dams would be the largest ever built in Central Asia and would potentially affect water deliveries and availability to the downstream nations of Uzbekistan, Turkmenistan, and Kazakhstan, all of which have expressed their concerns. By some estimates, the Rogun Dam would significantly reduce summer water availability and increase winter flows, depending on operations. There is concern that if the dams are built, economic and political tensions among the countries will escalate. Uzbekistan has the ability to cut off natural gas exports to both Kyrgyzstan and Tajikistan, and as noted above, it has actually done so. In September 2012, the leader of Uzbekistan, Islam Karimov, stated that the Rogun Dam could lead to "not just serious confrontation, but even wars" (*The Economist* 2012b). This regional dispute is further complicated by political relationships with Russia. Russia has strategic interests in the region and has supported Kyrgyzstan and Tajikistan in return for an agreement to host Russian military bases for several decades to come (Borisov 2012).

---

2. The Kambarata-1 hydropower plant, proposed by Kyrgyzstan for the Naryn River, a tributary of the Syr Darya; and the Rogun Dam, proposed by Tajikistan for the Vakhsh River, a tributary of the Amu Darya.

Other Regions in Africa

Several instances of violent clashes over access to water between pastoralists and farmers were reported in many parts of Africa in 2012. The worst cases occurred in Kenya, but additional conflicts were reported between communities in Mali and Burkina Faso, in Mali and Mauritania, in Kenya and Uganda, in Kenya and Somalia, and within Tanzania.

Somalia/Kenya

In September 2012, Kenyan military officials reported that Al-Shabaab insurgent fighters had poisoned a water well and damaged water infrastructure for another well near the port city of Kismayu in an effort to counter the participation of Kenyan troops as part of the African Union Mission in Somalia (AMISOM) (Wabala 2012).

Kenya

Extensive violence over water was reported in Kenya in 2012, with more than one hundred deaths in clashes between farmers and cattle herders over land and water (*New York Times* 2012). This conflict is part of a long-running dispute between the Pokomo people—mostly farmers near the Tana River—and the Orma, seminomadic cattle herders. In 2001, at least 130 people were killed in a string of clashes in the same district and between the same two communities over access to land and a river (Google News 2012a). There is some indication, however, that the current conflict is being exacerbated by Kenyan and foreign investment in vast tracts of land for food and biofuel cultivation, putting pressure on local resources (BBC 2012). Additional violence, including deaths, was reported in 2012 in disputes over access to water in the poorest slums around Nairobi (Njeru 2012).

Mali/Burkina Faso, Mali/Mauritania, Kenya/Uganda, Kenya/Somalia, and Tanzania

The past few years have seen a series of increasingly violent clashes over access to water and land between nomadic herders, pastoralists, and settled villagers. In May 2012, a clash along the border between Dogon tribesmen from Mali and nomadic Fulani herders from Burkina Faso killed at least thirty people after the revocation of an earlier agreement to share water and pastureland (Xinhua News 2012b).

The conflict in Mali is also having repercussions in neighboring Mauritania, with protests and violence over water shortages reaching the capital, Nouakchott. By July 2012, over 70,000 Malian refuges were seeking asylum in Mauritania, putting added pressure on scarce food and water supplies. This led to protests in the capital over "the lack of water" (Taha 2012).

In August 2012, fighting between two clans in the Lower Jubba region of South Somalia killed at least three people and wounded five. Reports from the village of Waraq (near the border with Kenya) indicated that the dispute originated with a disagreement over the ownership of new water wells (Shabelle Media Network 2012).

Cross-border tensions are also growing between Uganda and Kenya after Kenyan Pokot herdsman crossed the border in search of water and pasture. Borders are often ignored in the region, where traditional pastoralists move herds to follow rains and grazing opportunities. In October 2012, the Ugandan government deployed more than 5,000 soldiers to the region in order to contain armed conflict among pastoralists from the two countries (Bii 2012).

Violence between farmers and pastoralists has also expanded in Tanzania's southeastern Rufiji River valley, a region hit by drought. In May 2012, a farmer was killed in a conflict with a herdsman in a dispute over access to water in the southern regions of Lindi and Mtwara. Five more people died and many more were injured in subsequent violence. According to local sources, the violence has been worsened in the past five years by prolonged drought (Makoye 2012).

South Africa

Protesters in Cape Town, South Africa, rioted over local failure to deliver adequate water and power to poor communities. Hundreds of protesters burned tires, destroyed cars, and threw rocks at police in August 2012 after continued disappointment and anger over the lack of basic services (Xinhua News 2012a).

Asia-Pacific Region

Amnesty International reported in 2011 that women in the slums of the Solomon Islands are "continually harassed, attacked, and raped" when they are forced to walk long distances in unsafe areas in order to find clean water. A survey showed that 92 percent of households do not have a tap in their home and that local water sources are often polluted (Amnesty International 2011). Violence over access to a water source was reported in Maluku, Indonesia. Clashes in early 2012 led to several deaths and injuries (*Jakarta Globe* 2012).

Conclusions

Water-related violence has a long history and continues to be a global and regional problem. The past several years have seen an increase in the total number of reports of violent conflict over water. Unlike the situation in the early and middle parts of the twentieth century, an increasing proportion of reported cases have to do with subnational disputes, terrorism, and local violence rather than transnational incidents. Many are small in scale, involving local violence over water allocations and use or violence over local development decisions that affect environmental and economic conditions at the community scale. But more and more of the reported cases have their roots in water scarcity and competition for a fixed resource that is reaching peak limits (Gleick and Palaniappan 2010).

Much has been written in recent years about the nature of water-related conflicts and the need for improved international cooperation. Far less attention has been given to what appears to be the more significant risk: subnational violence and competition. Some of this violence has been, and will remain, local. But some may spill over into the international arena: the ongoing civil war in Syria (at the time of this writing) is a critical example in which water shortages, mismanagement, drought, and subsequent economic and population dislocations appear to have contributed directly to the political unraveling of that country.

The data suggest that the challenges of water conflicts are growing, not shrinking, especially at the subnational scale. Far better mechanisms and far greater efforts are needed to address these kinds of conflicts.

References

Abuwara, A. 2011. Can water end the Arab-Israeli conflict? *Al Jazeera English*, July 29. http://english.aljazeera.net/indepth/features/2011/07/20117278519784574.html (accessed March 2, 2013).
Al Arabiya News. 2012. Egypt, Sudan could seek military action over Nile: WikiLeaks. October 12. http://english.alarabiya.net/articles/2012/10/12/243396.html (accessed February 5, 2013).
Amnesty International. 2011. "Where Is the Dignity in That?" Women in Solomon Islands Slums Denied Sanitation and Safety. September. http://www.amnesty.org.nz/files/SolomonIslandsWEB.pdf.
———. 2012. Guatemala: Lives and Livelihoods at Stake in Mining Conflict. June 21. http://www.amnesty.org/en/news/guatemala-lives-and-livelihoods-stake-mining-conflict-2012-06-21 (accessed January 31, 2013).
Associated Press (AP). 2011. 9 killed in renewed violence in Yemeni capital. September 22. http://www.foxnews.com/world/2011/09/22/killed-in-renewed-violence-in-yemeni-capital/ (accessed March 2, 2013).
———. 2012. Brazil police investigate indigenous claim that creek in sacred land was poisoned. November 21. http://www.foxnews.com/world/2012/11/21/brazil-police-investigate-indigenous-claim-that-creek-in-sacred-land-was/#ixzz2Jae2OGLn (accessed January 31, 2013).
BBC. 2012. Mass graves found in Kenya's Tana River Delta region. BBC News Africa, September 18. http://www.bbc.co.uk/news/world-africa-19634076 (accessed February 6, 2013).
Bii, B. 2012. Tension at Kenya-Uganda border over rustling. *Daily Nation*, October 3. http://www.nation.co.ke/News/-/1056/1523908/-/xtc5qbz/-/index.html (accessed February 6, 2013).
Borisov, A. 2012. Water tensions overflow in ex-Soviet Central Asia. Phys.org, November 21. http://phys.org/news/2012-11-tensions-ex-soviet-central-asia.html#jCp (accessed February 5, 2013).
Bsharat, M., and S. A. Ramadan. 2011. Fears of violence mount in West Bank on eve of Palestinian UN bid. Xinhua News Agency, September 21. http://news.xinhuanet.com/english2010/world/2011-09/21/c_131150218.htm (accessed March 2, 2013).
Catholic Online. 2012. Severe drought in Brazil destroying herds and lives. October 12. http://www.catholic.org/international/international_story.php?id=48869 (accessed January 31, 2013).
Center for Strategic and International Studies (CSIS) and Sandia National Laboratories. 2005. Addressing Our Global Water Future. September 30. Washington, DC: Center for Strategic and International Studies; Albuquerque: Sandia National Laboratories. http://csis.org/files/media/csis/pubs/csis-snl_ogwf_sept_28_2005.pdf (accessed August 15, 2013).
Circle of Blue. 2011. Water as a tool of war: Qaddafi loyalists turn off tap for half of Libya. *Circle of Blue Water News*, August 30. http://www.circleofblue.org/waternews/2011/world/water-as-a-tool-of-war-qaddafi-loyalists-turn-off-tap-for-half-of-libya/ (accessed March 2, 2013).
———. 2012. Protests break out after India's Supreme Court rules in favor of downstream state in Cauvery River dispute. *Circle of Blue Water News*, October 3. http://www.circleofblue.org/waternews/2012/world/protests-break-out-after-indias-supreme-court-rules-in-favor-of-downstream-state-in-cauvery-river-dispute/ (accessed February 5, 2013).
The Economist. 2012a. Bolivia and Chile: Trickle-down diplomacy; Evo Morales tries to swap a stream for a piece of Chilean seafront. November 17. http://www.economist.com/news/americas/21566673-evo-morales-tries-swap-stream-piece-chilean-seafront-trickle-down-diplomacy (accessed January 31, 2013).
———. 2012b. Water wars in Central Asia: Dammed if they do; Spats over control of water roil an already unstable region. September 29. http://www.economist.com/node/21563764 (accessed February 5, 2013).
Egypt Independent. 2012. Abu Simbel farmers release tourists held over water shortage. June 3. http://www.egyptindependent.com/news/farmers-hold-tourists-abu-simbel-over-water-shortage-news2 (accessed February 5, 2013).

Ferrie, J. 2012. South Sudan refugee water shortage killing up to 10 people daily. Bloomberg News, June 13. http://www.businessweek.com/news/2012-06-13/south-sudan-refugee-water-shortage-killing-up-to-10-people-daily (accessed February 6, 2013).

Gleick, P. H. 1998. Conflict and cooperation over fresh water. In *The World's Water 1998–1999: The Biennial Report on Freshwater Resources*, edited by P. H. Gleick, 105. Washington, DC: Island Press.

———. 2006. Water and terrorism. *Water Policy* 8 (6): 481–503.

Gleick, P. H., and M. Palaniappan. 2010. Peak water limits to freshwater withdrawal and use. *Proceedings of the National Academy of Sciences* 107 (25): 11155–11162. http://www.pnas.org/cgi/doi/10.1073/pnas.1004812107.

Google News. 2012a. Kenya to hold peace meeting after 52 killed. AFP, August 23. http://www.google.com/hostednews/afp/article/ALeqM5jZdyUiAcUA7p94cH8PjEtslQUK3A?docId=CNG.4fe0cb53fa3afbb412c2a959ae285e36.211 (accessed February 6, 2013).

Hamid, M. 2012. Afghan schoolgirls poisoned in anti-education attack. Reuters, April 17. http://www.reuters.com/article/2012/04/17/us-afghanistan-women-idUSBRE83G0PZ20120417 (accessed January 31, 2013).

Hassan, I. 2012. Militants sabotage works on J&K lake. *dna*, August 30. http://www.dnaindia.com/india/report_militants-sabotage-works-on-j-and-k-lake_1734411 (accessed February 6, 2013).

Indian Express. 2012. Stopped releasing Cauvery water to TN, says Karnataka CM. December 10. http://www.indianexpress.com/news/stopped-releasing-cauvery-water-to-tn-says-karnataka-cm/1043124/ (accessed February 5, 2013).

Jakarta Globe. 2012. Bloody clashes in Indonesia's Maluku claim another life. March 8. http://www.thejakartaglobe.com/news/bloody-clashes-in-indonesias-maluku-claim-another-life/503346 (accessed February 6, 2013).

Kozhevnikov, R. 2012. Uzbekistan resumes gas supplies to Tajikistan. Reuters, April 16. http://www.reuters.com/article/2012/04/16/tajikistan-gas-idUSL6E8FG3YL20120416 (accessed February 5, 2013).

Kramer, A., A. T. Wolf, A. Carius, and G. D. Dabelko. 2013. The key to managing conflict and cooperation over water. *A World of Science* 11 (1) (January–March): 4–12. http://unesdoc.unesco.org/images/0021/002191/219156E.pdf.

Leurig, S. 2012. Water Ripples: Expanding Risks for U.S. Water Providers. Boston, MA: Ceres. http://www.ceres.org/resources/reports/water-ripples-expanding-risks-for-u.s.-water-providers/view.

Makoye, K. 2012. Drought drives Tanzanian herders into conflict with farmers. Reuters, June 12. http://www.trust.org/alertnet/news/drought-drives-tanzanian-herders-into-conflict-with-farmers/ (accessed February 6, 2013).

Mashal, M. 2012. What Iran and Pakistan want from the Afghans: Water. *Time*, December 2. http://world.time.com/2012/12/02/what-iran-and-pakistan-want-from-the-afghans-water/ (accessed January 31, 2013).

McNeish, H. 2012. Refugees in South Sudan face water crisis. AFP, March 31. http://www.google.com/hostednews/afp/article/ALeqM5ho-X1B6kIDKnLCHEUnlrW9665trw?docId%3DCNG.91f54df3baaf54aa7554fca3c3d379df.41 (accessed February 5, 2013).

New York Times. 2012. Kenya: Clashes over land and water leave 12 dead. September 7. http://www.nytimes.com/2012/09/08/world/africa/kenya-clashes-over-land-and-water-leave-12-dead.html?partner=rssnyt&emc=rss&_moc.semityn&_r=1& (accessed February 6, 2013).

Njeru, G. 2012. Water shortages driving growing thefts, conflicts in Kenya. Reuters, August 6. http://www.trust.org/alertnet/news/water-shortages-driving-growing-thefts-conflicts-in-kenya/ (accessed February 6, 2013).

Ooska News. 2012. Egypt water shortage protests are widespread and violent. July 25. http://www.ooskanews.com/daily-water-briefing/egypt-water-shortage-protests-are-widespread-and-violent_23549 (accessed February 5, 2013).

Phys.org. 2012. Trouble at Brazil mega-dam stops construction for now. November 12. http://phys.org/news/2012-11-brazil-mega-dam.html#jCp (accessed January 31, 2013).

Reuters. 2012a. Power cuts spark protests as India swelters. July 4. http://www.reuters.com/article/2012/07/04/uk-india-electricity-protest-idUSLNE86300Q20120704. (accessed February 5, 2013.)

———. 2012b. Update 3—Peru clash over Newmont mine kills three. July 3. http://www.reuters.com/article/2012/07/04/peru-newmont-idUSL2E8I3GB820120704 (accessed February 5, 2013).

———. 2012c. Update 3—Peru uses emergency rules to try to end anti-mining protest. May 28. http://www.reuters.com/article/2012/05/29/xstrata-peru-idUSL1E8GS34F20120529 (accessed February 5, 2013).

Shabelle Media Network. 2012. Somalia: Ethnic fighting renews in lower Jubba region, South Somalia. August 15. http://allafrica.com/stories/201208160039.html (accessed February 6, 2013).

Shah, A. 2012. Officials say 7 Afghan children killed by bomb. Associated Press, July 25. http://news.yahoo.com/officials-7-afghan-children-killed-bomb-110922657.html (accessed January 31, 2013).

Sudan Tribune. 2012. Egypt: Govt denies deal with Sudan to strike Ethiopian dam. September 24. http://allafrica.com/stories/201209250218.html (accessed February 5, 2013).

Taha, R. M. 2012. Mali conflict spilling over: Influx of refugees from Mali exacerbates political instability in neighbouring Mauritania. *Daily News of Egypt,* July 1.http://www.dailynewsegypt.com/2012/07/01/mali-conflict-spilling-over/ (accessed February 6, 2013).

United Press International (UPI). 2011. Gadhafi turns water project into a weapon. UPI.com, September 2. http://www.upi.com/Business_News/Energy-Resources/2011/09/02/Gadhafi-turns-water-project-into-a-weapon/UPI-66451314984213/?spt=hs&or=er (accessed March 2, 2013).

United States Department of Defense, Defense Intelligence Agency (US DIA). 2012. Global Water Security. Intelligence Community Assessment ICA 2012-08. February 2. Washington, DC: Office of the Director of National Intelligence. http://www.dni.gov/files/documents/Newsroom/Press%20Releases/ICA_Global%20Water%20Security.pdf (accessed January 31, 2013).

Wabala, D. 2012. Kenya: KDF airlifts water to troops as wells poisoned. *AllAfrica, The Star,* September 24. http://allafrica.com/stories/201209250173.html (accessed February 6, 2013).

Water Politics. 2012. Central Asia's looming conflict over water: The upriver countries. November 12. http://www.waterpolitics.com/2012/11/12/central-asias-looming-conflict-over-water-the-upriver-countries/ (accessed February 5, 2013).

Xinhua News. 2012a. Cape Town hit by continued protest against poor service delivery. August 14. http://news.xinhuanet.com/english/world/2012-08-14/c_131782882.htm (accessed February 6, 2013).

———. 2012b. Tribal clash kills 30 along Mali–Burkina Faso border. May 25. http://news.xinhuanet.com/english/world/2012-05-25/c_131611541.htm (accessed February 6, 2013).

Yeager, C. 2012. When gold loses its luster: Four dead in Peru mining protests. *Circle of Blue* Water News, July 6. http://www.circleofblue.org/waternews/2012/commentary/editorial-in-the-circle-fresh-focus/when-gold-loses-its-luster-four-dead-in-peru-mining-protests/ (accessed February 5, 2013).

Zakaria, F. 2013. The coming water wars. CNN World, Global Public Square, February 25. http://globalpublicsquare.blogs.cnn.com/2013/02/25/the-coming-water-wars/?iref=allsearch (accessed March 2, 2013).

WATER BRIEF 4

Water Conflict Chronology

Peter H. Gleick and Matthew Heberger

The full Pacific Institute Water Conflict Chronology, updated for 2010–2012, appears below. A summary of recent events is provided in Water Brief 3 in this volume. As noted in volume 7 of *The World's Water*, a detailed, interactive online database now provides integrated Google Maps and filters to permit readers and researchers to view entries filtered by time period, location, and subject. It also includes a separate searchable bibliography.

The Water Conflict Chronology has appeared in every volume of *The World's Water* since 1998. It continues to be one of the most popular features of this report, and we continue to seek out and include additions sent by readers and researchers around the world. The chronology is increasingly being used by media and academics, and in 2013 it was cited by Fareed Zakaria (2013) of CNN.

The current categories or types of conflicts include the following:

Military tool (state actors): water resources, or water systems themselves, are used by a nation or state as a weapon during a military action.

Military target (state actors): water resources or water systems are targets of military actions by nations or states.

Terrorism or domestic violence, including cyberterrorism (nonstate actors): water resources or water systems are the targets or tools of violence or coercion by nonstate actors. A distinction is drawn between environmental terrorism and ecoterrorism (see Gleick 2006).

Development disputes (state and nonstate actors): water resources or water systems are a major source of contention and dispute in the context of economic and social development.

Please email any contributions with full citations and supporting information to pgleick@pipeline.com.

Water Conflict Chronology
May 2013 Update

Date	Parties Involved	Basis of Conflict	Violent or in Violent Context?	Description	Sources
3000 BC	Ea, Noah	Religious account	No: Threat	Ancient Sumerian legend recounts the deeds of the deity Ea, who punished humanity for its sins by inflicting the earth with a six-day storm. The Sumerian myth parallels the biblical account of Noah and the deluge, although some details differ.	Hatami and Gleick (1994)
2500 BC	Lagash, Umma	Military tool	Yes	The dispute over the "Gu'edena" (edge of paradise) region begins. Urlama, King of Lagash from 2450 to 2400 BC, diverts water from this region to boundary canals, drying up boundary ditches to deprive Umma of water. His son Il cuts off the water supply to Girsu, a city in Umma.	Hatami and Gleick (1994)
1790 BC	Hammurabi	Development dispute	No	The Code of Hammurabi for the State of Sumer lists several laws pertaining to irrigation systems and water theft.	Hatami and Gleick (1994)
1720–1684 BC	Abi-Eshuh, Iluma-Ilum	Military tool	Yes	A grandson of Hammurabi, Abish or Abi-Eshuh, dams the Tigris to prevent the retreat of rebels led by Iluma-Ilum, who declared the independence of Babylon. This failed attempt marks the decline of the Sumerians, who had reached their apex under Hammurabi.	Hatami and Gleick (1994)
circa 1300 BC	Sisera, Barak, God	Religious account; Military tool	Yes	The Old Testament gives an account of the defeat of Sisera and his "nine hundred chariots of iron" by the unmounted army of Barak on the fabled Plains of Esdraelon. God sends heavy rainfall in the mountains, and the Kishon River overflows the plain and immobilizes or destroys Sisera's technologically superior forces ("... the earth trembled, and the heavens dropped, and the clouds also dropped water," Judges 5:4; "... The river of Kishon swept them away, that ancient river, the river Kishon," Judges 5:21).	Scofield (1967)
1200 BC	Moses, Egypt	Military tool; Religious account	Yes	When Moses and the retreating Jews find themselves trapped between the Pharaoh's army and the Red Sea, Moses miraculously parts the waters of the Red Sea, allowing his followers to escape. The waters close behind them and cut off the Egyptians.	Hatami and Gleick (1994)
720–705 BC	Assyria, Armenia	Military tool	Yes	After a successful campaign against the Halidians of Armenia, Sargon II of Assyria destroys their intricate irrigation network and floods their land.	Hatami and Gleick (1994)

Date	Parties	Basis	Violent Conflict	Description	Sources
705–682 BC	Sennacherib, Babylon	Military tool; Military target	Yes	In quelling rebellious Assyrians in 695 BC, Sennacherib razes Babylon and diverts one of the principal irrigation canals so that its waters wash over the ruins.	Hatami and Gleick (1994)
701 BC	Israel (Judah), Assyria	Military tool; Military maneuvers	Yes	When King Hezekiah of Judah sees that Sennacherib of Assyria is coming in war, he has springs and a brook outside Jerusalem stopped to keep water from the Assyrians. ("So there was gathered much people together, who stopped all the fountains, and the brook that ran through the midst of the land, saying, Why should the kings of Assyria come, and find much water?", 2 Chronicles 32:1–4).	Scofield (1967)
681–699 BC	Assyria, Tyre	Military tool; Religious account	Yes	Esarhaddon, an Assyrian, refers to an earlier period when gods, angered by insolent mortals, created destructive floods. According to inscriptions recorded during his reign, Esarhaddon besieges Tyre, cutting off food and water.	Hatami and Gleick (1994)
669–626 BC	Assyria, Arabia, Elam	Military tool; Military target	Yes	Assurbanipal's inscriptions also refer to a siege against Tyre, although scholars attribute it to Esarhaddon. In campaigns against both Arabia and Elam in 645 BC, Assurbanipal, son of Esarhaddon, dries up wells to deprive Elamite troops. He also guards wells from Arabian fugitives in an earlier Arabian war. On his return from victorious battle against Elam, Assurbanipal floods the city of Sapibel, an ally of Elam. According to inscriptions, he dams the Ulai River with the bodies of dead Elamite soldiers and deprives dead Elamite kings of their food and water offerings.	Hatami and Gleick (1994)
612 BC	Egypt, Persia, Babylon, Assyria	Military tool	Yes	A coalition of Egyptian, Median (Persian), and Babylonian forces attacks and destroys Nineveh, the capital of Assyria. Nebuchadnezzar's father, Nebopolassar, leads the Babylonians. The converging armies divert the Khosr River to create a flood, which allows them to elevate their siege engines on rafts.	Hatami and Gleick (1994)
605–562 BC	Babylon	Military tool	No	Nebuchadnezzar builds immense walls around Babylon, using the Euphrates and canals as defensive moats surrounding the inner castle.	Hatami and Gleick (1994); Drower (1954)
6th century BC	Assyria	Military target; Military tool	Yes	Assyrians poison the wells of their enemies with rye ergot.	Eitzen and Takafuji (1997)
590–600 BC	Cirrha, Delphi	Military tool	Yes	Athenian legislator Solon reportedly has roots of *Helleborus* thrown into a small river or aqueduct leading from the Pleistrus River to Cirrha during a siege of this city. The enemy forces become violently ill and are defeated as a result. Some accounts have Solon building a dam across the Plesitus River, cutting off the city's water supply. Such practices were widespread.	Absolute Astronomy (2006)

continues

Water Conflict Chronology *continued*

Date	Parties Involved	Basis of Conflict	Violent Conflict or in the Context of Violence?	Description	Sources
558–528 BC	Babylon	Military tool	Yes	On his way from Sardis to defeat Nabonidus at Babylon, Cyrus faces a powerful tributary of the Tigris, probably the Diyalah. According to Herodotus's account, the river drowns his royal white horse and presents a formidable obstacle to his march. Cyrus, angered by the "insolence" of the river, halts his army and orders them to cut 360 canals to divert the river's flow. Other historians argue that Cyrus needed the water to maintain his troops on their southward journey, while another asserts that the construction was an attempt to win the confidence of the locals.	Hatami and Gleick (1994)
539 BC	Babylon	Military tool	Yes	According to Herodotus, Cyrus invades Babylon by diverting the Euphrates above the city and marching troops along the dry riverbed. This popular account describes a midnight attack that coincided with a Babylonian feast.	Hatami and Gleick (1994)
430 BC	Athens	Military tool	Yes	During the second year of the Peloponnesian War, a plague breaks out in Athens. The Spartans are accused of poisoning the cisterns of the Piraeus, the source of most of Athens's water.	Strategy Page (2006)
355–323 BC	Babylon	Military tool	Yes	Returning from the razing of Persepolis, Alexander proceeds to India. After the Indian campaigns, he heads back to Babylon via the Persian Gulf and the Tigris, where he tears down defensive weirs that the Persians had constructed along the river. Arrian describes Alexander's disdain for the Persians' attempt to block navigation, which he saw as "unbecoming to men who are victorious in battle."	Hatami and Gleick (1994)
210–209 BC	Rome and Carthage	Military tool	Yes	In 210 BC, Scipio crosses the Ebro to attack New Carthage. During a short siege, Scipio leads a breaching column through a supposedly impregnable lagoon located on the landward side of the city; a strong northerly wind combined with the natural ebb of the tide leaves the lagoon shallow enough for the Roman infantry to wade through. New Carthage is quickly taken.	Fonner (1996); Gowan (2004)
52 BC	Rome, Gaul	Military tool	Yes	Caesar constructs water-filled ditches as a blockade during the Siege of Alesia in Gaul, site of modern-day Alise-Sainte-Reine in Côte d'Or, near Dijon, France.	*Wikipedia* (2011a)

Date	Parties Involved	Basis	Violence	Description	Sources
51 BC	Rome, Gaul	Military target	Yes	Caesar attacks water supplies during the siege of Uxellodunum by undermining one of the local springs and placing attackers near the other. Shortage of water leads to the surrender of the Gauls.	History of War Online (2011)
49 BC	France, Rome	Military tool	Yes	During the first year of the Great Roman Civil War, Julius Caesar's troops lay siege to the walled city of Massilia (modern-day Marseille) using siege towers and battering rams and by digging "mines" or tunnels to undermine the city walls. Massillians defend their city with "dogged determination"; tactics include directing water through pipes to wash down on attackers, which the Romans counter by covering siege buildings with bricks and "several coatings of stucco." Defenders also excavate a large basin inside the walls, filling it with water. As the Roman miners reach the walls, the tunnels are flooded with water and collapse. Massilia ultimately surrenders after a five-month siege.	Illustrated History of the Roman Empire (2011); Caesar (1906); Wikipedia (2013a)
AD 30	Roman Empire (Pontius Pilate), Jews	Development dispute	Yes	Roman Procurator Pontius Pilate uses sacred money to divert a stream to Jerusalem. The Jews are angered by the diversion, and tens of thousands gather to protest. Pilate's soldiers mingle among the crowd and, with daggers hidden in their garments, attack the protesters. "A great number" are slain and wounded and the "sedition" is ended.	Josephus (AD 90)
537	Goths and Rome	Military tool; Military target	Yes	In the sixth century AD, as the Roman Empire begins to decline, the Goths besiege Rome and cut almost all of the aqueducts leading into the city. In AD 537, this siege succeeds. The only aqueduct that continues to function is the Aqua Virgo, which runs almost entirely underground.	Rome Guide (2004); InfoRoma (2004)
1187	Saladin and the Crusaders	Military tool	Yes	Saladin is able to defeat the Crusaders at the Horns of Hattin in 1187 by denying them access to water. In some reports, Saladin fills all the wells along the way with sand and destroys the villages of the Maronite Christians, who would supply the Christian army with water.	Lockwood (2006); Priscoli (1998)
1503	Florence and Pisa	Military tool	No: Plan only	Leonardo da Vinci and Machiavelli plan to divert the Arno River away from Pisa during a conflict between Pisa and Florence.	Honan (1996)
1573–1574	Holland and Spain	Military tool	Yes	In 1573, at the beginning of the Eighty Years' War against Spain, the Dutch flood the land to break the siege of Spanish troops on the town Alkmaar. The same defense is used to protect Leiden in 1574. This strategy becomes known as the Dutch Water Line and is used frequently for defense in later years.	Dutch Water Line (2002)

continues

Water Conflict Chronology continued

Date	Parties Involved	Basis of Conflict	Violent Conflict or in the Context of Violence?	Description	Sources
1626–1629	Spain, Dutch Republic	Development dispute; Military tool	No	The Spanish Habsburgs attempt to prevent ship traffic on the River Rhine from reaching the Dutch Republic in order to damage the Dutch economy. Plans are also made to divert water from the Rhine to lands under Spanish control, to dry up downstream cities in Holland. The first stage is a canal between the Rhine and Meuse Rivers, between the cities of Rheinberg and Venlo. Plans for a later stage call for a connection between the Meuse and the Scheldt to circumvent the Scheldt Estuary, controlled by the Dutch. Although some 60 kilometers of canal are constructed, the plan fails because of changed military conditions and lack of funding. Parts of the canal are still visible in present-day Germany.	Israel (1997); Bachiene (1791)
1642	China; Ming Dynasty	Military tool	Yes	The Huang He's dikes are breached for military purposes. In 1642, "toward the end of the Ming dynasty (1368–1644), General Gao Mingheng used the tactic near Kaifeng in an attempt to suppress a peasant uprising."	Hillel (1991)
1672	French, Dutch	Military tool	Yes	Louis XIV starts the third of the Dutch Wars in 1672, in which the French overrun the Netherlands. In defense, the Dutch open their dikes and flood the country, creating a watery barrier that is virtually impenetrable.	*Columbia Encyclopedia* (2000a)
1748	United States	Development dispute; Terrorism	Yes	Ferry house on Brooklyn shore of East River burns down. New Yorkers accuse Brooklynites of having set the fire as revenge for unfair East River water rights.	MCNY (n.d.)
1777	United States	Military tool	Yes	British and Hessians attack the water system of New York. "The enemy wantonly destroyed the New York water works" during the War for Independence.	Thatcher (1827)
1804	France, Holland	Development dispute; Military tool	No	Napoleon orders the construction of a canal between Neuss and Venlo to connect the Rhine and Meuse Rivers, in order to divert trade from the Batavia Republic to the Southern Netherlands, then under French control. Three-quarters of the canal is completed, but work stops because of lack of funds.	Israel (1997)
1841	Canada	Development dispute; Terrorism	Yes	A reservoir in Ops Township, Upper Canada (now Ontario), is destroyed by neighbors who consider it a hazard to health.	Forkey (1998)

1844	United States	Development dispute; Terrorism	Yes	A reservoir in Mercer County, Ohio, is destroyed by a mob that considers it a health hazard.	Scheiber (1969)
1850s	United States	Development dispute; Terrorism	Yes	A New Hampshire dam that had impounded water for factories downstream is attacked by local residents unhappy about its effect on water levels.	Steinberg (1990)
1853–1861	United States	Development dispute; Terrorism	Yes	The banks and reservoirs of the Wabash and Erie Canal in southern Indiana are repeatedly destroyed by mobs regarding it as a health hazard.	Fatout (1972); Fickle (1983)
1860–1865	United States	Military tool; Military target	Yes	General William T. Sherman's memoirs contain an account of Confederate soldiers poisoning ponds by dumping the carcasses of dead animals into them. Other accounts suggest this tactic was used by both sides.	Eitzen and Takafuji (1997)
1862	United States Union and Confederate armies	Military tool	Yes	During the US Civil War, Confederate forces near Yorktown use dams to flood the Warwick River and cut off Union troops. "The enemy is pushed behind a . . . branch of the Warwick river in which they control the depths of water by dams. McClellan did not intend to pass that stream at that time, or at that point where the skirmish took place. But the troops, finding the stream fordable went over (under whose immediate orders does not appear) and the water was then deepened so that they were measurably cut off."	Hitchcock (1862)
1863	United States	Military tool	Yes	General Ulysses S. Grant, during the Civil War campaign against Vicksburg, cuts levees in the battle against the Confederates.	Grant (1885); Barry (1997)
1870s	China	Development dispute	No	An unauthorized dam in Hubei, China, is constructed by locals and removed by government (twice).	Rowe (1988)
1870s–1881	United States	Development dispute	Yes	Recurrent friction and eventual violent conflict over water rights in the vicinity of Tularosa, New Mexico, involves villagers, ranchers, and farmers.	Rasch (1968)
1887	United States	Development dispute; Terrorism	Yes	A canal reservoir in Paulding County, Ohio, is dynamited by a mob regarding it as a health hazard. State militia are called out to restore order.	Walters (1948)
1890	Canada	Development dispute; Terrorism	Yes	A partly successful attempt is made to destroy a lock on the Welland Canal in Ontario, Canada, either by Fenians protesting English policy in Ireland or by agents of Buffalo, New York, grain handlers unhappy about the diversion of trade through the canal.	Styran and Taylor (2001)

continues

Water Conflict Chronology *continued*

Date	Parties Involved	Basis of Conflict	Violent Conflict or in the Context of Violence?	Description	Sources
1898	Egypt, France, Britain	Military tool; Political tool	Military maneuvers	Military conflict nearly ensues between Britain and France when a French expedition attempts to gain control of the headwaters of the White Nile. While the parties ultimately negotiate a settlement of the dispute, the incident has been characterized as having "dramatized Egypt's vulnerable dependence on the Nile, and fixed the attitude of Egyptian policy-makers ever since."	Moorehead (1960)
1907–1913	Owens Valley, Los Angeles, California	Terrorism; Development dispute	Yes	The Los Angeles Valley aqueduct/pipeline suffers repeated bombings in an effort to prevent diversions of water from the Owens Valley to Los Angeles.	Reisner 1993
1908–1909	United States	Development dispute	Yes	Violence, including a murder, is directed against agents of a land company that had claimed title to Reelfoot Lake in northwestern Tennessee; the agents had attempted to levy charges for fish taken and threatened to drain the lake for agriculture.	Vanderwood (1969)
1915	German Southwest Africa	Military tool	Yes	Union of South African troops capture Windhoek, capital of German Southwest Africa, in May. Retreating German troops poison wells—"a violation of the Hague convention."	Daniel (1995)
1935	California, Arizona	Development dispute	Military maneuvers	Arizona calls out the National Guard and militia units to its border with California to protest the construction of Parker Dam and diversions from the Colorado River; the dispute ultimately is settled in court. The militia mount machine guns on commandeered ferries and patrol the river, earning the nickname the "Arizona Navy."	Reisner (1986/1993)
1937	Republican government of Spain, Spanish Nationalists	Military target	Yes	During the Spanish Civil War, two concrete gravity dams, at Burguillo and Ordunte, are attacked by the Nationalist army, with a two-and-one-half-ton charge placed in an inspection gallery at Ordunte. There is some limited damage, which is repaired in 1938–1939.	Pagan (2005)

Date	Parties	Basis	Yes/No	Description	Sources
1938	China and Japan	Military tool; Military target	Yes	Chiang Kai-shek orders the destruction of flood control dikes of the Huayuankou section of the Huang He (Yellow) River, in order to flood areas threatened by the Japanese army. West of Kaifeng, dikes are destroyed with dynamite, spilling water across the flat plain. Even though the flood destroys part of the invading army and mires its equipment in mud, Wuhan, headquarters of the Nationalist government, is taken by the Japanese in October. Floodwaters cover an area variously estimated as between 3,000 and 50,000 square kilometers, killing Chinese estimated in numbers between "tens of thousands" and "one million."	Hillel (1991); Yang Lang (1989/1994)
1939–1940	Netherlands, Germany	Military tool	Yes	During the mobilization of the Dutch at the beginning of World War II, 1939–1940, the Dutch attempt to flood the Gelderse Vallei with the New Dutch Water Defence Line, which had been completed in 1885. During the German invasion in May 1940, large areas are inundated.	IDG (1996)
1939–1942	Japan, China	Military target; Military tool	Yes	Japanese chemical and biological weapons activities reportedly include tests by "Unit 731" against military and civilian targets in which water wells and reservoirs are laced with typhoid and other pathogens.	Harris (1994)
1940	Finland, Soviet Union	Military tool	Yes	Partisan Finns manipulate the waters of the Saimaan Canal (Finland) in order to flood surrounding land and hinder Soviet troop movements during the Soviet-Finnish conflict.	Malik (2005)
1940–1945	Multiple parties	Military target	Yes	Hydroelectric dams are routinely bombed as strategic targets during World War II.	Gleick (1993)
1941	USSR and Germany	Military target	Yes	The strategically important Dnieper hydropower plant in the Ukraine is targeted by both Soviet and German troops during World War II. On August 18, 1941, the dam and power plant are dynamited by Soviet troops retreating before advancing German forces. The facility is bombed again in 1943 by retreating German troops.	Pagan (2005); *New York Times* (1941); Makarov (2005); AHF (2004)
1941	Germany, Soviet Union	Military tool	Yes	In November, Soviet troops flood the area south of the Istra Reservoir near Moscow in an effort to slow the German advance. Just a few weeks later, German troops used the same tactic to create a water barrier to halt advances by the Soviet 16th Army.	Malik (2005)
1941–1943	Germany, USSR	Military target	Yes	World War II inflicts enormous harm to hydroelectric systems in the Soviet Union. Over two-thirds of the hydroelectric power stations are lost.	Malik (2005)

continues

Water Conflict Chronology *continued*

Date	Parties Involved	Basis of Conflict	Violent Conflict or in the Context of Violence?	Description	Sources
1943	Britain, Germany	Military target	Yes	The British Royal Air Force bombs dams on the Möhne, Sorpe, and Eder Rivers in Germany (May 16 and 17). A Möhne Dam breach kills 1,200 and destroys all downstream dams for fifty kilometers. The flood that occurs after the breaching of the Eder Dam reaches a peak discharge of 8,500 cubic meters per second, nine times higher than the highest flood observed. Many houses and bridges are destroyed, and sixty-eight are killed.	Kirschner (1949); Semann (1950)
1944	Germany, Italy, Britain, United States	Military tool	Yes	German forces use waters from the Isoletta Dam (Liri River) in January and February to successfully destroy British assault forces crossing the Garigliano River (downstream of the Liri River). The German army then dams the Rapido River, flooding a valley occupied by the American army.	US ACE (1953)
1944	Germany, Italy, Britain, United States	Military tool	Yes	The German army floods the Pontine Marshes, on the coast of Italy southeast of Rome, by stopping pumps and opening dikes, in order to disrupt Allied forces who had established a beachhead at Anzio. Allied forces are surrounded by German artillery and pinned down for months by heavy shelling. The purpose of the flooding is to bring mosquitoes and malaria, "deliberately introduced as an act of biological warfare." This has a limited military impact but devastates the local population, which the Third Reich wished to punish for disloyalty.	US ACE (1953); Evans (2008)
1944	Germany, Allied forces	Military tool	Yes	Germans flood the Ay River in France in July, creating a lake two meters deep and several kilometers wide, slowing an advance on Saint Lô, a German communications center in Normandy.	US ACE (1953)
1944	Germany, Allied forces	Military tool	Yes	Germans flood the Ill River Valley during the Battle of the Bulge (winter 1944–1945), creating a lake sixteen kilometers long, three to six kilometers wide, and one to two meters deep, greatly delaying the American army's advance toward the Rhine.	US ACE (1953)
1944	United States, Japan	Military target	Yes	The US bombardment of a Japanese-occupied island in June targets water supply points, resulting in severe shortages.	Stewart (n.d.)
1944	Finland, USSR	Military target	Yes	In June, the Soviet air force attacks the Svir River Dam, near Leningrad, then under the control of the Finnish military.	Orlenko (1981); AHF (2004)

Date	Parties	Basis of Conflict	Violent Conflict or in the Context of Violence?	Description	Sources
1945	Romania, Germany	Military target	Yes	In one of the few verified German tactical uses of biological warfare, German forces pollute a large reservoir in northwestern Bohemia, Czechoslovakia, with sewage. See also the entry for the Pontine Marshes in 1944.	SIPRI (1971)
1947 onward	Bangladesh, India	Development dispute	No	A partition divides the Ganges River between Bangladesh and India; construction of the Farakka barrage by India, beginning in 1962, increases tension; short-term agreements settle disputes in 1977–1982, 1982–1984, and 1985–1988; and a thirty-year treaty is signed in 1996.	Butts (1997); Samson and Charrier (1997)
1947–1960s	India, Pakistan	Development dispute	No	A partition leaves the Indus River basin divided between India and Pakistan; disputes over irrigation water ensue, during which India stems the flow of water into irrigation canals in Pakistan. The Indus Waters Agreement is reached in 1960, after twelve years of World Bank–led negotiations.	Bingham et al. (1994)
1948	Arabs, Israelis	Military tool	Yes	Arab forces cut off West Jerusalem's water supply in the first Arab-Israeli war.	Wolf (1995); Wolf (1997)
1948	Arabs, Israelis	Military tool	Yes	Water and food supplies are cut off during an Arab siege of Jerusalem from December 1, 1947, to July 10, 1948. Arab forces block the road to Jerusalem in an attempt to defeat Jewish Jerusalem. Shortages cause Israelis to begin rationing water on May 12, limiting each person to two gallons (eight liters) per day, of which four pints (two liters) are for drinking.	Collins and LaPierre (1972); Joseph (1960); *Wikipedia* (2011b)
1950s	Korea, United States, others	Military target	Yes	Centralized dams on the Yalu (Amnok) River serving North Korea and China are attacked during the Korean War.	Gleick (1993)
1951	Korea, United Nations	Military tool; Military target	Yes	North Korea releases flood waves from the Hwachon Dam, damaging floating bridges operated by UN troops in the Pukhan Valley. The US Navy then sends planes to destroy the spillway crest gates.	US ACE (1953)
1951	Israel, Jordan, Syria	Military tool; Development dispute	Yes	Jordan makes public its plans to irrigate the Jordan Valley by tapping the Yarmouk River. Israel responds by commencing drainage of the Huleh swamps, located in the demilitarized zone between Israel and Syria; border skirmishes ensue between Israel and Syria.	Wolf (1997); Samson and Charrier (1997)
1951	United States, Korea	Military target	Yes	The Hwacheon Dam in Korea, completed in 1944, becomes both a target and a tool of opposing forces during the Korean War. In 1951, North Korea opens the dam to flood downstream areas and slow advancing UN forces. In response, the US Navy sends aircraft to bomb the dam, in the last known American use of aerial torpedoes in war.	Calcagno (2004)

continues

183

Water Conflict Chronology *continued*

Date	Parties Involved	Basis of Conflict	Violent Conflict or in the Context of Violence?	Description	Sources
1953	Israel, Jordan, Syria	Development dispute; Military target	Yes	Israel begins construction of its National Water Carrier to transfer water from the north of the Sea of Galilee out of the Jordan River basin to the Negev Desert for irrigation. Syrian military actions along the border and international disapproval lead Israel to move its intake to the Sea of Galilee.	Naff and Matson (1984); Samson and Charrier (1997)
1958	Egypt, Sudan	Military tool; Development dispute	Yes	Egypt sends an unsuccessful military expedition into disputed territory amid pending negotiations over Nile waters, Sudanese general elections, and an Egyptian vote on Sudan-Egypt unification; the Nile Water Treaty is signed when a pro-Egyptian government is elected in Sudan.	Wolf (1997)
1960s	North Vietnam, United States	Military target	Yes	Irrigation water supply systems in North Vietnam are bombed during the Vietnam War. An estimated 661 sections of dikes are damaged or destroyed.	IWCT (1967); Gleick (1993); Zemmali (1995)
1962	Israel, Syria	Development dispute; Military target	Yes	Israel destroys irrigation ditches in the lower Tarfiq in the demilitarized zone. Syria complains.	Naff and Matson (1984)
1962–1967	Brazil, Paraguay	Military tool; Development dispute	Military maneuvers	Negotiations between Brazil and Paraguay over development of the Paraná River are interrupted in 1962 by a unilateral show of military force by Brazil, which invades the area and claims control over the Guaira Falls site. Military forces are withdrawn in 1967 following an agreement for a joint commission to examine development in the region.	Murphy and Sabadell (1986)
1963–1964	Ethiopia, Somalia	Development dispute; Military tool	Yes	Creation of boundaries in 1948 leaves Somali nomads under Ethiopian rule; border skirmishes occur over disputed territory in the Ogaden Desert, where critical water and oil resources are located. A cease-fire is negotiated only after several hundred are killed.	Wolf (1997)
1964	Cuba, United States	Military tool	No	On February 6, the Cuban government cuts off the normal water supply to the US Naval Base at Guantanamo Bay in retaliation for the US seizure of four Cuban fishing boats days earlier near Florida.	*Guantanamo Bay Gazette* (1964)
1964	Israel, Syria	Military target	Yes	Headwaters of the Dan River on the Jordan River are bombed at Tell El-Qadi in a dispute about sovereignty over the source of the Dan.	Naff and Matson (1984)

Date	Parties	Basis of conflict	Violent conflict or in the context of violence?	Description	Sources
1965	Zambia, Rhodesia, Great Britain	Military target	No	President Kenneth Kaunda calls on the British government to send troops to Kariba Dam to protect it from possible saboteurs from the Rhodesian government.	Chenje (2001)
1965	Israel, Palestinians	Terrorism	Yes	The first attack ever by the Palestinian National Liberation Movement Al-Fatah is made on the diversion pumps for the Israeli National Water Carrier. The attack fails.	Naff and Matson (1984); Dolatyar (1995)
1965–1966	Israel, Syria	Military tool; Development dispute	Yes	Fire is exchanged over an "all-Arab" plan to divert the Jordan River headwaters (Hasbani and Banias) and presumably preempt the Israeli National Water Carrier; Syria halts construction of its diversion in July 1966.	Wolf (1995); Wolf (1997)
1967	Israel, Syria	Military target; Military tool	Yes	Israel destroys the Arab diversion works on the Jordan River headwaters. During the Arab-Israeli War, Israel occupies Golan Heights, with the Banias tributary to the Jordan; Israel occupies the West Bank.	Gleick (1993); Wolf (1995); Wolf (1997); Wallensteen and Swain (1997)
1967–1972	Vietnam, United States	Military tool	Yes	The US military, in "Operation Popeye," uses silver iodide for cloud seeding over Indochina (Vietnam) in an attempt to extend the monsoon season and stop the flow of materiel along Ho Chi Minh Trail. "Continuous rainfall was intended to slow down the truck traffic and was relatively successful."	Plant (1995); Troung and Cooper (2003)
1969	Israel, Jordan	Military target; Military tool	Yes	Israel, suspicious that Jordan is overdiverting the Yarmouk, leads two raids to destroy the newly built East Ghor Canal. Secret negotiations, mediated by the United States, lead to an agreement in 1970.	Samson and Charrier (1997)
1970	United States	Terrorism	No: Threat	The Weathermen, a group opposed to American imperialism and the Vietnam War, allegedly attempt to obtain biological agents to contaminate the water supply systems of US urban centers.	Kupperman and Trent (1979); Eitzen and Takafuji (1997); Purver (1995)
1970	Chinese citizens	Development dispute	Yes	Conflicts over excessive water withdrawals from, and subsequent water shortages in, China's Zhang River had been worsening for over three decades between villages in Shenxian and Linzhou counties. In the 1970s, militias from competing villages fight over withdrawals. (See also entries for 1976, 1991, 1992, and 1999.)	*China Water Resources Daily* (2002)

continues

Water Conflict Chronology *continued*

Date	Parties Involved	Basis of Conflict	Violent Conflict or in the Context of Violence?	Description	Sources
1970s	Argentina, Brazil, Paraguay	Development dispute	No	Brazil and Paraguay announce plans to construct a dam at Itaipu on the Paraná River, causing concern in Argentina about downstream environmental repercussions and the efficacy of Argentina's own planned dam project downstream. Argentina demands to be consulted during the planning of Itaipu, but Brazil refuses. An agreement is reached in 1979 that provides for the construction of both Brazil's and Paraguay's dams at Itaipu and Argentina's Yacyreta Dam.	Wallensteen and Swain (1997)
1972	United States	Terrorism	No: Threat	Two members of the right-wing "Order of the Rising Sun" are arrested in Chicago with thirty to forty kilograms of typhoid cultures, with which they allegedly planned to poison the water supply in Chicago, St. Louis, and other cities. Experts say the plan would have been unlikely to cause health problems, given that the water is chlorinated.	Eitzen and Takafuji (1997)
1972	United States	Terrorism	No: Threat	There is a reported threat to contaminate the water supply of New York City with nerve gas.	Purver (1995)
1972	North Vietnam	Military target	Yes	The United States bombs dikes in the Red River delta, rivers, and canals during a massive bombing campaign.	*Columbia Encyclopedia* (2000b)
1973	Germany	Terrorism	No: Threat	A German biologist threatens to contaminate water supplies with anthrax bacilli and botulinum toxin unless he is paid $8.5 million.	Jenkins and Rubin (1978); Kupperman and Trent (1979)
1974	Iraq, Syria	Military target; Military tool; Development dispute	Military maneuvers	Iraq threatens to bomb the al-Thawra (Tabaqah) Dam in Syria and masses troops along the border, alleging that the dam has reduced the flow of Euphrates River water to Iraq.	Gleick (1994)

Date	Parties	Type	Violent conflict or in context of violence	Description	Sources
1975	Iraq, Syria	Development dispute; Military tool	Military maneuvers	As upstream dams are filled during a low-flow year on the Euphrates, Iraqis claim that flow reaching its territory is "intolerable" and ask the Arab League to intervene. Syrians claim they are receiving less than half the river's normal flow and pull out of an Arab League technical committee formed to mediate the conflict. In May, Syria closes its airspace to Iraqi flights and both Syria and Iraq reportedly transfer troops to their mutual border. Saudi Arabia successfully mediates the conflict.	Gleick (1993); Gleick (1994); Wolf (1997)
1975	Angola, South Africa	Military goal; Military target	Yes	South African troops move into Angola to occupy and defend the Ruacana hydropower complex, including the Gové Dam on the Kunene River. The goal is to take possession of and defend the water resources of southwestern Africa and Namibia.	Meissner (2000)
1976	China	Development dispute	Yes	A local militia chief is shot to death in a clash over the damming of the Zhang River. Conflicts over excessive water withdrawals and subsequent water shortages from China's Zhang River had been worsening for over three decades. (See also entries for 1970, 1991, 1992, 1999.)	*China Water Resources Daily* (2002)
1977	United States	Terrorism	Yes	A North Carolina reservoir is contaminated with unknown materials. According to Clark, "Safety caps and valves were removed, and poison chemicals were sent into the reservoir. . . . Water had to be brought in."	Clark (1980); Purver (1995)
1978 onward	Egypt, Ethiopia	Development dispute; Political tool	No	There are long-standing tensions over the Nile, especially the Blue Nile, originating in Ethiopia. Ethiopia's proposed construction of dams on the headwaters of the Blue Nile leads Egypt to repeatedly declare the vital importance of water. "The only matter that could take Egypt to war again is water" (Anwar Sadat, 1979). "The next war in our region will be over the waters of the Nile, not politics" (Boutros Boutrous-Ghali, 1988).	Gleick (1991); Gleick (1994)
1978–1984	Sudan	Development dispute; Military target; Terrorism	Yes	Demonstrations in Juba, Sudan, in 1978 opposing the construction of the Jonglei Canal lead to the deaths of two students. Construction of the canal is suspended in 1984 following a series of attacks on the construction site.	Suliman (1998); Keluel-Jang (1997)
1980s	Mozambique, Rhodesia/Zimbabwe, South Africa	Military target; Terrorism	Yes	Power lines from Cahora Bassa Dam are repeatedly destroyed during a fight for independence in the region. The dam is targeted by RENAMO (the Mozambican National Resistance).	Chenje (2001)
1980–1988	Iran, Iraq	Military tool	Yes	Iran diverts water to flood Iraqi defense positions.	Plant (1995)
1981	Iran, Iraq	Military target; Military tool	Yes	Iran claims to have bombed a hydroelectric facility in Kurdistan, thereby blacking out large portions of Iraq, during the Iran-Iraq War.	Gleick (1993)

continues

Water Conflict Chronology *continued*

Date	Parties Involved	Basis of Conflict	Violent Conflict or in the Context of Violence?	Description	Sources
1981–1982	Angola, Namibia	Military target; Military tool	Yes	Water infrastructure, including dams and the major Cunene-Cuvelai pipeline, is targeted during the conflicts in Namibia and Angola in the 1980s.	Turton (2005)
1982	United States	Terrorism	No: Threat	Los Angeles police and the FBI arrest a man who is preparing to poison the city's water supply with a biological agent.	Livingston (1982); Eitzen and Takafuji (1997)
1982	Israel, Lebanon, Syria	Military tool	Yes	Israel cuts off the water supply to Beirut during a siege.	Wolf (1997)
1982	Guatemala	Development dispute	Yes	In Río Negro, 177 civilians are killed over opposition to the Chixoy hydroelectric dam.	Levy (2000)
1983	Lebanon	Terrorism	Yes	An explosives-laden truck disguised as a water delivery vehicle destroys a barracks in a US military compound, killing more than 300 people. The attack is blamed on Hezbollah, with the support of the Iranian government.	BBC (2007)
1983	Israel	Terrorism	No	The Israeli government reports that it has uncovered a plot by Israeli Arabs to poison the water in Galilee with "an unidentified powder."	Douglass and Livingstone (1987)
1984	United States	Terrorism	Yes	Members of the Rajneeshee religious cult contaminate a city water supply tank in The Dalles, Oregon, using salmonella bacteria. A community outbreak of over 750 cases occurs in a county that normally reports fewer than 5 cases per year.	Clark and Deininger (2000)
1985	United States	Terrorism	No	Law enforcement authorities discover that a small survivalist group in the Ozark Mountains of Arkansas known as The Covenant, the Sword, and the Arm of the Lord (CSA) has acquired a drum containing thirty gallons of potassium cyanide, with the apparent intent to poison water supplies in New York, Chicago, and Washington, DC. CSA members believe that such attacks would make the Messiah return more quickly by punishing unrepentant sinners. The objective appears to be mass murder in the name of a divine mission rather than a change in governmental policy.	Tucker (2000); NTI (2005)

1986	North Korea, South Korea	Military tool	No	North Korea's announcement of its plans to build the Kumgansan hydroelectric dam on a tributary of the Han River upstream of Seoul raises concerns in South Korea that the dam could be used as a tool for ecological destruction or war.	Gleick (1993)
1986	Lesotho, South Africa	Development dispute; Military goal	Yes	South Africa supports a bloodless coup by Lesotho's defense forces. Immediately afterward, the two countries agree to share water from the Highlands of Lesotho, following thirty years of unsuccessful negotiations. There is disagreement over the degree to which water is a motivating factor for either party.	Mohamed (2001)
1986	Lesotho, South Africa	Military goal; Development dispute	Yes	South Africa supports a coup in Lesotho over support for the African National Congress, opposition to apartheid, and water. The new government in Lesotho quickly signs the Lesotho Highlands water agreement.	American University (2000b)
1988	Angola, South Africa, Cuba	Military goal; Military target	Yes	Cuban and Angolan forces launch an attack on Calueque Dam via land and then air. Considerable damage is inflicted on the dam's wall, and the power supply to the dam is cut. The water pipeline to Owamboland is cut and destroyed.	Meissner (2000)
1990	South Africa	Development dispute	No	A pro-apartheid council cuts off water to the Wesselton township of 50,000 blacks following protests over miserable sanitation and living conditions.	Gleick (1993)
1990	Iraq, Syria, Turkey	Development dispute; Military tool	No	The flow of the Euphrates is interrupted for a month as Turkey finishes construction of the Ataturk Dam, part of the Grand Anatolia Project. Syria and Iraq protest that Turkey now has a weapon of war. In mid-1990, Turkish president Turgut Ozal threatens to restrict water flow to Syria to force it to withdraw support for Kurdish rebels operating in southern Turkey.	Gleick (1993); Gleick (1995)
1991	Iraq, Kuwait, United States	Military target	Yes	During the Gulf War, Iraq destroys much of Kuwait's desalination capacity during its retreat.	Gleick (1993)
1991	Canada	Terrorism	No: Threat	A threat is made via an anonymous letter to contaminate the water supply of the city of Kelowna, British Columbia, with "biological contaminates." The motive is apparently "associated with the Gulf War." The security of the water supply is increased in response, and no group is identified as the perpetrator.	Purver (1995)
1991	Iraq, Turkey, United Nations	Military tool	No: Threat	The United Nations discusses use of the Ataturk Dam in Turkey to cut off flows of the Euphrates to Iraq.	Gleick (1993)

continues

189

Water Conflict Chronology *continued*

Date	Parties Involved	Basis of Conflict	Violent Conflict or in the Context of Violence?	Description	Sources
1991	Iraq, Kuwait, United States	Military target	Yes	During the Persian Gulf War, Allied Coalition forces damage Baghdad's modern water supply and sanitation system—intentionally and unintentionally. "Four of seven major pumping stations were destroyed, as were 31 municipal water and sewerage facilities—20 in Baghdad—resulting in sewage pouring into the Tigris. Water purification plants were incapacitated throughout Iraq" (Arbuthnot 2000). Following the damage, the *New England Journal of Medicine* reports that during the first eight months of 1991, childhood deaths in Iraq increase by 47,000 and the country's infant mortality rate doubles, to 92.7 per 1,000 live births.	Gleick (1993); Arbuthnot (2000); Barrett (2003)
1991	China	Development dispute	Yes	In December 1991, Huanglongkou village and Qianyu village exchange mortar fire over the construction of new water diversion facilities. Conflicts over excessive water withdrawals and subsequent water shortages from China's Zhang River had been worsening for over three decades. (See also entries for 1970, 1976, 1992, and 1999.)	*China Water Resources Daily* (2002)
1991–2001	United States, Iraq	Military target; Military tool	No	The United States deliberately pursues a policy of destroying Iraq's water systems through sanctions and withholding contracts.	Nagy (2001)
1991–2007	Karnataka, India	Development dispute	Yes	Violence erupts when Karnataka rejects an interim order handed down by the Cauvery Waters Tribunal, set up by the Supreme Court of India. The tribunal had been established in 1990 to settle two decades of dispute between Karnataka and Tamil Nadu over irrigation rights to the Cauvery River.	Gleick (1993); Butts (1997); American University (2000a)
1992		Political tool; Development dispute	Military maneuvers	Citing environmental concerns, Hungary abrogates a 1977 treaty with Czechoslovakia concerning construction of the Gabcikovo/Nagymaros project. Slovakia continues construction unilaterally, completes the dam, and diverts the Danube into a canal inside the Slovak Republic. Massive public protest and movement of military to the border ensue; the issue is taken to the International Court of Justice.	Gleick (1993)
1992	Turkey	Terrorism	Yes	Lethal concentrations of potassium cyanide are reportedly discovered in the water tanks of a Turkish Air Force compound in Istanbul. The Kurdish Workers' Party (PKK) claims credit.	Chelyshev (1992)

1992	Bosnia, Bosnian Serbs	Military tool	Yes	The Serbian siege of Sarajevo, Bosnia and Herzegovina, includes a cutoff of all electrical power and the water feeding the city from the surrounding mountains. The lack of power cuts the two main pumping stations inside the city, despite pledges from Serbian nationalist leaders to United Nations officials that they would not use their control of Sarajevo's utilities as a weapon. Bosnian Serbs take control of water valves regulating flow from wells that provide more than 80 percent of water to Sarajevo; the reduced water flow to the city is used to "smoke out" Bosnians.	Burns (1992); Husarska (1995)
1992	China	Development dispute	Yes	In August 1992, bombs are set off along a Zhang River distribution canal, collapsing part of the canal and causing flooding and economic losses. Violence continues in the late 1990s with confrontations, mortar attacks, and bombings. Conflicts over excessive water withdrawals and subsequent water shortages from China's Zhang River had been worsening for over three decades. (See also entries for 1970, 1976, 1991, and 1999.)	*China Water Resources Daily* (2002)
1992	Moldova, Russia	Military target	Yes	In June, hostilities between Moldova and Russia in a short but intense conflict include a rocket-artillery attack on the hydroelectric turbines at the Dubossary power station on the Nistru (or Dniester) River.	Malik (2005); Belitser et al. (2009)
1993	Iran	Terrorism	No	A report suggests that proposals were made at a meeting of fundamentalist groups in Tehran, under the auspices of the Iranian Foreign Ministry, to poison water supplies of major cities in the West "as a possible response to Western offensives against Islamic organizations and states."	Haeri (1993)
1993	Yugoslavia	Military target; Military tool	Yes	The sixty-five-meter-high Peru a Dam on the Cetina River was Yugoslavia's second-largest hydroelectric facility before the country's breakup with the Croatian War beginning in 1991. On January 28, 1993, Serbian/Yugoslav army forces detonate explosives at the dam in an attempt to wipe out Croatian villages and the port city of Omiš. A successful Croatian counterattack allows military engineers to reach the dam and release water in time to prevent it from bursting, saving an estimated twenty to thirty thousand civilians. Credit for preventing a dam burst is also given to British Marine Captain Mark Gray, a UN observer, for opening gates to reduce water levels prior to the attack.	Gleick (1993); Rathfelder (2007)

continues

Water Conflict Chronology *continued*

Date	Parties Involved	Basis of Conflict	Violent Conflict or in the Context of Violence?	Description	Sources
1993–2003	Iraq	Military tool	No	To quell opposition to his government, Saddam Hussein reportedly poisons and drains the water supplies of southern Shiite Muslims, the Ma'dan. The marshes of southern Iraq are intentionally targeted. The European Parliament and UN Human Rights Commission deplore the use of water as a weapon in the region.	Gleick (1993); American University (2000c); National Geographic News (2001)
1994	Moldova, Russia	Terrorism	No: Threat	Moldavian general Nikolay Matveyev reportedly threatens to contaminate the water supply of the Russian 14th Army in Tiraspol, Moldova, with mercury.	Purver (1995)
1998/1994	United States	Cyberterrorism	No	The *Washington Post* reports that a twelve-year-old computer hacker has broken into the SCADA computer system that runs Arizona's Roosevelt Dam, gaining complete control of the dam's massive floodgates. The cities of Mesa, Tempe, and Phoenix, Arizona, are downstream of this dam. This report turns out to be incorrect. A hacker did break into the computers of an Arizona water facility, the Salt River Project in the Phoenix area. But he was twenty-seven, not twelve, and the incident occurred in 1994, not 1998. And while clearly trespassing in critical areas, the hacker never could have had control of any dams, leading investigators to conclude that no lives or property were ever threatened.	Gellman (2002); Lemos (2002)
1995	Ecuador, Peru	Military tool; Political tool	Yes	Armed skirmishes arise in part because of disagreement over the control of the headwaters of the Cenepa River. Wolf argues that this is primarily a border dispute simply coinciding with the location of a water resource.	Samson and Charrier (1997); Wolf (1997)
1997	Singapore, Malaysia	Political tool	No	Malaysia, which supplies about half of Singapore's water, threatens to cut off that supply in retribution for criticisms by Singapore of policy in Malaysia.	Zachary (1997)
1998	Tajikistan	Terrorism; Political tool	No: Threat	Tajik guerrilla commander Makhmud Khudoberdyev threatens to blow up a dam on the Kairakkhum channel if his political demands are not met.	*WRR* (1998)
1998	Angola	Military tool; Political tool	Yes	In September, fierce fighting between UNITA and Angolan governmental forces breaks out at Gove Dam on the Kunene River for control of the installation.	Meissner (2001)

1998	Democratic Republic of Congo	Military target; Terrorism	Yes	Attacks are made on Inga Dam during efforts to topple President Kabila. Electricity supplies from Inga Dam and water supplies to Kinshasa are disrupted.	Chenje (2001); Human Rights Watch (1998)
1998–1999	Kosovo	Terrorism; Political tool	Yes	Water supplies and wells are contaminated by Serbs disposing of the bodies of Kosovar Albanians. There are also reports of Yugoslav federal forces poisoning wells with carcasses and hazardous materials.	CNN (1999); Hickman (1999)
1998–2000	Eritrea and Ethiopia	Military target	Yes	Water-pumping plants and pipelines in the border town of Adi Quala are destroyed during the civil war between Eritrea and Ethiopia.	ICRC (2003)
1999	Lusaka, Zambia	Terrorism; Political tool	Yes	A bomb blast destroys Lusaka's main water pipeline, cutting off water for the city, population 3 million.	FTGWR (1999)
1999	Yugoslavia	Military target	Yes	Belgrade reports that NATO planes have targeted a hydroelectric plant during the Kosovo campaign.	Reuters (1999c)
1999	Bangladesh	Development dispute; Political tool	Yes	Fifty people are hurt during strikes called to protest power and water shortages. Protests are led by former prime minister Begum Khaleda Zia over a decrease in public services and deterioration of law and order.	Ahmed (1999)
1999	Yugoslavia	Military target	Yes	NATO targets utilities and shuts down water supplies in Belgrade. NATO bombs bridges on Danube, disrupting navigation.	Reuters (1999b)
1999	Yugoslavia	Political tool	Yes	Yugoslavia refuses to clear war debris (downed bridges) on the Danube unless financial aid for reconstruction is provided; European countries on the Danube fear that flooding due to winter ice dams will result. Diplomats decry environmental blackmail.	Simons (1999)
1999	Kosovo	Political tool	Yes	Serbian engineers shut down the water system in Pristina prior to occupation by NATO.	Reuters (1999a)
1999	South Africa	Terrorism	Yes	A homemade bomb is discovered at a water reservoir at Wallmansthal, near Pretoria. It is thought to have been meant to sabotage water supplies to farmers.	Pretoria Dispatch (1999)
1999	Angola	Terrorism; Political tool	Yes	One hundred bodies are found in four drinking water wells in central Angola.	International Herald Tribune (1999)
1999	Puerto Rico, United States	Political tool	No	Protesters block water intake to Roosevelt Roads Navy Base in opposition to US military presence and the navy's use of the Blanco River, following chronic water shortages in neighboring towns.	New York Times (1999)

continues

Water Conflict Chronology *continued*

Date	Parties Involved	Basis of Conflict	Violent Conflict or in the Context of Violence?	Description	Sources
1999	China	Development dispute; Terrorism	Yes	Around the Chinese New Year, farmers from Hebei and Henan Provinces fight over limited water resources. Heavy weapons, including mortars and bombs, are used and nearly one hundred villagers are injured. Houses and facilities are damaged, and the total loss reaches US$1 million.	*China Water Resources Daily* (2002)
1999	East Timor	Military tool; Terrorism	Yes	Militia opposing East Timor independence kills pro-independence supporters and throws their bodies in a water well.	BBC (1999)
1999	Yemen	Development dispute	Yes	Yemen sends 700 soldiers to quell fighting that claimed six lives and injured sixty in clashes that erupted between two villages fighting over a local spring near Ta'iz. The village of Al-Marzuh believed it was entitled to exclusive use of a spring because the spring was located on its land; the neighboring village of Quradah believed its right to the water had been affirmed in a fifty-year-old court verdict. The dispute erupted in violence. President Ali Abdullah Saleh intervenes by summoning the sheikhs of the two villages to the capital and sorts out the problem by dividing the water into halves.	Al-Qadhi (2003)
1999–2000	Namibia, Botswana, Zambia	Military goal; Development dispute	No	On Sedudu/Kasikili Island, in the Zambezi/Chobe River, a dispute arises about the border and access to water. The conflict is presented to the International Court of Justice.	ICJ (1999)
2000	Ethiopia	Development dispute	Yes	A man is stabbed to death during a fight over clean water during the famine in Ethiopia.	Sandrasagra (2000)
2000	Central Asia: Kyrgyzstan, Kazakhstan, Uzbekistan	Development dispute	No	Kyrgyzstan cuts off water to Kazakhstan until coal is delivered; Uzbekistan cuts off water to Kazakhstan for nonpayment of debt.	Pannier (2000)
2000	France, Belgium, Netherlands	Terrorism	Yes	In July, workers at the Cellatex chemical plant in northern France dump 5,000 liters of sulfuric acid into a tributary of the Meuse River after they are denied workers' benefits. A French analyst points out that this is the first time "the environment and public health were made hostage in order to exert pressure, an unheard-of situation until now."	*Christian Science Monitor* (2000)

2000	Hazarajat, Afghanistan	Development dispute	Yes	Violent conflicts break out over water resources in the villages of Burma Legan and Taina Legan, and in other parts of the region, as drought depletes local resources.	Cooperation Center for Afghanistan (2000)
2000	India: Gujarat	Development dispute	Yes	Water riots are reported in some areas of Gujarat in protest of authority's failure to arrange an adequate supply of tanker water. Police are reported to have shot into a crowd at Falla village near Jamnagar, resulting in the deaths of three and injuries to twenty following protests against the diversion of water from the Kankavati Dam to Jamnagar town.	*FTGWR* (2000)
2000	Bolivia	Development dispute	Yes	Massive protests, riots, and violence result from efforts to privatize the water system of Cochabamba, Bolivia.	Shultz and Draper (2009)
2000	United States	Terrorism	No	A drill simulating a terrorist attack on the Nacimiento Dam in Monterey County, California, gets out of hand when two radio stations report it as a real attack.	Gaura (2000)
2000	Kenya	Development dispute	Yes	A clash between villagers and thirsty monkeys leaves eight monkeys dead and ten villagers wounded. The duel started after water tankers brought water to a drought-stricken area and the monkeys, desperate for water, attacked the villagers.	BBC (2000); Okoko (2000)
2000	Australia	Cyberterrorism	Yes	In Queensland, Australia, police arrest a man for using a computer and radio transmitter to take control of the Maroochy Shire wastewater system and release sewage into parks, rivers, and property.	Gellman (2002)
2000	China	Development dispute	Yes	Civil unrest erupts over the use and allocation of water from Baiyangdian Lake, the largest natural lake in northern China. Several people die in riots by villagers in July in Shandong after officials cut off water supplies. In August, six die when officials in the southern province of Guangdong blow up a water channel to prevent a neighboring county from diverting water.	Pottinger (2000)
2001	Israel, Palestine	Terrorism; Military target	Yes	Palestinians destroy water supply pipelines to the West Bank settlement of Yitzhar and to Kibbutz Kisufim. The Agbat Jabar refugee camp near Jericho disconnects from its water supply after Palestinians loot and damage local water pumps. Palestinians accuse Israel of destroying a water cistern, blocking water tanker deliveries, and attacking materials for a wastewater treatment project.	*Israel Line* (2001a); *Israel Line* (2001b); ENS (2001)

continues

Water Conflict Chronology *continued*

Date	Parties Involved	Basis of Conflict	Violent Conflict or in the Context of Violence?	Description	Sources
2001	Pakistan	Development dispute; Terrorism	Yes	Civil unrest erupts over severe water shortages caused by the long-term drought. Protests begin in March and April and continue into summer, involving riots, four bombs in Karachi (June 13), one death, twelve injuries, and thirty arrests. Ethnic conflicts arise as some groups "accuse the government of favoring the populous Punjab province [over Sindh Province] in water distribution."	Nadeem (2001); Soloman (2001)
2001	Macedonia	Terrorism; Military target	Yes	Water flow to Kumanovo (population 100,000) is cut off for twelve days in a conflict between ethnic Albanians and Macedonian forces. Valves of Glaznja and Lipkovo Lakes are damaged.	AFP (2001); Macedonia Information Agency (2001)
2001	China	Development dispute	Yes	In an act to protest the destruction of fisheries by uncontrolled water pollution, fishermen in northern Jiaxing City, Zhejiang Province, dam a large industrial wastewater canal for twenty-three days. The wastewater discharges into the neighboring Shengze Town, Jiangsu Province, killing fish and threatening public health.	China Ministry of Water Resources (2001)
2001	Philippines	Terrorism	No	The militant Islamist separatist group Abu Sayyaf threatens to poison the water supply in Isablela, a mainly Christian town on Basilan Island in the country's south. In October, residents in six nearby villages suspect contamination because the water smells like gasoline. Local officials respond by closing pipelines and bringing in drinking water by truck. In the months following the 9/11 attacks on New York, numerous false alarms of terrorist activity are reported around the world.	World Environment News (2001); Fenton et al. (2001)
2001	Afghanistan	Military target	Yes	American forces bomb the hydroelectric facility at Kajaki Dam in Helmand Province, cutting off electricity for the city of Kandahar. The dam itself is not targeted.	BBC (2001); Parry (2001)
2001	Pokomo farmers and Orma cattle herders	Development dispute	Yes	At least 130 people are killed in a string of clashes between Pokomo farmers and Orma people, seminomadic cattle herders, over access to land and river water.	AFP (2012)

2002	Nepal	Terrorism; Political tool	Yes	The Khumbuwan Liberation Front (KLF) blows up a 250-kilowatt hydroelectric powerhouse in Nepal's Bhojpur District, cutting off power to Bhojpur and surrounding areas. The damage takes six months to repair and costs 10 million Rs (US$120,000). During 2002, Maoist rebels destroy more than seven micro-hydro projects, a water supply intake, and supply pipelines to Khalanga, in western Nepal.	*Kathmandu Post* (2002); *FTGWR* (2002)
2002	Rome, Italy	Terrorism	No: Threat	Italian police arrest four Moroccans allegedly planning to contaminate the water supply system in Rome with a cyanide-based chemical, targeting buildings that include the United States embassy. Ties to Al-Qaeda are suggested.	BBC (2002)
2002	Kashmir, India	Development dispute	Yes	Two people are killed and twenty-five others are injured in Kashmir when police fire at a group of villagers clashing over water sharing. The incident takes place in Garend village in a dispute over sharing water from an irrigation stream.	*Japan Times* (2002)
2002	United States	Terrorism	No: Threat	Papers seized during the arrest of a Lebanese imam at a mosque in Seattle include "instructions on poisoning water sources" from a London-based Al-Qaeda recruiter. The FBI issues a bulletin to computer security experts around the country indicating that Al-Qaeda terrorists may have been studying American dams and water supply systems in preparation for new attacks. "U.S. law enforcement and intelligence agencies have received indications that al-Qaeda members have sought information on Supervisory Control And Data Acquisition (SCADA) systems available on multiple SCADA-related Web sites," reads the bulletin, according to SecurityFocus. "They specifically sought information on water supply and wastewater management practices in the U.S. and abroad."	McDonnell and Meyer (2002); MSNBC (2002)
2002	Colombia	Terrorism	Yes	Colombian rebels in January damage a gate valve in the dam that supplies most of Bogotá's drinking water. The Revolutionary Armed Forces of Colombia (FARC) detonates an explosive device planted on a German-made gate valve located inside a tunnel in the Chingaza Dam.	*WaterWeek* (2002)
2002	Karnataka, Tamil Nadu, India	Development dispute	Yes	There is continuing violence over allocation of water from the Cauvery (Kaveri) River between Karnataka and Tamil Nadu, with riots, property destruction, more than thirty injuries, and arrests through September and October.	*The Hindu* (2002a); *The Hindu* (2002b); *Times of India* (2002)

continues

Water Conflict Chronology *continued*

Date	Parties Involved	Basis of Conflict	Violent Conflict or in the Context of Violence?	Description	Sources
2002	United States	Terrorism	No: Threat	The Earth Liberation Front threatens the water supply for the town of Winter Park, Co. Previously, this group claimed responsibility for the destruction of a ski lodge in Vail, Colorado, that threatened lynx habitat.	Crecente (2002); AP (2002)
2002	Botswana, Bushmen	Development dispute	Yes	Botswana's president, Festus Mogae, sends troops to the Kalahari Desert to destroy wells and empty water sources of indigenous Khoisan people (also known as Bushmen), ostensibly in an effort to remove them from their ancestral lands and assimilate them into modern society. Critics accuse the government of taking away water rights in favor of mining interests and label the government's actions a "siege"; Botswana is condemned by international observers. Against expectations, a band of Bushmen retreat into the desert and survive for years with little outside assistance.	Workman (2009)
2002	Bolivian irrigators, townsmen	Development dispute	Yes	Violence erupts between irrigators and the townsmen of Tarata over the use of water from the Laka Laka Dam. After local authorities begin diverting some water to cultivate gardens in the town, irrigators in the downstream area of Arbieto protest that their water rights are being violated. Irrigators destroy a portion of the town's pipeline, and townsmen retaliate by vandalizing part of an irrigation canal.	Bustamante et al. (2004)
2003	United States	Terrorism	No: Threat	Al-Qaeda threatens US water systems via a call to a Saudi Arabian magazine. Al-Qaeda does not "rule out . . . the poisoning of drinking water in American and Western cities."	AP (2003b); Waterman (2003); NewsMax (2003); *US Water News* (2003)
2003	United States	Terrorism	Yes	Four incendiary devices are found in the pumping station of a Michigan water-bottling plant. The Earth Liberation Front claims responsibility, accusing Ice Mountain Water Company of "stealing" water for profit. Ice Mountain is a subsidiary of Nestle Waters.	AP (2003a)
2003	Colombia	Terrorism; Development dispute	Yes	A bomb blast at the Cali Drinking Water Treatment Plant kills three workers on May 8. The workers were members of a trade union involved in intense negotiations over privatization of the water system.	PSI (2003)

2003	Jordan	Terrorism	No: Threat	Jordanian authorities arrest Iraqi agents in connection with a botched plot to poison the water supply that serves American troops in the eastern Jordanian desert near the border with Iraq. The scheme involves the poisoning of a water tank that supplies American soldiers at a military base in Khao, which lies in an arid region of the eastern frontier near the industrial town of Zarqa.	*MJS* (2003)
2003	Iraq, United States, others	Military target	Yes	During the US-led invasion of Iraq, water systems are reportedly damaged and destroyed by different parties, and major dams are military objectives of the US forces. Damage directly attributable to the war includes vast segments of the water distribution system and the Baghdad water system, damaged by a missile.	UNICEF (2003); ARC (2003)
2003	Iraq	Terrorism	Yes	Insurgents bomb a main water pipeline in Baghdad. City engineers say this is the first strike against Baghdad's water system during the Iraq War, which began in March 2003. The bombing occurs when an explosive is fired at the six-foot-wide water main in the northern part of Baghdad, says Hayder Muhammad, chief engineer for the city's water treatment plants.	Tierney and Worth (2003)
2003–2007	Sudan, Darfur	Military tool; Military target; Terrorism	Yes	The ongoing civil war in Sudan includes violence against water resources. In 2003, villagers from around Tina say that bombings have destroyed water wells. In Khasan Basao, they allege that water wells have been poisoned. In 2004, wells in Darfur are intentionally contaminated as part of a strategy of harassment against displaced populations.	Amnesty International (2004); Reuters (2004b)
2004	Mexico	Development dispute	Yes	Two Mexican farmers had argued for years over water rights to a small spring used to irrigate a small corn plot near the town of Pihuamo. In March, these farmers shoot and kill each other.	*Guardian* (2004)
2004	Pakistan	Terrorism	Yes	In military action aimed at Islamic terrorists, including Al-Qaeda and the Islamic Movement of Uzbekistan, homes, schools, and water wells are damaged and destroyed.	Reuters (2004a)
2004	India, Kashmir	Terrorism	Yes	Twelve Indian security forces are killed by an improvised explosive device planted in an underground water pipe during a "counter-insurgency operation in Khanabal area in Anantnag district."	TNN (2004)
2004	China	Development dispute	Yes	Tens of thousands of farmers stage a sit-in against the construction of the Pubugou Dam on the Dadu River in Sichuan Province. Riot police are deployed to quell the unrest, and one person is killed. Witnesses also report the deaths of a number of residents. (See China, 2006, for a follow-up.)	BBC (2004a); VOA (2004)

Water Conflict Chronology *continued*

Date	Parties Involved	Basis of Conflict	Violent Conflict or in the Context of Violence?	Description	Sources
2004	China, United States	Military target	No	A 2004 Pentagon report on China's military capacity raises the concept of Taipei adopting military systems capable of being used as a tool for deterring Chinese military coercion by "presenting credible threats to China's urban population or high-value targets, such as the Three Gorges Dam." China promptly denounces "a U.S. suggestion" that Taiwan's military target the Three Gorges Dam, leading the United States to deny that it had so urged.	*China Daily* (2004); US DOD (2004)
2004	South Africa	Development dispute	Yes	Poor delivery of water and sanitation services in Phumelela Township leads to several months of protests. No one is killed during the protests, but a few people are seriously injured, and municipal property is damaged.	CDE (2007)
2004	Gaza Strip	Terrorism; Development dispute	Yes	The United States halts two water development projects as punishment to the Palestinian Authority for its failure to find those responsible for a deadly attack on a US diplomatic convoy in October 2003.	AP (2004)
2004	India	Development dispute	Yes	Four people are killed in October and more than thirty are injured in November in ongoing protests by farmers over allocations of water from the Indira Gandhi Irrigation Canal in Sriganganagar District, which borders Pakistan. Authorities impose curfews on the towns of Gharsana, Raola, and Anoopgarh.	Indo-Asian News Service (2004)
2004–2006	Ethiopia	Development dispute	Yes	At least 250 people are killed and many more are injured in clashes over water wells and pastoral lands. Villagers call the clashes the "War of the Well" and describe "well warlords, well widows, and well warriors." A three-year drought has led to extensive violence over limited water resources, worsened by the lack of effective governmental and central planning.	BBC (2004b); AP (2005); Wax (2006)
2005	Kenya	Development dispute	Yes	Police are sent to the northwestern part of Kenya to control a major violent dispute between Kikuyu and Maasai groups over water. More than twenty people are killed in fighting in January. By July, the death toll exceeds ninety, principally in the rural center of Turbi. The tensions arose over grazing and water. Maasai herdsmen accused a local Kikuyu politician of diverting a river to irrigate his farm, depriving downstream livestock. Fighting displaces more than 2,000 villages, reflecting tensions between nomadic and settled communities.	BBC (2005); Ryu (2005); Lane (2005)

2005	Ukraine	Terrorism	Yes	On April 13, the Kiev Hydropower Station on the Dnieper River receives a threat that forty rail cars filled with explosives have been placed on a portion of levees holding back the reservoir.	Levitsky (2005)
2006	Yemen	Development dispute	Yes	Local media report a struggle between the Hajja and Amran tribes over a well located between the two governorates in Yemen. According to news reports, armed clashes between the two sides force many families to leave their homes and migrate. News reports confirm that authorities have arrested twenty people in an attempt to stop the fighting.	Al-Ariqi (2006)
2006	China	Development dispute	Yes	Chinese authorities execute a man who took part in protests against the Pubugou Dam in Sichuan Province in 2004 (see the 2004 China entry). Chen Tao was convicted of killing a policeman but is executed before legal appeals are completed.	BBC (2006a); Coonan (2006)
2006	Ethiopia	Development dispute	Yes	At least twelve people die and over twenty are wounded in clashes over water and pasture in the Somali border region.	BBC (2006b)
2006	Ethiopia and Kenya	Development dispute	Yes	At least forty people die in Kenya and Ethiopia in continuing clashes over water, livestock, and grazing land. Fighting takes place in southern Ethiopia in the region of Oromo and in the northern Kenya Marsabit District.	Reuters (2006)
2006	Sri Lanka	Military tool; Military target; Terrorism	Yes	Tamil Tiger rebels cut the water supply to government-held villages in northeastern Sri Lanka. Sri Lankan governmental forces then launch attacks on the reservoir, declaring the Tamil actions to be terrorism. As of August, conflict around the water blockade had claimed over 425 lives.	BBC (2006c); BBC (2006d); Gutierrez (2006)
2006	Israel, Lebanon	Military target; Terrorism	Yes	Hezbollah rockets damage a wastewater plant in Israel. Israeli counterattacks damage water systems throughout southern Lebanon, including tanks, pipes, pumping stations, and facilities along the Litani River.	Science (2006); Amnesty International (2006); Murphy (2006)
2006	Sudan	Development dispute	Yes	Militia of the Merowe Dam Militia Implementation Unit in Sudan attack a gathering of villagers concerned about the community impacts of the dam at a school in Amri village, killing three farmers and injuring more than fifty others.	Bosshard (2009)
2007	India	Development dispute	Yes	Thousands of farmers breach security and storm the area of Hirakud Dam to protest allocation of water to industry. Minor injuries are reported during the conflict between the farmers and police.	Statesman News Service (2007)

continues

Water Conflict Chronology *continued*

Date	Parties Involved	Basis of Conflict	Violent Conflict or in the Context of Violence?	Description	Sources
2007	Afghanistan	Military target; Terrorism	Yes	The Kajaki Dam is the scene of major fighting between Taliban and NATO forces, mainly British and Dutch. The Taliban is attempting to make it impossible to work on reconstruction of the dam and power lines to boost output.	Friel (2007)
2007	Canada	Terrorism	No	A Toronto man previously accused of attempted murder and illegal possession of explosives is charged with eight more counts of attempted murder after allegedly tampering with bottled water, which he injected with an unspecified liquid.	*Toronto Star* (2007)
2007	Burkina Faso, Ghana, Côte d'Ivoire	Development dispute	Yes	Declining rainfall has led to growing fights between animal herders and farmers with competing needs. In August, 2007 people are forced to flee their homes by fighting in Zounweogo Province.	UN OCHA (2007)
2007	Israel, Palestine	Development dispute	No	Israel's sanctions against Gaza cause water shortages and a growing public health risk. In particular, restrictions on fuel, spare parts, and maintenance equipment threaten the functioning of Gaza's already limited water and sanitation system.	Oxfam (2007)
2007	Sydney residents	Development dispute	Yes	A thirty-six-year-old Australian is charged with murder after killing a man during a fight over water restrictions in Sydney. A number of incidents have been reported following ten years of drought and water restrictions, leading scholars to suggest a "link between persistent urban water restrictions and civil unrest."	ABC News (2007); Crase (2009)
2007	Sudan	Development dispute	Yes	Angry villagers in Sudan stage protests against Kajbar Dam; four villagers are killed by governmental militia.	Bosshard (2009)
2008	Nigeria	Development dispute	Yes	A protest over the price of water in Nyanya, Abuja, Nigeria, results in violence, including the beating of water vendors.	Yakubu (2008)
2008	China, Tibet	Military target; Development dispute	Yes	China launches a political crackdown in Tibet. At least some observers note the importance of Tibet for the water resources of China, though the political complications between Tibet and China extend far beyond water. As noted: "Tibet is referred to in some circles as the 'world's water tower'; the Tibetan plateau is home to vast reserves of glaciated water, the sources of 10 of the largest rivers in Asia, including the Yellow, Yangtze, Mekong, Brahmaputra, Salween, Hindus and Sutlej among others. By some estimates, the Tibetan plateau is the source of fresh water for fully a quarter of the world's population."	Sharife (2008)

2008	Pakistan	Terrorism	Yes	In October, the Taliban threatens to blow up Warsak Dam, the main water supply for Peshawar, during a governmental offensive in the region.	Perlez and Shah (2008)
2008	Murulle, Garre clans	Development dispute	Yes	Fighting over boreholes in arid northern Kenya kills at least four people as competition for resources grows in the drought-hit region between the Murulle and Garre clans in Elwak, Mandera District.	Reuters (2008)
2009	China, India	Development dispute; Military tool	No	China claims a part of historical Tibet that is now under Indian control as part of the state of Arunachal Pradesh. To influence this territorial dispute, China tries to block a $2.9 billion Asian Development Bank loan to India because it would help finance Indian water projects in the disputed area.	Wong (2009)
2009	Ethiopian Oromia, Somali regions	Development dispute	Yes	Ethiopian Somalis attack a Borana community in the Oromia region over ownership of a new borehole being drilled on the disputed border between them. Three people from the Oromia village of Kafa are killed and seven are injured, and the entire community is driven from their homes. The drilling rig is destroyed.	BBC (2009)
2009	Indian citizens	Development dispute	Yes	A family in Madhya Pradesh State in India is killed by a small mob for illegally drawing water from a municipal pipe. Others run to collect water for themselves before the water in the pipe runs out. Drought and inequality in water distribution lead to increasing conflict in the region. Indian media report more than fifty violent clashes in the state capital, Bhopal, during May alone. Since January, twelve people have been killed and even more have been injured.	Singh (2009)
2009	Mumbai residents, police	Development dispute	Yes	Police clash with hundreds of Mumbai residents protesting water cuts. One man is killed, and a dozen others are injured. Mumbai authorities are forced to ration supplies after the worst monsoon season in decades.	Chandran (2009)
2009	North Korea, South Korea	Political tool	Yes	Without previous warning, North Korea releases 40 million cubic meters of water from the Hwanggag Dam, causing a flash flood on the Imjin River. In South Korea, at least six fishermen and campers drown. North Korea claims that the water had to be urgently released and promises to warn the South of future releases. South Korea fears that North Korea could use the water of the dam as a weapon during a violent conflict.	Choe (2009)
2010	Pakistani tribes	Development dispute; Military tool	Yes	More than one hundred are dead and scores are injured following two weeks of tribal fighting in Parachinar, in the Kurram region of Pakistan near the Afghanistan border. The conflict over irrigation water began as the Shalozan Tangi tribe cut off supplies to the Shalozan tribe. Some report that Al-Qaeda may be involved; others claim that sectarian violence is to blame, as one group is Sunni Muslim and the other Shiite.	*Express Tribune* (2010); AP (2010)

continues

Water Conflict Chronology *continued*

Date	Parties Involved	Basis of Conflict	Violent Conflict or in the Context of Violence?	Description	Sources
2010	Afghanistan	Terrorism	Yes	A remote-controlled bomb hidden in a water truck kills three people, including two children, in the eastern Afghan province of Khost, which borders Pakistan.	AP (2009)
2010	Pakistani Mangal, Tori tribes	Development dispute	Yes	A water dispute in Pakistan's tribal region leads to 116 deaths. In early September, the Mangal tribe cuts off the supply of irrigation water to lands used by the neighboring Tori tribe, leading to fighting.	CNN (2010)
2010	India	Development dispute	Yes	A protest over water shortages in the National Capital Territory of Delhi in India leads to violence. Erratic water supply and cutoffs in the Kondli area of Mayur Vihar in East Delhi cause a violent protest and several injuries.	Gosh (2010)
2010	Guatemala	Development dispute	Yes	Two unidentified gunmen on a motorbike shoot an activist protesting the impacts of mining on water quality and local water rights.	Amnesty International (2012)
2010	Protestors, authorities in India	Development dispute	Yes	At least three deaths and dozens of injuries are reported during protests over land and water given away for a power plant in Sompeta, in Srikakulam District in Andhra Pradesh, India.	*The Hindu* (2010)
2011	Yemen	Military target	Yes	Violence in Yemen's capital, Sana'a, leads to "acute water and power shortages, forcing residents to rely on power generators and buy water extracted from wells and sold on a thriving black market." The violence arises during the Yemeni uprising during the Arab Spring protests across the Middle East. During the violence, governmental soldiers shell neighborhoods and destroy many rooftop water tanks.	Shah (2012)
2011	Solomon Islands	Development dispute	Yes	Amnesty International reports that women in the slums of the Solomon Islands must walk more than a kilometer to fetch clean water and are "continually harassed, attacked, and raped." A survey shows that 92 percent of households do not have a tap in their home and that local water sources are often polluted, forcing women to walk through unsafe areas.	Amnesty International (2011)
2011	Israel, Palestinians	Development dispute; Military target	Yes	Israel's military destroys nine water tanks in the Bedouin village of Amniyr in the South Hebron Hills in the West Bank, Palestine. Later, soldiers destroy pumps and wells in the Jordan Valley villages of Al-Nasaryah, Al-Akrabanyah, and Beit Hassan.	Aburawa (2011)
2011	Israel, Palestine	Development dispute; Military target	Yes	Israelis from nearby settlements attack Qasra, a West Bank village of 6,000, destroying crops and a water well. Attackers previously burned down a mosque and damaged hundreds of olive trees.	Bsharat and Ramadan (2011)

2011–2012	Syria	Development dispute; Military goal	Yes	Severe political conflict in Syria has been aggravated by the multiyear drought gripping the region. More than 1.5 million people—mostly farmers and their families—have moved to cities and their outskirts. In 2008, US diplomats in Syria warned that the influx of rural people to cities "could act as a multiplier on social and economic pressures already at play and undermine stability in Syria." Political unrest begins in March 2011 in Dara'a and soon escalates into civil war as ousters seek to overturn the regime of President Bashar al-Assad and the ruling Ba'ath Party.	Mohtadi (2012); Worth (2010); Femia and Werrell (2012)
2012	Libya	Military tool	Yes	During the 2011 Libyan Civil War, forces loyal to dictator Muammar Gaddafi gain control of a water operations center and cut off water supply to the capital. The system controls Libya's "Great Manmade River"—a system of pumps, pipes, and canals that brings water from distant aquifers to Tripoli and other cities. Half the country is left without running water, prompting the United Nations and neighboring countries to mobilize tanker ships to deliver water to coastal cities.	Circle of Blue (2011); UPI (2011)
2012	Afghanistan	Terrorism	Yes	Up to 150 schoolgirls are reported sickened by poison in a school water supply in an intentional attack thought to have been carried out by religious conservatives opposed to the education of women.	Hamid (2012)
2012	Afghanistan	Terrorism	Yes	Seven children are killed by a bomb thought to have been intended for Afghan police and planted at a freshwater spring in Ghor Province.	Shah (2012)
2012	Afghanistan	Military target; Terrorism	Yes	Islamist militants execute militia members defending the Machalgho Dam in eastern Afghanistan. The dam is being developed for irrigation and local power supply. This dispute is one of several surrounding the international waters of Afghanistan, Iran, and Pakistan, which share several rivers.	Mashal (2012)
2012	India	Development dispute	Yes	Thousands of farmers in Karnataka try to prevent the release of water from two dams (Krishna Raja Sagar and Kabini) on the Cauvery River. Injuries to protestors and police are reported. The water releases were ordered by the Supreme Court of India, which required Karnataka to deliver water to the downstream state of Tamil Nadu despite severe drought. The dispute continues later in the year when Karnataka again halts releases.	Circle of Blue (2012); *Indian Express* (2012)
2012	India	Development dispute	Yes	Scuffles and protests break out around New Delhi during the summer as residents surround water delivery trucks and fight over water. The summer is the hottest in thirty-three years, leading to extensive energy and water shortages.	Reuters (2012a)

continues

Water Conflict Chronology *continued*

Date	Parties Involved	Basis of Conflict	Violent Conflict or in the Context of Violence?	Description	Sources
2012	India, Pakistan	Development dispute; Military target	Yes	Violence erupts in the latest event in the dispute between Pakistan and India over the waters of the Indus River basin. Pakistani militants attack and sabotage water systems, flood protection works, and dams in the Wullar Lake region of northern Kashmir. They attack engineers and workers and detonate explosives at the unfinished Tulbul Navigation Lock/Wullar Dam. Pakistan claims the new dam violates the Indus Water Treaty by cutting flows to Pakistan.	Ul Hassan (2012)
2012	Brazil	Development dispute	Yes	Brazil's federal police respond to reports that water used by the indigenous Guarani-Kaiowa tribe has been poisoned by nearby landowners attempting to gain control over disputed land. Since 2009, the dispute has led to the deaths of three tribesmen; tribesmen say the water runs through sacred land.	AP (2012)
2012	Brazil	Development dispute	Property damage	Work on the controversial $13 billion Belo Monte Dam is halted after protesters burn buildings at three dam sites.	Phys.org (2012)
2012	Brazil	Development dispute	Yes	Northeastern Brazil sees growing conflicts after severe drought reduces water availability. News agencies report that one person a day is being killed in "water wars" that involve locals fighting over scarce supplies.	Catholic Online (2012)
2012	Peru	Development dispute	Yes	Several incidents of protests, injuries, and deaths are reported in regions of Peru where residents oppose large mines because of concerns over water quality and water rights. Police kill four protestors in clashes over the proposed Canadian-operated $5 billion Minas Conga gold mine.	Reuters (2012b); Yeager (2012)
2012	Peru	Development dispute	Yes	Protests because of concerns over water quality and water rights around the Xstrata Tintaya copper mine lead to two deaths and fifty injuries.	Reuters (2012c)
2012	Egypt	Development dispute	Yes	Farmers from the Abu Simbel region in Egypt hold over 200 tourists hostage to protest inadequate irrigation water. The farmers had captured the tourists after they visited nearby monuments. They release them after officials agree to a temporary release of water.	*Egypt Independent* (2012)
2012	Egypt	Development dispute	Yes	Public protests over drinking and irrigation water shortages take place across Egypt. Several protests turn violent: in Beni Suef, one person is killed and many are injured during a conflict over irrigation water; in Minya, villagers clash with officials over water shortages and water pollution; in Fayyoum, hundreds of people protesting water shortages block a highway and set fires.	Ooska News (2012)

Year	Parties	Basis of Conflict	Violent Conflict or in the Context of Violence	Description	Sources
2012	Somalia, Kenya	Military target	Yes	Somali Al-Shabaab insurgents poison a well and damage water infrastructure near the port city of Kismayo, Somalia. Insurgents are fighting Kenyan peacekeeping troops participating in the African Union mission in Somalia.	Wabala (2012)
2012	Kenya	Development dispute	Yes	Violence, including several deaths, occurs during disputes over access to water in the poorest slums around Nairobi, Kenya.	Njeru (2012)
2012	Tajikistan, Uzbekistan	Development dispute	Yes	Uzbekistan cuts natural gas deliveries to Tajikistan in retaliation over a Tajik hydroelectric dam that Uzbeks say will disrupt water supplies. Gas flows resume after a new contract is signed.	Kozhevnikov (2012)
2012	Kyrgyzstan, Tajikistan, Uzbekistan, Turkmenistan, Kazakhstan	Development dispute	No	Tensions escalate over two proposed dams in central Asia: Kambarata-1 in Kyrgyzstan and the Rogun Dam in Tajikistan. These dams could affect water supplies in the downstream nations of Uzbekistan, Turkmenistan, and Kazakhstan. Uzbekistan's president, Islam Karimov, says the dams could cause "not just serious confrontation, but even wars."	*The Economist* (2012)
2012	Sudan, South Sudan	Development dispute	Yes	Violence breaks out at water points in the Jamam refugee camp in South Sudan. Médicins Sans Frontières reports that as many as ten refugees die every day because of water shortages at refugee camps in South Sudan.	McNeish (2012); Ferrie (2012)
2012	Egypt, Ethiopia, Sudan	Development dispute; Military target	Yes	Information is leaked about an alleged secret agreement that would allow Egypt to build an air base in Sudan to attack the Grand Ethiopian Renaissance Dam (GERD). Egypt is concerned that the dam, under construction in Ethiopia just upstream of Sudan on the Blue Nile, would reduce flows into its territory. The news reports, strongly denied by Egypt, claim that Sudan would allow Egypt to launch attacks if diplomatic efforts fail to resolve water-sharing disputes between Egypt and Ethiopia. The allegations are based on an internal 2010 e-mail made available by WikiLeaks.	*Sudan Tribune* (2012); Al Arabiya News (2012)
2012	Mali, Burkina Faso	Development dispute	Yes	A clash along the border between Dogon villagers from Mali and nomadic Fulani herders from Burkina Faso kills at least thirty people after an earlier agreement to share water and pastureland is revoked. Chaos following a military coup in March is partly responsible for the breakdown in law and order in Mali.	Xinhua News (2012b)
2012	Mali, Mauritania	Development dispute	Yes	Protests and violence over water shortages erupt in the capital of Mauritania, Nouakchott. By July, over 70,000 Malian refugees are seeking asylum in Mauritania, putting pressure on scarce food and water supplies.	Taha (2012)

continues

Water Conflict Chronology *continued*

Date	Parties Involved	Basis of Conflict	Violent Conflict or in the Context of Violence?	Description	Sources
2012	Somalia	Development dispute	Yes	In August, fighting between two clans in the Lower Jubba region of south Somalia kills at least three people and wounds five. Reports from the village of Waraq (near the border with Kenya) indicate that the dispute began over the ownership of new water wells.	Shabelle Media Network (2012)
2012	Uganda, Kenya	Development dispute	Yes	Tensions lead to violence between Uganda and Kenya after Kenyan Pokot herdsmen cross the border seeking water and pasture. In October, the Ugandan government sends 5,000 soldiers to control violence among pastoralists from the two countries.	Bii (2012)
2012	Tanzania	Development dispute	Yes	Violence between farmers and pastoralists expands in Tanzania's southeastern Rufiji Valley, a region hit by drought. A farmer is killed in a conflict with a herdsman over access to water in the southern regions of Lindi and Mtwara. Five more people die and many more are injured in subsequent violence. According to local sources, violence has worsened during the prolonged drought.	Makoye (2012)
2012	South Africa	Development dispute	Yes	Protesters in poor communities of Cape Town, South Africa, riot over inadequate water and power. Hundreds burn tires, destroy cars, and throw rocks at police in anger over the lack of basic services.	Xinhua News (2012a)
2012	Syria	Military target	Yes	During the Syrian Civil War, the major pipeline delivering water to the city of Aleppo is badly damaged. The city of 3 million suffers severe shortages of drinking water.	BBC (2012a)
2012	Syria	Military target	Yes	In November, Syrian rebels fighting the government of President Bashar al-Assad overrun governmental forces and capture the Tishrin hydroelectric dam on the Euphrates River, after days of heavy clashes. The dam supplies electricity to part of Syria and is considered strategically important to the Syrian regime.	Mroue (2012)
2012	Indonesia	Development dispute	Yes	Violence breaks out over access to a water source in Maluku, Indonesia. Rival mobs from two villages attack each other "with sharp weapons, guns and explosives," causing several deaths and injuries.	Antara (2012)
2012–2013	Kenya	Development dispute	Yes	Extensive violence over water is reported in Kenya, with more than one hundred deaths in clashes between farmers and cattle herders. The conflict is part of a long-running dispute between Pokomo farmers and Orma people, seminomadic cattle herders, over land and water. The current conflict is exacerbated by Kenyan and foreign investment in vast tracts of land for food and biofuel cultivation, putting pressure on local resources. (See also the entry in 2001.)	AFP (2012); *Wikipedia* (2013b)

References

ABC News. 2007. Man charged with murder after lawn watering dispute. Australian Broadcasting Company (ABC), November 1. http://www.abc.net.au/news/stories/2007/11/01/2078076.htm (accessed February 11, 2011).

Absolute Astronomy. 2006. Incapacitating Agent. http://www.absoluteastronomy.com/reference/incapacitating_agent (accessed 2006).

Aburawa, A. 2011. Can water end the Arab-Israeli conflict? *Al Jazeera English*, July 29. http://english.aljazeera.net/indepth/features/2011/07/20117278519784574.html.

Agence France-Press (AFP). 2001. Macedonian troops fight for water supply as president moots amnesty. June 8. http://www.balkanpeace.org/hed/archive/june01/hed3454.shtml.

———. 2012. Kenya to hold peace meeting after 52 killed. August 23. http://www.google.com/hostednews/afp/article/ALeqM5jZdyUiAcUA7p94cH8PjEtslQUK3A?docId=CNG.4fe0cb53fa3afbb412c2a959ae285e36.211.

Ahmed, A. 1999. Fifty hurt in Bangladesh strike violence. Reuters, Dhaka, April 18.

Al Arabiya News. 2012. Egypt, Sudan could seek military action over Nile: WikiLeaks. October 12. http://english.alarabiya.net/articles/2012/10/12/243396.html.

Al-Ariqi, A. 2006. Water war in Yemen. *Yemen Times* 14 (932) (April 24). http://yementimes.com/article.shtml?i=932&p=health&a=1.

Al-Qadhi, M. 2003. Thirst for water and development leads to conflict in Yemen. *Choices* (United Nations Development Programme) 12 (1): 13–14. See also http://yementimes.com/article.shtml?i=642&p=health&a=1.

American Red Cross (ARC). 2003. Baghdad hospitals reopen but health care system strained. Mason Booth, staff writer, RedCross.org, April 24. http://www.redcross.org/news/in/iraq/030424baghdad.html.

American University. 2000a. Cauvery River dispute. Washington, DC: American University, Inventory of Conflict and Environment (ICE). http://www.american.edu/projects/mandala/TED/ice/cauvery.htm.

———. 2000b. Lesotho "Water Coup." Washington, DC: American University, Inventory of Conflict and Environment (ICE). http://www.american.edu/projects/mandala/ted/ice/leswater.htm.

———. 2000c. Marsh Arabs and Iraq. Washington, DC: American University, Inventory of Conflict and Environment (ICE). http://www.american.edu/projects/mandala/TED/ice/marsh.htm.

Amnesty International. 2004. Sudan: Darfur; "Too Many People Killed for No Reason." February 3. http://www.amnesty.org/en/library/info/AFR54/008/2004.

———. 2006. Lebanon: Deliberate Destruction or "Collateral Damage"? Israeli Attacks on Civilian Infrastructure. August 22. http://web.amnesty.org/library/Index/ENGMDE180072006.

———. 2011. "Where Is the Dignity in That?" Women in Solomon Islands Slums Denied Sanitation and Safety. September. http://www.amnesty.org.nz/files/SolomonIslandsWEB.pdf.

———. 2012. Guatemala: Lives and Livelihoods at Stake in Mining Conflict. June 21. http://www.amnesty.org/en/news/guatemala-lives-and-livelihoods-stake-mining-conflict-2012-06-21.

Antara, J. G. 2012. Bloody clashes in Indonesia's Maluku claim another life. *Jakarta Globe*, March 8. http://www.thejakartaglobe.com/news/bloody-clashes-in-indonesias-maluku-claim-another-life/503346.

Arbuthnot, F. 2000. Allies deliberately poisoned Iraq public water supply in Gulf War. *Sunday Herald*, Scotland, September 17.

Associated Press (AP). 2002. "Earth Liberation Front members threaten Colorado town's water." October 15.

———. 2003a. Incendiary devices placed at water plant. September 25.

———. 2003b. Water targeted, magazine reports. May 29.

———. 2004. US dumps water projects in Gaza over convoy bomb. May 6.

———. 2005. At least 16 killed in Somalia over water, pasture battles. June 8.

———. 2009. Bomb in water truck kills 3 in Afghanistan. November 23. http://www.signonsandiego.com/news/2009/nov/23/bomb-in-water-truck-kills-3-in-afghanistan/ (accessed February 10, 2011).

———. 2010. Fighting between tribes in Pakistan kills 102. September 17. http://www.google.com/hostednews/ap/article/ALeqM5hkiMxbHNH0BqgpWA2ZG6VD6wVTmAD9I9NH7G0.

———. 2012. Brazil police investigate indigenous claim that creek in sacred land was poisoned. *Fox News*, November 21. http://www.foxnews.com/world/2012/11/21/brazil-police-investigate-indigenous-claim-that-creek-in-sacred-land-was/.

Axis History Forum (AHF). 2004. Dnieper Dam blown up by retreating Russians. http://forum.axishistory.com/viewtopic.php?f=55&t=52940 (accessed February 10, 2011).

Bachiene, W. 1791. *Vaderlandsche geographie of nieuwe tegenwoordige staat en hedendaagsche historie der Nederlanden*. Amsterdam.

Barrett, G. 2003. Iraq's bad water brings disease, alarms relief workers. *Olympian*, Olympia, Washington, June 29.

Barry, J. M. 1997. *Rising Tide: The Great Mississippi Flood of 1927 and How It Changed America*, 67. New York: Simon and Schuster.

Belitser, N., S. Gerasymchuk, O. Grytsenko, Y. Dovgopol, Z. Zhminko, Y. Matiychyk, O. Sushko, and O. Chabala. 2009. Transnistrian Problem: A View from Ukraine. Kiev, Ukraine: Strategic and Security Studies Group (SSSG). http://www.irf.ua/files/eng/text_eng.pdf.

Bii, B. 2012. Tension at Kenya-Uganda border over rustling. *Daily Nation*, October 3. http://www.nation.co.ke/News/-/1056/1523908/-/xtc5qbz/-/index.html.

Bingham, G., A. Wolf, and T. Wohlegenant. 1994. Resolving water disputes: Conflict and cooperation in the United States, the Near East, and Asia. Washington, DC: United States Agency for International Development (USAID), Bureau for Asia and the Near East.

Bosshard, P. 2009. China dams the world. *World Policy Journal* (Winter 2009/2010): 43–51.

British Broadcasting Corporation (BBC). 1999. World: Asia-Pacific Timor atrocities unearthed. BBC News, September 22. http://news.bbc.co.uk/hi/english/world/asia-pacific/news id_455000/455030.stm.

———. 2000. Kenyan monkeys fight humans for water. BBC News, March 21. http://news.bbc.co.uk/1/hi/world/africa/685381.stm.

———. 2001. US "bombed Afghan power plant." BBC News, November 1. http://news.bbc.co.uk/1/hi/world/south_asia/1632304.stm.

———. 2002. "Cyanide attack" foiled in Italy. BBC News, February 20. http://news.bbc.co.uk/hi/english/world/europe/newsid_1831000/1831511.stm.

———. 2004a. China tries to calm dam protests. BBC News, Louisa Lim, November 18. http://news.bbc.co.uk/go/pr/fr/-/2/hi/asia-pacific/4021901.stm.

———. 2004b. "Dozens dead" in Somalia clashes. BBC News, December 6. http://news.bbc.co.uk/2/hi/africa/4073063.stm.

———. 2005. Thousands flee Kenyan water clash. BBC News, January 24. http://news.bbc.co.uk/1/hi/world/africa/4201483.stm.

———. 2006a. China "executes dam protester." BBC News, December 7. http://www.bbc.co.uk/go/pr/fr/-/2/hi/asia-pacific/6217148.stm.

———. 2006b. Somalis clash over scarce water. BBC News, February 17. http://news.bbc.co.uk/go/pr/fr/-/1/hi/world/africa/4723008.stm.

———. 2006c. Sri Lanka forces attack reservoir. BBC News, August 7. http://news.bbc.co.uk/2/hi/south_asia/5249884.stm?ls.

———. 2006d. Water and war in Sri Lanka. BBC News, August 3. http://news.bbc.co.uk/2/hi/south_asia/5239570.stm.

———. 2007. Iran faces $2.65 bn US bomb award. BBC News, September 7. http://news.bbc.co.uk/go/pr/fr/-/2/hi/middle_east/6984365.stm.

———. 2009. Water pipe sparks Ethiopian conflict. BBC News, March 13. http://news.bbc.co.uk/2/hi/africa/7929104.stm.

———. 2012a. Aleppo water supply cut as Syria fighting rages. BBC News, September 8. http://www.bbc.co.uk/news/world-middle-east-19533112.

———. 2012b. Mass graves found in Kenya's Tana River Delta region. BBC News, September 18. http://www.bbc.co.uk/news/world-africa-19634076.

Bsharat, M., and S. A. Ramadan. 2011. Fears of violence mount in West Bank on eve of Palestinian UN bid. Xinhua News, September 21. http://news.xinhuanet.com/english2010/world/2011-09/21/c_131150218.htm.

Burns, J. F. 1992. Tactics of the Sarajevo siege: Cut off the power and water." *New York Times*, September 25, A1.

Bustamante, R., J. Butterworth, M. Flierman, D. Herbas, M. den Hollander, S. van der Meer, P. Ravenstijn, M. Reynaga, and G. Zurita. 2004. Livelihoods in conflict: Disputes over water for household-level productive uses in Tarata, Bolivia. In Beyond Domestic: Case Studies on Poverty and Productive Uses of Water at the Household Level, edited by P. Moriarty, J. Butterworth, and B. van Koppen, 137–152. Technical Paper Series 41. Delft, Netherlands: IRC International Water and Sanitation Centre. http://www.irc.nl/content/download/6802/105351/file/preprints.pdf.

Butts, K., ed. 1997. *Environmental Change and Regional Security*. Carlisle, PA: US Army War College, Center for Strategic Leadership, Asia-Pacific Center for Security Studies.

Cable News Network (CNN). 1999. U.S.: Serbs destroying bodies of Kosovo victims. May 5. http://www.cnn.com/WORLD/europe/9905/05/kosovo.bodies.

———. 2010. Water conflict in Pakistan's tribal region leaves dozens dead. CNN International Edition. http://articles.cnn.com/2010-09-19/world/pakistan.water.dispute_1_tribal-region-water-dispute-mumtaz-zareen?_s=PM:WORLD (accessed February 10, 2011).

Caesar, J. 1906. *Caesar's Civil War with Pompeius*. Translated by F. P. Long. Rome: Clarendon Press, 1906. http://books.google.com/books?id=z-8JAQAAIAAJ.

Calcagno, F. 2004. Historical Overview—Licensee Assessments (Lessons Learned). January 27. San Francisco: US Department of Energy, Federal Energy Regulatory Commission. http://www.ferc.gov/EventCalendar/Files/20040309095131-1%20Historical%20Overview.pdf.

Catholic Online. 2012. Severe drought in Brazil destroying herds and lives. October 12. http://www.catholic.org/international/international_story.php?id=48869.

Centre for Development and Enterprise (CDE). 2007. Voices of anger: Phumelela and Khutsong; Protest and conflict in two municipalities. *CDE Focus* 10 (April). http://www.cde.org.za/index.php/component/content/article/85-4-south-africa-s-future-and-issues-of-national-importance/214-voices-of-anger-phumelela-and-khutsong-protest-and-conflict-in-two-municipalities?highlight=Y To2OntpOjA7czo2OiJ2b2lj ZXMiO2k6MTtz OjI6Im9 mIjtpOjI7czo1OiJhb mdlciI7aTozO3M6OToidm9pY2VzIG9mIjtpOjQ7czoxNToidm9pY2VzIG9mIGFuZ2VyIjtpO jU7czo4OiJvZiBhbm dlciI7fQ==.

Chandran, R. 2009. One killed in Mumbai water shortage protests. Reuters, December 3. http://www.reuters.com/article/idUSBOM105009.

Chelyshev, A. 1992. Terrorists poison water in Turkish army cantonment. Telegraph Agency of the Soviet Union (TASS), March 29.

Chenje, M. 2001. Hydro-politics and the quest of the Zambezi River Basin Organization. In *International Waters in Southern Africa*, edited by M. Nakayama. Tokyo: United Nations University.

China Daily. 2004. PLA general: Attempt to destroy dam doomed. June 16. http://www.chinadaily.com.cn/english/doc/2004-06/16/content_339969.htm.

China Ministry of Water Resources. 2001. Policy and Regulatory Department website. http://shuizheng.chinawater.com.cn/ssjf/20021021/200210160087.htm.

China Water Resources Daily. 2002. Villagers fight over water resources. October 24. Citation provided by Ma Jun, personal communication.

Choe, S.-H. 2009. South Korea demands apology from North over dam incident. *New York Times*, September 8. http://www.nytimes.com/2009/09/09/world/asia/09korea.html (accessed February 10, 2011).

Christian Science Monitor. 2000. Ecoterrorism as negotiating tactic. July 21, 8.

Circle of Blue. 2011. Water as a tool of war: Qaddafi loyalists turn off tap for half of Libya. *Circle of Blue Water News*, August 30. http://www.circleofblue.org/waternews/2011/world/water-as-a-tool-of-war-qaddafi-loyalists-turn-off-tap-for-half-of-libya/.

———. 2012. Protests break out after India's Supreme Court rules in favor of downstream state in Cauvery River dispute. *Circle of Blue Water News*, October 3. http://www.circleofblue.org/waternews/2012/world/protests-break-out-after-indias-supreme-court-rules-in-favor-of-downstream-state-in-cauvery-river-dispute/.

Clark, R. C. 1980. *Technological Terrorism*. Old Greenwich, CT: Devin-Adair.

Clark, R. M., and R. A. Deininger. 2000. Protecting the nation's critical infrastructure: The vulnerability of U.S. water supply systems. *Journal of Contingencies and Crisis Management* 8 (2): 73–80.

Collins, L., and D. LaPierre. 1972. *O Jerusalem*. New York: Simon and Schuster.

Columbia Encyclopedia. 2000a. Netherlands. 6th ed. http://www.bartleby.com/65/ne/Nethrlds.html.

———. 2000b. Vietnam: History. http://www.infoplease.com/ce6/world/A0861793.html.

Coonan, C. 2006. China secretly executes anti-dam protester. *The Independent*, London, December 7. http://www.independent.co.uk/news/world/asia/china-secretly-executes-antidam-protester-427401.html.

Cooperation Center for Afghanistan. 2000. The Social Impact of Drought in Hazarajat. http://www.ccamata.com/impact.html.

Crase, L. 2009. Water policy in Australia: The impact of drought and uncertainty. In *Policy and Strategic Behaviour in Water Resource Management*, edited by A. Dinar and J. Albiac, 91–107. London: Earthscan.

Crecente, B. D. 2002. ELF targets water: Group threatens eco-terror attack on Winter Park tanks. *Rocky Mountain News*, October 15. http://www.rockymountainnews.com/drmn/state/article/0,1299,DRMN_21_1479883,00.html.

Daniel, C., ed. 1995. *Chronicle of the 20th Century*. New York: Dorling Kindersley.

Dolatyar, M. 1995. Water diplomacy in the Middle East. In *The Middle Eastern Environment*, edited by E. Watson. London: John Adamson.

Douglass, J. D., and N. C. Livingstone. 1987. *America the Vulnerable: The Threat of Chemical and Biological Warfare*. Lexington, MA: Lexington Books.

Drower, M. S. 1954. Water-supply, irrigation, and agriculture. In *A History of Technology*, edited by C. Singer, E. J. Holmyard, and A. R. Hall. New York: Oxford University Press.

Dutch Water Line. 2002. Information on the historical use of water in defense of Holland. http://www.xs4all.nl/~pho/Dutchwaterline/dutchwaterl.htm.

The Economist. 2012. Water wars in Central Asia: Dammed if they do. September 29. http://www.economist.com/node/21563764.

Egypt Independent. 2012. Abu Simbel farmers release tourists held over water shortage. June 3. http://www.egyptindependent.com/news/farmers-hold-tourists-abu-simbel-over-water-shortage-news2.

Eitzen, E. M., and E. T. Takafuji. 1997. Historical overview of biological warfare. In *Medical Aspects of Chemical and Biological Warfare: Textbook of Military Medicine*. Bethesda, MD: United States Army, Office of the Surgeon General.

Environment News Service (ENS). 2001. Environment a weapon in the Israeli-Palestinian conflict. February 5. http://www.ens-newswire.com/ens/feb2001/2001-02-05-01.asp.

Evans, R. J. 2008. *The Third Reich at War*. New York: Penguin.

Express Tribune. 2010. Kurram tribal clash leaves 13 more dead. Lahore, Pakistan, September 19. http://tribune.com.pk/story/51998/kurram-tribal-clash-leaves-13-more-dead/.

Fatout, P. 1972. *Indiana Canals*, 158–162. West Lafayette, IN: Purdue University Studies.

Femia, F., and C. Werrell. 2012. Syria: Climate change, drought, and social unrest. Center for Climate and Security, February 29. http://climateandsecurity.org/2012/02/29/syria-climate-change-drought-and-social-unrest/.

Fenton, B., T. Helm, and B. Dutter. 2001. Anthrax letter spreads panic around world. *The Telegraph*, October 16. http://www.telegraph.co.uk/news/worldnews/northamerica/usa/1359605/Anthrax-letter-spreads-panic-around-world.html.

Ferrie, J. 2012. South Sudan refugee water shortage killing up to 10 people daily. Bloomberg News, June 13. http://www.businessweek.com/news/2012-06-13/south-sudan-refugee-water-shortage-killing-up-to-10-people-daily.

Fickle, J. E. 1983. The "people" versus "progress": Local opposition to the construction of the Wabash and Erie Canal. *Old Northwest* 8 (4): 309–328.

Financial Times Global Water Report (*FTGWR*). 1999. Zambia: Water cutoff. *FTGWR* 68 (March 19): 15.

———. 2000. Drought in India comes as no surprise. *FTGWR* 94 (April 28): 14.

———. 2002. Maoists destroy Nepal's infrastructure. *FTGWR* 146 (May 17): 4–5.

Fonner, D. K. 1996. Scipio Africanus. *Military History Magazine*, March. Cited at http://historynet.com/mh/blscipioafricanus/index1.html.

Forkey, N. S. 1998. Damning the dam: Ecology and community in Ops Township, Upper Canada. *Canadian Historical Review* 79 (1): 68–99.

Friel, T. 2007. Taliban flee battle using children as shields: NATO. Reuters.

Gaura, M. A. 2000. Disaster simulation too realistic. *San Francisco Chronicle*, October 27, A1.

Gellman, B. 2002. Cyber-attacks by Al Qaeda feared. *Washington Post*, June 27, A1.

Gleick, P. H. 1991. Environment and security: The clear connections. *Bulletin of the Atomic Scientists* (April): 17–21.

———. 1993. Water and conflict: Fresh water resources and international security. *International Security* 18 (1): 79–112.

———. 1994. Water, war, and peace in the Middle East. *Environment* 36 (3): 6–42. Washington, DC: Heldref Publishers.

———. 1995. Water and Conflict: Critical Issues. Presentation to the 45th Pugwash Conference on Science and World Affairs, Hiroshima, Japan, July 23–29.

Gleick, P.H. "Water and terrorism" in *The World's Water 2006-2007* (Washington, DC: Island Press, 2007), 1-28.

Gosh, D. 2010. Protest against water shortage turns violent. *Times of India*, July 7. http://timesofindia.indiatimes.com/articleshow/6136544.cms.

Gowan, H. 2004. Hannibal Barca and the Punic Wars. http://www.barca.fsnet.co.uk/ (accessed March 2005).

Grant, U. S. 1885. *Personal Memoirs of U.S. Grant*. New York: C. L. Webster.

Guantanamo Bay Gazette. 1964. The History of Guantanamo Bay: An Online Edition. Chapter 21: The 1964 water crisis. Archived version at archive.org: http://web.archive.org/web/20011121080500/www.nsgtmo.navy.mil/History/HISCHP21.HTM.

The Guardian. 2004. Water duel kills elderly cousins. London, March 11. http://www.guardian.co.uk/world/2004/mar/11/mexico.

Gutierrez, J. 2006. Sri Lanka rejects Norway deal with Tigers, battle for water resumes. Agence France-Press, August 6.

Haeri, S. 1993. Iran: Vehement reaction. *Middle East International*, March 19, 8.

Hamid, M. 2012. Afghan schoolgirls poisoned in anti-education attack. Reuters, April 17. http://www.reuters.com/article/2012/04/17/us-afghanistan-women-idUSBRE83G0PZ20120417.

Harris, S. H. 1994. *Factories of Death: Japanese Biological Warfare 1932–1945 and the American Cover-up*. New York: Routledge.

Hatami, H., and P. Gleick. 1994. Chronology of conflict over water in the legends, myths, and history of the ancient Middle East. In Water, war, and peace in the Middle East. *Environment* 36 (3): 6ff. Washington, DC: Heldref Publishers.

Hickman, D. C. 1999. A Chemical and Biological Warfare Threat: USAF Water Systems at Risk. Counterproliferation Paper No. 3. Maxwell Air Force Base, AL: United States Air Force Counterproliferation Center, Air War College.

Hillel, D. 1991. Lash of the dragon. *Natural History*, August, 28–37.

The Hindu. 2002a. Farmers go berserk; MLA's house attacked. October 30. http://www.hinduonnet.com/thehindu/2002/10/30/stories/2002103004870400.htm.

———. 2002b. Ryots on the rampage in Mandya. October 31. http://www.hinduonnet.com/thehindu/2002/10/31/stories/2002103106680100.htm.

———. 2010. Sompeta: Land allotment suspended. June 24. http://www.hindu.com/2011/06/24/stories/2011062457830100.htm.

History of War Online. 2011. Siege of Uxellodunum, Spring–Summer 51 B.C. http://www.historyofwar.org/articles/siege_uxellodunum.html (accessed February 11, 2011).

Hitchcock, E. A. 1862. Signed letter from Major General Ethan Allen Hitchcock to his niece Mary, Washington City. April 20. (During this period, Hitchcock served as chairman of the War Board serving President Lincoln and Secretary of War Stanton.) http://www.liveauctioneers.com/item/7093913.

Honan, W. H. 1996. Scholar sees Leonardo's influence on Machiavelli. *New York Times*, December 8, 18. http://www.nytimes.com/1996/12/08/us/scholar-sees-leonardo-s-influence-on-machiavelli.html.

Human Rights Watch. 1998. Human Rights Watch Condemns Civilian Killings by Congo Rebels. http://www.hrw.org/press98/aug/congo827.htm.

Husarska, A. 1995. Running dry in Sarajevo: Water fight. *New Republic*, July 17 and 24.

Illustrated History of the Roman Empire. 2011. http://www.roman-empire.net/army/leg-siege.html (accessed February 11, 2011).

Indian Express. 2012. Stopped releasing Cauvery water to TN, says Karnataka CM. December 10. http://www.indianexpress.com/news/stopped-releasing-cauvery-water-to-tn-says-karnataka-cm/1043124/.

Indo-Asian News Service. 2004. Curfew imposed in three Rajasthan towns. *Hindustan Times*, December 4. http://www.hindustantimes.com/news/181_1136315,000900010008.htm.

Information and Documentation Center for the Geography of the Netherlands (IDG). 1996. Water in, around, and under the Netherlands. IDG-Bulletin 1995/1996.

InfoRoma. 2004. Roman Aqueducts. http://www.inforoma.it/feature.php?lookup=aqueduct (accessed March 2005).

International Committee of the Red Cross (ICRC). 2003. Eritrea: ICRC repairs war-damaged health centre and water system. December 15. ICRC News No. 03/158. http://www.alertnet.org/thenews/fromthefield/107148342038.htm.

International Court of Justice (ICJ). 1999. International Court of Justice Press Communiqué 99/53, Kasikili Island/Sedudu Island (Botswana/Namibia). December 13. The Hague, Netherlands: International Court of Justice. http://www.icj-cij.org/icjwww/ipresscom/ipress1999/ipresscom9953_ibon.

International Herald Tribune. 1999. 100 bodies found in well. August 14–15, 4.

International War Crimes Tribunal (IWCT). 1967. Some Facts on Bombing of Dikes. http://www.infotrad.clara.co.uk/antiwar/warcrimes/index.html.

Israel, J. 1997. *Conflicts of Empires: Spain, the Low Countries, and the Struggle for World Supremacy, 1585–1713*. London: Hambledon Press.

Israel Line. 2001a. Palestinians loot water pumping center, cutting off supply to refugee camp. January 5. http://www.mfa.gov.il/mfa/go.asp?MFAH0iy50 (accessed January 5, 2001).

———. 2001b. Palestinians vandalize Yitzhar water pipe. January 9. http://www.mfa.gov.il/mfa/go.asp?MFAH0izu0.

Japan Times. 2002. Kashmir water clash. May 27, 3.

Jenkins, B. M., and A. P. Rubin. 1978. New vulnerabilities and the acquisition of new weapons by nongovernment groups. In *Legal Aspects of International Terrorism*, edited by A. E. Evans and J. F. Murphy, 221–276. Lexington, MA: Lexington Books.

Joseph, D. B. 1960. *The Faithful City: The Siege of Jerusalem, 1948*. New York: Simon and Schuster.

Josephus. AD 90. *Antiquities of the Jews*. Book 18, chapter 3, section 2. http://www.fullbooks.com/The-Antiquities-of-the-Jews20.html.

Kathmandu Post. 2002. KLF destroys micro hydro plant. January 28. http://www.nepalnews.com.np/contents/englishdaily/ktmpost/2002/jan/jan28/index.htm.

Keluel-Jang, S. A. 1997. Alier and the Jonglei Canal. *Southern Sudan Bulletin* 2 (3) (January). http://www.sufo.demon.co.uk/poli007.htm.

Kirschner, O. 1949. Destruction and Protection of Dams and Levees. Military Hydrology, Research and Development Branch, U.S. Corps of Engineers, Department of the Army, Washington District. From *Schweizerische Bauzeitung*, March 14, 1949. Translated by H. F. Schwarz, Washington, DC.

Kozhevnikov, R. 2012. Uzbekistan resumes gas supplies to Tajikistan. Reuters, April 16. http://www.reuters.com/article/2012/04/16/tajikistan-gas-idUSL6E8FG3YL20120416.

Kupperman, R. H., and D. M. Trent. 1979. *Terrorism: Threat, Reality, Response*. Stanford, CA: Hoover Institution Press.

Lane, M. 2005. Personal communication to P. Gleick regarding conflicts in northern Kenya, with reference to *Sunday Nation* newspaper reports of July 17, 2005.

Lemos, R. 2002. Safety: Assessing the infrastructure risk. CNET News.com, August 26. http://news.cnet.com/2009-1001_3-954780.html.

Levitskiy, Y. 2005. Kiev hydropower station threatened. *Evening News* (15): 123. In Russian only. Левицький Я. Київську ГЕС погрожували підірвати, Вечірні вісті.

Levy, K. 2000. Guatemalan dam massacre survivors seek reparations from financiers. *World Rivers Review*, December, 12–13. Berkeley, CA: International Rivers Network.

Livingston, N. C. 1982. *The War Against Terrorism*. Lexington, MA: Lexington Books.

Lockwood, R. P. 2006. The Battle over the Crusades. http://www.catholicleague.org/research/battle_over_the_crusades.htm (accessed April 2006).

Macedonia Information Agency. 2001. Humanitarian catastrophe averted in Kumanovo and Lipkovo. Republic of Macedonia Agency of Information Archive. June 18. http://wwww.reliefweb.int/w/rwb.nsf/0/dbd4ef105d93da4ac1256a6f005bc328?OpenDocument.

Makarov, P. 2005. To look in the shameless eyes of officials! In Russian only. Взглянуть в бесстыжие глаза чиновника! http://www.whp057.narod.ru/nikopol1-st1.htm and http://www.makarov.nikopol.net/doc/otnoshenie_vlasti.doc.

Makoye, K. 2012. Drought drives Tanzanian herders into conflict with farmers. Alert Net, June 12. http://www.trust.org/alertnet/news/drought-drives-tanzanian-herders-into-conflict-with-farmers/.

Malik, L. K. 2005. *Factors of Risk of Hydro Technical Buildings Damage: Problems of Safety*. Moscow: Nauka. In Russian only. Факторы риска повреждения гидротехнических сооружений: Проблемы безопасности.

Mashal, M. 2012. What Iran and Pakistan want from the Afghans: Water. *Time*, December 2. http://world.time.com/2012/12/02/what-iran-and-pakistan-want-from-the-afghans-water/.

McDonnell, P. J., and J. Meyer. 2002. Links to terrorism probed in Northwest. *Los Angeles Times*, July 13.

McNeish, H. 2012. Refugees in South Sudan face water crisis. Agence France-Press, March 31. http://www.google.com/hostednews/afp/article/ALeqM5ho-X1B6kIDKnLCHEUnlrW9665trw?docId%3DCNG.91f54df3baaf54aa7554fca3c3d379df.41.

Meissner, R. 2000. Hydropolitical hotspots in southern Africa: Will there be a water war? The case of the Kunene River." In *Water Wars: Enduring Myth or Impending Reality?*, edited by H. Solomon and A. Turton. Africa Dialogue Monograph Series No. 2. Umhlanga Rocks, South Africa: ACCORD.

———. 2001. Interaction and existing constraints in international river basins: The case of the Kunene River Basin. In *International Waters in Southern Africa*, edited by M. Nakayama. Tokyo: United Nations University.

Milwaukee Journal Sentinel (MJS). 2003. Jordan foils Iraqi plot to poison U.S. troops' water, officials say. April 1. http://www.jsonline.com/news/gen/apr03/130338.asp.

Mohamed, A. E. 2001. Joint development and cooperation in international water resources: The case of the Limpopo and Orange River Basins in southern Africa. In *International Waters in Southern Africa*, edited by M. Nakayama. Tokyo: United Nations University.

Mohtadi, S. 2012. Climate change and the Syrian uprising. *Bulletin of the Atomic Scientists*, August 16. http://thebulletin.org/web-edition/features/climate-change-and-the-syrian-uprising.

Moorehead, A. 1960. *The White Nile*. London: Penguin Books.

Mroue, B. 2012. Activists: Syrian rebels seize major dam in north. *Daily Star*, Lebanon, November 26. http://www.dailystar.com.lb/News/Middle-East/2012/Nov-26/196180-activists-syrian-rebels-seize-major-dam-in-north.ashx.

MSNBC. 2002. FBI says al-Qaida after water supply. Numerous wire reports; see, for example, http://www.ionizers.org/water-terrorism.html.

Murphy, I. L., and J. E. Sabadell. 1986. International river basins: A policy model for conflict resolution. *Resources Policy* 12 (1): 133–144.

Murphy, K. 2006. Old feud over Lebanese river takes new turn: Israel's airstrikes on canals renew enduring suspicions that it covets water from the Litani. *Los Angeles Times*, August 10. http://articles.latimes.com/2006/aug/10/world/fg-litani10.

Museum of the City of New York (MCNY). n.d. The Greater New York Consolidation Timeline. http://www.mcny.org/Exhibitions/GNY/timeline.htm.

Nadeem, A. 2001. Bombs in Karachi kill one." Associated Press, June 13.

Naff, T., and R. C. Matson, eds. 1984. *Water in the Middle East: Conflict or Cooperation?* Boulder, CO: Westview Press.

Nagy, T. J. 2001. The secret behind the sanctions: How the U.S. intentionally destroyed Iraq's water supply. *The Progressive* 165 (9): 22–26.

National Geographic News. 2001. Ancient Fertile Crescent almost gone, satellite images show." May 18. http://news.nationalgeographic.com/news/2001/05/0518_crescent.html.

NewsMax. 2003. Al-Qaida threat to U.S. water supply. NewsMax Wires, May 29. http://www.newsmax.com/archives/articles/2003/5/28/202658.shtml.

New York Times. 1941. Act laid to Stalin: Razing of dam reported ordered to bar Nazis' path with flood." August 21.

———. 1999. Puerto Ricans protest navy's use of water. October 31, 30.

———. 2012. Kenya: Clashes over land and water leave 12 dead. September 7. http://www.nytimes.com/2012/09/08/world/africa/kenya-clashes-over-land-and-water-leave-12-dead.html.

Njeru, G. 2012. Water shortages driving growing thefts, conflicts in Kenya. Alert Net, August 6. http://www.trust.org/alertnet/news/water-shortages-driving-growing-thefts-conflicts-in-kenya/.

Nuclear Threat Initiative (NTI). 2005. A Brief History of Chemical Warfare. http://www.nti.org/h_learnmore/cwtutorial/chapter02_02.html.

Okoko, T. O. 2000. Monkeys, humans fight over drinking water. Panafrican News Agency, March 21.

Ooska News. 2012. Egypt water shortage protests are widespread and violent. July 25. http://www.ooskanews.com/daily-water-briefing/egypt-water-shortage-protests-are-widespread-and-violent_23549.

Orlenko, I. P. 1981. We are "Tallinim." Eesti raamat, Tallinn, Estonia. In Russian only. Орленко И. Ф. Мы. 1981."Мы—'Таллиннские'." Ээсти раамат, Таллин, Эстония. http://www.bellabs.ru/51/Book1/Book1-00.html.

Oxfam. 2007. Gaza siege puts public health at risk as water and sanitation services deteriorate, warns Oxfam. http://www.reliefweb.int/rw/rwb.nsf/db900sid/AMMF-796CFU?OpenDocument&rc=3&emid=ACOS-635PFR.

Pagan, A. 2005. Catastrophic dam failures. CE News.com, December 2005. http://www.cenews.com/magazine-article-cenews-com-december-2005-catastrophic_dam_failures-4617.html.

Pannier, B. 2000. Central Asia: Water becomes a political issue. Radio Free Europe. http://www.rferl.org/nca/features/2000/08/F.RU.000803122739.html.

Parry, R. L. 2001. UN fears "disaster" over strikes near huge dam. *The Independent*, London, November 8.

Perlez, J., and P. Z. Shah. 2008. Confronting the Taliban at home, Pakistan finds itself at war. *New York Times*, October 3, 1.

Phys.org. 2012. Trouble at Brazil mega-dam stops construction for now. November 12. http://phys.org/news/2012-11-brazil-mega-dam.html.

Plant, G. 1995. Water as a Weapon in War. Water and War: Symposium on Water in Armed Conflicts, November 21–23, 1994, Montreux, Switzerland. Geneva, Switzerland: International Committee of the Red Cross.

Pottinger, M. 2000. Major Chinese lake disappearing in water crisis. Reuters Science News. http://us.cnn.com/2000/NATURE/12/20/china.lake.reut/.
Pretoria Dispatch. 1999. Dam bomb may be "aimed at farmers." July 21. http://www.dispatch.co.za/1999/07/21/southafrica/RESEVOIR.HTM.
Priscoli, J. D. 1998. Water and Civilization: Conflict, Cooperation, and the Roots of a New Eco Realism. Keynote address, 8th Stockholm World Water Symposium, August 10–13. http://www.genevahumanitarianforum.org/docs/Priscoli.pdf.
Public Services International (PSI). 2003. Urgent action: Bomb blast kills 3 workers at the Cali water treatment plant." http://www.world-psi.org and http://209.238.219.111/Water.htm.
Purver, R. 1995. Chemical and Biological Terrorism: The Threat According to the Open Literature. Ottawa, Ontario: Canadian Security Intelligence Service. http://www.csis.gc.ca/en/publications/other/c_b_terrorism01.asp.
Rasch, P. J. 1968. The Tularosa ditch war. *New Mexico Historical Review* 43 (3): 229–235.
Rathfelder, E. 2007. Dangerous forces: Dams, dikes, and nuclear stations, in *Crimes of War*. West Sussex, England: Random House. http://www.crimesofwar.org/thebook/dangerous-forces.html.
Reisner, M. 1986/1993. *Cadillac Desert: The American West and Its Disappearing Water*. New York: Penguin Books.
Reuters. 1999a. NATO builds evidence of Kosovo atrocities. June 17. http://dailynews.yahoo.com/headlines/ts/story.html?s=v/nm/19990617/ts/yugoslavia_leadall_171.html (accessed June 1999).
———. 1999b. NATO keeps up strikes but Belgrade quiet. June 5. http://dailynews.yahoo.com/headlines/wl/story.html?s=v/nm/19990605/wl/yugoslavia_strikes_129.html (accessed June 1999).
———. 1999c. Serbs say NATO hit refugee convoys. April 14. http://dailynews.yahoo.com/headlines/ts/story.html?s=v/nm/19990414/ts/yugoslavia_192.html. http://www.uia.ac.be/u/carpent/kosovo/messages/397.html.
———. 2004a. Al Qaeda spy chief killed in Pakistani raid. http://www.washingtonpost.com/wp-dyn/articles/A32878-2004Mar29_2.html.
———. 2004b. Darfur: "2.5 million people will require food aid in 2005." R. Schofield. November 22. http://www.medair.org/en_portal/medair_news?news=258.
———. 2006. Clashes over water, pasture kill 40 in east Africa. June 7. http://asia.news.yahoo.com/060606/3/2lk9x.html.
———. 2008. Clashes over water kill 4 in drought-hit Kenya. October 17. http://www.reuters.com/article/latestCrisis/idUSWAL762271.
———. 2012a. Power cuts spark protests as India swelters. July 4. http://www.reuters.com/article/2012/07/04/uk-india-electricity-protest-idUSLNE86300Q20120704.
———. 2012b. Update 3—Peru clash over Newmont mine kills three. July 3. http://www.reuters.com/article/2012/07/04/peru-newmont-idUSL2E8I3GB820120704.
———. 2012c. Update 3—Peru uses emergency rules to try to end anti-mining protest. May 28. http://www.reuters.com/article/2012/05/29/xstrata-peru-idUSL1E8GS34F20120529.
Rome Guide. 2004. Fontana di Trevi: History. http://web.tiscali.it/romaonlineguide/Pages/eng/rbarocca/sBMy5.htm (accessed March 2005).
Rowe, W. T. 1988. Water control and the Qing political process: The Fankou Dam controversy, 1876–1883. *Modern China* 14 (4): 353–387.
Ryu, A. 2005. Water rights dispute sparks ethnic clashes in Kenya's Rift Valley. Voice of America. http://www.voanews.com/english/archive/2005-03/2005-03-21-voa28.cfm.
Samson, P., and B. Charrier. 1997. International freshwater conflict: Issues and prevention strategies. Green Cross International. http://www.greencrossinternational.net/GreenCrossPrograms/waterres/gcwater/report.html.
Sandrasagra, M. J. 2000. Development Ethiopia: Relief agencies warn of major food crisis. Inter Press Service, April 11.
Scheiber, H. N. 1969. *Ohio Canal Era*, 174–175. Athens: Ohio University Press.
Science. 2006. Tallying Mideast Damage. *Science* 313 (5793): 1549.
Scofield, C. I., ed. 1967. *New Scofield Reference Bible, Authorized King James Version*. Oxford, England: Oxford University Press.
Semann, D. 1950. Die Kriegsbeschädigungen der Edertalsperrmauer, die Wiederherstellungsarbeiten und die angestellten Untersuchungen über die Standfestigkeit der Mauer. *Die Wasserwirtschaft*, 41 Jg., Nr. 1 u. 2.
Shabelle Media Network. 2012. Ethnic fighting renews in Lower Jubba region, South Somalia. Mogadishu, August 15. http://allafrica.com/stories/201208160039.html.

Shah, A. 2012. Officials say 7 Afghan children killed by bomb. Associated Press, July 25. http://news.yahoo.com/officials-7-afghan-children-killed-bomb-110922657.html.

Sharife, K. 2008. Tibet: Shifting climates on the roof of the world. *The Mail and Guardian Online*, Johannesburg, South Africa. http://www.thoughtleader.co.za/khadijasharife/2008/04/17/tibet-shifting-climates-on-the-roof-of-the-world-part-one/.

Shultz, J., and M. C. Draper, eds. 2009. *Dignity and Defiance: Stories from Bolivia's Challenge to Globalization*. Berkeley: University of California Press.

Simons, M. 1999. Serbs refuse to clear bomb-littered river. *New York Times*, October 24.

Singh, G. 2009. Water wars strike ahead of predictions. EcoWorldly.com, May 16. http://ecoworldly.com/2009/05/16/water-wars-strike-ahead-of-predictions/.

Soloman, A. 2001. Policeman dies as blasts rock strike-hit Karachi. Reuters, June 13. http://dailynews.yahoo.com/h/nm/20010613/ts/pakistan_strike_dc_1.html.

Statesman News Service. 2007. Clash takes place over water rights. *The Statesman*, Sambalpur, India. http://www.thestatesman.net/page.arcview.php?clid=9&id=202958&usrsess=1.

Steinberg, T. S. 1990. Dam-breaking in the nineteenth-century Merrimack Valley. *Journal of Social History* 24 (1): 25–45.

Stewart, W. H. n.d. An Analysis of the 1944 Japanese Defense of Saipan. http://saipanstewart.com/essays/Defense%20of%20Saipan.html (accessed February 11, 2011).

Stockholm International Peace Research Institute (SIPRI). 1971. *The Rise of CB Weapons: The Problem of Chemical and Biological Warfare*. New York: Humanities Press.

Strategy Page. 2006. Biotoxins in Warfare. http://www.strategypage.com/articles/biotoxin_files/BIOTOXINSINWARFARE.asp (accessed April 2006).

Styran, R. M., and R. R. Taylor. 2001. *"The Great Swivel Link": Canada's Welland Canal*. Toronto, Ontario: The Champlain Society.

Sudan Tribune. 2012. Government denies deal with Sudan to strike Ethiopian dam. September 24. http://allafrica.com/stories/201209250218.html.

Suliman, M. 1998. Resource Access: A Major Cause of Armed Conflict in the Sudan. The Case of the Nuba Mountains. London: Institute for African Alternatives. http://srdis.ciesin.org/cases/Sudan-Paper.html.

Taha, R. M. 2012. Mali conflict spilling over: Influx of refugees from Mali exacerbates political instability in neighbouring Mauritania." Daily News Egypt, July 1. http://www.dailynewsegypt.com/2012/07/01/mali-conflict-spilling-over/.

Thatcher, J. 1827. *A Military Journal during the American Revolutionary War, from 1775 to 1783.* 2nd ed., revised and corrected. Boston, MA: Cottons and Barnard. http://www.fortklock.com/journal1777.htm.

Tierney, J., and R. F. Worth. 2003. Attacks in Iraq may be signals of new tactics. *New York Times*, August 18, 1. http://www.nytimes.com/2003/08/18/international/worldspecial/18IRAQ.html?hp.

Times News Network (TNN). 2004. IED was planted in underground pipe. December 5. http://timesofindia.indiatimes.com/articleshow/947432.cms.

Times of India. 2002. Cauvery row: Farmers renew stir. October 20. http://timesofindia.indiatimes.com/cms.dll/html/uncomp/articleshow?art_id=26586125.

Toronto Star. 2007. Man faces attempted murder charge over water. November 6. http://www.thestar.com/printArticle/274128.

Troung, A. G., and T. Cooper. 2003. Laos, 1948–1989; Part 2. *Air Combat Information Group Journal*, November 3. http://www.acig.org/artman/publish/article_348.shtml.

Tucker, J. B., ed. 2000. *Toxic Terror: Assessing Terrorist Use of Chemical and Biological Weapons*. Cambridge, MA: MIT Press.

Turton, A. R. 2005. A Critical Assessment of the River Basins at Risk in the Southern African Hydropolitical Complex. Paper presented at the Workshop on the Management of International Rives and Lakes, hosted by the Third World Centre for Water Management and the Helsinki University of Technology, August 17–19, Helsinki, Finland. CSIR Report No. ENV-P-CONF 2005-001.

Ul Hassan, I. 2012. Militants sabotage works on J&K lake. *dna*, August 30. http://www.dnaindia.com/india/report_militants-sabotage-works-on-j-and-k-lake_1734411.

United Nations Children's Fund (UNICEF). 2003. Iraq: Cleaning Up Neglected, Damaged Water System, Clearing Away Garbage. News Note Press Release, May 27. http://www.unicef.org/media/media_6998.html.

United Nations Office for the Coordination of Humanitarian Affairs (UN OCHA). 2007. Burkina Faso: Innovation and Education Needed to Head Off Water War. http://www.irinnews.org/PrintReport.aspx?ReportId=74308.

United Press International (UPI). 2011. Gadhafi turns water project into a weapon. September 2. http://www.upi.com/Business_News/Energy-Resources/2011/09/02/Gadhafi-turns-water-project-into-a-weapon/UPI-66451314984213/.

United States Army Corps of Engineers (US ACE). 1953. Applications of Hydrology in Military Planning and Operations and Subject Classification Index for Military Hydrology Data. Washington, DC: Department of the Army, Corps of Engineers, Engineering Division, Military Hydrology R&D Branch.

United States Department of Defense (US DOD). 2004. FY04 Report to Congress on PRC Military Power, Pursuant to the FY2000 National Defense Authorization Act. Annual Report on the Military Power of the People's Republic of China 2004. http://www.defenselink.mil/pubs/d20040528PRC.pdf.

US Water News. 2003. Report suggests al-Qaida could poison U.S. water. June. http://www.uswaternews.com/archives/arcquality/3repsug6.html.

Vanderwood, P. J. 1969. *Night Riders of Reelfoot Lake.* Memphis, TN: Memphis State University Press.

Voice of America News (VOA). 2004. China's Sichuan Province tense in aftermath of violent anti-dam protests. Luis Ramirez. November 24.

Wabala, D. 2012. KDF airlifts water to troops as wells poisoned. *The Star*, Nairobi, September 24. http://allafrica.com/stories/201209250173.html.

Wallensteen, P., and A. Swain. 1997. International fresh water resources: Conflict or cooperation? In *Comprehensive Assessment of the Freshwater Resources of the World.* Stockholm: Stockholm Environment Institute.

Walters, E. 1948. *Joseph Benson Foraker: Uncompromising Republican*, 44–45. Columbus: Ohio History Press.

Waterman, S. 2003. Al-Qaida threat to U.S. water supply. United Press International (UPI), May 28.

Water Politics. 2012. Central Asia's Looming Conflict over Water: The Upriver Countries. November 12. http://www.waterpolitics.com/2012/11/12/central-asias-looming-conflict-over-water-the-upriver-countries/.

WaterWeek. 2002. Water facility attacked in Colombia. American Water Works Association. January. http://www.awwa.org/advocacy/news/020602.cfm.

Wax, E. 2006. Dying for water in Somalia's drought: Amid anarchy, warlords hold precious resource. *Washington Post*, April 14, A1. http://www.washingtonpost.com/wp-dyn/content/article/2006/04/13/AR2006041302116.html.

Wikipedia. 2011a. Battle of Alesia. http://en.wikipedia.org/wiki/Battle_of_Alesia (accessed February 10, 2011).

———. 2011b. Siege of Jerusalem (1948). http://en.wikipedia.org/wiki/Siege_of_Jerusalem_(1948) (accessed February 10, 2011).

———. 2013a. Siege of Massilia. http://en.wikipedia.org/wiki/Siege_of_Massilia (accessed August 15, 2013).

———. 2013b. 2012–2013 Tana River District Clashes. http://en.wikipedia.org/w/index.php?title=2012-2013_Tana_River_District_clashes&oldid=536856006 (accessed February 25, 2013).

Wolf, A. T. 1995. *Hydropolitics along the Jordan River: Scarce Water and Its Impact on the Arab-Israeli Conflict.* Tokyo: United Nations University Press.

———. 1997. "Water Wars" and Water Reality: Conflict and Cooperation along International Waterways. Presentation at NATO Advanced Research Workshop on Environmental Change, Adaptation, and Human Security, October 9–12, Budapest, Hungary.

Wong, E. 2009. Tibetan area a tinderbox for China-India tensions. *New York Times*, September 3. http://www.nytimes.com/2009/09/04/world/asia/04chinaindia.html.

Workman, J. G. 2009. *Heart of Dryness: How the Last Bushmen Can Help Us Endure the Coming Age of Permanent Drought.* New York: Walker & Company.

World Environment News. 2001. Philippine rebels suspected of water "poisoning." October 16. http://web.archive.org/web/20030708180932/http://www.planetark.org/avantgo/dailynewsstory.cfm?newsid=12807.

World Rivers Review (WRR). 1998. Dangerous dams: Tajikistan. *WRR* 13 (6) (December): 13. Berkeley, CA: International Rivers Network.

Worth, R. F. 2010. Earth is parched where Syrian farms thrived. *New York Times*, October 13. http://www.nytimes.com/2010/10/14/world/middleeast/14syria.html.

Xinhua News. 2012a. Cape Town hit by continued protest against poor service delivery. August 14. http://news.xinhuanet.com/english/world/2012-08/14/c_131782882.htm.

———. 2012b. Tribal clash kills 30 along Mali–Burkina Faso border. http://news.xinhuanet.com/english/world/2012-05/25/c_131611541.htm.

Yakubu, A. 2008. Water vendors protest at Nyanya. *Daily Trust*, Abuja, AllAfrica.com, March 4. http://allafrica.com/stories/printable/200803040513.html.

Yang Lang. 1989/1994. High dam: The sword of Damocles." In *Yangtze! Yangtze!*, edited by Dai Qing, 229–240. London: Probe International, Earthscan.

Yeager, C. 2012. When gold loses its luster: Four dead in Peru mining protests. *Circle of Blue Water News*, July 6. http://www.circleofblue.org/waternews/2012/commentary/editorial-in-the-circle-fresh-focus/when-gold-loses-its-luster-four-dead-in-peru-mining-protests/.

Zachary, G. P. 1997. Water pressure: Nations scramble to defuse fights over supplies. *Wall Street Journal*, December 4, A17.

Zakaria, F. 2013. "The Coming Water Wars." *Fareed Zakaria GPS*, CNN World. February 25, 2013. http://globalpublicsquare.blogs.cnn.com/2013/02/25/the-coming-water-wars/?iref=all search (Accessed March 2, 2013.)

Zemmali, H. 1995. International humanitarian law and protection of water. Water and War, Symposium on Water in Armed Conflicts, November 21–23, Montreux, Switzerland. Geneva, Switzerland: International Committee of the Red Cross.

DATA TABLE 1

Total Renewable Freshwater Supply by Country (2013 Update)

Description

Average annual renewable freshwater resources are listed by country, updating data table 1 in the previous versions of *The World's Water*. Data in this table from the AQUASTAT database of the Food and Agriculture Organization of the United Nations (FAO) were updated to reflect the most recent data in that database as of May 2013. However, because these data are typically produced by modeling or estimation, rather than measurement, we continue to use data from other sources where appropriate (even if older than data in AQUASTAT). For example, most data for European countries come from the Eurostat data set of the European Union. Data for Timor-Leste and for the Occupied Palestinian Territory (as defined by AQUASTAT) were added in this volume.

Data in this table typically comprise both renewable surface water and groundwater supplies, including surface inflows from neighboring countries. The FAO refers to this as total natural renewable water resources. Flows to other countries are not subtracted from these numbers. All quantities are in cubic kilometers per year (km^3/yr). These data represent average freshwater resources in a country—actual annual renewable supply will vary from year to year.

Limitations

These detailed country data should be viewed, and used, with caution. The data come from different sources and were estimated over different periods. Many countries do not directly measure or report internal water resources data, so some of these entries were produced using indirect methods.

Not all of the annual renewable water supply is available for use by the countries to which it is credited here; some flows are committed to downstream users. For example, under treaty requirements, Sudan must pass significant flows downstream to Egypt. Other countries, such as Turkey, Syria, and France, to name only a few, also pass significant amounts of water to other users. The annual average figures hide large seasonal, interannual, and long-term variations.

Sources

The 2013 update was compiled by P. H. Gleick, Pacific Institute.

When a recent source uses older data previously cited here, we have tried to leave the original citation date and source.

a. Total natural renewable surface water and groundwater. Typically includes flows from other countries. (FAO: "Natural total renewable water resources.")
b. Estimates from Belyaev, Institute of Geography, USSR (1987).
c. Frenken, K., ed. Estimates from UN FAO. 2005. Irrigation in Africa in Figures—Aquastat Survey 2005. Rome, Italy: Food and Agriculture Organization of the United Nations.
d. Estimates from WRI (1994). See this source for original data source.
e. Estimates from UN FAO. 1997. Irrigation in the Countries of the Former Soviet Union in Figures. Rome, Italy: Food and Agriculture Organization of the United Nations.
f. UN FAO. 1999. Irrigation in Asia in Figures. Rome, Italy: Food and Agriculture Organization of the United Nations.
g. Nix, H. 1995. Water/Land/Life: The Eternal Triangle. Canberra, Australia: Water Research Foundation of Australia.
h. UN FAO. 2000. Irrigation in Latin America and the Caribbean. Rome, Italy: Food and Agriculture Organization of the United Nations.
i. UN FAO. 2013. AQUASTAT database. Rome, Italy: Food and Agriculture Organization of the United Nations. http://www.fao.org (accessed May 1, 2013).
j. Margat, J./OSS. 2001. Les ressources en eau des pays de l'OSS. Evaluation, utilisation et gestion. UNESCO/Observatoire du Sahara et du Sahel (updating of 1995).
k. Estimates from UN FAO. 2003. Review of World Water Resources by Country. Rome, Italy: Food and Agriculture Organization of the United Nations (see specific references in this document for more information).
l. United States Geological Survey revised: conterminous US (2071); Alaska (980); Hawaii (18).
m. Eurostat, U. Wieland. 2003. Water Resources in the EU and in the Candidate Countries. Statistics in Focus, Environment and Energy, European Communities.
n. Margat, J., and D. Vallée. 2000. Blue Plan—Mediterranean Vision on Water, Population and the Environment for the 21st Century. France: Sophia Antipolis.
o. Geres, D. 1998. Water Resources in Croatia. International Symposium on Water Management and Hydraulic Engineering, Dubrovnic, Croatia, September 14–19, 1998.
p. AQUASTAT website as of November 2005.
q. Pearse, P. H., F. Bertrand, and J. W. MacLaren. 1985. Currents of Change. Final Report of Inquiry on Federal Water Policy. Ottawa, Canada: Environment Canada.
r. China Ministry of Water. Annual Report 2007–2008.
s. Eurostat. 2013. http://epp.eurostat.ec.europa.eu/tgm/table.do?tab=table&init=1&plugin=1&language=en&pcode=ten00001 (accessed May 1, 2013).
t. Australian Government National Water Commission. 2005. Australian Water Resources 2005.
u. Frenken, K., ed. Estimates from UN FAO. 2005. Irrigation in the Middle East Region in Figures. AQUASTAT Survey—2008.

DATA TABLE 1 Total Renewable Freshwater Supply by Country (2013 Update)

Region	Country	Annual Renewable Water Resources (km³/yr)	Year of Estimate	Source of Estimate
AFRICA	Algeria	11.6	2005	c
	Angola	148.0	2011	i
	Benin	25.8	2001	j
	Botswana	12.2	2011	i
	Burkina Faso	12.5	2011	i
	Burundi	12.5	2011	i
	Cameroon	285.5	2003	k
	Cape Verde	0.3	2005	c
	Central African Republic	144.4	2005	c
	Chad	43.0	1987	b
	Comoros	1.2	2005	c
	Congo	832.0	1987	b
	Congo, Democratic Republic (formerly Zaire)	1283	2001	j
	Côte d'Ivoire	81	2001	j
	Djibouti	0.3	2005	c
	Egypt	57.3	2011	i
	Equatorial Guinea	26	2001	j
	Eritrea	6.3	2001	j
	Ethiopia	122.0	2011	i
	Gabon	164.0	1987	b
	Gambia	8.0	2005	c
	Ghana	53.2	2001	j
	Guinea	226.0	1987	b
	Guinea-Bissau	31.0	2005	c
	Kenya	30.7	2005	c
	Lesotho	3.0	2011	i
	Liberia	232.0	1987	b
	Libya	0.6	2005	c
	Madagascar	337.0	2005	c
	Malawi	17.3	2001	j
	Mali	100.0	2005	c
	Mauritania	11.4	2005	c
	Mauritius	2.8	2005	c
	Morocco	29.0	2005	c
	Mozambique	217.1	2005	c
	Namibia	17.7	2005	c
	Niger	33.7	2005	c
	Nigeria	286.2	2005	c
	Réunion	5.0	1988	k
	Rwanda	9.5	2005	c
	Senegal	38.8	2011	i
	Sierra Leone	160.0	1987	b
	Somalia	14.2	2005	c
	South Africa	51.4	2011	i
	Sudan	64.5	2005	c
	Swaziland	4.5	1987	b
	Tanzania	96.3	2011	i

continues

DATA TABLE 1   *continued*

Region	Country	Annual Renewable Water Resources (km³/yr)	Year of Estimate	Source of Estimate
	Togo	14.7	2001	j
	Tunisia	4.6	2005	c
	Uganda	66.0	2005	c
	Zambia	105.2	2001	j
	Zimbabwe	20.0	1987	b
NORTH AND CENTRAL AMERICA	Antigua and Barbuda	0.1	2000	h
	Bahamas	0.0	2011	i
	Barbados	0.1	2003	k
	Belize	18.6	2000	h
	Canada	2902.0	2011	i
	Costa Rica	112.4	2000	h
	Cuba	38.1	2000	h
	Dominica	nd	nd	
	Dominican Republic	21.0	2000	h
	El Salvador	25.2	2001	j
	Grenada	nd	nd	
	Guatemala	111.3	2000	h
	Haiti	14.0	2000	h
	Honduras	95.9	2000	h
	Jamaica	9.4	2000	h
	Mexico	457.2	2000	h
	Nicaragua	196.7	2000	h
	Panama	148.0	2000	h
	St. Kitts and Nevis	0.02	2000	h
	Trinidad and Tobago	3.8	2000	h
	United States of America	3069.0	1985	l
SOUTH AMERICA	Argentina	814.0	2000	h
	Bolivia	622.5	2000	h
	Brazil	8233.0	2000	h
	Chile	922.0	2000	h
	Colombia	2132.0	2000	h
	Ecuador	424.0	2011	i
	Guyana	241.0	2000	h
	Paraguay	336.0	2000	h
	Peru	1913.0	2000	h
	Suriname	122.0	2003	k
	Uruguay	139.0	2000	h
	Venezuela	1233.2	2000	h
ASIA	Afghanistan	65.3	2011	i
	Bahrain	0.1	2008	u
	Bangladesh	1227.0	1999	f
	Bhutan	78.0	2011	i
	Brunei	8.5	2011	i
	Cambodia	476.1	1999	f
	China	2738.8	2008	r
	India	1911.0	2011	i
	Indonesia	2019.0	2011	i
	Iran	137.5	2008	u

225

	Iraq	89.9	2011	i
	Israel	1.8	2008	u
	Japan	430.0	1999	f
	Jordan	0.9	2008	u
	Korea Democratic People's Republic	77.1	1999	f
	Korea Republic	69.7	1999	f
	Kuwait	0.02	2008	u
	Laos	333.6	2003	k
	Lebanon	4.5	2008	u
	Malaysia	580.0	1999	f
	Maldives	0.03	1999	f
	Mongolia	34.8	1999	f
	Myanmar	1168.0	2011	i
	Nepal	210.2	1999	f
	Oman	1.4	2008	u
	Pakistan	247.0	2011	i
	Palestinian Territory, Occupied	0.8	2011	i
	Philippines	479.0	1999	f
	Qatar	0.1	2008	u
	Saudi Arabia	2.4	2008	u
	Singapore	0.6	1975	d
	Sri Lanka	52.8	2011	i
	Syria	16.8	2008	u
	Taiwan	67.0	2000	p
	Thailand	438.6	2011	i
	Turkey	211.6	2011	i
	United Arab Emirates	0.2	2008	u
	Vietnam	884.0	2011	i
	Yemen	2.1	2008	u
EUROPE	Albania	41.7	2001	n
	Austria	84.0	2007	s
	Belgium	20.0	2007	s
	Bosnia and Herzegovina	37.5	2003	k
	Bulgaria	107.2	2010	s
	Croatia	105.5	1998	m, o
	Cyprus	0.3	2007	s
	Czech Republic	16.0	2007	s
	Denmark	16.3	2007	s
	Estonia	12.3	2007	s
	Finland	110.0	2007	s
	France	186.3	2007	s
	Germany	188.0	2007	s
	Greece	72.0	2007	s
	Hungary	116.4	2007	s
	Iceland	170.0	2007	s
	Ireland	51.0	2013	s
	Italy	175.0	2007	s
	Luxembourg	1.6	2007	s
	Macedonia	6.4	2001	n
	Malta	0.07	2005	s
	Netherlands	89.7	2007	s

continues

DATA TABLE 1 *continued*

Region	Country	Annual Renewable Water Resources (km³/yr)	Year of Estimate	Source of Estimate
	Norway	384.0	2013	s
	Poland	63.1	2007	s
	Portugal	73.6	2007	s
	Romania	42.3	2013	s
	Serbia-Montenegro*	208.5	2003	k
	Slovakia	80.3	2013	s
	Slovenia	32.1	2007	s
	Spain	111.1	2007	s
	Sweden	186.2	2013	s
	Switzerland	53.5	2007	s
	United Kingdom	164.0	2013	s
FORMER SOVIET UNION	Armenia	7.8	2008	u
	Azerbaijan	34.7	2008	u
	Belarus	58.0	1997	e
	Estonia	12.8	1997	e
	Georgia	63.3	2008	u
	Kazakhstan	107.5	2011	i
	Kyrgyzstan	23.6	2011	i
	Latvia	33.7	2011	i
	Lithuania	24.5	2007	s
	Moldova	11.7	1997	e
	Russia	4508.0	2011	i
	Tajikistan	21.9	2011	i
	Turkmenistan	24.8	2011	i
	Ukraine	139.5	1997	e
	Uzbekistan	48.9	2011	i
OCEANIA	Australia	336.1	2005	t
	Fiji	28.6	1987	b
	New Zealand	397.0	1995	g
	Papua New Guinea	801.0	1987	b
	Solomon Islands	44.7	1987	b
	Timor-Leste	8.2	2011	i

*Referred to as Yugoslavia in previous volumes of *The World's Water*.

DATA TABLE 2

Freshwater Withdrawal by Country and Sector (2013 Update)

Description

As with data table 1, this table on freshwater withdrawals has appeared in some form in every volume of *The World's Water* and remains highly popular. The use of water varies greatly from country to country and from region to region. Data on water use by regions and by different economic sectors are among the most sought after in the water resources area. Ironically, these data are often the least reliable and most inconsistent of all water resources information. This table includes the data available on total freshwater withdrawals by country in cubic kilometers per year and cubic meters per person per year, using national population estimates from 2010. The table also gives the breakdown of that water use by the municipal, agricultural, and industrial sectors, in both percentage of total water use and cubic meters per person per year. Note that "municipal" withdrawals were previously called "domestic" withdrawals and now often include some estimates of commercial and institutional use in the single category.

Data for a number of countries have been updated in this table since the previous volume of *The World's Water*. The AQUASTAT database of the Food and Agriculture Organization of the United Nations (FAO) continues to provide the bulk of the data, but we also work to update water-use data using country-specific sources. The data sources are explicitly identified.

"Withdrawal" typically refers to water taken from a water source for use. It does not refer to water "consumed" in that use. The domestic sector typically includes household and municipal uses as well as commercial and governmental water use. The industrial sector includes water used for power plant cooling and industrial production. The agricultural sector includes water for irrigation and livestock.

Limitations

Extreme care should be used when applying these data. They come from a wide variety of sources and are collected using a wide variety of approaches, with few formal standards. As a result, this table includes data that are actually measured, estimated, modeled using different assumptions, or derived from other data. The data also come from different years, making direct intercomparisons difficult. For example, some water-use data are over twenty years old. Also note that the per capita water-use estimates are computed using withdrawals from different years, but population estimates are normalized for 2010.

As noted in past volumes of *The World's Water*, the FAO AQUASTAT data set, while the most comprehensive single database, contains inadequate information on sources and assumptions and often contains modeled rather than measured values. Data from this database should be used with great care and with appropriate caveats about their quality.

Another major limitation of these data is that they do not include the use of rainfall in agriculture—what is sometimes referred to as "green water." Many countries use a significant fraction of the rain falling on their territory for agricultural production, but this water use is neither accurately measured nor reported in this data set. We repeat our regular call for a systematic reassessment of water-use data and for national and international commitments to collect and standardize this information.

Sources

a. World Resources Institute. 1990 and 1994. *World Resources*. In collaboration with the United Nations Environment Programme and the United Nations Development Programme. New York: Oxford University Press.
b. Eurostat. 1997, 2005, and 2011. Statistics of the European Union. EC/C/6/Ser.26GT. Also, Statistics in Focus. http://europa.eu.int/comm/eurostat. Luxembourg.
c. UN FAO. 1999. Irrigation in Asia in Figures. Rome, Italy: Food and Agriculture Organization of the United Nations.
d. UN FAO. 2000. Irrigation in Latin America and the Caribbean. Rome, Italy: Food and Agriculture Organization of the United Nations.
e. Kenny, J. F., N. L. Barber, S. S. Hutson, K. S. Linsey, J. K. Lovelace, and M. A. Maupin. 2009. Estimated Use of Water in the United States in 2005. US Geological Survey Circular 1344. Reston, VA: US Geological Survey. http://pubs.usgs.gov/circ/1344/pdf/c1344.pdf.
f. UN FAO. 2013. AQUASTAT database. Rome, Italy: Food and Agriculture Organization of the United Nations. http://www.fao.org (accessed May 2013). See text for details.
g. Environment Canada. Withdrawal Uses. http://www.ec.gc.ca/eau-water/default.asp?lang=En&n=851B096C-1.
h. Hidalgo, H. 2010. Water Resources in Costa Rica: A Strategic View. Draft copy. Citing Ministerio de Salud et al. (2003).
i. Pink, B. 2010. Australian Bureau of Statistics. Water Account Australia 2008–2009. http://www.ausstats.abs.gov.au/Ausstats/subscriber.nsf/0/D2335EFFE939C9BCCA2577E700158B1C/$File/46100_2008-09.pdf.
j. China Ministry of Water. Annual Report 2007–2008. Ministry of Water Resources People's Republic of China, 2007–2008 Annual Report. http://www.mwr.gov.cn/english/2007-2008.doc (accessed June 2011).

DATA TABLE 2 Freshwater Withdrawal by Country and Sector (2013 Update)

Region	Country	Year	Total Freshwater Withdrawal (km³/yr)	Per Capita Withdrawal (m³/p/yr)	Domestic Use (%)	Industrial Use (%)	Agricultural Use (%)	Domestic Use m³/p/yr	Industrial Use m³/p/yr	Agricultural Use m³/p/yr	Source	2010 Population (millions)
AFRICA	Algeria	2001	5.70	161	24	15	61	39	24	98	f	35.42
	Angola	2005	0.70	37	45	34	21	17	13	8	f	18.99
	Benin	2001	0.13	14	32	23	45	5	3	6	f	9.21
	Botswana	2000	0.19	96	41	18	41	39	17	39	f	1.98
	Burkina Faso	2001	0.72	44	39	2	59	17	1	26	f	16.29
	Burundi	2000	0.29	34	17	6	77	6	2	26	f	8.52
	Cameroon	2000	0.97	49	17	7	76	8	3	37	f	19.96
	Cape Verde	2000	0.02	39	7	2	91	3	1	36	f	0.51
	Central African Republic	2005	0.07	16	83	17	1	13	3	0	f	4.51
	Chad	2005	0.88	76	12	12	76	9	9	58	f	11.51
	Comoros	1999	0.01	14	48	5	47	7	1	7	f	0.69
	Congo, Democratic Republic (formerly Zaire)	2005	0.68	10	68	21	11	7	2	1	f	67.83
	Congo, Republic of	2000	0.05	13	70	22	9	9	3	1	f	3.76
	Côte d'Ivoire	2005	1.50	70	41	21	38	29	15	26	f	21.57
	Djibouti	2000	0.02	23	84	0	16	19	0	4	f	0.88
	Egypt	2000	68.30	809	8	6	86	62	49	695	f	84.47
	Equatorial Guinea	2000	0.02	29	79	15	6	23	4	2	f	0.69
	Eritrea	2004	0.58	111	5	0	95	6	0	105	f	5.22
	Ethiopia	2002	5.60	66	6	0	94	4	0	62	f	84.98
	Gabon	2005	0.14	93	61	10	29	57	9	27	f	1.50
	Gambia	2000	0.09	51	38	19	43	20	10	22	f	1.75
	Ghana	2000	0.98	40	24	10	66	10	4	27	f	24.33
	Guinea	2001	0.55	53	38	9	53	20	5	28	f	10.32

continues

229

Data Table 2. *continued*

Region	Country	Year	Total Freshwater Withdrawal (km³/yr)	Per Capita Withdrawal (m³/p/yr)	Domestic Use (%)	Industrial Use (%)	Agricultural Use (%)	Domestic Use m³/p/yr	Industrial Use m³/p/yr	Agricultural Use m³/p/yr	Source	2010 Population (millions)
AFRICA, cont.	Guinea-Bissau	2000	0.18	109	13	5	82	14	5	90	f	1.65
	Kenya	2003	2.74	67	17	4	79	11	2	53	f	40.86
	Lesotho	2000	0.04	19	46	46	9	9	9	2	f	2.08
	Liberia	2000	0.13	32	54	36	9	17	11	3	f	4.10
	Libya	2000	4.30	657	14	3	83	92	20	545	f	6.55
	Madagascar	2000	16.50	819	1	1	98	8	8	803	f	20.15
	Malawi	2005	1.40	89	11	4	86	10	4	77	f	15.69
	Mali	2000	6.50	488	9	1	90	44	5	439	f	13.32
	Mauritania	2005	1.40	416	7	2	91	29	8	379	f	3.37
	Mauritius	2003	0.73	563	30	3	68	169	17	383	f	1.30
	Morocco	2000	12.60	389	10	3	87	39	12	339	f	32.38
	Mozambique	2001	0.88	38	19	3	78	7	1	29	f	23.41
	Namibia	2002	0.29	131	25	5	70	33	6	92	f	2.212
	Niger	2005	0.98	62	30	3	67	19	2	41	f	15.89
	Nigeria	2005	13.00	82	31	15	54	25	12	44	f	158.26
	Rwanda	2000	0.15	15	24	8	68	4	1	10	f	10.28
	Senegal	2002	2.22	173	4	3	93	7	5	161	f	12.86
	Sierra Leone	2005	0.21	36	52	26	22	19	9	8	f	5.84
	Somalia	2003	3.30	352	1	0	99	4	0	351	f	9.36
	South Africa	2000	12.50	248	31	6	63	77	15	156	f	50.49
	Sudan and South Sudan	2005	27.60	639	4	1	95	26	4	607	f	43.19
	Swaziland	2000	1.04	865	2	1	97	17	9	839	f	1.20
	Tanzania, Republic of	2002	5.20	115	10	0	89	12	0	103	f	45.04
	Togo	2002	0.17	25	53	2	45	13	1	11	f	6.78

	Tunisia	2001	2.85	275	13	4	82	36	11	225	f	10.37
	Uganda	2002	0.32	9	48	14	38	5	1	4	f	33.80
	Zambia	2002	1.50	113	18	8	73	20	9	83	f	13.26
	Zimbabwe	2002	4.21	333	14	7	79	47	23	263	f	12.64
NORTH	Antigua and Barbuda	2005	0.008	90	63	21	15	57	19	14	f	0.09
AND	Barbados	2005	0.10	390	20	26	54	78	101	210	f	0.26
CENTRAL	Belize	2000	0.22	703	5	49	46	35	344	323	f	0.313
AMERICA	Canada	2006	45.08	1,330	20	69	12	260	913	157	g	33.89
	Costa Rica	2003	0.54	116				0	0	0	h	4.64
	Cuba	2007	4.40	393	27	17	56	106	67	220	f	11.20
	Dominica	2004	0.02	301	—	—					d	0.07
	Dominican Republic	2005	5.50	538	26	1	72	140	5	387	f	10.23
	El Salvador	2007	1.80	291	22	14	64	64	41	186	f	6.19
	Guatemala	2006	3.50	243	15	31	54	37	75	131	f	14.38
	Haiti	2000	1.20	118	19	4	78	22	5	92	f	10.19
	Honduras	2006	2.10	276	16	23	60	44	63	165	f	7.62
	Jamaica	2000	0.41	150	34	17	49	51	26	73	f	2.73
	Mexico	2009	80.40	727	14	9	77	102	67	557	f	110.65
	Nicaragua	2001	1.40	240	13	2	85	31	5	204	f	5.82
	Panama	2000	0.91	259	23	2	76	60	5	197	f	3.51
	St. Lucia	2005	0.02	115	—	—	—				d	0.17
	St. Vincent and the Grenadines	1995	0.01	92	—	—	—				d	0.11
	Trinidad and Tobago	2000	0.23	171	66	25	9	113	43	15	f	1.34
	United States of America	2005	482.20	1,518	13	46	41	193	699	626	e	317.64
SOUTH	Argentina	2000	32.60	802	22	12	66	176	96	529	f	40.67
AMERICA	Bolivia	2000	2.60	259	21	12	67	54	31	174	f	10.03
	Brazil	2006	58.07	297	28	17	55	83	52	162	f	195.42
	Chile	2007	26.70	1,558	4	10	86	62	156	1340	f	17.13
	Colombia	2000	12.70	274	57	4	39	156	10	107	f	46.30
	Ecuador	2005	9.90	719	13	6	81	93	43	582	f	13.77

continues

Data Table 2. *continued*

Region	Country	Year	Total Freshwater Withdrawal (km³/yr)	Per Capita Withdrawal (m³/p/yr)	Domestic Use (%)	Industrial Use (%)	Agricultural Use (%)	Domestic Use (m³/p/yr)	Industrial Use (m³/p/yr)	Agricultural Use (m³/p/yr)	Source	2010 Population (millions)
SOUTH AMERICA, cont.	Guyana	2000	1.64	2,154	2	1	98	36	19	2111	f	0.76
	Paraguay	2000	0.49	76	20	8	71	15	6	54	f	6.46
	Peru	2000	19.30	654	7	8	85	46	52	556	f	29.50
	Suriname	2000	0.67	1,278	4	3	93	57	37	1183	f	0.52
	Uruguay	2000	3.70	1,097	11	2	87	121	22	955	f	3.37
	Venezuela	2000	9.10	313	49	8	44	154	25	138	f	29.04
ASIA	Afghanistan	2000	20.30	697	1	1	99	7	7	690	f	29.12
	Armenia	2007	2.90	938	30	4	66	281	41	616	f	3.09
	Azerbaijan	2005	12.21	1,367	4	19	76	55	260	1039	f	8.93
	Bahrain	2003	0.36	442	50	6	45	221	27	199	f	0.81
	Bangladesh	2008	35.87	253	10	2	88	25	5	222	f	141.82
	Bhutan	2008	0.34	480	5	1	94	24	5	451	f	0.71
	Brunei	1994	0.09	221	nd	nd	nd				c	0.41
	Cambodia	2006	2.20	146	4	2	94	6	3	137	f	15.05
	China	2007	578.9	425	12	23	64	52	99	272	j	1,361.76
	Cyprus	2000	0.21	239	27	1	71	65	3	170	b	0.88
	Georgia	2005	1.80	427	20	22	58	85	94	247	f	4.22
	India	2010	761.00	627	7	2	90	46	14	567	f	1,214.46
	Indonesia	2000	113.30	487	12	7	82	58	34	400	f	232.52
	Iran	2004	93.30	1,243	7	1	92	85	12	1143	f	75.08
	Iraq	2000	66.00	2,097	7	15	79	147	315	1657	f	31.47
	Israel	2004	1.95	268	36	6	58	97	16	156	f	7.29
	Japan	2001	90.00	709	19	18	63	135	127	446	f	127.00
	Jordan	2005	0.94	145	31	4	65	45	6	95	f	6.47
	Kazakhstan	2010	21.10	1,339	4	30	66	54	402	884	f	15.75

	Korea Democratic People's Republic	2005	8.70	363	10	13	76	36	47	276	f	23.99
	Korea Republic	2002	25.47	525	26	12	62	136	63	326	f	48.50
	Kuwait	2002	0.91	299	44	2	54	132	6	162	f	3.05
	Kyrgyzstan	2006	8.00	1,441	3	4	93	45	58	1340	f	5.55
	Laos	2005	3.50	544	4	5	91	23	27	495	f	6.44
	Lebanon	2005	1.31	308	29	11	60	89	34	185	f	4.25
	Malaysia	2005	11.20	401	35	43	22	140	173	88	f	27.91
	Maldives	2008	0.006	19	95	5	0	18	1	0	c	0.31
	Mongolia	2009	0.55	204	13	43	44	26	88	90	f	2.70
	Myanmar	2000	33.23	658	10	1	89	66	4	586	f	50.50
	Nepal	2006	9.50	318	2	0	98	6	0	312	f	29.85
	Oman	2003	1.32	455	10	1	88	45	5	400	f	2.91
	Pakistan	2008	183.50	993	5	1	94	52	8	933	f	184.75
	Palestine Territory, Occupied	2005	0.42	95	48	7	45	46	7	43	f	4.41
	Philippines	2009	81.6	872	8	10	82	70	87	715	f	93.62
	Qatar	2005	0.44	294	39	2	59	115	6	174	f	1.51
	Saudi Arabia	2006	23.67	902	9	3	88	81	27	793	f	26.25
	Singapore	1975	0.19	39	45	51	4	18	20	2	b	4.84
	Sri Lanka	2005	13.00	637	6	6	87	38	38	554	f	20.41
	Syria	2005	16.80	746	9	4	88	67	30	657	f	22.51
	Tajikistan	2006	11.50	1,625	6	4	91	98	65	1479	f	7.07
	Thailand	2007	57.31	841	5	5	90	40	41	760	f	68.14
	Turkey	2003	40.10	530	15	11	74	78	58	393	f	75.71
	Turkmenistan	2004	28.00	5,409	3	3	94	162	162	5085	f	5.18
	United Arab Emirates	2005	4.00	849	15	2	83	127	17	705	f	4.71
	Uzbekistan	2005	56.00	2,015	7	3	90	141	60	1813	f	27.79
	Vietnam	2005	82.00	921	1	4	95	9	37	875	f	89.03
	Yemen	2005	3.57	147	7	2	91	10	3	134	f	24.26
EUROPE	Albania	2006	1.30	410	43	18	39	176	74	160	f	3.17
	Austria	2002	3.70	441	18	79	3	79	348	13	f	8.39
	Belarus	2000	4.30	448	27	54	19	121	242	85	f	9.59

continues

Data Table 2. *continued*

Region	Country	Year	Total Freshwater Withdrawal (km³/yr)	Per Capita Withdrawal (m³/p/yr)	Domestic Use (%)	Industrial Use (%)	Agricultural Use (%)	Domestic Use m³/p/yr	Industrial Use m³/p/yr	Agricultural Use m³/p/yr	Source	2010 Population (millions)
EUROPE cont.	Belgium	2007	6.20	580	12	88	1	70	510	7	b, f	10.70
	Bosnia and Herzegovina	2009	0.34	90								3.76
	Bulgaria	2009	6.10	814	16	68	16	130	553	130	f	7.50
	Croatia	2010	0.63	143	85	14	1	121	20	1	f	4.41
	Czech Republic	2007	1.70	163	41	57	2	67	93	3	f	10.41
	Denmark	2009	0.66	120	58	5	36	70	6	43	f	5.48
	Estonia	2007	1.80	1,344	3	97	1	40	1304	13	f	1.34
	Finland	2005	1.60	299	25	72	3	75	215	8	f	5.35
	France	2007	31.60	504	18	69	12	91	348	61	f	62.64
	Germany	2007	32.30	394	12	68	20	47	268	78	b	82.06
	Greece	2007	9.50	849	9	2	89	76	17	756	f	11.18
	Hungary	2007	5.60	562	12	83	5	67	466	28	f	9.97
	Iceland	2005	0.17	516	50	8	42	258	41	217	f	0.33
	Ireland	1994	1.18	257	23	77	0	58	198	0	f	4.59
	Italy	2000	45.40	755	20	36	44	151	272	332	f	60.10
	Latvia	2002	0.40	179	38	49	13	68	87	23	f	2.24
	Lithuania	2007	2.40	737	7	90	3	52	664	22	f	3.26
	Luxembourg	1999	0.06	116	63	37	1	73	43	1	b	0.49
	Macedonia	2007	1.00	489	21	67	12	103	328	59	f	2.04
	Malta	2002	0.05	122	63	1	35	77	1	43	f	0.41
	Moldova	2007	1.10	308	14	83	3	43	255	9	f	3.58
	Netherlands	2008	10.60	637	12	87	1	76	554	6	f	16.65
	Norway	2006	2.90	597	28	43	29	167	257	173	b	4.86
	Poland	2009	12.00	315	31	60	10	98	189	32	f	38.04

	Portugal	2002	8.50	792	8	19	73	63	150	578	f	10.73
	Romania	2003	6.50	307	22	61	17	67	187	52	f	21.19
	Russian Federation	2000	76.68	546	20	60	20	109	328	109	b, f	140.37
	Serbia	2009	4.10	418	17	82	2	71	343	8	f	9.80
	Slovakia	2007	0.69	128	47	50	3	60	64	4	f	5.41
	Slovenia	2009	0.94	464	18	82	1	84	381	5	f	2.02
	Spain	2008	32.50	717	18	22	61	129	158	437	f	45.32
	Sweden	2007	2.60	280	37	59	4	103	165	11	f	9.29
	Switzerland	2000	2.60	342	40	58	2	137	199	7	f	7.59
	Ukraine	2005	19.20	423	24	70	6	101	296	25	f	45.43
	United Kingdom	2007	13.00	210	57	33	10	120	69	21	a, f	61.90
OCEANIA	Australia	2010	59.84	2,782	16	11	74	445	306	2058	i	21.51
	Fiji	2000	0.08	94	28	11	61	26	10	57	f	0.85
	New Zealand	2002	4.80	1,115	21	4	74	234	45	825	f	4.30
	Papua New Guinea	2005	0.39	57	57	43	0	32	24	0	a, f	6.89
	Solomon Islands	1987			40	20	40	0	0	0	a, f	0.54
	Timor-Leste	2004	1.17	1,064	8	1	91	85	11	968	f	1.10

Notes:
Figures may not add to totals because of independent rounding.
Population data from Population Division of the Department of Economic and Social Affairs of the United Nations Secretariat, World Population Prospects: The 2008 Revision. Total Population, Both Sexes. Estimates 2010 (New York: United Nations, 2009). http://esa.un.org/unpd/wpp2008/all-wpp-indicators_components.htm.
US data include freshwater only; previous data included saline water.
China data include Hong Kong.

DATA TABLES 3A AND 3B

Access to Improved Drinking Water by Country, 1970-2008 & 2011 Update

Description

Safe drinking water is one of the most basic human requirements, and one of the Millennium Development Goals (MDGs) by 2015 is to reduce by half the proportion of people unable to reach or afford safe drinking water. As a result, estimates of access to safe drinking water are a cornerstone of most international assessments of progress, or lack thereof, toward solving global and regional water problems.

Data are given here in two related tables.

Data table 3A includes the percentage of urban, rural, and total populations, by country, with access to improved drinking water for 1970, 1975, 1980, 1985, 1990, 1994, 2000, 2002, 2004, 2005, and 2008—similar to data table 3 in the previous volume of *The World's Water*. In the current volume, however, these previous data are sorted alphabetically by country rather than separated by regions.

Data table 3B provides an update of the percentage of national, urban, and rural populations with improved drinking water for 2011, the most recent year for which data are available. This table also includes the percentage of country populations in urban and rural categories for 2011.

The World Health Organization (WHO) / UNICEF Joint Monitoring Programme (JMP) for Water Supply and Sanitation collected the data presented here over various periods. Most of the data presented were drawn from responses by national governments to WHO questionnaires. Participants in data collection include the JMP, the United Nations Children's Fund, and the Water Supply and Sanitation Collaborative Council, which has continued sector monitoring and aims to support and strengthen the monitoring efforts of individual countries. The forty largest countries in the developing world account for 90 percent of population in these regions. As a result, the World Health Organization spent extra effort in collecting comprehensive data for these countries.

Data for 2000 and later reflect a significant change in definition. Data are now reported for populations with and without access to an "improved" drinking water supply. An improved drinking water source is defined as one that, "by nature of its construction or through active intervention, is protected from outside contamination, in particular from contamination with fecal matter."

According to the World Health Organization, the following technologies were included in the assessment as representing an improved water supply:

Household connection

Public standpipe

Borehole

Protected dug well

Protected spring

Rainwater collection

In comparison, an unimproved drinking water source refers to

Unprotected well

Unprotected spring

Rivers or ponds

Vendor-provided water

Bottled water

Tanker truck water

Limitations

A review of water and sanitation coverage data from the 1980s and 1990s shows that the definition of safe, or improved, water supply and sanitation facilities differs from one country to another and for a given country over time. Indeed, some of the data from individual countries often showed rapid and implausible changes in the level of coverage from one assessment to the next. This indicates that some of the data are also unreliable, irrespective of the definition used. Countries used their own definitions of "rural" and "urban."

For the 1996 data, two-thirds of the countries reporting indicated how they defined "access." At the time, the definition most commonly centered on walking distance or time from household to water source, such as a public standpipe, which varied from 50 to 2,000 meters and 5 to 30 minutes. Definitions sometimes included considerations of quantity, with the acceptable limit ranging from 15 to 50 liters per capita per day.

WHO/JMP assessments since 2000 have attempted to shift from gathering information from water providers only to including consumer-based information. The current approach uses household surveys in an effort to assess the actual use of facilities. "Reasonable access" was broadly defined as the availability of at least 20 liters per person per day from a source within 1 kilometer of the user's dwelling. A drawback of this approach is that household surveys are not conducted regularly in many countries. Thus, direct comparisons between countries, and across time within the same country, are difficult. Direct comparisons are additionally complicated by the fact that these data hide disparities between regions and socioeconomic classes.

Access to water, as reported by WHO/JMP assessments, does not imply that the level of service or quality of water is "adequate" or "safe." In addition, the household surveys and censuses on which the JMP relies also measure "use" and not "access." The assessment questionnaire did not include any methodology for discounting coverage figures to allow for intermittence of supply or poor quality of the water supplies. However, the instructions stated that piped systems should not be considered "functioning" unless they were operating at over 50 percent capacity on a daily basis and that hand pumps should not be considered functioning unless they were operating for at least 70 percent of the time, with a lag between breakdown and repair not exceeding two weeks. These aspects were taken into consideration when estimating coverage for countries for which national surveys had not been conducted. More details of the methods used beginning with the 2000 assessment, and their limitations, can be found at http://www.who.int/docstore/water_sanitation_health/Globassessment/GlobalTOC.htm.

Sources

United Nations Environment Programme (UNEP). 1989. *Environmental Data Report.* GEMS Monitoring and Assessment Research Centre. Oxford, England: Basil Blackwell.
———. 1993–1994. *Environmental Data Report.* GEMS Monitoring and Assessment Research Centre in cooperation with the World Resources Institute and the UK Department of the Environment. Oxford, England: Basil Blackwell.
World Health Organization (WHO). 1996. Water Supply and Sanitation Sector Monitoring Report: 1996 (Sector Status as of 1994). In collaboration with the Water Supply and Sanitation Collaborative Council and the United Nations Children's Fund, UNICEF, New York.
———. 2000. Global Water Supply and Sanitation Assessment 2000 Report. http://www.who.int/docstore/water_sanitation_health/Globassessment/GlobalTOC.htm.
World Health Organization (WHO) and United Nations Children's Fund (UNICEF). 2004. Meeting the MDG Drinking Water and Sanitation Target: A Mid-Term Assessment of Progress. http://www.who.int/water_sanitation_health/monitoring/jmp2004/en/index.html.
———. 2006. Meeting the MDG Drinking Water and Sanitation Target: The Urban and Rural Challenge of the Decade. http://www.who.int/water_sanitation_health/monitoring/jmpfinal.pdf.
———. 2010. WHO/UNICEF Joint Monitoring Programme (JMP) for Water Supply and Sanitation. http://www.wssinfo.org/data-estimates/table/.
———. 2013. WHO/UNICEF Joint Monitoring Programme (JMP) for Water Supply and Sanitation. http://www.wssinfo.org/data-estimates/table/_(2011 data were accessed May 29, 2013).
World Resources Institute (WRI). 1988. World Health Organization data, cited by the World Resources Institute, *World Resources 1988–89.* World Resources Institute and the International Institute for Environment and Development in collaboration with the United Nations Environment Programme. New York: Basic Books.

DATA TABLE 3A   Access to Safe Drinking Water by Country, 1970–2008

Percentage of Population with Access to Improved Drinking Water

Region and Country	URBAN													RURAL													TOTAL												
	1970	1975	1980	1985	1990	1994	2000	2002	2004	2005	2008			1970	1975	1980	1985	1990	1994	2000	2002	2004	2005	2008			1970	1975	1980	1985	1990	1994	2000	2002	2004	2005	2008			
Afghanistan	18	40	28	38	40	39	19	19	63	66	78			1	5	8	17	19	5	11	11	31	33	39			3	9	8	17	23	12	13	13	39	41	48			
Albania								99	99	98	96										95	94	97	98										97	96	97	97			
Algeria	84	100		85			88	92	88	88	85						55			94	80	80	81	79						68			94	87	85	85	83			
American Samoa						100														100												100							
Angola			85	87	73	69	34	70	75	54	60					10	15	20	15	40	40	40	39	38					26	33	35	32	38	50	53	47	50			
Anguilla									60																										60				
Antigua and Barbados									95	95	95											89	89											91	91				
Argentina	69	76	61	63			85	97	98	98	98			12	26	17	17			30		80	80	80			56	66	54	56			79		96	96	97			
Armenia								99	99	99	98										80	80	89	93										92	92	95	96			
Aruba								100	100	100	99										100	100	100	100										100	100	100	100			
Australia							100	100	100	100	100									100	100	100	100	100									100	100	100	100	100			
Azerbaijan								95	95	88	88										59	59	66	71										77	77	77	80			
Bahamas	100	100	100	100	98		99	98	98	98	98			12	13			75		86	86	86					65	65	100	100	90		96	97	97					
Bahrain	100	100		100	100			100	100	100	100			94	100		100	0									99	100		100										
Bangladesh	13	22	26	24	39	100	99	82	82	85	85			47	61	40	49	89	97	97	72	72	78	78			45	56	39	46	81	97	75	75	74	80	80			
Barbados	95	100	99	100	100		100	100	100	100	100			100	100	98	99	100		100	100	100	100	100			98	100	99	99	100		100	100	100	100	100			
Belize			99	100	95	96	83	100	100	97	99					36	26	53	82	69	82	82	94	100					68	64	74	89	76	91	91	95	99			
Benin	83	100	26	80	73	41	74	79	78	82	84			20	20	15	34	43	53	55	60	57	65	69			29	34	18	50	54	50	63	68	67	72	75			
Bhutan			50		60	75	86	86	86	99	99					5	19	30	54	60	60	60	88	88					7		32	64	62	62	62	91	92			
Bolivia	92	81	69	75	76	78	93	95	95	95	96			2	6	10	13	30	22	55	68	68	63	67			33	34	36	43	53	55	79	85	85	84	86			
Bosnia and Herzegovina								99	100	100	100										96	97		98										97	98	99			
Botswana	71	95		84	100		100	100	99	99	99			26	39	51	46	88	54	90	90	90	90				29	45		53	91			95	95	95	95			
Brazil	78	87	83	85	95	85	95	96	96	98	99			28		51	56	61	31	54	58	57	81	84			55		72	77	87	72	87	89	90	95	97			

continues

DATA TABLE 3A *continued*

Percentage of Population with Access to Improved Drinking Water

Region and Country	URBAN												RURAL												TOTAL											
	1970	1975	1980	1985	1990	1994	2000	2002	2004	2005	2008		1970	1975	1980	1985	1990	1994	2000	2002	2004	2005	2008		1970	1975	1980	1985	1990	1994	2000	2002	2004	2005	2008	
British Virgin Island					100			98		98	100						100			98		98	100									98		100	100	100
Brunei Darussalam			100												95																					
Bulgaria									100	100	100										97	100	100										99	100	100	
Burkina Faso	35	50	27	43			84	82	94	91	95		10		23		31	69	70	44	54	65	72		12		25	31			78	51	61	70	76	
Burundi	77		90	98	92	92	96	90	92	85	83				20	21	43	49		78	77	71	71					23	45	52		79	79	72	72	
Cambodia							53	58	64	75	81								25	29	35	51	56								30	34	41	56	61	
Cameroon	77			43	42		82	84	86	90	92		21			24			42	41	44	48	51		32			32			62	63	66	71	74	
Canada							100	100	100	100	100								99	99	99	99	99								100	100	100	100	100	
Cape Verde			100	83		70	64	86	86	86	85				21	50		34	89	73	73	82	82				25	52		51	74	80	80	84	84	
Cayman Islands			100	98						95	95																									
Central African Republic				13	19	18	80	93	93	89	92						26	18	43	61	61	50	51						23	18	60	75	75	65	67	
Chad	47	43				48	31	40	41	66	67		24	23			17		26	32	43	43	44		27	26				24	27	34	42	49	50	
Chile	67	78	100	98		94	99	100	100	99	99		13	28	17	29		37	66	59	58	75	75		56	70	84	87		85	94	95	95	96	96	
China					87	93	94	92	93	98	98						68	89	66	68	67	78	82						73	90	75	77	77	86	89	
Colombia	88	86	93	100	87	88	98	99	99	99	99		28	33	73	76	82	48	73	71	71	73	73		63	64	86	86	86	76	91	92	93	92	92	
Comoros							98	90	92	91	91								95	96	82	96	97								96	94	86	95	95	
Congo	63	81	42				71	72	84	95	95		6	9	7				17	17	27	34	34		27	38	20				51	46	58	71	71	
Congo,	33	38		52	68	37	89	83	82	82	80		4	12		21	24	23	26	29	29	28	28		11	19		32	36	27	45	46	46	45	46	
Cook Islands			100	99	100		100	98	98	97	101					88	100		100	88	88	88						92	100		100	95	94	94		
Costa Rica	98	100	100	100		85	98	100	100	100	100		59	56	82	83		99	98	92	92	90	91		74	72	90	91		92	98	97	97	96	97	
Côte d'Ivoire	98				57	59	90	98	97	92	93		29				80	81	65	74	74	67	68		44			71	72	77	84	84	79	80		
Cuba	82	96		100	100	96	99	95	95	95	96		15				91	85	82	78	78	83	89		56				98	93	95	91	91	92	94	
Cyprus	100	94		100	100	100	100	100	100	100	100								100	100	100	100	100								100	100	100	100	100	

Country																																	
Djibouti				50	50	77	100	82	76	95	98			20	20	100	100	67	59	55	52		43	45	90	100	80	73	89	92			
Dominica							100	100	100	96							100	90	90	93						100	97	97	95				
Democratic Republic																																	
Dominican Republic	72	88		85	82	74	83	98	97	89	87	14	27	34	33	45	67	70	85	91	83	84	37	55	60	62	67	71	79	93	95	87	86
East Timor								73	77	80	86							51	56	57	63							52	58	63	69		
Ecuador	76	67	79	81	63	82	81	92	97	96	97	7	8	20	31	44	55	51	77	89	86	88	34	36	50	57	55	70	71	86	94	92	94
Egypt	94		88	95		82	96	100	99	100		93		64		86	50	94	97	97	96	98	93		84		90	64		98	98	99	
El Salvador	71	89	67	87	68	78	88	91	94	94	100	20	28	40	40	15	37	61	68	70	73	76	40	53	50	51	47	55	74	82	84	86	87
Equatorial Guinea			47		65	88	45	45	45		94	20	28	40	40	18	100	42	42	42	42	76					32	95	43	44	43	43	
Eritrea							63	72	74	74	74						42	54	57	57	57	57						46	57	60	60	61	
Estonia								100	99	99								99	97	97									100	98	98		
Ethiopia	61	58		69			77	81	81	95	98	1			9		13	11	11	24	24	26	6	8		16		24	22	22	35	38	
Fiji	78	89	94				43	43	95			15	56	66		69	100	51	51	51			37	69	77		80	100	47	47	47		
French Guiana				96	100		88	88										71	71	71								84	84	84			
French Polynesia					100		100	100	100	100	100							100	100	100	100	100						100	100	100	100	100	
Gabon							73	95	95	95	95	3						55	47	47	43	41						70	87	88	86	87	
Gambia	97		85	97	100		80	95	94	94	96				50	48		53	77	77	83	86	12		59		60	62	82	82	89	92	
Gaza Strip									94	91	91								88	91	91									92	91	91	
Georgia								90	96	99	100							61	67	92	96								76	82	96	98	
Ghana	86	86	72	93	63	70	87	93	88	89	90		14	33	39		49	49	68	64	68	74	35	45		56	21	56	79	75	78	82	
Grenada	100	100					97	97	97	96	97	47	77					93	93	93								94	95	95			
Guadeloupe							94	98	98	98	98							94	93	93								94	98	98			
Guam							100	100	100	100	100							100	100	100	100	100						100	100	100	100	100	
Guatemala	88	85	90	72	92	97	97	99	99	97	98	12	14	18	14	43		88	92	92	88	90	38	39	46	37	62	92	95	95	92	94	
Guinea	68	69	69	41	100		72	78	78	89	89			2	12	37	62	36	38	35	57	61			15	18	53	48	51	50	68	71	
Guinea-Bissau			18	17			29	79	79	82	83			8	22		57	55	49	49	48	51			10	21		53	49	59	58	60	
Guyana	100	100	100	100	100	90	98	83	83	96	98	63	75	60	65	71	45	91	83	83	90	93	75	84	72	76	81	61	94	83	92	94	
Haiti		46	51	59	56	37	46	91	52	70	71	3		8	30	35	23	45	59	56	53	55		12	19	38	41	28	46	54	60	63	
Honduras	99	99	93	56	85	81	97	99	95	95	98	10	13	40	45	48	53	82	81	74	77	34	41	59	49	64	65	90	87	94	86		

continues

DATA TABLE 3A *continued*

Percentage of Population with Access to Improved Drinking Water

Region and Country	URBAN														RURAL														TOTAL													
	1970	1975	1980	1985	1990	1994	2000	2002	2004	2005	2008				1970	1975	1980	1985	1990	1994	2000	2002	2004	2005	2008				1970	1975	1980	1985	1990	1994	2000	2002	2004	2005	2008			
Hong Kong																																										
Hungary			100		100		100	100	100	100	100						95		96					97	100								100					99	100	100		
India	60	80	77	76	86	85	92	96	95	95	96				6	18	31	50	69	79	86	82	83	81	84				17	31	42	56	73	81	88	86	86	85	88			
Indonesia	10	41	35	43	35	78	91	89	87	90	89				1	4	19	36	33	54	65	69	69	70	71				3	11	23	38	34	62	76	78	77	80	80			
Iran	68	76	82		100	89	99	98	99	98	98				11	30	50		75	77	89	83	84						35	51	66		89	83	95	93	94		
Iraq	83	100		100	93		96	97	97	93	91				7	11	54	54	41		48	50	50	53	55				51	66		86	78	44	85	81	81	80	79
Israel								100	100	100	100										100	100	100	100	100										100	100	100	100	100		
Jamaica	100	100	55	99			81	98	98	98	98				48	79	46	93			59	87	88	88	89				62	86	51	96			71	93	93	93	94
Japan			100	100	100		100	100	100	100	100										100	100	100	100	100										100	100	100	100	100		
Jordan	98						100	91	99	98	98				59		65	88	97		84	91	91	91	91				77		86	96	99	89	96	91	97	96	96
Kazakhstan							98	96	97	99	99										82	72	73	91	90										91	86	86	96	95		
Kenya	100	100	85			67	87	89	83	85	83				2	4	15			49	31	46	46	48	52				15	17	26			53	49	62	61	56	59
Kiribati			93		91		82	77	77	78							25				25	53	53	54											47	64	65	65		
Korea Democratic People's Republic							100	100	100	100	100										100	100	100	100	100										100	100	100	100	100		
Korea Republic	84	95	86	90	100		97	97	97	99	100				38	33	61	48	76		71	71	71	83	88				58	66	75	75	93		92	92	92	96	98
Kuwait	60	100	86	97						99	99						100							100	100				51	89	87							99	99
Kyrgyzstan							98	98	98	99	99										66	66	66	80	85										77	76	77	87	90		
Laos	97	100	28			40	59	66	79	74	72				39	32	20		25	39	100	38	43	47	51				48	41	21		29	39	90	43	51	54	57
Latvia								100	100	100	100											96	96	96	96											99	99	99	99		
Lebanon					100		100	100	100	100	100										100	100	100	100	100								100		100	100	100	100	100		
Lesotho	100	65	37	65		14	98	88	92	96	97				1	14	11	30	64		88	74	76	79	81				3	17	15	36	52		91	76	79	83	85
Liberia	100			100		58		72	72	80	79				6		23		8			52	52	48	51				15		53	30			62	61	67	68
Libya	100	100	100				72	72							42	82	90		68		68	68							58	87	96				72	72			
Madagascar	67	76	80	81		83	85	75	77	71	71				1	14	7	17	10		31	34	35	27	29				11	25	21	31	29		47	45	50	40	41

Country																																
Malawi																																
Malaysia	100	100	77	97	96	52	95	96	98	94	95											29	34	41	56	45	57	67	73	74	80
Maldives			90	96	77	98	100	99	100	100	100	1	6	37	50	66	44	44	62	68	70	77		63	84	89	100	95	99	100	100
Mali	29		11	58	41	36	74	99	98	99	99			49	76	68	86	94	94	96	99	99		2	21			84	83	90	91
Marshall Islands		37	46						78	77	81			3	12	4	38	100	78	76	86	86			16	37	65	48	50	51	56
Martinique				100				80	82	93	92				10		61		35	36	40	44									
Mauritania	98		80	73		84	34	63	59	49	52	10	22	85		45	69	40	45	44	43	47	17		84	76	37	85	87	95	94
Mauritius	100	100	100	100	100	95	100	100	100	100	100	29	22	98	100		100	100	100	100			61	60	99	98	100	56	53	45	49
Mexico	71	70	90	99	94	91	94	97	100	96	100	29	49	40	47		62	63	72	87	83	87	54	62	73	83	86	91	97	93	94
Micronesia						100		95	95	94	96					38		94	94	94						100		94	94		
Mongolia					100		77	87	87	94	97				58		100		30	30	45	49					60	62	62	73	76
Montenegro										100											96	96								98	98
Montserrat							100	100	100	100	100								100	100	100	100						100	100	100	100
Morocco	92		100	100	100	98	100	99	99	97	98	28			25	18	14	58	56	56	60	60	51		59	52	82	80	81	80	81
Mozambique				38		17	86	76	72	76	77				9		40	43	24	26	29	29			15	32	60	42	43	45	47
Myanmar (Burma)	35	31	38	36	79	36	88	95	80	75	75	13	14	15	24	72	39	60	74	77	69	69	18	17	27	38	68	80	78	71	71
Namibia																															
Nauru					90	87	100	98	98	99	99					37	42	67	72	81	82	88				52	77	80	87	88	92
Nepal	53	85	83	70	66	66	85	93	96	93	93	5		7	25	34	41	80	82	89	85	87	2	8	11	28	38	44	81	86	88
Netherlands								100	100	100	100										100	100							99	100	100
New Zealand						100	100	100	100	100	100						100	100	100	100	100	100					100	100	100	100	100
Nicaragua	58	100	67	76		81	95	93	90	97	98	16	14	6	11	27	59	65	63	66	68	35	39	48	61	79	81	79	83	85
Niger	37	36	41	35	98	46	70	80	80	89	96	19	26	32	49	45	55	56	36	36	37	39	20	27	33	47	56	59	46	45	48
Nigeria				100	100	63	81	72	67	76	75				20	22	26	39	49	31	40	42			38	39	57	60	48	57	58
Niue					0	100	100	100	100	100	100						100	100	100	100	100	100					100	100	100	100	100
Northern Mariana Islands				100				98	98	98						0			97	97	90	100						98	99	97	98
Oman		100	90			41	81	95	90	92		48		49		52	30	72	89	76	77		52	53		63	39	79	90	86	88
Pakistan	77	75	72	83	82	77	96	95	96	95	95	4	5	20	27	42	84	87	89	86	87	21	25	35	44	60	88	90	91	89	90
Palau				100			100	79	78							97	20	94	94	98						79	84	85	84	

continues

DATA TABLE 3A *continued*

Percentage of Population with Access to Improved Drinking Water

Region and Country	URBAN												RURAL												TOTAL											
	1970	1975	1980	1985	1990	1994	2000	2002	2004	2005	2008		1970	1975	1980	1985	1990	1994	2000	2002	2004	2005	2008		1970	1975	1980	1985	1990	1994	2000	2002	2004	2005	2008	
Panama	100	100	100	100			88	99	99	97	97		41	54	62	64			86	79	79	83	83		69	77	81	82		83	87	91	90	93	93	
Papua New Guinea	44	30	55	95	94	84	88	88	88	87	87		72	19	10	15	20	17	32	32	32	33	33		70	20	16	26	32	28	42	39	39	40	40	
Paraguay	22	25	39	53	61		95	100	99	98	99		5	5	9	8	9		58	62	68	93	66		11	13	21	28	34		79	83	86	83	86	
Peru	58	72	68	73	68	74	87	87	89	90	90		8	15	18	17	24	24	51	66	65	58	61		35	47	50	55	55	60	77	81	83	81	82	
Philippines	67	82	49	49	93	93	92	90	87	93	93		20	31	43	54	72	77	80	77	82	85	87		36	50	45	52	81	85	87	85	85	90	91	
Qatar	100	100	76		100			100	100	100	100		75	83	43					100	100	100	99		95	97	71					100	100	100	100	
Republic of Moldova								97	97	96	96									88	88	85	85									92	92	90	90	
Romania							91	91												16	16										57	57			
Russian Federation							99	100	98	98									88	88	89	89								96	97	96	96		
Rwanda	81	84	48	79	84		60	92	92	80	77		66	68	55	48	67		40	69	69	63	62		67	68	55	50	68		41	73	74	66	65	
Samoa	86	100	97		100		95	91	90	90				23		94	77		100	88	87	87			17	43				99	88	88	88		
São Tomé and Príncipe							89	89	89	88	89				45	45				73	73	81	89				45				79	79	85	89	
Saudi Arabia	100	97	92	100			100	97	97	97	97		37	56	87	88			64						49	64	90	94			95					
Senegal	87	56	77	79	65	82	92	90	92	91	92				25	38	26	28	65	54	60	51	52				43	53	42	50	78	72	76	68	69	
Serbia									99	99	99										98	98										99	99		
Serbia and Montenegro								99	99											86	86										93	93			
Seychelles							100	100	100	94	100					95				75	75							95				87	88			
Sierra Leone	75		50	68	80	58	23	75	75	82	86		1		2	7	20	21	31	46	46	33	26		12		14	24	39	34	28	57	57	51	49	
Singapore			100	100	100		100	100	100	100	100																100	100	100	100	100	100	100	100	100	
Slovakia									100	100	100										97	100										99	99	100	100	
Solomon Islands		96		82			94	94	94	95					45		58		65	65	65	65							62		71	70	70	70		

Country																														
Somalia	17	77		58				32	32	58	67	14	22	22				27	27	11	9	15	38	34			29	29	28	30
South Africa	46	36	65	82	80	43	92	98	99	99	99	14	13	29	55	47	80	73	73	75	78	21	19		70	86	87	88	89	91
Sri Lanka							91	99	98	97	98		18					72	74	84	88		28	40	46	83	78	79	86	90
St. Kitts								99	99	100	100						80	99	99	99	100				60		99	100	100	100
St. Lucia								98	98	99	97						80	98	98	98	98						98	98	98	98
St. Vincent																			93											
Sudan	61	96	100			66		78	78	68	64	13	43		45		69	64	64	53	52	19	50	51	50	75	69	70	59	57
Suriname		100	71				94	98	98	97	97			94			96	73	73	78	81			88	83	95	92	92	92	93
Swaziland		83		100		41		87	87	90	92		29	7		44		42	54	56	61		37	31	43		52	62	64	69
Syria	98		98			92	94	94	98	95	94	50	54			78	64	64	87	82	84	71		74	85	80	79	93	89	89
Tajikistan								93	92	93	94				85			47	48	57	61						58	59	67	70
Tanzania	61	88	90				80	92	85	82	80	9	36	42		45	42	62	49	45	45	13	39	53		54	73	62	54	54
Thailand	60	69	65				89	95	98	99	99	10	16	66			77	80	100	97	98	17	25	63	64	80	85	99	98	98
Togo	100	49	70	100		74	85	80	80	86	87	5	10	31	58		38	36	36	40	41	17	16	38	54	54	51	52	58	60
Tokelau					100									41	100			89	88	83	84				63					
Tonga	100	100	86	99	92	100	100	100	100	98	100	53	71	70	98	100	100	100	100	99	100	63	83	17	99	100	100	100	99	100
Trinidad/Tobago	100	79	100	100	100	47		92	92	97	98	95	100	93	88			88	88	93	93	96	97	96	100	86	91	91	93	94
Tunisia	92	93	100	100		100		94	99	99	99	17		31		89		60	82	84	84	49		60	99		82	93	94	94
Turkey			95				82	96	98	99	100			62			84	87	93	92	96			70	76	83	93	96	97	99
Turkmenistan								93	93	97	97							54	54	72							71	72	84	
Turks/Caicos Islands			87						100	97	98			68				100	100	100	98			77			100	100	98	98
Tuvalu		100	100			100	100	94	94	100	100			100			95	92	92	99	99					98	93	100	100	100
Uganda	88	100	37	60		47	72	87	87	89	91	17	29	18	30	32	46	52	56	60	64	22	35	20	34	50	56	60	64	67
Ukraine								100	99	99	98							94	91	95	97						98	96	98	98
United Arab Emirates			95					100	100	100	100						100	100	100	100	100					100	100	100	100	100
United States of America							100	100	100	100	100						100	100	100	94	94					100	93	100	99	99
United States Virgin Islands									100										100									100		
Uruguay	100	100	96	95	100	98	98	98	98	100	100	59	87	2	27		93	93	93	95	100	92	98	81	85	89	98	100	100	100

continues

DATA TABLE 3A *continued*

Percentage of Population with Access to Improved Drinking Water

Region and Country	URBAN													RURAL													TOTAL												
	1970	1975	1980	1985	1990	1994	2000	2002	2004	2005	2008			1970	1975	1980	1985	1990	1994	2000	2002	2004	2005	2008			1970	1975	1980	1985	1990	1994	2000	2002	2004	2005	2008		
Uzbekistan							96	97		98	98									78	84		82	81									85	89		88	87		
Vanuatu		65		95			63	85	86	95	95									94	52	52	74	79									88	60	60	79	83		
Venezuela	92		93	93		80	88	85	85	94						53	54	36	75	58	70	70	75				75		86	89		79	84	83	83	93			
Vietnam				70	47	53	81	93	99	97	99			38			53	65	33	32	50	39	67	80	85	92			45	36	36	56	73	85	88	94			
Wallis and Futuna Islands																							100	100	99									100				
Western Samoa			97	75												94	67												69									
Yemen Arab Republic	45		100	100			85	74	71	75	72			2		18	25			64	68	65	58	57					31	40			69	69	67	63	62		
Yemen Dem	88		85				85							43		25				64							57		52				69						
Zambia	70	86		76		64	88	90	90	87	87			22	16		41		27	48	36	40	42	46			37	42		58		43	64	55	58	58	60		
Zimbabwe					95		100	100	98	99	99						32	80		77	74	72	72	72							84		85	83	81	82	82		

Sources: UNEP (1989); WRI (1988) UNEP (1989); WRI (1988) UNEP (1989); WRI (1988) UNEP (1989); WRI (1988) UNEP (1993–1994) WHO (1996) WHO (2000) WHO/UNICEF (2004) WHO/UNICEF (2006) WHO/UNICEF (2010) UNEP (1989); WRI (1988) UNEP (1989); WRI (1988) UNEP (1993–1994) WHO (1996) WHO (2000) WHO/UNICEF (2004) WHO/UNICEF (2006) WHO/UNICEF (2010) UNEP (1989); WRI (1988) UNEP (1989); WRI (1988) UNEP (1989); WRI (1988) Calculated from UNEP (1993–1994) WHO (1996) WHO (2000) WHO/UNICEF (2004) WHO/UNICEF (2006) WHO/UNICEF (2010) WHO/UNICEF (2010)

Note: The United Nations considers all European countries, except those shown, to have 100 percent water supply and sanitation coverage.

DATA TABLE 3B   Access to Improved Drinking Water by Country, 2011 Update

	2011 Population		2011 Fraction of Population with Access to Improved Drinking Water		
Country	Urban %	Rural %	Urban Improved Total Improved (%)	Rural Improved Total Improved (%)	National Improved Total Improved (%)
Afghanistan	24	76	85	53	61
Albania	53	47	95	94	95
Algeria	73	27	85	79	84
American Samoa	93	7	100	100	100
Andorra	87	13	100	100	100
Angola	59	41	66	35	53
Anguilla	100	0	95		95
Antigua and Barbuda	30	70	98	98	98
Argentina	93	7	100	95	99
Armenia	64	36	100	98	99
Aruba	47	53	98	98	98
Australia	89	11	100	100	100
Austria	68	32	100	100	100
Azerbaijan	54	46	88	71	80
Bahamas	84	16	96	96	96
Bahrain	89	11	100	100	100
Bangladesh	28	72	85	82	83
Barbados	44	56	100	100	100
Belarus	75	25	100	99	100
Belgium	97	3	100	100	100
Belize	45	55	97	100	99
Benin	45	55	85	69	76
Bermuda	100	0			
Bhutan	36	64	100	96	97
Bolivia (Plurinational State of)	67	33	96	72	88
Bosnia and Herzegovina	48	52	100	98	99
Botswana	62	38	99	93	97
Brazil	85	15	100	84	97
British Virgin Islands	41	59			
Brunei Darussalam	76	24			
Bulgaria	73	27	100	99	99
Burkina Faso	27	73	96	74	80
Burundi	11	89	82	73	74
Cambodia	20	80	90	61	67
Cameroon	52	48	95	52	74
Canada	81	19	100	99	100
Cape Verde	63	37	91	86	89
Cayman Islands	100	0	96		96
Central African Republic	39	61	92	51	67
Chad	22	78	71	44	50
Channel Islands	31	69			
Chile	89	11	100	90	98
China	51	49	98	85	92
China, Hong Kong SAR	100	0			
China, Macao SAR	100	0			
Colombia	75	25	100	72	93

continues

DATA TABLE 3B *continued*

	2011 Population		2011 Fraction of Population with Access to Improved Drinking Water		
Country	Urban %	Rural %	Urban Improved Total Improved (%)	Rural Improved Total Improved (%)	National Improved Total Improved (%)
Comoros	28	72		97	
Congo	64	36	95	32	72
Cook Islands	74	26	100	100	100
Costa Rica	65	35	100	91	96
Côte d'Ivoire	51	49	91	68	80
Croatia	58	42	100	97	99
Cuba	75	25	96	86	94
Cyprus	70	30	100	100	100
Czech Republic	73	27	100	100	100
Democratic People's Republic of Korea	60	40	99	97	98
Democratic Republic of the Congo	34	66	80	29	46
Denmark	87	13	100	100	100
Djibouti	77	23	100	67	92
Dominica	67	33	96		
Dominican Republic	70	30	82	81	82
Ecuador	67	33	96	82	92
Egypt	43	57	100	99	99
El Salvador	65	35	94	81	90
Equatorial Guinea	39	61			
Eritrea	21	79			
Estonia	69	31	99	97	99
Ethiopia	17	83	97	39	49
Faeroe Islands	41	59			
Falkland Islands (Malvinas)	74	26			
Fiji	52	48	100	92	96
Finland	84	16	100	100	100
France	86	14	100	100	100
French Guiana	76	24	95	75	90
French Polynesia	51	49	100	100	100
Gabon	86	14	95	41	88
Gambia	57	43	92	85	89
Georgia	53	47	100	96	98
Germany	74	26	100	100	100
Ghana	52	48	92	80	86
Greece	61	39	100	99	100
Greenland	85	15	100	100	100
Grenada	39	61			
Guadeloupe	98	2	99	100	99
Guam	93	7	99	99	99
Guatemala	50	50	99	89	94
Guinea	35	65	90	65	74
Guinea-Bissau	44	56	94	54	72
Guyana	28	72	98	93	95
Haiti	53	47	77	48	64
Honduras	52	48	96	81	89

	2011 Population		2011 Fraction of Population with Access to Improved Drinking Water		
Country	Urban %	Rural %	Urban Improved Total Improved (%)	Rural Improved Total Improved (%)	National Improved Total Improved (%)
Hungary	69	31	100	100	100
Iceland	94	6	100	100	100
India	31	69	96	89	92
Indonesia	51	49	93	76	84
Iran (Islamic Republic of)	69	31	98	90	95
Iraq	66	34	94	67	85
Ireland	62	38	100	100	100
Isle of Man	51	49			
Israel	92	8	100	100	100
Italy	68	32	100	100	100
Jamaica	52	48	97	89	93
Japan	91	9	100	100	100
Jordan	83	17	97	90	96
Kazakhstan	54	46	99	90	95
Kenya	24	76	83	54	61
Kiribati	44	56	87	50	66
Kuwait	98	2	99	99	99
Kyrgyzstan	35	65	96	85	89
Lao People's Democratic Republic	34	66	83	63	70
Latvia	68	32	100	96	98
Lebanon	87	13	100	100	100
Lesotho	28	72	91	73	78
Liberia	48	52	89	60	74
Libyan Arab Jamahiriya	78	22			
Liechtenstein	14	86			
Lithuania	67	33	98		
Luxembourg	85	15	100	100	100
Madagascar	33	67	78	34	48
Malawi	16	84	95	82	84
Malaysia	73	27	100	99	100
Maldives	41	59	100	98	99
Mali	35	65	89	53	65
Malta	95	5	100	100	100
Marshall Islands	72	28	93	97	94
Martinique	89	11	100	100	100
Mauritania	41	59	52	48	50
Mauritius	42	58	100	100	100
Mayotte	50	50			
Mexico	78	22	96	89	94
Micronesia (Fed. States of)	23	77	95	88	89
Monaco	100	0	100		100
Mongolia	69	31	100	53	85
Montenegro	63	37	100	95	98
Montserrat	14	86	99	99	99
Morocco	57	43	98	61	82
Mozambique	31	69	78	33	47
Myanmar	33	67	94	79	84
Namibia	38	62	99	90	93

continues

DATA TABLE 3B *continued*

	2011 Population		2011 Fraction of Population with Access to Improved Drinking Water		
Country	Urban %	Rural %	Urban Improved Total Improved (%)	Rural Improved Total Improved (%)	National Improved Total Improved (%)
Nauru	100	0	96		96
Nepal	17	83	91	87	88
Netherlands	83	17	100	100	100
Netherlands Antilles	93	7			
New Caledonia	62	38	98	98	98
New Zealand	86	14	100	100	100
Nicaragua	58	42	98	68	85
Niger	18	82	100	39	50
Nigeria	50	50	75	47	61
Niue	38	62	99	99	99
Northern Mariana Islands	91	9	97	97	97
Norway	79	21	100	100	100
Occupied Palestinian Territory	74	26	82	82	82
Oman	73	27	95	85	92
Pakistan	36	64	96	89	91
Palau	84	16	97	86	95
Panama	75	25	97	86	94
Papua New Guinea	12	88	89	33	40
Paraguay	62	38	99		
Peru	77	23	91	66	85
Philippines	49	51	93	92	92
Poland	61	39	100		
Portugal	61	39	100	100	100
Puerto Rico	99	1			
Qatar	99	1	100	100	100
Republic of Korea	83	17	100	88	98
Republic of Moldova	48	52	99	93	96
Réunion	94	6	99	98	99
Romania	53	47	99		
Russian Federation	74	26	99	92	97
Rwanda	19	81	80	66	69
Saint Kitts and Nevis	32	68	98	98	98
Saint Lucia	18	82	98	93	94
Saint Vincent and the Grenadines Samoa	49	51	95	95	95
	20	80	97	98	98
San Marino	94	6			
São Tomé and Príncipe	63	37	99	94	97
Saudi Arabia	82	18	97	97	97
Senegal	43	57	93	59	73
Serbia	56	44	99	99	99
Seychelles	54	46	96	96	96
Sierra Leone	39	61	84	40	57
Singapore	100	0	100		100
Slovakia	55	45	100	100	100
Slovenia	50	50	100	99	100
Solomon Islands	20	80	93	76	79

	2011 Population		2011 Fraction of Population with Access to Improved Drinking Water		
Country	Urban %	Rural %	Urban Improved Total Improved (%)	Rural Improved Total Improved (%)	National Improved Total Improved (%)
Somalia	38	62	66	7	30
South Africa	62	38	99	79	91
South Sudan	18	82	63	55	57
Spain	77	23	100	100	100
Sri Lanka	15	85	99	92	93
Sudan	33	67	66	50	55
Suriname	70	30	97	81	92
Swaziland	21	79	93	67	72
Sweden	85	15	100	100	100
Switzerland	74	26	100	100	100
Syrian Arab Republic	56	44	93	87	90
Tajikistan	27	73	92	57	66
TFYR Macedonia	59	41	100	99	100
Thailand	34	66	97	95	96
Timor-Leste	28	72	93	60	69
Togo	38	62	90	40	59
Tokelau	0	100		97	97
Tonga	23	77	99	99	99
Trinidad and Tobago	14	86	98	93	94
Tunisia	66	34	100	89	96
Turkey	72	28	100	99	100
Turkmenistan	49	51	89	54	71
Turks and Caicos Islands	94	6			
Tuvalu	51	49	98	97	98
Uganda	16	84	91	72	75
Ukraine	69	31	98	98	98
United Arab Emirates	84	16	100	100	100
United Kingdom	80	20	100	100	100
United Republic of Tanzania	27	73	79	44	53
United States of America	82	18	100	94	99
United States Virgin Islands	95	5	100	100	100
Uruguay	93	7	100	98	100
Uzbekistan	36	64	98	81	87
Vanuatu	25	75	98	88	91
Venezuela (Bolivarian Republic of)	94	6			
Vietnam	31	69	99	94	96
Western Sahara	82	18			
Yemen	32	68	72	47	55
Zambia	39	61	86	50	64
Zimbabwe	39	61	97	69	80

DATA TABLES 4A AND 4B

Access to Improved Sanitation by Country, 1970-2008 & 2011 Update

Description

Adequate sanitation is also a fundamental requirement for basic human well-being, and improving access to sanitation is one of the Millennium Development Goals (MDGs). Data are given here in two related tables.

Data table 4A includes the percentage of urban, rural, and total populations, by country, with access to improved sanitation for 1970, 1975, 1980, 1985, 1990, 1994, 2000, 2002, 2004, 2005, and 2008—similar to data table 4 in the previous volume of *The World's Water*. In the current volume, however, these previous data are sorted alphabetically by country rather than separated by regions.

Data table 4B provides an update of the percentage of national, urban, and rural populations with improved sanitation for 2011, the most recent year for which data are available. This table also includes the percentage of country populations in urban and rural categories for 2011.

The World Health Organization (WHO) / UNICEF Joint Monitoring Programme (JMP) for Water Supply and Sanitation collected the data presented here over various periods. Most of the data presented were drawn from responses by national governments to questionnaires. Participants in data collection include the JMP, the United Nations Children's Fund, and the Water Supply and Sanitation Collaborative Council, which has continued sector monitoring and aims to support and strengthen the monitoring efforts of individual countries. Countries used their own definitions of "rural" and "urban."

For all WHO assessments since 2000, new definitions were provided for "improved" sanitation with allowance for acceptable local technologies. For MDG monitoring, an improved sanitation facility is defined as one that hygienically separates human excreta from human contact.

The following technologies were included in the 2000 assessment as representing improved sanitation:

Connection to a public sewer

Connection to a septic system

Pour-flush latrine

Simple pit latrine

Ventilated improved pit latrine

In comparison, unimproved sanitation facilities refer to the following:

Public or shared latrine

Open pit latrine

Bucket latrine

Limitations

As is the case with drinking water data, definitions for access to sanitation vary from country to country and from year to year within the same country. Countries generally regard sanitation facilities that break the fecal-oral transmission route as adequate. In urban areas, adequate sanitation may be provided by connections to public sewers or by household systems such as pit privies, flush latrines, septic tanks, and communal toilets. In rural areas, pit privies, pour-flush latrines, septic tanks, and communal toilets are considered adequate. Direct comparisons between countries and across time within the same country are difficult and are additionally complicated by the fact that these data hide disparities between regions and socioeconomic classes.

WHO assessments since 2000 have attempted to shift from gathering information from water providers only to including consumer-based information. The current approach uses household surveys in an effort to assess the actual use of facilities. Access to sanitation services, as reported by the World Health Organization, does not imply that the level of service is "adequate" or "safe." The assessment questionnaire did not include any methodology for discounting coverage figures to allow for intermittence or poor quality of the service provided. More details of the methods used, and their limitations, can be found at http://www.who.int/docstore/water_sanitation_health/Globassessment/GlobalTOC.htm.

Sources

United Nations Environment Programme (UNEP). 1989. *Environmental Data Report*. GEMS Monitoring and Assessment Research Centre. Oxford, England: Basil Blackwell.

———. 1993–1994. *Environmental Data Report*. GEMS Monitoring and Assessment Research Centre in cooperation with the World Resources Institute and the UK Department of the Environment. Oxford, England: Basil Blackwell.

World Health Organization (WHO). 1996. Water Supply and Sanitation Sector Monitoring Report: 1996 (Sector Status as of 1994). In collaboration with the Water Supply and Sanitation Collaborative Council and the United Nations Children's Fund, UNICEF, New York.

———. 2000. Global Water Supply and Sanitation Assessment 2000 Report. http://www.who.int/docstore/water_sanitation_health/Globassessment/GlobalTOC.htm.

World Health Organization (WHO) and United Nations Children's Fund (UNICEF). 2004. Meeting the MDG Drinking Water and Sanitation Target: A Mid-Term Assessment of Progress. http://www.who.int/water_sanitation_health/monitoring/jmp2004/en/index.html.

———. 2006. Meeting the MDG Drinking Water and Sanitation Target: The Urban and Rural Challenge of the Decade. http://www.who.int/water_sanitation_health/monitoring/jmpfinal.pdf.

———. 2010. WHO/UNICEF Joint Monitoring Programme (JMP) for Water Supply and Sanitation. http://www.wssinfo.org/data-estimates/table.

———. 2013. WHO/UNICEF Joint Monitoring Programme (JMP) for Water Supply and Sanitation. http://www.wssinfo.org/data-estimates/table/ (2011 data were accessed May 29, 2013).

World Resources Institute (WRI). 1988. World Health Organization data, cited by the World Resources Institute, *World Resources 1988–89*. World Resources Institute and the International Institute for Environment and Development in collaboration with the United Nations Environment Programme. New York: Basic Books.

DATA TABLE 4A Access to Sanitation by Country, 1970–2008

Percentage of Population with Access to Improved Sanitation

Region and Country	URBAN											RURAL											TOTAL										
	1970	1975	1980	1985	1990	1994	2000	2002	2004	2005	2008	1970	1975	1980	1985	1990	1994	2000	2002	2004	2005	2008	1970	1975	1980	1985	1990	1994	2000	2002	2004	2005	2008
Afghanistan	69	63		5	13	38	25	16	49	56	60	16	15				1	8	5	29	29	30	21	21				8	12	8	34	35	37
Albania								99	99										81	84										89	91		
Algeria	13	100		80			90	99	99	98	98	6	50		40			47	82	82	86	88	9	67		57			73	92	92	94	95
Angola			40	29	25	34	70	56	56	80	86			15	16	20	8	30	16	16	15	18			20	19	21	16	44	30	31	50	57
Anguilla								98	99	99	99																			99		
Antigua and Barbuda	87							98	98	98	98								94	99										95	95		
Argentina	87	100	80	75			89	92	91	91	91	79	83	35	35			48		83	77	77	85	97		69			85	84	83	89	90
Armenia								96	96	95	95								61	61	79	80								84	83	89	90
Australia							100	100	100	100	100							100	100	100	100	100							100	100	100	100	100
Azerbaijan								73	73	55	51								36	36	34	39								55	54	45	45
Bahamas	100	100	88	100	98		93	100	100	100	100	13	13			2		94	100	100	100	100	66	65	88	100	63		93	100	100	100	100
Bahrain				100	100			100	100	100	100				100	0										100							
Bangladesh	87	40	21	24	40	77	82	75	51	57	56			1	3	4	30	44	39	35	48	52	6	5	3	5	10	35	53	48	39	50	53
Barbados	100	100	100	100	100		100	99	99	100	100	13	13			2		94	100	100	100	100	66	65	88	100	63		93	100	100	100	100
Belarus								93											61											84			
Belize			62	87	76	23	59	71	71	89	93			75	45	22	87	21	25	25	84	86			69	66	50	57	42	47	47	86	90
Benin	83		48	58	60	54	46	58	59	22	24	1		4	20	35	6	6	12	11	3	4	14		16	33	45	20	23	32	33	11	12
Bhutan					80	66	65	65	65	87	87					3	18	70	70	70	54	54						7	41	69	70	64	65
Bolivia	25		37	33	38	58	82	58	60	33	34	4	9	4	10	14	16	38	23	22	9	9	12		18	21	26	41	66	45	46	24	25
Bosnia and Herzegovina								99											92											95			
Botswana				93	100		85	57	57	72	74				28	85		40	25	25	36	39				40	89		77	41	42	57	60
Brazil	85		86	84	55		83	83	86	87		24		1	1	32	3	40	35	37	37	37	58		63	71	44		75	75	78	80	
British Virgin Islands				100			100	100	100	100	100	13	13			2		94	100	100	100	100	66	65	88	100	63		93	100	100	100	100

continues

DATA TABLE 4A continued

Percentage of Population with Access to Improved Sanitation

Region and Country	URBAN													RURAL													TOTAL												
	1970	1975	1980	1985	1990	1994	2000	2002	2004	2005	2008			1970	1975	1980	1985	1990	1994	2000	2002	2004	2005	2008			1970	1975	1980	1985	1990	1994	2000	2002	2004	2005	2008		
Bulgaria									100													96												99					
Burkina Faso	49	47	38	44		42	88	45	42	32	33					5	6		11	16	5	6	6	6			4	4	7	9		18	29	12	13	11	11		
Burundi	96		40	84	64	60	79	47	47	49	49					35	56	16	50		35	35	46	46					35	58	18	51		36	36	46	46		
Cambodia			100				58	53	53	60	67					76				10	8	8	15	18									18	16	17	24	29		
Cameroon				100			99	63	58	58	56						1			85	33	43	35	35									92	48	51	47	47		
Canada							100	100	100	100	100									99	99	99	99	99							43		100	100	100	100	100		
Cape Verde			34	32		40	95	61	61	65	65					10	9		10	32	19	19	34	38					11	10		24	71	42	43	52	54		
Cayman Islands			94	96						96	96						94																						
Central African Republic	64	100			45		43	47	47	39	43			96				46		23	12	12	23	28							46	46	31	27	27	29	34		
Chad	7	9				73	81	30	24	23	23				1				7	13	0	4	4	4				1				21	29	8	9	9	9		
Chile	33	36	100	100		82	98	96	95	98	98			10	11	10	4			93	64	62	83	83			29	32		84			97	92	91	96	96		
China					100	58	68	69	69	58	58							81	7	24	29	28	50	52					83		86		38	44	44	53	55		
Colombia	75	73	93	96	84	76	97	96	96	81	81			8	13	4	13	18	33	51	54	54	53	55			47	48	61		64	63	85	86	86	74	74		
Comoros							98	38	41	49	50									98	15	29	29	30									98	23	33	35	35		
Congo	8	10					14	14	28	31	31			6	9						2	25	29	29			6	9						9	27	30	30		
Congo, Democratic Republic	5	65			46	23	53	43	42	23	23			5	6		9	11	4	6	23	25	19	23			5	22			21	9	20	29	30	20	23		
Cook Islands			100	100	100		100	100	100	100	100									100	100	100	100	100									100	100	100	100	100		
Costa Rica	66	94	99	99		85	98	89	89	95	95			43	93	84	89		99	95	97	97	95	96			52	93	91	95	92	92	96	92	92	95	95		
Côte d'Ivoire	23				81	59	61	46	37	36								100	51	24	23	29	11	11			5					54	38	40	37	23	23		
Cuba	57	100			100	71	96	99	99	92	94						68		51	91	95	95	78	81					91		92	66	95	98	98	89	91		
Cyprus	100	94		100	96		100	100	100	100	100			92	95	100	100			100	100	100	100	100			95	95		100	98		100	100	100	100	100		
Czech Republic								99													97												98					

Country																													
Djibouti			43	78		77	99	55	88	65	63		54		20	17		50	27	50	18	10				39		90	91	50	82	58	56
Dominica									86	80									75	75	84									83	84	80	
Dominican Republic	63	74	25	41	95	76	75	67	81	86	87			16	4	10	75	83	64	43	73	72	74	58		42	15	23	87	78	71	57	78	81	83
East Timor							65	66	68	86									30	33	35	40									33	36	44	50
Ecuador			73	98	56	87	70	80	94	95	96		7	17	29	38	34	37	59	82	81	84				43	65	48	64	59	72	89	90	92
Egypt					80	20	98	84	86	97	97				10	26	5	91	56	58	90	92						50	11	94	68	70	93	94
El Salvador	66	71	48	82	85	78	88	78	77	89	89	18	17	17	43	38	59	78	40	39	79	83	37		39	35	58	59	94	83	78	85	87
Equatorial Guinea					54	61	60	60	60	60					26	24	48	46	46	46	46						33	54	53	53	62	70	51	
Eritrea											52		8	8					1	3	3	4	4								13	9	13	14
Estonia								66	32	52											96											97		
Ethiopia	67	56	96				58	34	44	28	29			8	96				6	4	7	7	8	14	14				15		6	13	10	12
Fiji	100	100	85		91	100	75	19	87			87	93		60			85	12	98	55			91	96		70	75	92	43	98	72		
French Guiana							85	99											57		85									79		78		
French Polyneisa							99	99	97	99	99							95	97	97	99	97	97							98	98	98	98	98
Gabon							25	37	37	34	33								4	30	30	30	30	55						21	36	36	33	33
Gambia						83	41	72	72	67	68					27	23		35	46	46	63	65			44			37	37	53	53	65	67
Gaza Strip				100					78	91	91									61	61	84	84									73	89	89
Georgia								96	96	96	96								69		91	94	93								83	94	95	95
Ghana	92	95	47	51	63	53	62	74	27	17	18	40	40	17	16	60	36	64	46	11	6	7			56	26	30	61	42	63	58	18	11	13
Grenada								96	96	96	96							97	97	97	97	97								97	97	96	97	98
Guadeloupe								61	61										61		64									61		64		
Guam							98	99	98	99	99	11	16	20		52		76	98	99	98	98	98					60		85	99	99	98	99
Guatemala	70		45	41	72		98	72	90	88	89	2		1	12	0		41	52	6	69	73	30	13		30	24		70	58	61	86	78	81
Guinea			54				94	25	31	31	34		13	18			17		6	11	10	11				11			20	13	13	18	17	19
Guinea-Bissau			21	29		32	88	57	57	47	49			79	80	81		34	23	23	8	9				15	21			47	34	35	20	21
Guyana	95	99	73	100	97		97	86	86	85	85	92	94	10	13			81	60	78	80	93	96	78	86	86	87	70	80	81
Haiti			42	42	44	42	50	52	57	28	24	43	1	13	17	16	14	16	23	12	10	19	21	25	24	28	34	30	19	17
Honduras	64	53	49	24	89	81	94	89	87	78	80	9	13	26	34	42	53	57	52	54	56	62	24	26	35	30	63	65	77	68	69	66	71
Hong Kong				90												50						88							
Hungary									100	100									85	85											95	95	

continues

DATA TABLE 4A *continued*

Percentage of Population with Access to Improved Sanitation

Region and Country	URBAN													RURAL													TOTAL												
	1970	1975	1980	1985	1990	1994	2000	2002	2004	2005	2008			1970	1975	1980	1985	1990	1994	2000	2002	2004	2005	2008			1970	1975	1980	1985	1990	1994	2000	2002	2004	2005	2008		
India	85	87	27	31	44	70	73	58	59	54	54			1	2	1	2	3	14	14	18	22	18	21			18	20	7	9	14	29	31	30	33	28	31		
Indonesia	50	60	29	33	79	73	87	71	73	66	67			4	5	21	38	30	40	52	38	40	33	36			12	15	23	37	44	51	66	52	55	49	52		
Iran	100	100	96		100	89	86	86						48	59	43		35	37	74	78						70	78	69		72	67	81	84					
Iraq	82	75		100	96		93	95	95	76	76				1		11			31	48	48	61	66			47	47				36	79	80	79	71	73		
Israel								100	100	100	100												100	100											100	100		
Jamaica	100	100	12	92		98	98	90	91	82	82			92	91	2	90			66	68	69	84	84			94	94	7	91			84	80	80	83	83		
Japan								100	100	100	100												100	100											100	100		
Jordan			94	92	100	100	100	94	94	98	98					34		100		98	85	87	96	97				70		100	95	99	93	93	98	98		
Kazakhstan							100	87	87	97	97									98	52	52	98	98								99	72	72	97	97		
Kenya	85	98	89			69	96	56	46	27	27			45	48	19			81	81	43	41	31	32			50	55	30			77	86	48	43	30	31		
Kiribati					91	100	54	59	59	49								49	100	44	22	22	22								100	48	39	40	35			
Korea Demo-cratic People's Republic							99	58	58										100	60	60										99	59	59				
Korea Republic	59	80	100	100	67		76			100	100												100	100											100	100		
Kuwait		100	100							100	100												100	100											100	100		
Kyrgyzstan							100	75	75	94	94									100	51	51	93	93									100	60	59	93	93		
Laos		10	13		30	70	84	61	67	77	86				2	4		8	13	34	14	20	30	38				3	5		12	24	46	24	30	43	53		
Latvia									82													71												78				
Lebanon						100	100	100	100	100	100								100	92	87	87	87					100			99	98	98	98		
Lesotho	44	51	13	22		1	93	61	61	39	40			10	12	14	14		7	96	32	32	25	25			11	13	14	15		6	92	37	37	28	29		
Liberia	100			6		38		49	49	24	25			9			2		2		7	7	4	4			19					18		26	27	16	17		
Libya	100	100	100					97	97	97	97			54	69	72			30	96	96	96	96			67	79	88				97	97	97	97	97		
Madagascar	88	9		55		50	70	49	48	15	15				9				3		70	27	26	9	10								15	42	33	34	11	11		
Malawi			100			70	96	66	62	51	51					81			51	98	42	61	55	57					83			53	77	46	54	61	56		
Malaysia	100	100	100	100	94			95	95	96	96			43	43	55	60	94		41	98	93	95	95			59	60	70	75	94		94	96	94	96	96		

258

Country																														
Maldives		21	60	100	95	100	100	100	100	100	45				1	2	4	26	58	42	42	88	96		8	3	13	22	44	56	58	59	92	98
Mali	63		79	90	81	58	93	59	59	44						3	10	21	100	38	39	31	32					19	31	69	45	46	35	36
Marshall Islands					100		93	93	93	82	83							45		59	58	51	53					27			82	82	72	73
Martinique	100									95	95																							
Mauritania			5	8			44	64	49	45	50					86		19	9	8	9			7						33	42	34	24	26
Mauritius	51	63	100	100	100	100	100	100	95	93	93	99	100	90	13	100	26	99	99	94	90	90	77	82	94	92	100	99	99	94	91	91
Mexico			77	77	85	81	87	90	91	88	90	13	14	12	13		26	32	39	41	61	68			55	58	66	73	77	79	82	85
Micronesia					99	100		61	61	61							100		14	14	15									28	28	26	
Mongolia				100			46	75	75	64	64					46		2	37	37	30	32					78		30	59	59	49	50
Montserrat								96	96	96	96					47	100		96	96	96	96								96	100	100	100
Morocco	75		62	100	69		83	88	83	83	83	4				16	18	42	31	52	50	52	29					40	75	61	73	68	69
Mozambique			53		70		51	53	37	38						12	70	26	14	19	4	4					20		43	27	32	15	17
Myanmar (Burma)	45	38	33	50	42		96	88	86	86	75						13	40	39	63	72	79	79	35	33	20	24	22	41	46	73	77	81	81
Namibia				24				66	50	61	60						11		17	14	13	15	17						15		41	30	25	31	33
Nauru										50	50																						
Nepal	14	14	16	17	34	51	75	68	62	50	51				1	1	3	16	20	20	30	24	27	1	1	1		6	20	27	27	35	28	31
New Zealand												8	24																					
Nicaragua			34	35		34	96	78	56	62	63		1		16		27	27	68	51	34	35	37					27		31	84	66	47	50	52
Niger	10	30	36		71	71	79	43	43	32	34			3		4	4	4	5	4	4	4	4	1	3	7		17	15	20	12	13	9	9
Nigeria					80	61	85	48	53	36	36					5	11	21	45	30	36	29	28					35	36	63	38	44	32	32
Niue				0	100	100	100	100	100	100	100						71		100	100	100	100	100							100	100	100	100	100
Northern Mariana Islands			100					94	96	96									92	96	94	96	96								94	95	94	
Oman	100	100		88		98	97	97	97	97	97			5		25		61	61					3	12		31			76	92	89		
Pakistan	12	21	42	51	53	94	92	92	92	72	72			2		6	12	19	42	35	41	25	29		6	13	19	25	30	61	54	59	41	45
Palau			95		100	96	100	96	96	96	96						100		100	52	52	52								100	83	80	79	
Panama	87	78	83		71	87	89	89	89	75	75			59		61		94	51	51	51	51		78	77	71	81	86	99	72	73	68	69
Papua New Guinea	100	100	99	57	80	85	67	67	73	73	71		69	76		35		80	41	41	42	41		14	18	15	44	22	82	45	44	46	45
Paraguay	16	28	95	89	31	95	94	94	94	88	90		5	80		83	60	95	58	61	39	40	6	10	86	85	46	95	78	80	69	70

continues

DATA TABLE 4A *continued*

Percentage of Population with Access to Improved Sanitation

Region and Country	URBAN													RURAL													TOTAL												
	1970	1975	1980	1985	1990	1994	2000	2002	2004	2005	2008			1970	1975	1980	1985	1990	1994	2000	2002	2004	2005	2008			1970	1975	1980	1985	1990	1994	2000	2002	2004	2005	2008		
Peru	52	57	67	76	62	90	72	74	79	81				16		0	12	20	10	40	33	32	33	36			36		36	49	59	44	76	62	63	66	68		
Philippines	90	76	81	83	79		92	81	80	78	80			40	44	67	56	63		71	61	59	65	69			57	56	75	67	70		83	73	72	73	76		
Qatar	100	100			100		100	100	100	100	100			16	100						100	100	100	100			83	100					100	100	100	100	100		
Republic of Moldova								86	86												52	52												68	68				
Romania								86	89												10												51					
Russian Federation								93	93												70	70	70											87	87				
Rwanda	83	87	60	77	88		12	56	56	47	50			52	56	50	55	17		8	38	38	49	55			53	57	51	56	21		8	41	42	49	54		
Samoa	100	100	86		100		95	100	100	100	100									100	100	100	100	100									100	100	100	100	100		
São Tomé and Príncipe								32	32	29	30						15				20	20	17	19						15				24	25	24	26		
Saudi Arabia	67	91	81	100			100	100	100	100	100			11	35	50	33			100							21	47	70	82			100						
Senegal			100	87	57	83	94	70	79	68	69					2		38	40	48	34	34	36	38					36		46	58	70	52	57	49	51		
Serbia and Montenegro								97	97												77	77											87	87					
Seychelles										96	97										100	100																
Sierra Leone			31	60	55	17	23	53	53	23	24					6	10	31	8	31	30	30	6	6					12	24	39	11	28	39	39	12	13		
Singapore			80	99	99		100	100	100	100	100																		80	99			100	100	100				
Slovakia									100													98												99					
Solomon Islands					73		98	98	98	98	98									18	18	18	18	18						13		34	31	31	32			
Somalia		77		44				47	48	50	52				35	21	5	2		14	14	14	7	6				47		18			25	26	22	23		
South Africa			80	65		79	99	86	79	83	84								12	73	44	46	64	65								46	86	67	65	75	77		
Sri Lanka	76	68	80		68	33	91	98	98	87	88			61	55	63	39	45	58	80	89	89	88	92			64	59	67	44	50	52	83	91	91	88	91		
St. Kitts								96	96	96	96										96	96	96	96									96	95	96	96		
St. Lucia								89	89												89	89											89	89					

Country																													
St. Vincent	100	100																											
Sudan			73	73	79	87	50	50	56	55	4	10	4	48		96	96	96	96	96	16	22		62	34	34	34	34	34
Suriname																									83	93	94	84	84
Swaziland		99	100		36	100	99	78	90	90		79		34		24	24	19	18	66			88		52	48	53	55	
Syria			74		77	98	97	78	61	61	25	28	25		81	76	44	66	53		36	36	45	36	77	90	93	96	
Tajikistan								71	95	96				37		56	44	50	90	95			50	56					
Tanzania		88					71	70	95	95	14		58			47	81	92	92	94	17				53	51	93	94	
Thailand	65	58	93		98	98	54	53	31	32					86	41	45	43	22	21	40	17		90	46	47	24	24	
Togo	4	64	78			97	97	98	95	95	8	41	46	36	86	100	96	99	96	96		45	52	96	99	99	96	96	
Tokelau		24	31		57	69	71	71	24	24		10	9	12	17	15	15	4	3		15	13	14	34	34	35	12	12	
Tonga	100	100		100	100							41									1								
Trinidad/Tobago			97		100		98	98	98	98	100	94	40	100		96	74	82	92	92	100	100	52	26		78			78
Tunisia	100			88			100	100	92	92	100		78	100		100	96	96	96	96		19	97		97	96	96	97	97
Turkey		84		100			90	96	96	96	34		16	85		62	62	64	64		62	100		88	80	100	92	92	
Turkmenistan		56			98		94	96	97	97					70	62	65	73	75	66			55	96		85	88	89	85
Turks/Caicos Islands	51	83	96	100			77	77	99	99						50	72	50	75	53					91	83	89	90	85
Tuvalu		100	81		90		98	94	98	98	96	88	95			94	98	95	97	97	81	92	93	98			62	98	90
Uganda	84	82	32	32	75	100	92	93	87	88	76	80	73	85	85	83	84	80	81	81	76		55	87	62	98	98	98	
Ukraine					96	53	54	37	38		95	30	60	55	100	39	41	48	49	48	94		30	57	100	88	90	82	81
United Arab Emirates		93			100	98	100	98	98	98	22			72	97	93	93	93				80		75	75	41	43	47	48
United States of America					100	98										100	95	95	95	95					99	98	98	97	97	
United States Virgin Islands					100	100	100	100	100					100	100	100	100	99	99					100	100	100	100	100	
Uruguay	97	59	59		96	95	99	99	99	89	13	6	59		100	85	99	96	96		82	83	51	59	95	94	100	99	100
Uzbekistan				100	100	73	78	99	99	100	17				100	48	61	96	96	100					100	57	67	97	100
Vanuatu	95	86		100	100	78	78	63	66		68	25			100	42	42	43	48				40	100	100	50	50	48	53
Venezuela	60	57		64	86	71	71	94	94	45	12	5	30	69		48	48	57				52	50	58	74	68	68	91	
Vietnam	100			23	43	87	84	92	88	98		55	72	15	70	26	50	61	67	13	26			21	73	41	61	68	75
Wallis and Futuna Islands						80	94	94		2		10						96	96							80			

continues

DATA TABLE 4A continued

Percentage of Population with Access to Improved Sanitation

Region and Country	URBAN													RURAL													TOTAL											
	1970	1975	1980	1985	1990	1994	2000	2002	2004	2005	2008			1970	1975	1980	1985	1990	1994	2000	2002	2004	2005	2008			1970	1975	1980	1985	1990	1994	2000	2002	2004	2005	2008
Western Samoa			86	88												83	83												84								
Yemen Arab Republic			60	83			99	76	86	89	94									31	14	28	29	33								45	30	43	46	52
Yemen Dem			70				99									15													35			45				
Zambia	12	87		76		40	99	68	59	59	59			18	16		34		10	64	32	52	41	43						55		23	78	45	55	47	49
Zimbabwe					95		99	69	63	57	56						15	22		51	51	47	37	37						43			68	57	53	44	44

Sources: UNEP (1989); WRI (1988) UNEP (1989); WRI (1988) UNEP (1989); WRI (1988) UNEP (1989); WRI (1988) UNEP (1989); WRI (1988) UNEP (1989) WHO (1996) WHO (2000) WHO/UNICEF (2004) WHO/UNICEF (2006) WHO/UNICEF (2010) WHO/UNICEF (2010) UNEP (1989); WRI (1988) UNEP (1989); WRI (1988) UNEP (1989); WRI (1988) UNEP (1989); WHO (2000) WHO/UNICEF (2004) WHO/UNICEF (2006) WHO/UNICEF (2010) UNEP (1989); WRI (1988) UNEP (1989); WRI (1988) UNEP (1989); WRI (1988) UNEP (1989); WRI (1988) Calculated from UNEP (1993) WHO (1996) WHO (2000) WHO/UNICEF (2004) WHO/UNICEF (2006) WHO/UNICEF (2010) WHO/UNICEF (2010)

DATA TABLE 4B Access to Improved Sanitation by Country, 2011 Update

	2011 Population		2011 Fraction of Population with Access to Improved Sanitation		
Country	Urban %	Rural %	Urban Improved Total Improved (%)	Rural Improved Total Improved (%)	National Improved Total Improved (%)
Afghanistan	24	76	46	23	28
Albania	53	47	95	93	94
Algeria	73	27	98	88	95
American Samoa	93	7	97	97	97
Andorra	87	13	100	100	100
Angola	59	41	86	19	59
Anguilla	100	0	98		98
Antigua and Barbuda	30	70	91	91	91
Argentina	93	7	96	98	96
Armenia	64	36	96	81	90
Aruba	47	53	98	98	98
Australia	89	11	100	100	100
Austria	68	32	100	100	100
Azerbaijan	54	46	86	78	82
Bahamas	84	16			
Bahrain	89	11	99	99	99
Bangladesh	28	72	55	55	55
Barbados	44	56			
Belarus	75	25	92	97	93
Belgium	97	3	100	100	100
Belize	45	55	93	87	90
Benin	45	55	25	5	14
Bermuda	100	0			
Bhutan	36	64	74	29	45
Bolivia (Plurinational State of)	67	33	57	24	46
Bosnia and Herzegovina	48	52	100	92	96
Botswana	62	38	78	42	64
Brazil	85	15	87	48	81
British Virgin Islands	41	59	98	98	97
Brunei Darussalam	76	24			
Bulgaria	73	27	100	100	100
Burkina Faso	27	73	50	6	18
Burundi	11	89	45	51	50
Cambodia	20	80	76	22	33
Cameroon	52	48	58	36	48
Canada	81	19	100	99	100
Cape Verde	63	37	74	45	63
Cayman Islands	100	0	96		96
Central African Republic	39	61	43	28	34
Chad	22	78	31	6	12
Channel Islands	31	69			
Chile	89	11	100	89	99
China	51	49	74	56	65
China, Hong Kong SAR	100	0			
China, Macao SAR	100	0			
Colombia	75	25	82	65	78
Comoros	28	72			

continues

DATA TABLE 4B *continued*

	2011 Population		2011 Fraction of Population with Access to Improved Sanitation		
Country	Urban %	Rural %	Urban Improved Total Improved (%)	Rural Improved Total Improved (%)	National Improved Total Improved (%)
Congo	64	36	19	15	18
Cook Islands	74	26	95	95	95
Costa Rica	65	35	95	92	94
Côte d'Ivoire	51	49	36	11	24
Croatia	58	42	99	98	98
Cuba	75	25	94	87	92
Cyprus	70	30	100	100	100
Czech Republic	73	27	100	100	100
Democratic People's Republic of Korea	60	40	88	73	82
Democratic Republic of the Congo	34	66	29	31	31
Denmark	87	13	100	100	100
Djibouti	77	23	73	22	61
Dominica	67	33			
Dominican Republic	70	30	86	74	82
Ecuador	67	33	96	86	93
Egypt	43	57	97	93	95
El Salvador	65	35	79	53	70
Equatorial Guinea	39	61			
Eritrea	21	79		4	
Estonia	69	31	100	94	98
Ethiopia	17	83	27	19	21
Faeroe Islands	41	59			
Falkland Islands (Malvinas)	74	26			
Fiji	52	48	92	82	87
Finland	84	16	100	100	100
France	86	14	100	100	100
French Guiana	76	24	95	76	90
French Polynesia	51	49	97	97	97
Gabon	86	14	33	30	33
Gambia	57	43	70	65	68
Georgia	53	47	96	91	93
Germany	74	26	100	100	100
Ghana	52	48	19	8	13
Greece	61	39	99	97	99
Greenland	85	15	100	100	100
Grenada	39	61			
Guadeloupe	98	2	97	90	97
Guam	93	7	97	97	97
Guatemala	50	50	88	72	80
Guinea	35	65	32	11	18
Guinea-Bissau	44	56	33	8	19
Guyana	28	72	88	82	84
Haiti	53	47	34	17	26
Honduras	52	48	86	74	81
Hungary	69	31	100	100	100
Iceland	94	6	100	100	100

	2011 Population		2011 Fraction of Population with Access to Improved Sanitation		
Country	Urban %	Rural %	Urban Improved Total Improved (%)	Rural Improved Total Improved (%)	National Improved Total Improved (%)
India	31	69	60	24	35
Indonesia	51	49	73	44	59
Iran (Islamic Republic of)	69	31	100	99	100
Iraq	66	34	86	80	84
Ireland	62	38	100	98	99
Isle of Man	51	49			
Israel	92	8	100	100	100
Italy	68	32			
Jamaica	52	48	78	82	80
Japan	91	9	100	100	100
Jordan	83	17	98	98	98
Kazakhstan	54	46	97	98	97
Kenya	24	76	31	29	29
Kiribati	44	56	51	30	39
Kuwait	98	2	100	100	100
Kyrgyzstan	35	65	94	93	93
Lao People's Democratic Republic	34	66	87	48	62
Latvia	68	32			
Lebanon	87	13	100		
Lesotho	28	72	32	24	26
Liberia	48	52	30	7	18
Libyan Arab Jamahiriya	78	22	97	96	97
Liechtenstein	14	86			
Lithuania	67	33	95		
Luxembourg	85	15	100	100	100
Madagascar	33	67	19	11	14
Malawi	16	84	50	53	53
Malaysia	73	27	96	95	96
Maldives	41	59	97	98	98
Mali	35	65	35	14	22
Malta	95	5	100	100	100
Marshall Islands	72	28	84	55	76
Martinique	89	11	94	73	92
Mauritania	41	59	51	9	27
Mauritius	42	58	92	90	91
Mayotte	50	50			
Mexico	78	22	87	77	85
Micronesia (Fed. States of)	23	77	83	47	55
Monaco	100	0	100		100
Mongolia	69	31	64	29	53
Montenegro	63	37	92	87	90
Montserrat	14	86			
Morocco	57	43	83	52	70
Mozambique	31	69	41	9	19
Myanmar	33	67	84	74	77
Namibia	38	62	57	17	32
Nauru	100	0	66		66
Nepal	17	83	50	32	35

continues

DATA TABLE 4B *continued*

Country	2011 Population Urban %	2011 Population Rural %	2011 Fraction of Population with Access to Improved Sanitation		
			Urban Improved Total Improved (%)	Rural Improved Total Improved (%)	National Improved Total Improved (%)
Netherlands	83	17	100	100	100
Netherlands Antilles	93	7			
New Caledonia	62	38	100	100	100
New Zealand	86	14			
Nicaragua	58	42	63	37	52
Niger	18	82	34	4	10
Nigeria	50	50	33	28	31
Niue	38	62	100	100	100
Northern Mariana Islands	91	9	98	98	98
Norway	79	21	100	100	100
Occupied Palestinian Territory	74	26	95	93	94
Oman	73	27	97	95	97
Pakistan	36	64	72	34	47
Palau	84	16	100	100	100
Panama	75	25	77	54	71
Papua New Guinea	12	88	57	13	19
Paraguay	62	38			
Peru	77	23	81	38	72
Philippines	49	51	79	69	74
Poland	61	39	96		
Portugal	61	39	100	100	100
Puerto Rico	99	1	99	99	99
Qatar	99	1	100	100	100
Republic of Korea	83	17	100	100	100
Republic of Moldova	48	52	89	83	86
Réunion	94	6	98	95	98
Romania	53	47			
Russian Federation	74	26	74	59	70
Rwanda	19	81	61	61	61
Saint Kitts and Nevis	32	68			
Saint Lucia	18	82	70	64	65
Saint Vincent and the Grenadines	49	51			
Samoa	20	80	93	91	92
San Marino	94	6			
São Tomé and Príncipe	63	37	41	23	34
Saudi Arabia	82	18	100	100	100
Senegal	43	57	68	39	51
Serbia	56	44	98	96	97
Seychelles	54	46	97	97	97
Sierra Leone	39	61	22	7	13
Singapore	100	0	100		100
Slovakia	55	45	100	100	100
Slovenia	50	50	100	100	100
Solomon Islands	20	80	81	15	29
Somalia	38	62	52	6	24

	2011 Population		2011 Fraction of Population with Access to Improved Sanitation		
Country	Urban %	Rural %	Urban Improved Total Improved (%)	Rural Improved Total Improved (%)	National Improved Total Improved (%)
South Africa	62	38	84	57	74
South Sudan	18	82	16	7	9
Spain	77	23	100	100	100
Sri Lanka	15	85	83	93	91
Sudan	33	67	44	13	24
Suriname	70	30	90	66	83
Swaziland	21	79	63	55	57
Sweden	85	15	100	100	100
Switzerland	74	26	100	100	100
Syrian Arab Republic	56	44	96	94	95
Tajikistan	27	73	95	94	95
TFYR Macedonia	59	41	97	83	91
Thailand	34	66	89	96	93
Timor-Leste	28	72	68	27	39
Togo	38	62	26	3	11
Tokelau	0	100		93	93
Tonga	23	77	99	89	92
Trinidad and Tobago	14	86	92	92	92
Tunisia	66	34	97	75	90
Turkey	72	28	97	75	91
Turkmenistan	49	51	100	98	99
Turks and Caicos Islands	94	6			
Tuvalu	51	49	86	80	83
Uganda	16	84	34	35	35
Ukraine	69	31	96	89	94
United Arab Emirates	84	16	98	95	98
United Kingdom	80	20	100	100	100
United Republic of Tanzania	27	73	24	7	12
United States of America	82	18	100	99	100
United States Virgin Islands	95	5	96	96	96
Uruguay	93	7	99	98	99
Uzbekistan	36	64	100	100	100
Vanuatu	25	75	65	55	58
Venezuela (Bolivarian Republic of)	94	6			
Vietnam	31	69	93	67	75
Western Sahara	82	18			
Yemen	32	68	93	34	53
Zambia	39	61	56	33	42
Zimbabwe	39	61	52	33	40

DATA TABLE 5

MDG Progress on Access to Safe Drinking Water by Region

Description

The Millennium Development Goals (MDGs)—adopted by the United Nations in 2000—established a set of targets for improving the lives of the world's poor, ranging from eradicating extreme hunger to reducing child mortality and ensuring environmental sustainability. These targets, agreed to by all countries and leading development institutions throughout the world, consist of eight goals and twenty-one targets. This table presents regional summaries of progress for the years 1990, 2004, 2008, and 2011 (from the 2013 update) as well as the 2015 target levels, measured as the proportion of the population using an improved water source (see data tables 3 and 4 for definitions).

In 2010, the human right to safe water and sanitation was formally recognized by the UN General Assembly and the UN Human Rights Council, but this basic right is not being met universally. In many parts of the world, particularly sub-Saharan Africa and Oceania, a lack of clean water adversely affects human health and development. Using 1990 as a baseline, one target of goal 7 of the MDGs seeks to reduce by half the proportion of people without sustainable access to safe drinking water by 2015.

At the global level, we are on track to meet the target for improving access to safe drinking water; however, not all regions are performing as well as others.

Europe, Latin America, the Caribbean, and much of Asia have met or are on track to meet the established targets. But in Oceania, there has been no progress or conditions have worsened; it is possible that the targets for Northern and Sub-Saharan Africa and South-Eastern Asia can still be met, but time is running out. The global community must continue to intensify efforts in these regions if we hope to achieve the established 2015 targets.

Limitations

These data give a good picture of the current lack of access to improved water and sanitation services, but comparison with different assessments should be done with extreme care, or not at all, because of changing definitions. For the 2011 update (released in 2013), some regional changes in country assignments slightly change the data from previous years. In particular, the region formerly noted as the Commonwealth of Independent States is now Caucasus and Central Asia.

Country-reported data may reflect national definitions of "improved," unlike survey data, which were standardized as much as possible. For example, in many African countries the population "without access" to improved sanitation means people with no access to any sanitary facility. In Latin America and the Caribbean, however, it is more likely that those "without access" in fact have a sanitary facility, but the facility is deemed unsatisfactory by local or national authorities. Low coverage figures for Latin America and the Caribbean may in part be a reflection of the comparatively narrow definitions used within that region.

Changes in the source of data also complicate comparisons over time. Prior to 2000, for example, data collected by the World Health Organization was provider based and was collected from service providers, such as utilities, ministries, and water agencies. The data shown here, however, are sometimes user based and were collected from household surveys and censuses. User-based data are more likely to include improvements installed by households or local communities and give a more complete picture of water supply and sanitation coverage.

Sources

United Nations. 2007a. The Millennium Development Goals Report. http://mdgs.un.org/unsd/mdg/Resources/Static/Data/2007%20Stat%20Annex%20current%20indicators.pdf.

———. 2007b. Millennium Development Goals: 2007 Progress Chart. http://mdgs.un.org/unsd/mdg/Resources/Static/Products/Progress2007/MDG_Report_2007_Progress_Chart_en.pdf.

———. 2010a. The Millennium Development Goals Report. http://www.un.org/millenniumgoals/pdf/MDG%20Report%202010%20En%20r15%20-low%20res%2020100615%20-.pdf.

———. 2010b. Millennium Development Goals: 2010 Progress Chart. http://unstats.un.org/unsd/mdg/Resources/Static/Products/Progress2010/MDG_Report_2010_Progress_Chart_En.pdf.

World Health Organization (WHO) and United Nations Children's Fund (UNICEF). 2013. Progress on Sanitation and Drinking-Water—2013 Update. Geneva, Switzerland.

DATA TABLE 5 MDG Progress on Access to Safe Drinking Water by Region (Proportion of Population Using an Improved Water Source)

	1990			2004			2008			2011			2015 Target			On target?
	Urban	Rural	Total	Urban	Rural	Total	Urban	Rural	Total	Urban	Rural	Total	Urban	Rural	Total	
Northern Africa	95	82	89	96	86	91	98	87	92	95	89	92	98	91	95	Progress insufficient to meet target
Sub-Saharan Africa	82	36	49	80	42	56	83	47	60	84	76	72	91	68	75	Might meet target
Latin America and the Caribbean	93	60	83	96	73	91	97	80	93	97	85	94	97	80	92	Target met or close to being met
Eastern Asia	99	59	71	93	67	78	98	82	89	98	85	92	100	80	86	Target met or close to being met
Southern Asia	90	66	72	94	81	85	95	83	87	95	88	90	95	83	86	Target met or close to being met
South-Eastern Asia	93	68	76	89	77	82	81	92	86	94	84	89	97	84	88	Target met or close to being met
Western Asia	94	70	85	97	79	91	96	78	90	96	78	90	97	85	93	Might meet target
Oceania	92	39	51	80	40	51	92	37	50	95	45	56	96	70	76	No progress or deterioration
Commonwealth of Independent States/ Caucasus and Central Asia	97	84	92	99	80	92	98	87	94	96	79	86	99	92	96	Europe: Progress sufficient to meet target. Asia: Progress insufficient to meet target.

DATA TABLE 6

MDG Progress on Access to Sanitation by Region

Description

The Millennium Development Goals (MDGs)—adopted by the United Nations in 2000—established a set of targets for improving the lives of the world's poor, ranging from eradicating extreme hunger to reducing child mortality and ensuring environmental sustainability. These targets, agreed to by all countries and leading development institutions throughout the world, consist of eight goals and twenty-one targets. This table presents regional summaries of progress for the years 1990, 2004, 2008, and 2011 (from the 2013 update) as well as the 2015 target levels, measured as the proportion of the population using an improved sanitation system (see data tables 3 and 4 for definitions).

Adequate sanitation is also now a recognized human right, but, as with access to safe drinking water, this basic right is not universal. In many parts of the world, particularly for the poor in rural and peri-urban areas, a lack of basic sanitation adversely affects human health and development. Using 1990 as a baseline, one target of goal 7 of the MDGs seeks to reduce by half the proportion of people without sustainable access to sanitation by 2015.

Meeting the sanitation targets has proven to be more challenging than meeting the water target. While an estimated 2.6 billion people lacked access to basic sanitation in 1990, experts predict that well over 2 billion will still lack access to this basic right by 2015 (Ki Moon 2008). Some areas are performing better than others, highlighting a growing regional disparity in access to sanitation. A majority of regions, however, are not on track to meet established targets, and in Oceania and many former Soviet Union countries, there has been no progress or conditions have worsened.

Limitations

These data give a good picture of the current lack of access to improved water and sanitation services, but comparison with different assessments should be done with extreme care, or not at all, because of changing definitions. For the 2011 update (released in 2013), some regional changes in country assignments slightly change the data from previous years. In particular, the region formerly noted as the Commonwealth of Independent States is now Caucasus and Central Asia.

Country-reported data may reflect national definitions of "improved," unlike survey data, which were standardized as much as possible. For example, in many African countries the population "without access" to improved sanitation means people with no access to any sanitary facility. In Latin America and the Caribbean, however, it is more likely that those "without access" in fact have a sanitary facility, but the facility is deemed unsatisfactory by local or national authorities. Low coverage figures for Latin America and the Caribbean may in part be a reflection of the comparatively narrow definitions used within that region. For these data, we have not included in the totals the proportion of the population using "shared" facilities—just "improved" ones. Those data can be found in the original source (WHO and UNICEF 2013).

Changes in the source of data also complicate comparisons over time. Prior to 2000, for example, data collected by the World Health Organization came from service providers, such as utilities, ministries, and water agencies. The data shown here, however, are sometimes user based and were collected from household surveys and censuses. User-based data are more likely to include improvements installed by households or local communities and give a more complete picture of water supply and sanitation coverage.

Sources

Ki-Moon, B. 2008. United Nations Secretary-General, in message for World Water Day, March 5. New York: United Nations.

United Nations. 2007a. The Millennium Development Goals Report. http://mdgs.un.org/unsd/mdg/Resources/Static/Data/2007%20Stat%20Annex%20current%20indicators.pdf.

———. 2007b. Millennium Development Goals: 2007 Progress Chart. http://mdgs.un.org/unsd/mdg/Resources/Static/Products/Progress2007/MDG_Report_2007_Progress_Chart_en.pdf.

———. 2010a. The Millennium Development Goals Report. http://www.un.org/millenniumgoals/pdf/MDG%20Report%202010%20En%20r15%20-low%20res%2020100615%20-.pdf.

———. 2010b. Millennium Development Goals: 2010 Progress Chart. http://unstats.un.org/unsd/mdg/Resources/Static/Products/Progress2010/MDG_Report_2010_Progress_Chart_En.pdf.

World Health Organization (WHO) and United Nations Children's Fund (UNICEF). 2013. Progress on Sanitation and Drinking-Water—2013 Update. Geneva, Switzerland.

DATA TABLE 6 MDG Progress on Access to Sanitation by Region (proportion of population using an improved sanitation facility)

	1990			2004			2008			2011			2015 Target			On target?
	Urban	Rural	Total	Urban	Rural	Total	Urban	Rural	Total	Urban	Rural	Total	Urban	Rural	Total	
Northern Africa	84	47	65	91	62	77	94	83	89	94	84	90	92	74	83	Already met the target or very close to meeting the target.
Sub-Saharan Africa	52	24	32	53	28	37	44	24	31	42	24	30	76	62	66	Progress insufficient to reach the target if prevailing trends persist.
Latin America and the Caribbean	81	36	68	86	49	77	86	55	80	87	63	82	91	68	84	Progress insufficient to reach the target if prevailing trends persist.
Eastern Asia	64	7	24	69	28	45	61	53	56	76	57	67	82	54	62	Already met the target or very close to meeting the target.
Southern Asia	54	8	20	63	27	38	57	26	36	64	30	41	77	54	60	Progress insufficient to reach the target if prevailing trends persist.
South-Eastern Asia	70	40	49	81	56	67	79	60	69	81	62	71	85	70	75	Progress sufficient to reach the target if prevailing trends persist.
Western Asia	97	55	81	96	59	84	94	67	85	96	71	88	99	78	91	Progress insufficient to reach the target if prevailing trends persist.
Oceania	80	46	54	80	43	53	81	45	53	78	24	36	90	73	77	No progress or deterioration.
Commonwealth of Independent States/ Caucasus and Central Asia	92	63	82	92	67	83	93	83	—	96	95	96	96	82	91	Europe: No progress or deterioration. Asia: Progress insufficient to reach the target if prevailing trends persist.

DATA TABLE 7

Monthly Natural Runoff for the World's Major River Basins, by Flow Volume

Description

These data come from work done by Arjen Hoekstra and Mesfin Mekonnen (2011) to evaluate the regional, national, and global water footprint of human activity. As part of that effort, "natural runoff" was estimated for around 400 river basins, covering some two-thirds of the global land area (excluding Antarctica) and about 65 percent of the global population in 2000. Units for runoff are million cubic meters per month. Data are also provided for the area of the river basin or watershed, in square kilometers.

The natural (undepleted) runoff was estimated by adding the actual runoff data from the Composite Runoff V1.0 database (of Fekete et al. 2002) to the estimated total blue water footprint within the river basin. Because runoff varies naturally over time, these are average estimates. River basin names come from the Global Runoff Data Centre database (GRDC 2007). Note the substantial extremes: just two rivers, the Amazon and Congo Rivers, account for 28 percent of the natural runoff in the river basins listed here.

Limitations

These data are a combination of measured and estimated or modeled flows. Accurate long-term measurements of river flow are not available for a variety of reasons, including the lack of monitoring stations, the complication of estimating undepleted flows in basins where human withdrawals greatly reduce natural flows, and natural variability of high and low runoff over time. These estimates should be considered only approximate averages, but they provide estimates using a consistent method described in the original sources.

The land areas not covered include Greenland; the Sahara Desert in North Africa; the Arabian Peninsula; the Iranian, Afghan, and Gobi Deserts in Asia; the Mojave Desert in North America; and the Australian desert. Also excluded are many smaller pieces of land, often along the coasts, that do not fall within major river basins.

Sources

Fekete, B. M., C. J. Vörösmarty, and W. Grabs. 2002. High-resolution fields of global runoff combining observed river discharge and simulated water balances. *Global Biogeochemical Cycles* 16 (3). doi:10.1029/1999GB001254. Data available at http://grdc.sr.unh.edu/.

Global Runoff Data Centre (GRDC). 2007. Major River Basins of the World. Koblenz, Germany: Federal Institute of Hydrology, Global Runoff Data Centre. Data available at http://grdc.bafg.de/.

Hoekstra, A. Y., and M. M. Mekonnen. 2011. Global Water Scarcity: Monthly Blue Water Footprint Compared to Blue Water Availability for the World's Major River Basins. Value of Water Research Report Series No. 53. Delft, Netherlands: UNESCO-IHE Institute for Water Education. http://www.waterfootprint.org/Reports/Report53-GlobalBlueWaterScarcity.pdf.

DATA TABLE 7 Monthly Natural Runoff for the World's Major River Basins, by Flow Volume

Natural runoff (Mm³/month)

Basin ID	Basin Name	Area (km²)	Jan	Feb	Mar	Apr	May	Jun	Jul	Aug	Sep	Oct	Nov	Dec	Average
259	Amazonas	5,880,855	950,375.6	705,085.8	813,922.6	857,876.6	713,184.6	564,388.1	424,106.1	299,107.1	238,076.7	243,680.0	298,076.1	455,711.2	546,965.9
243	Congo	3,698,918	193,908.5	92,837.8	126,684.3	138,968.9	93,475.1	62,522.2	55,481.6	71,460.1	90,395.6	111,123.4	108,901.0	123,157.3	105,743.0
237	Orinoco	952,173	73,559.9	9,908.6	16,110.2	48,197.5	96,502.6	137,370.7	156,922.8	139,130.4	112,036.3	103,445.6	77,389.5	46,830.9	84,783.7
177	Yangtze (Chang Jiang)	1,745,094	40,746.3	20,673.1	48,340.7	86,583.8	114,954.9	136,756.4	119,291.0	113,382.9	101,024.6	67,829.6	41,278.9	25,519.8	76,365.2
195	Brahmaputra	518,011	28,402.7	549.4	2,782.9	16,852.2	46,782.8	103,699.0	127,521.0	128,246.8	107,215.1	58,846.2	31,386.4	18,647.3	55,911.0
122	Mississippi	3,196,605	79,924.0	66,571.6	111,936.2	102,147.6	83,552.9	56,877.9	36,359.7	27,329.8	13,097.4	12,199.9	20,013.0	38,932.9	54,495.2
5	Yenisei	2,558,237	16,135.2	742.2	452.4	7,752.7	162,714.9	162,047.8	97,190.0	64,240.9	43,091.7	24,440.8	14,455.9	8,735.1	50,583.3
302	Paraná	2,640,486	105,979.7	72,853.1	68,346.4	50,136.2	39,988.6	34,546.3	22,504.9	18,037.0	19,546.7	24,410.8	32,882.5	57,492.3	45,560.4
353	Ganges	1,024,463	32,182.1	10,981.6	16,447.4	12,896.7	12,922.0	27,823.6	78,624.5	128,519.9	96,972.9	47,842.1	32,621.7	19,626.1	43,121.7
199	Irrawaddy	411,516	26,908.9	342.4	994.5	4,301.3	10,976.0	59,825.7	98,546.8	111,837.4	93,122.6	59,380.0	30,238.9	17,673.9	42,845.7
187	Mekong	787,257	30,683.5	405.8	584.3	939.8	5,846.3	34,283.6	85,153.9	107,075.0	107,961.6	64,294.1	35,434.1	20,497.8	41,096.7
7	Lena	2,425,551	15,771.8	655.1	396.4	361.9	87,091.9	124,907.2	84,650.4	63,046.2	53,636.1	23,839.3	14,390.4	8,692.2	39,786.6
25	Ob	2,701,041	6,930.1	198.1	135.9	80,842.2	148,619.4	68,228.3	36,839.8	23,540.4	18,275.7	14,074.8	6,865.0	4,161.9	34,059.3
273	Tocantins	774,718	82,937.0	65,826.8	71,926.3	45,721.6	24,110.4	14,567.5	8,930.8	5,585.4	3,923.1	3,979.0	16,536.2	44,543.4	32,382.3
90	Amur	2,023,520	13,098.2	58.9	59.3	41,918.1	57,294.3	53,865.4	45,446.7	49,807.8	50,357.7	28,455.4	14,176.3	8,572.4	30,259.2
117	St. Lawrence	1,055,021	13,835.1	351.1	29,605.6	132,375.7	51,230.9	31,947.4	19,602.9	13,031.9	15,304.9	21,038.1	22,898.9	9,715.3	30,078.1
194	Nile	3,078,088	20,724.2	2,310.3	6,660.0	16,300.5	18,779.0	21,757.3	48,384.2	80,541.4	66,624.1	36,444.6	22,470.9	14,322.2	29,609.9
207	Niger	2,117,889	21,778.0	99.0	256.2	1,297.8	4,833.0	14,794.5	38,348.0	80,634.7	90,704.8	47,626.3	23,829.5	14,278.6	28,206.7
293	Zambezi	1,388,572	73,279.9	82,019.0	68,514.8	32,327.8	18,268.1	10,936.3	6,663.6	4,203.7	2,618.2	1,621.8	1,305.4	21,500.8	26,938.3
19	Mackenzie	1,752,002	5,637.8	86.9	53.5	9,063.7	79,403.5	79,172.3	46,002.0	24,877.4	15,732.9	10,447.5	5,767.0	3,484.1	23,310.7
64	Volga	1,408,279	3,091.7	147.2	747.6	140,132.1	51,088.9	28,581.8	17,230.0	10,747.1	6,887.4	6,486.8	3,165.6	1,902.0	22,517.4
168	Indus	1,139,075	11,918.9	9,642.7	18,198.4	21,870.2	18,514.4	18,264.8	32,379.1	40,736.0	31,344.3	19,300.0	10,858.9	6,757.3	19,982.1
201	Xi Jiang	362,894	8,372.9	2,682.6	4,932.3	10,762.2	25,796.8	44,596.4	41,263.7	42,397.2	25,138.3	14,767.9	8,958.4	5,447.9	19,593.0
107	Columbia	668,562	11,960.9	11,259.5	20,903.3	38,188.0	55,190.6	36,849.4	18,551.8	11,683.5	7,290.7	5,462.2	6,682.2	7,923.4	19,328.8
235	Magdalena	261,205	27,118.0	3,452.3	6,430.2	14,175.3	21,211.1	18,633.1	15,055.6	15,479.8	18,291.2	31,846.5	31,789.4	20,916.3	18,699.9

118	Danube	793,705	15,369.2	12,969.9	30,056.7	34,399.1	27,077.0	19,150.4	14,083.1	11,248.0	10,213.5	12,685.9	15,065.0	12,575.3	17,907.8
16	Yukon	829,632	4,850.3	252.5	152.7	943.7	53,166.5	48,766.9	29,776.4	16,607.0	12,906.1	7,648.1	4,212.2	2,544.3	15,152.2
334	Uruguay	265,505	15,702.5	5,633.7	8,112.2	13,990.4	16,949.4	19,158.1	16,344.5	15,620.3	18,876.7	20,160.7	12,864.3	9,587.2	14,416.7
386	Sungai Kapuas	84,902	29,491.3	13,356.0	15,227.7	15,516.0	13,513.3	10,291.5	7,393.4	6,608.1	8,936.3	13,775.6	17,293.6	17,622.4	14,085.4
265	Ogooué	222,663	22,726.7	7,569.4	15,733.8	20,776.5	19,147.2	7,950.2	4,618.0	2,791.3	2,505.1	8,651.3	23,592.1	17,443.6	12,792.1
22	Pechora	312,763	2,459.1	31.2	19.2	1,449.0	58,305.9	38,794.9	16,270.2	9,630.9	7,855.7	4,450.1	2,542.8	1,536.1	11,945.4
354	Salween	258,475	8,366.0	95.5	592.3	1,511.4	2,846.6	12,055.4	24,649.1	32,068.1	27,586.6	18,159.4	9,737.1	5,522.3	11,932.5
155	Tigris and Euphrates	832,579	15,514.6	16,609.9	22,140.0	25,647.8	18,735.0	10,281.7	7,368.3	5,738.1	3,544.0	2,291.3	3,575.7	7,623.3	11,589.1
124	Aral Drainage	1,233,148	2,546.9	4,161.6	13,784.1	20,140.9	23,937.1	21,375.9	17,965.9	12,930.4	8,104.3	3,850.2	1,674.4	1,541.5	11,001.1
263	Rajang	49,944	20,497.2	8,513.0	9,657.7	10,211.2	9,935.6	7,769.6	6,882.1	6,855.9	9,242.6	11,129.0	11,855.9	12,276.9	10,402.2
385	Sungai Mahakam	75,823	18,606.0	7,876.5	10,134.8	13,756.5	12,600.1	9,668.4	6,831.4	5,649.9	5,951.6	7,402.7	11,152.8	12,616.4	10,187.3
380	Mamberamo	75,416	15,446.4	9,453.9	12,610.4	11,467.8	9,723.6	7,842.7	8,385.4	7,964.6	8,697.3	6,547.6	7,173.3	8,630.9	9,495.3
281	Sepik	81,120	15,375.9	9,555.7	12,665.7	12,180.9	9,397.5	7,423.9	6,795.9	6,812.5	7,707.7	8,100.4	8,084.8	9,182.5	9,440.3
290	São Francisco	628,629	27,716.3	15,251.7	13,981.8	7,511.4	4,656.8	5,028.4	4,947.7	3,343.7	1,665.4	1,268.7	4,929.8	18,730.9	9,086.0
13	Kolyma	652,850	3,721.1	160.9	97.2	58.8	5,342.4	25,441.2	35,941.4	16,207.8	10,517.4	5,575.9	3,367.8	2,034.2	9,038.8
222	Grisalva	127,675	11,859.4	1,204.2	664.4	610.2	1,646.9	8,622.6	11,809.7	13,230.5	20,505.5	19,300.8	10,998.3	7,877.1	9,027.5
356	Lake Chad	2,391,219	6,870.6	135.4	144.7	180.2	263.0	1,227.0	7,989.1	36,416.6	27,951.2	14,050.2	7,408.8	4,490.9	8,927.3
213	Godavari	311,699	6,805.4	626.4	1,240.0	1,477.3	1,571.4	891.7	13,461.6	27,528.4	26,621.8	12,095.3	7,347.2	4,834.4	8,708.4
83	Nelson	1,099,380	1,698.4	40.6	35.8	29,311.7	20,864.5	15,143.9	8,074.5	5,014.9	4,579.4	4,110.0	1,799.8	1,090.5	7,647.0
355	Hong (Red River)	157,657	4,779.9	80.7	104.5	254.5	1,566.3	7,439.6	18,447.1	22,644.5	16,383.8	9,602.6	5,433.7	3,149.0	7,490.5
88	Fraser	239,678	3,955.1	1,048.3	2,360.7	15,356.9	26,897.1	15,496.8	7,432.2	4,439.9	3,254.0	3,618.6	3,140.5	2,506.4	7,458.9
48	Northern Dvina (Severnaya Dvina)	323,573	1,021.4	35.1	22.2	44,567.2	19,378.9	9,541.8	5,592.3	3,376.1	2,058.6	2,097.6	971.8	587.9	7,437.6
245	Cuyuni	85,635	9,798.1	3,136.9	2,829.9	4,268.1	9,871.7	13,477.9	13,021.6	10,134.4	5,445.8	3,587.1	3,665.6	6,571.4	7,150.7
110	Rhine	190,522	13,179.1	7,657.6	8,094.3	9,602.3	8,013.0	6,055.1	4,779.6	4,183.7	3,971.9	4,514.6	6,546.6	8,363.7	7,080.1
384	Sungai Kajan	33,172	10,769.4	4,092.4	5,488.6	6,619.3	6,870.7	5,795.5	5,104.8	4,917.8	6,431.1	6,999.8	7,819.9	6,829.4	6,478.2

continues

DATA TABLE 7 *continued*

								Natural runoff (Mm³/month)							
Basin ID	Basin Name	Area (km²)	Jan	Feb	Mar	Apr	May	Jun	Jul	Aug	Sep	Oct	Nov	Dec	Average
219	Krishna	269,869	4,249.9	610.7	1,255.2	1,355.6	1,400.2	3,256.5	16,600.9	15,185.1	11,795.9	6,710.3	4,670.2	3,427.8	5,876.5
249	Sanaga	134,252	4,812.3	3.0	236.5	2,004.2	4,068.5	5,840.0	7,929.0	9,616.8	13,725.9	13,358.0	5,383.1	3,156.1	5,844.5
149	Huang He (Yellow River)	988,063	2,702.4	608.3	2,320.1	5,166.0	8,177.4	9,102.2	10,309.3	9,805.2	9,930.4	5,872.9	3,052.0	1,802.8	5,737.4
383	Uwimbu	29,374	8,952.7	5,395.2	6,379.0	5,992.2	6,010.0	5,432.3	5,263.6	5,126.4	5,387.6	4,552.8	4,357.4	5,401.7	5,687.6
244	Atrato	34,619	8,908.3	2,297.1	2,736.4	4,317.7	5,624.4	5,876.9	5,976.7	6,096.9	6,689.5	7,140.9	7,032.0	5,537.1	5,686.2
93	Grande Rivière	111,718	2,729.2	0.9	0.5	0.3	20,738.4	10,192.2	6,544.7	6,144.0	7,187.4	7,682.4	2,959.2	1,787.3	5,497.2
114	Nottaway	118,709	2,188.3	0.2	0.2	13,451.3	15,987.7	7,640.1	5,188.1	4,271.3	5,083.6	6,291.3	2,438.0	1,434.9	5,331.3
373	Rovuma	151,949	9,106.2	15,265.1	18,092.1	9,319.0	4,605.3	2,779.5	1,678.9	1,014.2	612.7	370.2	223.6	516.4	5,298.6
252	Essequibo	68,788	5,069.1	1,976.6	2,110.6	2,932.1	7,163.7	13,057.2	11,720.0	7,957.4	3,936.3	2,438.8	1,874.9	3,105.6	5,278.5
75	Caniapiscau	105,691	2,297.2	0.9	0.5	0.3	16,170.7	12,475.4	6,805.5	6,229.8	6,669.5	6,024.1	2,490.3	1,504.1	5,055.7
56	Neva	223,310	1,195.2	10.6	8.8	32,169.2	9,560.0	5,468.9	3,272.4	1,984.2	1,703.3	2,586.4	1,686.6	773.9	5,034.9
210	Mahanadi (Mahahadi)	135,061	3,751.2	108.6	158.0	151.1	158.2	157.6	5,209.0	21,461.7	14,877.9	6,698.9	4,119.4	2,571.8	4,951.9
248	Cross	52,820	3,986.4	1.4	608.2	1,184.8	2,346.5	4,731.8	7,828.2	9,133.6	11,624.3	10,815.8	4,452.2	2,614.4	4,944.0
257	Maroni	65,945	3,849.5	4,583.6	5,326.3	7,656.9	11,080.5	10,145.5	7,092.8	4,342.7	2,242.1	1,349.5	815.1	565.6	4,920.8
388	Batang Hari	42,872	10,918.4	4,536.0	5,450.1	6,252.7	4,790.1	2,820.9	1,767.2	1,540.6	2,438.0	4,411.4	6,272.8	7,176.5	4,864.6
218	Chao Phraya	188,419	4,183.6	318.1	519.8	537.6	493.6	1,640.2	5,541.8	9,607.1	16,009.4	10,072.0	5,811.4	3,059.8	4,816.2
1	Khatanga	294,908	1,571.7	58.7	35.4	21.4	221.2	25,398.5	12,884.8	6,944.9	4,357.0	2,419.0	1,461.0	882.4	4,688.0
180	Alabama River and Tombigbee	113,117	9,397.8	10,770.1	12,232.8	8,612.8	4,829.3	2,684.6	1,677.7	1,055.0	671.1	421.7	465.0	3,348.5	4,680.5
112	Saguenay (Rivière)	91,367	1,984.2	1.5	1.5	11,245.1	11,251.7	8,137.7	5,058.6	4,100.4	4,665.6	5,748.5	2,173.4	1,301.6	4,639.2
27	Taz	152,086	989.9	6.0	3.6	2.2	12,922.1	23,163.4	7,321.6	4,370.1	3,193.2	1,738.0	1,049.7	634.0	4,616.2
254	Corantijn	65,528	1,313.6	829.3	1,710.9	3,554.5	11,297.6	13,180.8	9,711.4	6,084.0	3,012.4	1,811.6	1,094.1	692.7	4,524.4
126	Rhone	97,485	7,316.8	3,365.1	5,895.8	6,325.3	5,588.7	4,446.6	2,690.2	2,212.7	2,277.1	3,477.1	5,232.4	5,154.7	4,498.5

ID	Name	Area												
378	Min Jiang	60,040	1,710.0	2,261.2	6,444.7	6,263.4	9,721.2	10,643.9	4,850.2	3,725.5	2,882.6	2,049.4	1,340.9	880.7	4,397.8
337	Rio Jacuí	70,798	5,005.1	2,492.1	2,901.2	3,920.9	4,897.0	5,793.0	5,339.7	5,145.1	5,743.1	5,136.4	3,405.9	2,794.2	4,381.1
96	Dniepr	510,661	849.5	45.1	9,860.5	20,513.6	7,860.1	4,484.4	2,811.4	1,783.2	1,063.7	891.9	1,449.4	588.9	4,350.1
366	Cuanza	141,391	10,360.7	7,565.1	10,296.0	8,746.9	3,552.8	2,127.7	1,286.3	779.0	472.4	307.3	304.1	5,458.5	4,271.4
286	Purari	32,140	7,089.0	4,200.2	5,204.7	5,335.2	4,459.8	3,535.0	2,952.7	3,064.9	3,853.1	3,615.9	3,635.9	4,215.8	4,263.5
95	Churchill, Fleuve (Labrador)	84,984	1,702.6	0.0	0.0	0.0	15,223.9	10,971.0	6,028.7	4,689.5	4,563.7	4,517.0	1,848.4	1,116.4	4,221.8
266	Rio Araguari	33,771	4,727.0	5,377.5	6,935.9	8,461.5	8,118.0	7,123.8	4,272.9	2,486.6	1,376.3	831.0	501.9	398.5	4,217.6
21	Anadyr	171,276	1,182.1	1.3	0.8	0.5	2,452.2	21,998.7	10,380.5	5,461.5	4,105.7	2,116.1	1,278.1	771.9	4,145.8
6	Indigirka	341,228	1,902.5	179.7	108.5	65.6	450.9	16,809.8	14,507.5	6,786.0	3,631.1	2,176.2	1,314.4	793.9	4,060.5
258	San Juan (Columbia-Pacific)	13,898	6,064.5	2,203.5	2,616.1	3,456.3	4,124.0	3,912.7	3,904.9	3,971.2	4,103.4	4,529.7	4,529.1	3,905.9	3,943.4
51	Kuskokwim	118,114	1,269.5	1.4	0.9	0.5	16,620.3	7,998.3	5,470.1	5,473.3	5,291.7	2,398.6	1,372.5	829.0	3,893.8
71	Stikine	51,147	1,826.9	603.3	508.0	1,656.9	9,699.8	11,933.0	5,766.5	3,728.6	3,815.5	3,379.2	1,948.6	1,370.5	3,853.1
128	Po	73,067	4,276.6	2,000.0	3,536.1	5,530.3	6,397.4	4,452.6	2,941.3	2,394.4	2,314.0	2,947.8	3,482.3	2,857.2	3,594.2
232	San Juan	41,659	5,223.3	533.6	282.7	261.5	1,036.0	3,952.8	4,778.7	4,793.9	6,119.6	7,350.0	4,839.9	3,921.9	3,591.2
208	Narmada	95,818	2,407.1	430.8	900.5	1,131.7	1,138.1	445.1	7,352.7	11,663.1	8,745.2	3,757.6	2,295.3	1,586.4	3,487.8
382	Eilanden	20,077	5,431.2	3,334.4	3,835.7	3,751.0	3,567.3	3,244.7	3,235.3	3,090.1	3,371.3	2,701.3	2,824.9	3,278.9	3,472.2
317	Tsiribihina	61,992	9,631.8	9,804.7	9,049.7	4,154.5	2,354.3	1,436.8	893.2	545.1	328.9	198.3	299.2	2,841.4	3,461.5
54	Susitna	49,470	1,270.7	2.2	1.4	3,654.1	8,319.0	8,905.7	5,269.7	4,095.0	4,842.3	2,738.5	1,370.5	827.8	3,441.4
215	Sittang	34,265	2,254.3	3.3	6.3	6.6	16.9	3,852.3	7,852.6	10,043.8	8,332.3	4,900.8	2,492.4	1,478.4	3,436.7
396	Lake Turkana	181,536	2,082.5	9.0	100.4	1,988.3	3,300.7	4,415.2	6,718.5	7,827.8	6,549.2	4,135.2	2,501.6	1,394.5	3,418.6
345	Negro (Argentina)	130,062	2,461.3	98.8	347.1	1,009.8	4,025.3	5,859.2	6,210.5	6,075.1	5,030.4	4,368.0	3,162.3	1,740.6	3,365.7
262	Oyapock	27,076	4,282.0	4,043.5	4,817.3	6,249.8	6,602.9	5,755.6	3,526.6	2,069.9	1,131.7	683.3	412.7	586.7	3,346.8
272	Rio Capim	54,888	1,537.9	5,606.4	8,799.8	7,926.1	5,924.9	3,782.3	2,616.7	1,569.4	878.6	525.8	317.7	205.5	3,307.6
227	Volta	414,004	2,522.1	7.7	80.1	291.8	828.4	2,616.6	3,570.8	8,402.5	11,296.3	5,427.0	2,744.7	1,654.9	3,286.9
314	Rio Doce	86,086	12,563.7	5,298.4	4,242.2	2,461.5	1,301.4	784.3	483.1	302.0	187.0	117.8	2,334.5	9,099.0	3,264.6
58	Copper	64,960	1,090.8	0.0	0.0	17.3	7,546.7	10,214.6	7,440.5	4,185.8	3,963.1	2,215.5	1,184.1	715.2	3,214.5

continues

DATA TABLE 7 *continued*

								Natural runoff (Mm³/month)							
Basin ID	Basin Name	Area (km²)	Jan	Feb	Mar	Apr	May	Jun	Jul	Aug	Sep	Oct	Nov	Dec	Average
116	Moose (Trib. Hudson Bay)	105,615	1,000.2	0.6	0.6	11,334.0	9,979.0	4,300.8	2,527.6	1,594.4	2,136.5	3,024.6	1,085.8	656.0	3,136.7
31	Pur	111,351	740.7	0.3	0.3	0.3	7,186.7	15,161.5	4,615.2	2,795.1	2,896.5	1,331.2	804.1	485.8	3,001.5
72	Churchill	298,505	789.5	20.1	12.3	3,400.0	13,120.8	7,724.3	3,761.6	2,131.7	1,960.1	1,724.5	774.9	468.2	2,990.7
148	Sacramento	77,209	5,375.8	6,127.3	6,249.1	5,067.8	3,136.9	2,369.0	2,064.4	1,708.0	1,170.4	511.3	343.5	1,439.3	2,963.6
82	Skeena	42,944	1,078.2	0.2	1,041.9	4,491.6	8,819.6	7,160.8	3,082.5	2,010.7	1,937.6	2,313.5	1,518.6	707.1	2,846.9
98	Wisla	193,764	947.0	37.8	12,812.8	6,810.9	4,190.9	2,577.5	1,781.3	1,243.2	858.0	779.6	1,157.5	847.9	2,837.0
183	Huai He	174,310	1,271.4	1,031.3	2,377.3	4,348.7	4,430.9	4,375.0	4,755.1	3,800.5	3,689.6	1,900.5	1,153.0	823.3	2,829.7
99	Don	425,630	308.1	29.4	4,138.3	14,368.8	5,709.0	3,280.6	2,272.7	1,517.1	801.0	444.2	431.0	212.1	2,792.7
204	Dong Jiang	32,103	861.2	105.2	1,116.0	3,077.8	5,758.0	6,467.3	4,881.8	4,699.8	3,338.0	1,558.5	934.6	567.8	2,780.5
143	Susquehanna	69,080	2,091.9	1,240.2	8,814.7	5,917.2	3,887.6	2,447.9	1,522.7	1,002.4	867.5	1,337.0	2,499.4	1,658.8	2,774.0
127	Saint John	55,152	1,543.0	1.9	1.9	13,364.2	4,478.5	3,060.9	1,929.0	1,260.9	1,384.3	2,179.7	2,871.4	1,012.4	2,757.4
106	Manicouagan (Rivière)	54,205	1,252.8	0.3	0.2	524.6	8,666.6	6,924.6	3,691.1	2,983.3	3,220.2	3,479.9	1,358.9	820.8	2,743.6
202	Bei Jiang	52,915	629.5	353.9	1,921.9	4,546.4	6,874.5	7,013.0	3,830.8	3,230.7	2,159.0	1,100.4	668.5	413.5	2,728.5
268	Esmeraldas	19,796	2,922.0	4,238.0	5,676.0	6,743.2	4,668.6	2,504.1	1,398.8	876.1	589.2	579.1	755.5	966.5	2,659.8
284	Rufiji	204,639	1,836.2	3,827.1	7,683.8	8,804.4	4,027.1	2,129.0	1,285.4	779.2	473.6	288.1	176.3	418.8	2,644.1
105	Eastmain	48,838	1,182.9	0.0	0.0	2,558.1	9,188.6	4,899.7	2,922.3	2,536.0	2,987.8	3,379.4	1,284.2	775.6	2,642.9
111	Albany	123,081	813.0	0.1	0.1	8,411.8	9,204.2	3,827.2	2,148.3	1,306.1	1,776.4	2,432.0	882.5	533.1	2,611.2
2	Olenek	208,522	1,231.0	159.8	96.5	58.3	658.3	16,431.9	5,510.3	3,070.3	1,870.4	1,106.6	668.3	403.7	2,605.5
346	Biobio	24,109	1,512.9	29.1	298.7	919.1	3,736.5	4,786.4	5,042.5	4,670.4	4,131.1	2,973.2	1,809.7	1,045.2	2,579.6
240	Sassandra	68,097	2,261.5	1.4	3.2	129.1	309.3	2,250.3	3,322.8	4,638.5	8,322.9	5,416.1	2,603.4	1,487.3	2,562.2
402	Lake Taymur	138,783	723.7	0.0	0.0	0.0	0.0	14,735.9	6,387.8	3,298.4	2,513.9	1,300.5	785.5	474.4	2,518.3
80	Kamchatka	54,104	689.9	0.0	0.0	0.0	9,603.9	7,199.7	5,399.2	2,604.2	1,918.9	1,546.8	749.0	452.4	2,513.7
399	Lake Titicaca	107,215	7,124.5	5,963.3	4,974.8	2,951.8	1,893.3	1,052.7	658.3	552.1	469.3	688.1	678.7	2,594.8	2,466.8
94	Winisk	106,470	890.2	2.1	1.3	3,470.5	10,190.2	4,433.5	2,449.4	1,479.2	2,350.9	2,486.7	957.9	578.6	2,440.9

224	Mae Klong	28,004	1,554.4	20.9	34.2	33.5	1,076.4	3,580.0	5,254.6	5,831.8	5,582.5	3,551.6	1,715.1	1,031.1	2,438.9
357	Okavango	705,056	4,075.2	6,488.8	8,619.1	3,971.7	2,041.0	1,233.1	745.9	452.1	274.8	167.1	100.7	882.0	2,421.0
125	Loire	115,944	5,691.9	3,966.2	3,912.1	3,192.9	2,280.7	1,475.0	937.1	710.2	546.1	804.6	1,839.3	3,262.7	2,384.9
398	Suriname	24,868	2,186.8	1,915.0	2,088.7	2,922.6	4,729.6	5,013.8	3,973.9	2,420.7	1,213.0	729.1	440.4	526.6	2,346.7
276	Rio Parnaíba	336,584	2,039.6	3,746.9	7,333.0	6,991.8	3,124.0	1,723.6	1,047.6	641.5	394.0	242.4	157.1	558.8	2,333.4
260	Pahang	28,437	5,776.0	1,292.5	1,416.5	1,962.1	1,937.4	1,260.3	863.4	823.6	1,395.6	2,714.8	3,612.5	4,175.9	2,269.2
322	Paraiba Do Sul	58,027	7,384.0	4,105.7	3,823.9	2,180.1	1,239.1	759.3	470.0	308.5	264.6	590.0	1,633.4	4,400.3	2,263.2
331	Murray	1,059,508	2,868.3	1,379.7	1,501.2	1,110.5	1,279.6	2,153.4	2,511.6	3,164.9	3,371.2	3,337.6	2,314.5	2,000.4	2,249.4
220	Senegal	436,981	1,629.1	10.0	15.2	11.7	12.0	447.8	3,062.5	8,464.3	6,863.4	3,270.9	1,792.3	1,076.7	2,221.3
377	Fuchun Jiang	37,698	1,253.4	1,966.9	3,249.4	3,037.5	4,025.8	5,573.0	2,420.0	1,437.2	1,282.3	886.7	707.7	572.7	2,201.1
45	Thelon	238,839	478.4	0.1	0.0	0.0	6,190.9	10,749.7	3,370.2	2,026.8	1,686.6	859.5	519.1	313.5	2,182.9
274	Kouilou	60,000	4,007.3	2,400.3	3,950.9	5,306.6	3,068.4	1,426.7	862.7	522.5	316.7	191.7	1,130.5	3,001.1	2,182.1
81	Nass	21,212	1,054.0	0.0	810.7	3,441.7	6,020.4	4,634.1	2,047.3	1,515.8	1,839.7	2,607.5	1,420.3	691.1	2,173.5
209	Brahmani River (Bhahmani)	51,973	1,562.1	33.6	53.2	42.6	38.7	549.7	3,888.5	7,939.2	6,035.6	3,059.0	1,695.6	1,049.7	2,162.3
246	Cavally	30,665	2,295.7	105.8	221.1	532.9	1,553.0	3,418.3	2,447.3	1,941.8	4,204.8	4,188.2	2,935.7	1,654.9	2,125.0
101	Elbe	139,348	2,858.7	2,594.5	4,899.4	3,701.9	2,272.5	1,577.0	1,147.9	892.7	725.8	865.4	1,454.7	1,857.4	2,070.7
39	Mezen	76,715	372.9	4.5	2.7	3,916.9	10,521.7	3,855.7	2,081.0	1,250.6	799.5	867.4	386.3	233.3	2,024.4
340	Negro (Uruguay)	70,756	1,301.6	86.9	492.1	1,499.0	2,274.1	3,311.9	3,222.8	3,290.1	3,379.1	2,823.2	1,553.8	846.5	2,006.8
270	Daule and Vinces	41,993	2,730.9	4,080.6	5,260.2	4,696.3	2,419.3	1,482.0	915.4	644.8	472.3	447.2	458.1	400.9	2,000.7
255	Coppename	24,750	1,551.1	1,394.4	1,540.4	2,071.4	4,023.6	4,578.3	3,861.0	2,331.4	1,160.6	695.4	420.0	321.1	1,995.7
371	Lurio	61,172	4,604.7	6,185.0	6,089.2	2,522.2	1,435.2	866.8	523.5	316.3	191.1	115.4	69.7	468.0	1,948.9
4	Yana	233,479	896.1	72.8	44.0	26.6	159.2	7,535.9	6,820.7	3,807.3	1,832.6	1,106.8	668.5	403.8	1,947.9
140	Ebro	85,159	4,220.1	2,820.2	2,785.8	2,876.1	2,619.8	1,631.9	1,192.1	853.9	509.9	561.7	873.7	2,424.5	1,947.5
364	Coco	25,502	2,767.8	143.3	70.7	43.3	48.4	1,698.7	3,244.2	2,959.2	3,280.5	3,939.1	2,788.7	2,041.2	1,918.8
74	George	39,054	758.1	0.0	0.0	0.0	4,958.0	5,039.7	3,961.6	2,616.4	2,618.4	1,701.2	823.0	497.1	1,914.5
221	Papaloapan	39,885	1,877.2	12.4	16.0	15.8	11.5	477.3	2,373.0	4,411.2	5,780.4	4,426.4	2,194.2	1,261.7	1,904.8
351	Baker	30,760	1,928.9	630.2	1,099.0	1,638.6	2,259.5	2,536.2	2,753.8	2,648.6	2,149.0	1,882.5	1,579.4	1,282.9	1,865.7
142	Douro	96,125	3,913.4	3,031.8	4,104.3	2,983.9	2,057.4	1,252.2	1,050.6	820.4	404.1	215.0	539.3	1,650.2	1,835.2

continues

DATA TABLE 7 *continued*

Natural runoff (Mm³/month)

Basin ID	Basin Name	Area (km²)	Jan	Feb	Mar	Apr	May	Jun	Jul	Aug	Sep	Oct	Nov	Dec	Average
87	Neman	97,300	740.2	3.4	2,540.2	8,813.4	3,141.2	1,745.1	1,037.9	641.8	406.1	467.3	1,597.2	486.5	1,801.7
97	Ural	339,084	324.2	5.7	165.2	10,759.3	3,817.4	2,162.0	1,470.8	948.2	527.8	322.1	587.8	214.5	1,775.4
387	Batang Kuantan	16,739	3,820.3	1,460.7	1,731.9	2,316.9	1,825.4	1,075.7	640.6	525.6	838.5	1,724.9	2,579.6	2,602.7	1,761.9
392	Issyk-Kul	191,032	300.5	48.6	2,132.1	3,385.1	4,295.9	3,990.2	3,002.3	1,671.5	1,072.0	585.2	384.5	205.0	1,756.1
137	Hudson	36,893	1,000.4	26.3	4,477.0	4,970.6	2,745.5	1,656.2	1,063.5	725.6	742.5	1,044.8	1,764.9	728.3	1,745.5
138	Colorado (Pacific Ocean)	640,464	323.3	99.7	738.0	3,046.4	5,903.7	4,320.1	2,390.6	1,653.7	1,126.8	740.4	370.5	231.4	1,745.4
231	Cauvery	91,159	2,091.4	159.7	385.4	347.3	350.9	1,080.8	3,669.7	3,305.4	2,574.1	2,347.6	2,777.6	1,849.0	1,744.9
214	Tapti	65,096	1,243.3	168.4	306.4	365.6	395.0	149.1	3,365.4	5,169.5	5,115.9	2,124.1	1,333.7	886.1	1,718.5
35	Back	141,352	327.2	0.0	0.0	0.0	5,571.3	8,215.3	2,637.8	1,590.9	990.4	588.0	355.2	214.5	1,707.6
271	Rio Gurupi	32,335	491.7	2,048.7	4,647.2	4,513.5	3,426.0	2,151.5	1,412.8	764.6	450.8	272.3	164.5	103.1	1,703.9
264	Ntem	33,527	2,055.6	8.3	442.8	1,623.3	2,595.7	1,808.1	791.1	470.6	1,558.6	4,430.8	3,192.8	1,428.3	1,700.5
352	Santa Cruz	30,600	1,652.2	385.0	560.0	1,181.8	1,590.7	1,850.7	1,577.2	2,464.8	3,221.0	3,226.9	1,605.8	1,042.5	1,696.6
160	Han-Gang (Han River)	24,772	814.3	12.5	1,372.3	1,838.7	1,166.9	1,200.5	4,312.9	3,914.3	2,697.0	1,388.7	1,010.5	540.7	1,689.1
379	Han Jiang	30,742	429.5	189.6	1,241.1	2,010.5	3,792.4	4,718.0	2,447.5	2,125.6	1,654.9	782.8	466.3	283.9	1,678.5
100	Oder	116,536	1,032.5	1,852.6	5,896.3	3,319.4	2,160.7	1,462.2	947.3	682.0	505.5	478.7	627.0	1,046.2	1,667.5
89	Severn (Trib. Hudson Bay)	98,590	593.2	0.0	0.0	969.8	7,964.8	3,325.3	1,813.8	1,158.6	1,429.9	1,689.2	644.0	388.9	1,664.8
139	Klamath	40,040	3,010.5	3,574.1	3,546.4	3,046.4	2,001.6	1,051.5	698.7	455.6	276.0	146.5	394.9	1,495.2	1,641.5
184	Apalachicola	51,413	2,889.5	3,847.6	4,385.4	3,016.0	1,572.5	912.7	651.6	531.5	402.8	235.8	202.2	1,034.8	1,640.2
76	Western Dvina (Daugava)	89,340	772.7	2.1	2.1	9,180.8	2,977.1	1,713.6	1,016.3	619.6	425.3	849.3	1,548.2	507.4	1,634.5
212	Panuco	82,929	1,540.6	95.9	186.5	197.5	137.9	362.8	2,013.6	2,491.4	6,156.6	3,524.7	1,770.8	1,046.5	1,627.1
52	Vuoksi	62,707	334.8	1.2	1.2	10,528.9	2,968.3	1,728.3	1,039.2	643.3	560.3	960.4	387.0	220.0	1,614.4

306	Mitchell (N. Au)	1,612.8	71,725	1,080.2	6,217.0	5,876.8	2,616.6	1,435.1	859.9	521.2	317.3	194.5	119.8	72.3	43.0
53	Onega	65,894	224.9	2.8	1.8	10,633.8	3,402.5	1,902.0	1,125.9	680.1	436.9	556.4	234.8	141.1	1,611.9
61	Vaenern-Goeta	51,791	1,198.3	3.2	2,259.8	5,377.8	2,297.5	1,379.5	819.2	707.0	791.8	1,475.7	1,616.9	1,037.2	1,580.3
132	Connecticut	27,468	934.3	7.8	3,116.2	4,979.1	2,528.5	1,556.0	985.4	660.3	761.0	1,049.6	1,699.4	665.0	1,578.5
393	Balkhash	423,657	273.5	5.8	2,199.0	4,046.6	4,089.9	2,907.0	2,102.5	1,368.5	905.0	412.3	445.2	181.7	1,578.1
37	Nadym	54,625	383.2	0.2	0.1	0.1	4,504.7	7,346.2	2,354.4	1,461.2	1,502.8	687.7	415.4	250.9	1,575.6
133	Liao He	194,436	678.2	20.4	220.6	1,657.1	2,266.7	2,171.0	2,661.2	3,988.1	2,606.5	1,383.3	782.0	454.6	1,574.1
282	Rio Mearim	56,687	356.4	2,713.7	5,445.6	4,657.5	2,321.9	1,253.1	745.7	450.8	272.8	165.1	99.9	61.7	1,545.3
238	Bandama	98,751	1,337.1	4.0	5.7	118.4	306.8	1,520.8	1,055.3	2,949.0	5,574.4	3,186.6	1,473.0	881.5	1,534.4
59	Gloma	42,863	563.2	1.3	12.0	3,439.3	3,368.1	3,615.4	2,048.4	1,487.9	1,400.5	1,344.1	677.6	369.7	1,527.3
18	Kemijoki	55,825	487.2	3.2	2.0	594.1	8,987.0	2,448.1	1,458.1	934.7	1,082.1	1,366.4	516.4	312.0	1,515.9
134	Garonne	55,807	3,122.4	1,918.1	2,169.4	2,289.6	1,911.6	1,113.4	759.2	594.0	474.6	657.6	1,074.9	1,915.8	1,500.0
280	Rio Pindare	39,112	253.4	2,304.7	4,716.9	4,525.5	2,674.7	1,415.4	814.8	488.9	295.4	178.6	108.0	65.3	1,486.8
77	Aux Melezes	41,384	637.5	0.1	0.1	0.0	5,433.5	3,250.9	1,841.6	1,727.2	1,787.2	1,696.1	691.6	417.7	1,457.0
145	Kura	182,283	559.4	187.2	708.9	3,035.7	4,111.4	2,768.8	1,901.8	1,396.6	913.7	766.2	692.7	413.7	1,454.7
267	Mira	13,265	2,073.0	1,259.4	1,291.0	1,430.0	2,032.3	1,959.3	1,248.7	1,229.1	1,355.3	1,165.3	1,283.2	912.8	1,436.6
375	Pyasina	63,889	470.7	4.9	3.1	2.0	1.3	8,582.4	2,828.9	1,874.0	1,803.8	814.2	491.9	297.2	1,431.2
230	Grande De Matagalpa	17,992	1,788.5	87.6	46.4	30.2	37.4	1,350.9	2,694.3	2,443.9	2,701.8	2,950.2	1,748.2	1,248.2	1,427.3
73	Feuilles (Rivière Aux)	37,425	673.6	-	-	-	2,539.5	5,199.5	2,051.0	1,757.3	1,931.8	1,773.6	731.3	441.7	1,424.9
206	Damodar	43,096	1,257.9	111.7	160.8	44.8	15.0	253.7	1,494.8	4,751.2	4,657.8	2,147.4	1,316.0	850.4	1,421.8
147	Delaware	26,713	1,640.3	801.3	4,062.4	2,255.7	1,789.8	1,043.1	712.5	568.2	587.3	762.0	1,406.7	1,294.0	1,410.3
217	Ca	28,747	1,447.9	56.1	26.8	23.3	154.3	584.1	1,876.9	2,678.4	4,330.1	2,767.9	1,652.1	965.9	1,380.3
103	Weser	43,140	3,511.3	2,008.3	1,895.7	1,397.8	902.7	617.5	504.3	484.0	501.8	819.6	1,601.5	2,294.1	1,378.2
367	Cunene	110,545	1,828.9	2,579.9	5,240.5	2,925.0	1,322.8	798.9	482.7	291.8	176.4	112.8	98.1	642.6	1,375.0
84	Hayes (Trib. Hudson Bay)	105,372	339.8	7.0	4.3	2.6	7,011.9	3,778.4	2,054.7	1,111.5	798.0	731.1	339.8	205.3	1,365.4
241	Shebelle	805,077	1,126.2	49.5	54.4	2,532.2	1,755.0	1,025.3	1,594.0	2,005.1	1,944.7	1,681.9	1,610.5	791.5	1,347.5
336	Colorado (Argentina)	390,631	3,500.8	346.2	220.9	135.4	373.1	707.2	841.7	929.8	908.5	2,371.9	3,152.6	2,575.4	1,338.6

continues

DATA TABLE 7 continued

Basin ID	Basin Name	Area (km²)	Jan	Feb	Mar	Apr	May	Jun	Jul	Aug	Sep	Oct	Nov	Dec	Average
261	Nyong	34,626	1,269.1	0.1	386.2	1,153.1	1,817.1	1,504.5	698.1	677.8	2,434.6	3,389.9	1,699.9	832.3	1,321.9
349	Chubut	145,352	837.4	70.0	172.1	336.2	1,273.9	2,263.3	2,596.8	3,116.1	2,246.4	1,454.2	895.8	580.3	1,320.2
152	Tejo	70,352	2,713.3	2,202.2	3,261.5	2,062.1	1,343.3	834.4	721.8	559.4	292.8	124.7	136.5	1,182.1	1,286.2
253	Kelantan	14,420	3,574.8	439.2	342.5	417.5	434.1	481.7	494.1	561.2	1,352.9	2,084.8	2,431.3	2,719.0	1,277.8
362	Ulua	26,392	1,735.8	77.7	42.4	31.4	19.3	510.5	1,791.6	1,997.4	3,135.9	2,748.5	1,916.8	1,262.0	1,272.4
46	Angerman	32,372	343.5	0.5	0.4	4,027.0	3,524.6	2,685.5	1,232.5	833.4	764.3	954.8	371.0	224.1	1,246.8
119	Seine	74,228	3,426.9	2,491.0	2,183.7	1,663.9	1,005.1	594.9	402.4	304.5	202.7	234.2	696.1	1,700.9	1,242.2
256	Kinabatangan	14,102	2,820.1	862.7	675.4	672.8	675.3	1,148.9	787.8	1,142.6	1,468.5	1,388.6	1,249.3	1,909.0	1,233.4
205	Mahi	36,238	884.2	157.9	246.0	223.2	137.5	37.8	2,640.9	4,426.7	3,152.3	1,389.2	865.8	573.5	1,227.9
67	Nushagak	29,514	517.3	0.0	0.0	1,308.4	4,936.8	1,770.2	1,042.6	1,293.2	1,548.5	1,382.8	561.6	339.2	1,225.1
66	Arnaud	44,932	564.3	-	-	-	632.5	5,641.8	2,064.4	1,543.3	1,878.6	1,362.9	612.6	370.0	1,222.5
165	Shinano, Chikuma	11,159	1,100.6	807.7	800.2	1,837.3	1,864.2	1,290.9	1,242.1	1,006.9	1,172.1	1,174.8	1,150.4	1,044.7	1,207.7
171	Pee Dee	46,531	2,824.5	2,381.4	2,346.1	1,609.2	944.6	589.6	563.3	455.2	460.3	379.9	545.7	1,384.2	1,207.0
135	Ishikari	13,783	859.4	2.4	2.4	4,390.2	1,918.3	1,080.5	793.8	776.5	1,178.1	1,434.9	1,481.8	564.3	1,206.9
108	Little Mecatina	17,903	444.5	0.0	0.0	0.0	6,156.3	2,178.5	1,505.7	1,078.9	1,004.0	1,263.0	482.6	291.5	1,200.4
151	Yongding He	214,406	174.7	403.8	1,472.8	2,529.4	2,501.8	1,275.4	1,729.3	2,397.6	1,071.8	442.9	185.1	137.8	1,193.5
23	Lule	25,128	313.1	1.0	0.6	1,119.1	3,851.3	3,784.3	1,895.4	1,112.2	951.7	683.0	335.9	202.9	1,187.5
131	Kuban	58,936	1,008.9	1,275.5	1,664.7	2,123.9	1,943.7	1,594.5	1,393.7	826.2	621.9	471.0	528.0	626.0	1,173.2
129	Penobscot	21,169	655.1	0.6	482.1	5,554.7	1,878.0	1,224.9	733.4	451.5	421.6	684.5	1,327.0	429.8	1,153.6
70	Narva	58,147	710.7	1.2	1.2	6,018.5	1,937.0	1,079.4	639.7	420.0	372.9	731.8	1,459.3	466.4	1,153.2
358	Tarim	1,051,731	241.9	77.0	269.6	593.6	1,657.0	2,845.3	3,326.7	2,360.3	1,351.4	540.9	295.2	164.1	1,143.6
234	Corubal	24,258	882.9	0.4	0.5	0.5	0.4	293.0	1,767.6	3,594.2	3,165.9	2,080.6	964.3	579.0	1,110.8
347	Waikato	15,359	1,209.9	436.0	382.1	601.3	1,271.4	1,642.5	1,617.1	1,551.8	1,388.8	1,356.2	1,080.3	781.6	1,109.9
309	Jequitinhonha	68,549	4,207.8	1,648.8	1,430.8	860.3	471.5	297.0	203.6	126.2	74.1	60.4	675.7	3,152.4	1,100.7
225	Tranh (Nr Thu Bon)	9,460	2,007.0	45.2	24.6	16.4	71.3	290.8	883.2	1,278.8	1,770.6	2,724.1	2,418.9	1,615.2	1,095.5

289	Solo (Bengawan Solo)	15,146	2,855.5	2,415.3	2,433.6	1,722.1	949.3	531.7	332.1	222.4	171.0	106.2	250.0	1,099.8	1,090.8
316	Burdekin	130,427	690.0	3,679.3	3,662.4	1,887.1	1,001.1	597.8	363.8	227.1	146.3	95.8	59.5	32.7	1,036.9
120	Dniestr	72,108	407.9	13.3	3,147.7	2,602.1	1,528.5	1,130.4	792.2	626.7	510.8	629.4	702.3	272.1	1,030.3
320	Limpopo	415,623	1,880.1	3,058.9	2,803.2	1,433.2	767.0	501.7	359.2	334.0	330.6	246.8	159.6	308.4	1,015.2
161	Guadalquivir	56,955	677.2	1,306.6	3,139.9	1,977.6	1,055.0	1,003.9	1,110.6	955.3	481.2	185.6	65.2	160.4	1,009.9
169	Tone	15,739	936.2	260.8	691.8	1,057.7	983.0	973.7	1,058.8	1,240.9	1,561.5	1,541.0	979.8	683.0	997.3
228	Lempa	18,088	888.1	2.9	7.6	10.3	11.7	547.6	1,582.1	1,968.2	3,042.2	2,295.6	976.3	585.2	993.2
329	Rio Ribeira Do Iguape	25,698	2,174.8	1,657.5	1,425.9	887.3	735.4	782.1	538.1	442.8	601.1	860.6	773.8	987.3	988.9
326	Orange	972,388	1,857.2	2,095.4	2,246.2	1,402.3	807.0	489.5	342.5	319.8	311.7	362.3	534.8	884.1	971.1
277	Rio Itapecuru	52,672	84.3	1,224.8	3,260.0	3,253.3	1,651.3	883.6	520.8	314.3	190.4	115.4	69.7	42.3	967.5
167	Naktong	23,325	508.2	274.7	755.4	983.6	697.6	959.3	1,987.4	1,879.2	1,673.3	882.4	545.1	344.9	957.6
342	Salado	266,264	1,011.5	24.7	73.2	787.5	1,228.1	1,234.8	1,105.6	959.3	1,261.2	1,567.9	1,418.0	767.6	953.3
190	Pearl	22,423	1,984.2	2,012.8	2,131.6	1,713.4	1,094.2	598.0	392.6	256.9	162.9	99.8	148.7	752.9	945.7
229	Gambia	69,874	750.8	0.3	0.3	0.3	0.3	145.1	922.2	3,078.2	3,466.5	1,537.8	816.1	492.4	934.2
92	Grande Rivière De La Baleine	24,257	482.1	0.0	0.0	0.0	3,057.6	1,978.3	1,140.9	1,097.4	1,241.0	1,368.9	523.3	316.1	933.8
211	Santiago	126,222	681.2	116.1	243.6	265.3	168.1	205.1	1,027.8	2,651.8	3,112.7	1,413.6	791.4	492.4	930.8
315	Save	114,958	2,203.3	3,356.1	2,440.7	1,065.6	627.7	386.7	242.8	173.8	130.9	85.0	41.4	348.7	925.2
78	Baleine, Grande Rivière De La	22,136	379.8	-	-	-	4,270.1	1,505.8	1,122.0	983.1	1,130.8	994.3	412.3	249.0	920.6
36	Kem	42,081	235.3	0.5	0.4	2,642.5	3,902.2	1,352.6	781.6	490.1	429.8	696.2	253.6	153.2	911.5
69	Taku	17,968	430.1	0.0	0.0	806.8	2,508.0	2,296.9	1,033.2	783.7	1,027.7	1,263.6	467.0	282.0	908.3
43	Oulujoki	30,555	230.9	1.1	0.8	5,398.3	1,702.8	939.7	562.7	373.1	416.6	710.7	247.3	149.5	894.5
50	Nizhny Vyg (Soroka)	31,334	206.5	0.1	0.1	5,451.2	1,622.9	926.7	553.6	334.4	420.4	626.2	224.2	135.5	875.1
186	Altamaha	37,117	1,328.3	2,248.8	2,575.1	1,555.3	829.5	474.6	313.9	232.8	182.4	112.6	72.0	424.7	862.5
363	Patacua	24,232	1,710.9	118.6	53.8	32.7	19.7	70.9	693.4	892.9	1,438.3	2,176.2	1,802.4	1,293.7	858.6

continues

DATA TABLE 7 continued

Natural runoff (Mm³/month)

Basin ID	Basin Name	Area (km²)	Jan	Feb	Mar	Apr	May	Jun	Jul	Aug	Sep	Oct	Nov	Dec	Average
63	Alsek	28,422	286.1	0.0	0.0	584.4	2,611.7	2,770.9	1,100.4	738.6	961.8	680.0	310.6	187.6	852.7
162	San Joaquin	34,366	707.1	829.3	1,285.7	1,335.3	1,136.9	1,098.6	1,267.9	1,193.7	818.4	322.2	74.9	102.3	847.7
109	Natashquan (Rivière)	16,948	291.0	0.0	0.0	156.2	3,726.8	1,620.1	1,208.4	808.1	691.2	792.3	315.9	190.8	816.7
156	Potomac	32,381	1,355.9	1,163.5	1,705.2	1,374.1	948.5	624.9	370.6	255.1	188.4	213.6	392.6	752.3	778.7
3	Anabar	85,016	525.8	85.3	51.5	31.1	18.8	4,128.1	2,082.3	1,011.7	602.5	354.9	214.4	129.5	769.6
136	Merrimack	12,645	381.1	8.4	3,157.8	1,789.4	1,035.9	634.7	377.8	233.8	212.7	361.4	783.0	252.8	769.1
115	Rupert	16,063	311.4	0.0	0.0	794.7	3,201.0	1,189.1	796.1	677.5	758.7	894.3	338.0	204.2	763.8
166	Roanoke	26,801	1,844.9	1,467.9	1,460.0	1,085.8	725.1	448.3	316.0	238.3	184.2	159.1	332.5	853.9	759.7
291	Brantas	10,823	1,795.8	1,741.1	1,762.4	1,247.5	696.8	395.2	243.3	163.6	122.2	71.2	160.7	531.2	744.2
350	Clutha	17,119	1,029.1	421.2	482.8	693.8	684.7	694.4	642.0	723.7	898.4	957.1	762.4	659.5	720.8
158	Kitakami	9,652	690.1	172.5	1,357.6	1,082.6	802.0	551.9	581.1	590.1	631.9	775.8	828.3	528.7	716.1
344	Rapel	15,689	1,119.4	178.1	113.7	66.3	510.5	1,373.7	1,356.5	1,185.4	876.8	701.4	412.7	681.5	714.7
324	Mangoky	43,141	1,857.9	2,203.9	1,852.5	898.9	518.2	330.4	220.7	136.7	83.0	50.1	56.6	333.7	711.9
150	Kizilirmak	77,874	199.1	836.5	1,201.8	2,397.1	1,578.7	804.4	537.4	393.4	241.0	132.0	83.4	128.5	711.1
55	Kymijoki	33,623	186.3	1.0	1.0	4,570.7	1,319.9	758.1	456.3	277.2	216.4	308.4	311.4	122.5	710.7
174	Tenryu	5,769	584.8	291.0	511.3	839.0	748.4	861.4	816.6	724.3	1,026.6	888.4	635.8	438.9	697.2
181	Savannah	27,740	1,501.0	1,570.4	1,658.6	1,058.3	583.9	360.8	272.1	190.7	187.5	165.1	238.5	572.9	696.6
178	Yodo	8,424	1,063.4	536.0	654.4	713.4	583.0	743.8	675.6	510.1	768.0	820.6	652.7	632.7	696.1
192	Suwannee	26,401	1,043.0	1,262.4	1,316.2	797.7	415.7	300.4	613.1	756.9	714.7	385.8	211.7	527.5	695.4
176	Kiso	5,419	652.3	218.1	352.4	848.3	836.4	896.9	961.9	750.1	967.8	780.0	608.0	420.8	691.1
123	Skagit	7,961	951.3	342.8	1,452.8	1,747.8	975.0	491.1	287.0	173.6	108.5	409.0	751.3	600.5	690.9
173	Cape Fear	22,653	1,592.2	1,301.8	1,225.2	817.3	528.8	353.8	389.9	381.5	348.4	265.8	359.7	725.3	690.8
300	Daly	53,415	783.7	2,684.1	2,482.2	919.0	552.0	333.8	202.0	122.6	74.6	45.3	27.1	16.9	686.9
47	Thjorsa	7,527	422.1	33.1	91.6	1,123.0	1,723.6	1,322.4	662.2	584.0	668.7	840.8	478.5	288.8	686.6
285	Rio Jaguaribe	72,804	64.1	11.7	1,619.3	2,744.3	1,562.6	861.9	515.6	319.8	213.3	142.0	92.4	60.4	683.9

ID	Name													
163	James	23,528	1,600.5	1,239.3	1,248.0	954.0	670.9	429.1	258.7	169.2	122.7	160.2	381.4	857.5	674.3
159	Mogami	6,853	969.3	827.6	1,039.4	774.9	556.2	376.7	392.9	356.7	430.5	519.9	740.8	1,023.9	667.4
239	Oueme	59,843	458.8	1.0	1.1	6.9	240.8	976.7	1,265.1	1,320.9	1,893.3	1,037.9	499.0	301.2	666.9
164	Bravo	510,056	263.4	84.8	204.6	657.0	1,336.3	937.5	913.3	1,079.5	1,202.6	640.0	304.7	204.1	652.3
185	Brazos	117,853	747.2	874.4	717.7	986.3	929.9	587.4	962.4	828.0	508.3	179.4	80.2	367.2	647.4
65	Dramselv	17,364	259.8	0.5	97.0	1,233.4	1,458.1	1,401.1	734.0	665.9	720.2	633.0	304.9	170.5	639.9
296	Rio Paraguacu	54,607	274.1	185.7	300.7	705.2	1,494.5	1,165.9	1,466.7	878.0	455.0	267.7	194.3	163.9	629.3
191	Sabine	25,612	1,109.9	1,334.8	1,286.5	1,307.3	936.9	456.1	276.7	165.9	101.9	62.1	86.8	295.2	618.3
372	Messalo	24,811	941.3	1,818.6	2,197.0	1,098.8	548.0	330.8	199.8	120.7	72.9	44.1	26.6	16.1	617.9
233	Geba	12,774	537.0	3.5	4.3	4.3	3.5	79.2	413.4	1,814.0	2,305.1	1,207.1	582.6	353.4	608.9
60	Kokemaenjoki	26,616	236.7	1.6	1.5	3,662.2	1,092.9	616.1	371.5	225.5	147.2	152.9	501.9	154.8	597.1
68	Seal	53,440	126.4	1.6	1.0	0.6	3,534.7	1,434.9	728.5	432.3	415.7	241.7	130.6	78.9	593.9
318	Buzi	27,905	1,304.7	1,917.2	1,888.8	752.6	437.8	265.0	160.8	98.8	61.6	38.5	22.6	110.9	588.3
236	Comoe	78,507	447.9	3.4	4.6	105.1	306.0	923.2	676.9	1,018.0	1,509.8	1,021.9	540.3	296.4	571.1
179	Sebou	36,201	1,058.8	1,081.1	1,248.4	934.9	533.2	346.0	284.6	201.0	141.9	76.4	197.1	563.6	555.6
49	Oelffusa	5,678	378.5	0.0	323.1	1,726.6	690.3	685.2	565.6	398.0	471.2	658.7	434.6	320.1	554.3
157	Guadiana	66,020	246.7	715.8	1,619.0	1,022.7	571.6	571.0	703.0	614.9	301.8	105.4	30.3	15.9	543.2
275	Nyanga	12,369	1,159.8	691.5	941.1	1,044.4	523.6	267.4	161.5	97.5	58.9	35.6	692.8	805.1	539.9
10	Colville	57,545	185.8	1.4	0.8	0.5	45.5	1,938.0	1,977.6	1,034.6	579.1	324.5	196.0	118.4	533.5
325	Fitzroy	142,915	72.4	1,909.6	2,134.9	885.3	493.9	300.8	192.9	132.5	101.8	79.8	54.4	35.2	532.8
250	Pra	23,480	378.6	3.2	102.6	282.2	657.8	1,316.6	691.0	327.9	645.8	1,039.5	483.0	247.6	514.6
141	Rogue	14,527	1,090.2	1,088.4	909.0	858.5	622.6	302.4	190.5	120.0	73.0	41.5	212.2	514.3	501.9
403	Daryacheh-Ye Orumieh	30,336	156.1	103.1	506.3	1,527.4	1,539.9	737.8	467.8	347.1	213.8	126.4	137.4	116.8	498.3
20	Noatak	32,320	156.5	0.8	0.5	0.3	401.9	1,980.3	1,099.7	624.3	629.7	275.6	166.5	100.6	453.1
247	Tano	15,656	321.4	0.2	32.5	186.6	547.1	1,356.6	700.0	337.1	480.9	810.9	421.8	218.4	451.1
154	Eel (California)	7,450	1,366.1	1,258.1	927.1	540.6	304.0	173.7	105.4	63.9	38.7	23.2	15.6	579.5	449.6
15	Muonio	37,347	143.9	13.1	8.0	213.9	2,005.5	1,110.4	750.1	388.4	294.9	187.3	101.6	61.4	439.9
189	Trinity (Texas)	46,168	588.6	822.5	766.2	973.4	839.8	365.7	234.6	152.1	98.0	66.6	52.7	255.9	434.7
299	Roper	79,908	434.9	1,537.1	1,682.2	599.2	358.4	216.7	131.2	79.7	48.5	29.6	17.7	10.5	428.8
28	Kobuk	30,242	211.1	31.8	19.2	11.6	2,063.7	1,100.6	619.7	373.5	338.6	159.6	96.4	58.2	423.7

continues

DATA TABLE 7 *continued*

			Natural runoff (Mm³/month)												
Basin ID	Basin Name	Area (km²)	Jan	Feb	Mar	Apr	May	Jun	Jul	Aug	Sep	Oct	Nov	Dec	Average
153	Sakarya	62,483	348.5	1,016.9	1,126.0	888.3	508.1	345.2	273.7	242.4	156.9	68.9	26.5	75.1	423.0
330	Incomati	46,296	1,039.9	1,118.5	1,027.4	517.1	276.9	173.5	113.5	83.8	67.7	43.3	129.4	456.6	420.6
391	Escaut (Schelde)	21,499	1,351.5	705.9	604.0	472.4	284.9	168.8	111.6	79.8	54.8	63.7	373.3	746.1	418.1
9	Tana (NO, FI)	14,518	71.8	0.0	0.0	0.0	2,851.4	778.3	457.0	276.1	217.4	149.0	77.9	47.1	410.5
144	Luan He	71,071	174.3	47.2	147.9	279.9	408.1	214.3	964.0	1,219.0	725.2	357.8	190.0	115.8	403.6
188	Colorado (Caribbean Sea)	110,640	466.3	688.6	553.1	621.2	522.3	343.0	497.7	455.0	315.6	115.8	35.9	228.7	403.6
223	Verde	18,343	378.7	2.9	6.5	6.9	4.5	10.1	365.4	850.4	1,648.0	893.8	414.1	250.8	402.7
175	Santee	40,035	804.8	794.4	752.0	496.5	298.4	193.4	174.2	169.8	140.6	124.1	150.6	395.6	374.5
12	Anderson	66,492	54.7	0.4	0.2	0.1	2,548.6	797.3	434.9	262.7	158.7	95.8	57.9	35.0	370.5
104	Attawapiskat	30,457	121.7	0.0	0.0	0.0	2,195.5	613.1	355.5	214.7	350.3	329.0	132.1	79.8	366.0
401	Great Salt Lake	74,114	65.3	237.4	354.3	782.1	915.8	634.2	481.7	350.5	213.3	121.5	68.5	66.7	357.6
17	Palyavaam	31,113	106.3	0.0	0.0	0.0	0.0	1,878.7	1,036.4	526.4	355.2	191.1	115.4	69.7	356.6
404	Van Golu	17,737	117.9	0.6	127.0	1,118.2	1,378.3	489.2	294.9	184.2	112.7	115.2	230.6	77.6	353.9
121	Southern Bug	60,121	38.8	7.9	1,840.9	967.0	519.4	311.3	205.6	138.9	77.0	43.3	27.5	18.1	349.6
307	Majes	18,612	893.9	994.8	875.9	436.0	232.5	139.9	84.7	52.7	34.4	31.7	29.3	363.0	347.4
381	Lorentz	4,299	493.1	445.0	515.7	475.1	346.5	253.8	280.8	253.8	345.8	199.3	277.0	262.3	345.7
313	Mucuri	16,732	1,331.8	412.5	315.0	254.2	167.2	113.9	88.6	50.0	28.9	21.6	212.3	1,046.2	336.8
196	St. Johns	22,489	379.1	187.4	226.0	126.3	72.7	67.3	331.1	475.2	796.6	731.2	335.3	225.9	329.5
376	Popigay	48,954	84.9	0.8	0.5	0.3	0.2	2,052.3	754.1	410.5	251.2	146.9	88.7	53.6	320.3
40	Iijoki	16,163	94.2	0.1	0.1	1,326.2	997.0	396.0	232.9	151.1	155.9	299.9	102.3	61.8	318.1
335	Tugela	30,079	758.7	786.8	752.8	372.3	204.5	126.4	86.3	74.3	66.9	64.9	86.0	376.1	313.0
312	Gilbert	46,429	183.1	1,374.9	1,126.1	428.1	256.1	154.7	93.5	56.5	34.2	20.8	12.5	7.6	312.4
333	Maputo	30,938	924.9	719.0	618.4	324.5	179.7	114.6	75.2	58.1	46.2	33.2	112.3	467.1	306.1
298	Rio De Contas	56,526	297.0	120.1	258.0	556.9	550.1	447.9	457.9	292.5	179.2	110.9	165.1	224.8	305.0

395	Lake Mar Chiquita	154,330	490.2	594.7	1,026.2	504.2	261.0	162.6	112.7	87.1	80.9	85.5	95.2	137.7	303.2
62	Thlewiaza	64,400	91.9	10.3	6.2	3.7	1,834.1	666.0	339.5	204.2	198.6	94.1	56.8	34.3	295.0
33	Ponoy	13,186	127.2	0.0	0.0	0.0	1,585.1	438.1	255.8	178.3	268.5	394.1	138.1	83.4	289.1
11	Alazeya	85,493	184.1	31.2	18.9	11.4	6.9	1,896.8	555.0	314.2	189.8	114.6	69.2	41.8	286.2
182	Gono (Go)	3,949	457.3	244.7	265.1	275.8	218.2	354.0	370.2	191.3	273.3	275.9	229.3	269.2	285.4
44	Lagarfljot	3,285	112.9	0.0	0.0	22.3	1,190.7	742.9	324.1	231.6	257.6	309.8	128.1	74.0	282.9
130	St. Croix	4,639	170.5	0.1	0.1	1,441.8	475.5	301.5	171.5	101.8	88.6	166.9	352.7	111.8	281.9
24	Kalixaelven	17,158	78.3	6.6	4.0	388.6	1,237.8	757.3	351.5	203.1	137.7	118.6	57.6	34.8	281.3
278	Rio Acarau	14,473	22.3	31.8	810.4	1,076.9	646.7	301.3	179.6	110.0	67.9	42.1	26.2	16.4	277.6
304	Rio Prado	31,674	696.0	283.3	357.3	417.6	243.4	187.0	168.8	106.4	58.7	36.6	137.5	549.1	270.1
14	Tuloma	26,058	94.6	11.6	7.1	4.4	1,732.1	547.3	300.0	180.4	119.2	121.3	55.2	33.5	267.2
200	Fuerte	36,420	340.7	90.6	48.7	39.8	35.5	39.1	166.0	764.0	806.9	406.5	207.8	249.3	266.2
397	Dead Sea	35,444	605.0	681.7	454.0	289.3	257.0	203.5	196.8	168.4	96.6	58.6	31.9	117.9	263.4
226	Penner	54,976	568.7	34.4	52.1	45.1	43.8	41.5	158.5	151.2	182.9	360.5	1,017.5	485.3	261.8
287	Ruvu	17,541	121.7	122.6	361.9	952.4	722.0	287.6	174.0	106.2	64.9	39.6	23.9	73.4	254.2
269	Tana	95,715	290.5	14.7	38.9	512.9	706.1	333.2	183.6	113.4	69.9	103.6	260.5	305.1	244.4
292	Santa	11,883	455.4	563.1	650.9	399.4	210.8	125.2	78.5	57.1	44.0	79.5	128.6	132.8	243.8
203	San Pedro	29,359	192.9	2.6	4.4	6.2	7.1	3.4	137.0	732.2	987.8	369.0	206.3	130.3	231.6
113	Thames	12,359	726.2	447.7	361.1	237.6	136.9	78.4	49.7	32.4	21.8	18.4	131.8	395.2	219.8
102	Trent	9,053	691.4	368.2	303.2	213.3	137.6	80.5	56.5	51.0	49.0	67.1	198.0	402.5	218.2
374	Galana	51,922	187.2	7.8	34.8	597.5	654.3	327.5	177.5	104.7	64.5	40.0	203.0	196.1	216.2
365	Ocona	16,064	539.4	582.1	517.1	245.1	134.9	80.9	48.8	30.6	19.9	49.2	71.2	235.0	212.9
348	South Esk	10,843	186.5	8.9	10.5	32.6	76.3	208.6	392.9	471.5	403.2	358.1	217.0	135.7	208.5
91	Tweed	4,771	575.8	239.8	225.1	158.2	110.7	72.3	54.4	61.3	91.1	172.0	311.6	366.0	203.2
41	Joekulsa A Fjoellum	7,311	75.0	0.0	0.0	5.7	754.0	592.9	236.8	148.0	147.0	221.4	82.3	49.2	192.7
242	Mono	23,899	122.1	0.3	14.5	50.7	126.1	330.7	356.9	319.3	472.0	289.3	132.5	80.1	191.2
400	Lake Vattern	11,337	109.1	0.5	645.9	535.8	246.0	136.0	81.3	60.0	40.2	82.5	196.1	85.7	184.9
38	Quoich	28,218	41.6	-	-	-	-	1,216.3	349.6	199.5	128.2	74.8	45.2	27.3	173.5
79	Spey	2,942	478.1	203.5	174.9	136.5	98.8	58.1	43.8	52.5	81.3	157.4	246.1	292.1	168.6

continues

DATA TABLE 7 continued

								Natural runoff (Mm³/month)							
Basin ID	Basin Name	Area (km²)	Jan	Feb	Mar	Apr	May	Jun	Jul	Aug	Sep	Oct	Nov	Dec	Average
251	Davo	8,460	133.7	0.2	0.2	2.0	9.6	542.8	283.0	126.7	238.1	288.6	192.8	89.5	158.9
57	Ferguson	15,200	40.3	0.0	0.0	0.0	0.0	1,053.5	296.1	171.1	142.1	72.5	43.8	26.4	153.8
295	Rio Itapicuru	37,593	56.5	2.3	2.1	7.2	99.2	263.0	645.3	362.8	169.7	102.3	62.2	38.4	150.9
288	Rio Parafba	18,969	37.7	2.0	3.5	89.5	253.8	514.2	396.9	208.3	115.4	73.2	46.6	29.6	147.6
29	Coppermine	43,016	28.9	0.1	0.1	0.0	390.0	714.8	246.4	139.9	84.5	51.0	30.8	18.6	142.1
34	Kovda	10,228	30.6	0.0	0.0	0.0	881.9	280.3	152.3	92.5	71.2	78.0	33.2	20.1	136.7
26	Ellice	12,600	30.5	-	-	-	-	958.3	248.8	150.3	90.8	54.8	33.1	20.0	132.2
42	Svarta, Skagafiroi	3,430	54.9	0.0	29.3	123.6	363.8	392.6	148.0	89.0	76.1	152.2	68.9	36.0	127.9
86	Skjern A	2,818	351.5	132.9	121.1	89.0	50.8	28.9	18.4	22.4	70.4	165.2	192.5	228.6	122.6
283	Chira	16,700	76.0	208.7	379.5	341.5	138.2	86.6	62.2	51.9	37.8	21.8	23.1	17.1	120.4
32	Varzuga	8,182	47.2	0.0	0.0	0.0	683.9	188.7	110.1	66.6	112.4	140.1	51.2	30.9	119.3
172	Chelif	45,249	234.5	283.5	281.9	175.3	103.2	88.2	81.1	66.7	44.4	17.2	8.8	39.5	118.7
279	Pangani	50,365	34.7	22.7	34.9	203.1	493.3	239.9	152.6	87.1	59.8	38.7	28.7	27.5	118.6
30	Hayes (Trib. Arctic Ocean)	22,993	27.0	-	-	-	-	834.5	225.2	133.1	80.4	48.5	29.3	17.7	116.3
343	Blackwood	22,585	79.6	0.6	0.7	0.4	0.1	29.9	254.5	407.6	301.6	166.9	86.8	52.5	115.1
311	Fitzroy	94,044	5.5	446.6	490.3	167.3	101.0	61.0	36.9	22.3	13.5	8.1	4.9	3.0	113.4
297	Canete	5,755	257.6	233.3	231.1	128.4	67.0	41.2	25.4	17.5	12.9	43.0	62.3	113.9	102.8
146	Dalinghe	22,823	58.3	3.2	6.3	33.3	66.4	80.0	85.0	386.7	210.1	123.7	64.5	39.6	96.4
8	Omoloy	38,871	26.9	0.7	0.4	0.2	0.2	426.1	277.1	128.9	73.6	43.7	26.4	15.9	85.0
85	Gudena	2,861	259.3	115.5	103.3	77.1	41.9	24.7	16.2	10.7	14.8	58.0	124.8	157.3	83.6
368	Doring	48,856	44.0	20.3	25.5	14.8	4.2	92.2	148.6	176.0	136.2	105.1	61.5	43.1	72.6
301	Drysdale	26,016	26.2	283.6	270.5	96.7	58.3	35.2	21.3	12.8	7.8	4.7	2.8	1.7	68.5
216	Armeria	9,639	60.6	8.1	17.0	28.8	27.6	10.1	3.3	40.7	292.2	147.3	71.6	48.0	62.9
369	Gamka	45,676	65.3	14.4	28.8	40.5	41.1	51.1	45.0	62.9	111.5	105.9	87.0	55.8	59.1

359	Horton	23,926	12.4	0.3	0.2	0.1	78.8	314.9	93.6	54.9	33.2	20.0	12.1	7.3	52.3
294	Rio Vaza-Barris	15,314	17.1	1.4	1.7	1.6	61.4	151.1	177.6	102.7	49.8	30.8	18.8	11.8	52.2
360	Hornaday	14,778	12.4	0.0	0.0	0.0	0.0	181.3	145.7	70.2	37.1	22.3	13.5	8.1	40.9
193	Yaqui	76,182	15.8	30.2	63.2	72.9	36.0	24.2	49.0	61.5	44.6	29.0	14.8	15.8	38.1
339	Limari	11,780	118.1	101.5	40.8	21.1	12.6	27.9	18.7	17.6	14.6	15.5	11.4	31.8	36.0
170	Salinas	12,655	11.8	34.5	58.5	36.9	32.6	47.2	66.4	68.2	42.9	9.6	2.7	1.6	34.4
405	Ozero Sevan	4,765	8.6	0.7	3.8	76.9	106.9	55.9	31.2	21.0	12.9	13.1	13.9	5.9	29.2
197	Nueces	43,878	4.2	8.4	21.4	28.3	35.8	41.5	66.6	49.3	24.4	10.5	5.4	3.8	25.0
341	Groot-Vis	30,441	10.7	24.1	23.0	13.9	11.7	9.5	9.9	13.2	21.7	27.3	18.9	18.9	16.9
338	Huasco	9,872	92.1	16.5	10.1	6.1	3.7	2.4	1.6	1.4	1.6	1.5	0.6	46.0	15.3
198	San Antonio	10,952	12.3	13.2	14.4	23.4	25.7	19.4	25.1	18.3	9.9	5.7	4.2	8.8	15.0
389	Flinders	110,041	0.4	89.8	28.2	15.5	9.4	5.7	3.5	2.2	1.4	0.9	0.6	0.4	13.2
390	Leichhardt	33,399	31.9	34.8	11.9	7.2	4.4	2.6	1.6	1.0	0.6	0.4	0.2	0.1	8.1
310	Macarthur	19,674	0.4	1.1	55.6	14.5	8.8	5.3	3.2	1.9	1.2	0.7	0.4	0.3	7.8
370	Groot-Kei	18,678	2.7	6.5	16.9	12.9	8.5	5.9	5.6	6.2	7.9	8.3	6.0	4.4	7.6
308	Ord	55,686	0.0	3.9	1.5	2.7	4.2	5.2	6.4	7.5	8.1	6.6	3.6	0.2	4.2
305	Victoria	78,462	0.1	20.4	14.0	5.5	3.3	2.0	1.2	0.7	0.4	0.3	0.2	0.1	4.0
361	Conception	25,569	0.6	1.8	3.6	4.8	3.8	4.7	4.3	5.9	5.5	4.2	1.5	1.1	3.5
394	Eyre Lake	1,188,841	0.2	0.3	0.6	0.5	0.5	0.4	0.4	0.6	0.7	0.7	0.5	0.4	0.5
303	Durack	29,363	0.0	1.9	1.6	0.6	0.4	0.2	0.1	0.1	0.1	0.0	0.0	0.0	0.4
319	Loa	50,206	0.3	0.4	0.3	0.3	0.3	0.3	0.3	0.3	0.3	0.3	0.3	0.3	0.3

DATA TABLE 8

Monthly Natural Runoff for the World's Major River Basins, by Basin Name

Description

These data are the same as those in data table 7 but sorted by river basin name. The data come from work done by Arjen Hoekstra and Mesfin Mekonnen (2011) to evaluate the regional, national, and global water footprint of human activity. As part of that effort, "natural runoff" was estimated for around 400 river basins, covering some two-thirds of the global land area (excluding Antarctica) and about 65 percent of the global population in 2000. Units for runoff are million cubic meters per month. Data are also provided for the area of the river basin or watershed, in square kilometers.

The natural (undepleted) runoff was estimated by adding the actual runoff data from the Composite Runoff V1.0 database (of Fekete et al. 2002) to the estimated total blue water footprint within the river basin. Because runoff varies naturally over time, these are average estimates. River basin names come from the Global Runoff Data Centre database (GRDC 2007). Note the substantial extremes: just two rivers, the Amazon and Congo Rivers, account for 28 percent of the natural runoff in the river basins listed here.

Limitations

These data are a combination of measured and estimated or modeled flows. Accurate long-term measurements of river flow are not available for a variety of reasons, including the lack of monitoring stations, the complication of estimating undepleted flows in basins where human withdrawals greatly reduce natural flows, and natural variability of high and low runoff over time. These estimates should be considered only approximate averages, but they provide estimates using a consistent method described in the original sources.

The land areas not covered include Greenland; the Sahara Desert in North Africa; the Arabian Peninsula; the Iranian, Afghan, and Gobi Deserts in Asia; the Mojave Desert in North America; and the Australian desert. Also excluded are many smaller pieces of land, often along the coasts, that do not fall within major river basins.

Sources

Fekete, B. M., C. J. Vörösmarty, and W. Grabs. 2002. High-resolution fields of global runoff combining observed river discharge and simulated water balances. *Global Biogeochemical Cycles* 16 (3). doi:10.1029/1999GB001254. Data available at http://grdc.sr.unh.edu/.

Global Runoff Data Centre (GRDC). 2007. Major River Basins of the World. Koblenz, Germany: Federal Institute of Hydrology, Global Runoff Data Centre. Data available at http://grdc.bafg.de/.

Hoekstra, A. Y., and M. M. Mekonnen. 2011. Global Water Scarcity: Monthly Blue Water Footprint Compared to Blue Water Availability for the World's Major River Basins. Value of Water Research Report Series No. 53. Delft, Netherlands: UNESCO-IHE Institute for Water Education. http://www.waterfootprint.org/Reports/Report53-GlobalBlueWaterScarcity.pdf.

DATA TABLE 8 Monthly Natural Runoff for the World's Major River Basins, by Basin Name

Natural runoff (Mm³/month)

Basin ID	Basin Name	Area (km²)	Jan	Feb	Mar	Apr	May	Jun	Jul	Aug	Sep	Oct	Nov	Dec	Average
180	Alabama River and Tombigbee	113,117.4	9,397.8	10,770.1	12,232.8	8,612.8	4,829.3	2,684.6	1,677.7	1,055.0	671.1	421.7	465.0	3,348.5	4,680.5
11	Alazeya	85,493.3	184.1	31.2	18.9	11.4	6.9	1,896.8	555.0	314.2	189.8	114.6	69.2	41.8	286.2
111	Albany	123,081.0	813.0	0.1	0.1	8,411.8	9,204.2	3,827.2	2,148.3	1,306.1	1,776.4	2,432.0	882.5	533.1	2,611.2
63	Alsek	28,422.0	286.1	0.0	0.0	584.4	2,611.4	2,770.9	1,100.4	738.6	961.8	680.0	310.6	187.6	852.7
186	Altamaha	37,117.5	1,328.3	2,248.8	2,575.1	1,555.3	829.5	474.6	313.9	232.8	182.4	112.6	72.0	424.7	862.5
259	Amazonas	5,880,854.9	950,375.6	705,085.8	813,922.6	857,876.6	713,184.6	564,388.1	424,106.1	299,107.1	238,076.7	243,680.0	298,076.1	455,711.2	546,965.9
90	Amur	2,023,520.4	13,098.2	58.9	59.3	41,918.1	57,294.3	53,865.4	45,446.7	49,807.8	50,357.7	28,455.4	14,176.3	8,572.4	30,259.2
3	Anabar	85,015.5	525.8	85.3	51.5	31.1	18.8	4,128.1	2,082.3	1,011.7	602.5	354.9	214.4	129.5	769.6
21	Anadyr	171,275.8	1,182.1	1.3	0.8	0.5	2,452.2	21,998.7	10,380.5	5,461.5	4,105.7	2,116.1	1,278.1	771.9	4,145.8
12	Anderson	66,491.7	54.7	0.4	0.2	0.1	2,548.6	797.3	434.9	262.7	158.7	95.8	57.9	35.0	370.5
46	Angerman	32,372.0	343.5	0.5	0.4	4,027.0	3,524.6	2,685.5	1,232.5	833.4	764.3	954.8	371.0	224.1	1,246.8
184	Apalachicola	51,412.9	2,889.5	3,847.6	4,385.4	3,016.0	1,572.5	912.7	651.6	531.5	402.8	235.8	202.2	1,034.8	1,640.2
124	Aral Drainage	1,233,148.5	2,546.9	4,161.6	13,784.1	20,140.9	23,937.1	21,375.9	17,965.9	12,930.4	8,104.3	3,850.2	1,674.4	1,541.5	11,001.1
216	Armeria	9,639.1	60.6	8.1	17.0	28.8	27.6	10.1	3.3	40.7	292.2	147.3	71.6	48.0	62.9
66	Arnaud	44,931.9	564.3				632.5	5,641.8	2,064.4	1,543.3	1,878.6	1,362.9	612.6	370.0	1,222.5
244	Atrato	34,619.5	8,908.3	2,297.1	2,736.4	4,317.7	5,624.4	5,876.9	5,976.7	6,096.9	6,689.5	7,140.9	7,032.0	5,537.1	5,686.2
104	Attawapiskat	30,457.4	121.7	0.0	0.0	0.0	2,195.5	613.1	355.5	214.7	350.3	329.0	132.1	79.8	366.0
77	Aux Melezes	41,384.1	637.5	0.1	0.1	0.0	5,433.5	3,250.9	1,841.6	1,727.2	1,787.2	1,696.1	691.6	417.7	1,457.0
35	Back	141,351.9	327.2	0.0	0.0	0.0	5,571.3	8,215.3	2,637.8	1,590.9	990.4	588.0	355.2	214.5	1,707.6
351	Baker	30,760.3	1,928.9	630.2	1,099.0	1,638.6	2,259.5	2,536.2	2,753.8	2,648.6	2,149.0	1,882.5	1,579.4	1,282.9	1,865.7
78	Baleine, Grande Rivière De La	22,136.1	379.8				4,270.1	1,505.8	1,122.0	983.1	1,130.8	994.3	412.3	249.0	920.6
393	Balkhash	423,657.4	273.5	5.8	2,199.0	4,046.6	4,089.9	2,907.0	2,102.5	1,368.5	905.0	412.3	445.2	181.7	1,578.1
238	Bandama	98,751.1	1,337.1	4.0	5.7	118.4	306.8	1,520.8	1,055.3	2,949.0	5,574.4	3,186.6	1,473.0	881.5	1,534.4

294

ID	Name												
388	Batang Hari	42,872.4	10,918.4	4,536.0	5,450.1	6,252.7	4,790.1	2,820.9	1,767.2	1,540.6	2,438.0	4,411.4	6,272.8	7,176.5	4,864.6
387	Batang Kuantan	16,739.0	3,820.3	1,460.7	1,731.9	2,316.9	1,825.4	1,075.7	640.6	525.6	838.5	1,724.9	2,579.6	2,602.7	1,761.9
202	Bei Jiang	52,915.4	629.5	353.9	1,921.9	4,546.4	6,874.5	7,013.0	3,830.8	3,230.7	2,159.0	1,100.4	668.5	413.5	2,728.5
346	Biobio	24,108.6	1,512.9	29.1	298.7	919.1	3,736.5	4,786.4	5,042.5	4,670.4	4,131.1	2,973.2	1,809.7	1,045.2	2,579.6
343	Blackwood	22,584.8	79.6	0.6	0.7	0.4	0.1	29.9	254.5	407.6	301.6	166.9	86.8	52.5	115.1
209	Brahmani River (Bhahmani)	51,973.4	1,562.1	33.6	53.2	42.6	38.7	549.7	3,888.5	7,939.2	6,035.6	3,059.0	1,695.6	1,049.7	2,162.3
195	Brahmaputra	518,011.4	28,402.7	549.4	2,782.9	16,852.2	46,782.8	103,699.0	127,521.0	128,246.8	107,215.1	58,846.2	31,386.4	18,647.3	55,911.0
291	Brantas	10,822.6	1,795.8	1,741.1	1,762.4	1,247.5	696.8	395.2	243.3	163.6	122.2	71.2	160.7	531.2	744.2
164	Bravo	510,056.3	263.4	84.8	204.6	657.0	1,336.3	937.5	913.3	1,079.5	1,202.6	640.0	304.7	204.1	652.3
185	Brazos	117,853.1	747.2	874.4	717.7	986.3	929.9	587.4	962.4	828.0	508.3	179.4	80.2	367.2	647.4
316	Burdekin	130,426.5	690.0	3,679.3	3,662.4	1,887.1	1,001.1	597.8	363.8	227.1	146.3	95.8	59.5	32.7	1,036.9
318	Buzi	27,904.7	1,304.7	1,917.2	1,888.8	752.6	437.8	265.0	160.8	98.8	61.6	38.5	22.6	110.9	588.3
217	Ca	28,747.0	1,447.9	56.1	26.8	23.3	154.3	584.1	1,876.9	2,678.4	4,330.1	2,767.9	1,652.1	965.9	1,380.3
297	Canete	5,755.2	257.6	233.3	231.1	128.4	67.0	41.2	25.4	17.5	12.9	43.0	62.3	113.9	102.8
75	Caniapiscau	105,690.6	2,297.2	0.9	0.5	0.3	16,170.7	12,475.4	6,805.5	6,229.8	6,669.5	6,024.1	2,490.3	1,504.1	5,055.7
173	Cape Fear	22,652.7	1,592.2	1,301.8	1,225.2	817.3	528.8	353.8	389.9	381.5	348.4	265.8	359.7	725.3	690.8
231	Cauvery	91,159.4	2,091.4	159.7	385.4	347.3	350.9	1,080.8	3,669.7	3,305.4	2,574.1	2,347.6	2,777.6	1,849.0	1,744.9
246	Cavally	30,665.2	2,295.7	105.8	221.1	532.9	1,553.0	3,418.3	2,447.3	1,941.8	4,204.8	4,188.2	2,935.7	1,654.9	2,125.0
218	Chao Phraya	188,419.1	4,183.6	318.1	519.8	537.6	493.6	1,640.2	5,541.8	9,607.1	16,009.4	10,072.0	5,811.4	3,059.8	4,816.2
172	Chelif	45,249.3	234.5	283.5	281.9	175.3	103.2	88.2	81.1	66.7	44.4	17.2	8.8	39.5	118.7
283	Chira	16,699.6	76.0	208.7	379.5	341.5	138.2	86.6	62.2	51.9	37.8	21.8	23.1	17.1	120.4
349	Chubut	145,351.9	837.4	70.0	172.1	336.2	1,273.9	2,263.3	2,596.8	3,116.1	2,246.4	1,454.2	895.8	580.3	1,320.2
72	Churchill	298,505.0	789.5	20.1	12.3	3,400.0	13,120.8	7,724.3	3,761.6	2,131.7	1,960.1	1,724.5	774.9	468.2	2,990.7
95	Churchill, Fleuve (Labrador)	84,984.3	1,702.6	0.0	0.0	0.0	15,223.9	10,971.0	6,028.7	4,689.5	4,563.7	4,517.0	1,848.4	1,116.4	4,221.8
350	Clutha	17,118.9	1,029.1	421.2	482.8	693.8	684.7	694.4	642.0	723.7	898.4	957.1	762.4	659.5	720.8
364	Coco	25,502.0	2,767.8	143.3	70.7	43.3	48.4	1,698.7	3,244.2	2,959.2	3,280.5	3,939.1	2,788.7	2,041.2	1,918.8

continues

DATA TABLE 8 *continued*

							Natural runoff (Mm³/month)							
Basin ID	Basin Name	Area (km²)	Jan	Feb	Mar	Apr	May	Jun	Jul	Aug	Sep	Oct	Nov	Dec	Average
336	Colorado (Argentina)	390,631.1	3,500.8	346.2	220.9	135.4	373.1	707.2	841.7	929.8	908.5	2,371.9	3,152.6	2,575.4	1,338.6
188	Colorado (Caribbean Sea)	110,640.3	466.3	688.6	553.1	621.2	522.3	343.0	497.7	455.0	315.6	115.8	35.9	228.7	403.6
138	Colorado (Pacific Ocean)	640,463.6	323.3	99.7	738.0	3,046.4	5,903.7	4,320.1	2,390.6	1,653.7	1,126.8	740.4	370.5	231.4	1,745.4
107	Columbia	668,561.9	11,960.9	11,259.5	20,903.3	38,188.0	55,190.6	36,849.4	18,551.8	11,683.5	7,290.7	5,462.2	6,682.2	7,923.4	19,328.8
10	Colville	57,544.7	185.8	1.4	0.8	0.5	45.5	1,938.0	1,977.6	1,034.6	579.1	324.5	196.0	118.4	533.5
236	Comoe	78,506.9	447.9	3.4	4.6	105.1	306.0	923.2	676.9	1,018.0	1,509.8	1,021.9	540.3	296.4	571.1
361	Conception	25,569.5	0.6	1.8	3.6	4.8	3.8	4.7	4.3	5.9	5.5	4.2	1.5	1.1	3.5
243	Congo	3,698,918.1	193,908.5	92,837.8	126,684.3	138,968.9	93,475.1	62,522.2	55,481.6	71,460.1	90,395.6	111,123.4	108,901.0	123,157.3	105,743.0
132	Connecticut	27,468.3	934.3	7.8	3,116.2	4,979.1	2,528.5	1,556.0	985.4	660.3	761.0	1,049.6	1,699.4	665.0	1,578.5
255	Coppename	24,750.2	1,551.1	1,394.4	1,540.4	2,071.4	4,023.6	4,578.3	3,861.0	2,331.4	1,160.6	695.4	420.0	321.1	1,995.7
58	Copper	64,959.7	1,090.8	0.0	0.0	17.3	7,546.7	10,214.6	7,440.5	4,185.8	3,963.1	2,215.5	1,184.1	715.2	3,214.5
29	Coppermine	43,016.4	28.9	0.1	0.1	0.0	390.0	714.8	246.4	139.9	84.5	51.0	30.8	18.6	142.1
254	Corantijn	65,527.6	1,313.6	829.3	1,710.9	3,554.5	11,297.6	13,180.8	9,711.4	6,084.0	3,012.4	1,811.6	1,094.1	692.7	4,524.4
234	Corubal	24,258.0	882.9	0.4	0.5	0.5	0.4	293.0	1,767.6	3,594.2	3,165.9	2,080.6	964.3	579.0	1,110.8
248	Cross	52,820.2	3,986.4	1.4	608.2	1,184.8	2,346.5	4,731.8	7,828.2	9,133.6	11,624.3	10,815.8	4,452.2	2,614.4	4,944.0
366	Cuanza	141,391.1	10,360.7	7,565.1	10,296.0	8,746.9	3,552.8	2,127.7	1,286.3	779.0	472.4	307.3	304.1	5,458.5	4,271.4
367	Cunene	110,545.5	1,828.9	2,579.9	5,240.5	2,925.0	1,322.8	798.9	482.7	291.8	176.4	112.8	98.1	642.6	1,375.0
245	Cuyuni	85,635.0	9,798.1	3,136.9	2,829.9	4,268.1	9,871.7	13,477.9	13,021.6	10,134.4	5,445.8	3,587.1	3,665.6	6,571.4	7,150.7
146	Dalinghe	22,823.1	58.3	3.2	6.3	33.3	66.4	80.0	85.0	386.7	210.1	123.7	64.5	39.6	96.4
300	Daly	53,414.6	783.7	2,684.1	2,482.2	919.0	552.0	333.8	202.0	122.6	74.6	45.3	27.1	16.9	686.9
206	Damodar	43,096.1	1,257.9	111.7	160.8	44.8	15.0	253.7	1,494.8	4,751.2	4,357.8	2,147.4	1,316.0	850.4	1,421.8
118	Danube	793,704.8	15,369.2	12,969.9	30,056.7	34,399.1	27,077.0	19,150.4	14,083.1	11,248.0	10,213.5	12,685.6	15,065.0	12,575.3	17,907.8
403	Daryacheh-Ye Orumieh	30,335.7	156.1	103.1	506.3	1,527.4	1,539.9	737.8	467.8	347.1	213.8	126.4	137.4	116.8	498.3

270	Daule and Vinces	41,993.5	2,730.9	4,080.6	5,260.2	4,696.3	2,419.3	1,482.0	915.4	644.8	472.3	447.2	458.1	400.9	2,000.7
251	Davo	8,460.3	133.7	0.2	0.2	2.0	9.6	542.8	283.0	126.7	238.1	288.6	192.8	89.5	158.9
397	Dead Sea	35,444.0	605.0	681.7	454.0	289.3	257.0	203.5	196.8	168.4	96.6	58.6	31.9	117.9	263.4
147	Delaware	26,713.4	1,640.3	801.3	4,062.4	2,255.7	1,789.8	1,043.1	712.5	568.2	587.3	762.0	1,406.7	1,294.0	1,410.3
96	Dniepr	510,661.3	849.5	45.1	9,860.5	20,513.6	7,860.1	4,484.4	2,811.4	1,783.2	1,063.7	891.9	1,449.4	588.9	4,350.1
120	Dniestr	72,108.2	407.9	13.3	3,147.7	2,602.1	1,528.5	1,130.4	792.2	626.7	510.8	629.4	702.3	272.1	1,030.3
99	Don	425,629.6	308.1	29.4	4,138.3	14,368.8	5,709.0	3,280.6	2,272.7	1,517.1	801.0	444.2	431.0	212.1	2,792.7
204	Dong Jiang	32,102.9	861.2	105.2	1,116.0	3,077.8	5,758.0	6,467.3	4,881.8	4,699.8	3,338.0	1,558.5	934.6	567.8	2,780.5
368	Doring	48,855.5	44.0	20.3	25.5	14.8	4.2	92.2	148.6	176.0	136.2	105.1	61.5	43.1	72.6
142	Douro	96,125.4	3,913.4	3,031.8	4,104.3	2,983.9	2,057.4	1,252.2	1,050.6	820.4	404.1	215.0	539.3	1,650.2	1,835.2
65	Dramselv	17,364.0	259.8	0.5	97.0	1,233.4	1,458.1	1,401.1	734.0	665.9	720.2	633.0	304.9	170.5	639.9
301	Drysdale	26,015.9	26.2	283.6	270.5	96.7	58.3	35.2	21.3	12.8	7.8	4.7	2.8	1.7	68.5
303	Durack	29,363.2	0.0	1.9	1.6	0.6	0.4	0.2	0.1	0.1	0.1	0.0	0.0	0.0	0.4
105	Eastmain	48,837.5	1,182.9	0.0	0.0	2,558.1	9,188.6	4,899.7	2,922.3	2,536.0	2,987.8	3,379.4	1,284.2	775.6	2,642.9
140	Ebro	85,158.6	4,220.1	2,820.2	2,785.8	2,876.1	2,619.8	1,631.9	1,192.1	853.9	509.9	561.7	873.7	2,424.5	1,947.5
154	Eel (California)	7,449.9	1,366.1	1,258.1	927.1	540.6	304.0	173.7	105.4	63.9	38.7	23.2	15.6	579.5	449.6
382	Eilanden	20,077.5	5,431.2	3,334.4	3,835.7	3,751.0	3,567.3	3,244.7	3,235.3	3,090.1	3,371.3	2,701.3	2,824.9	3,278.9	3,472.2
101	Elbe	139,347.6	2,858.7	2,594.5	4,899.4	3,701.9	2,272.5	1,577.0	1,147.9	892.7	725.8	865.4	1,454.7	1,857.4	2,070.7
26	Ellice	12,599.6	30.5	-	-	-	-	958.3	248.8	150.3	90.8	54.8	33.1	20.0	132.2
391	Escaut (Schelde)	21,498.7	1,351.5	705.9	604.0	472.4	284.9	168.8	111.6	79.8	54.8	63.7	373.3	746.1	418.1
268	Esmeraldas	19,796.2	2,922.0	4,238.0	5,676.0	6,743.2	4,668.6	2,504.1	1,398.8	876.1	589.2	579.1	755.5	966.5	2,659.8
252	Essequibo	68,788.3	5,069.1	1,976.6	2,110.6	2,932.1	7,163.7	13,057.2	11,720.0	7,957.4	3,936.3	2,438.8	1,874.9	3,105.6	5,278.5
394	Eyre Lake	1,188,841.3	0.2	0.3	0.6	0.5	0.5	0.4	0.4	0.6	0.7	0.7	0.5	0.4	0.5
57	Ferguson	15,200.4	40.3	0.0	0.0	0.0	0.0	1,053.5	296.1	171.1	142.1	72.5	43.8	26.4	153.8
73	Feuilles (Rivière Aux)	37,425.3	673.6	-	-	-	2,539.5	5,199.5	2,051.0	1,757.3	1,931.8	1,773.6	731.3	441.7	1,424.9
325	Fitzroy	142,915.3	72.4	1,909.6	2,134.9	885.3	493.9	300.8	192.9	132.5	101.8	79.8	54.4	35.2	532.8
311	Fitzroy	94,043.9	5.5	446.6	490.3	167.3	101.0	61.0	36.9	22.3	13.5	8.1	4.9	3.0	113.4

continues

DATA TABLE 8 *continued*

								Natural runoff (Mm³/month)							
Basin ID	Basin Name	Area (km²)	Jan	Feb	Mar	Apr	May	Jun	Jul	Aug	Sep	Oct	Nov	Dec	Average
389	Flinders	110,041.3	0.4	89.8	28.2	15.5	9.4	5.7	3.5	2.2	1.4	0.9	0.6	0.4	13.2
88	Fraser	239,678.4	3,955.1	1,048.3	2,360.7	15,356.9	26,897.1	15,496.8	7,432.2	4,439.9	3,254.0	3,618.6	3,140.5	2,506.4	7,458.9
377	Fuchun Jiang	37,697.9	1,253.4	1,966.9	3,249.4	3,037.5	4,025.8	5,573.0	2,420.0	1,437.2	1,282.3	886.7	707.7	572.7	2,201.1
200	Fuerte	36,419.8	340.7	90.6	48.7	39.8	35.5	39.1	166.0	764.0	806.9	406.5	207.8	249.3	266.2
374	Galana	51,921.7	187.2	7.8	34.8	597.5	654.3	327.5	177.5	104.7	64.5	40.0	203.0	196.1	216.2
229	Gambia	69,874.3	750.8	0.3	0.3	0.3	0.3	145.1	922.2	3,078.2	3,466.5	1,537.8	816.1	492.4	934.2
369	Gamka	45,676.2	65.3	14.4	28.8	40.5	41.1	51.1	45.0	62.9	111.5	105.9	87.0	55.8	59.1
353	Ganges	1,024,462.6	32,182.1	10,981.6	16,447.4	12,896.7	12,922.0	27,823.6	78,624.5	128,519.9	96,972.9	47,842.1	32,621.7	19,626.1	43,121.7
134	Garonne	55,807.2	3,122.4	1,918.1	2,169.4	2,289.6	1,911.6	1,113.4	759.2	594.0	474.6	657.6	1,074.9	1,915.8	1,500.0
233	Geba	12,774.4	537.0	3.5	4.3	4.3	3.5	79.2	413.4	1,814.0	2,305.1	1,207.1	582.6	353.4	608.9
74	George	39,054.1	758.1	0.0	0.0	0.0	4,958.0	5,039.7	3,961.6	2,616.4	2,618.4	1,701.2	823.0	497.1	1,914.5
312	Gilbert	46,429.1	183.1	1,374.9	1,126.1	428.1	256.1	154.7	93.5	56.5	34.2	20.8	12.5	7.6	312.4
59	Gloma	42,862.7	563.2	1.3	12.0	3,439.3	3,368.1	3,615.4	2,048.4	1,487.9	1,400.5	1,344.1	677.6	369.7	1,527.3
213	Godavari	311,698.7	6,805.4	626.4	1,240.0	1,477.3	1,571.4	891.7	13,461.6	27,528.4	26,621.8	12,095.3	7,347.2	4,834.4	8,708.4
182	Gono (Go)	3,949.5	457.3	244.7	265.1	275.8	218.2	354.0	370.2	191.3	273.3	275.9	229.3	269.2	285.4
230	Grande De Matagalpa	17,991.9	1,788.5	87.6	46.4	30.2	37.4	1,350.9	2,694.3	2,443.9	2,701.8	2,950.2	1,748.2	1,248.2	1,427.3
93	Grande Rivière	111,718.4	2,729.2	0.9	0.5	0.3	20,738.4	10,192.2	6,544.7	6,144.0	7,187.4	7,682.4	2,959.2	1,787.3	5,497.2
92	Grande Rivière De La Baleine	24,256.7	482.1	0.0	0.0	0.0	3,057.6	1,978.3	1,140.9	1,097.4	1,241.0	1,368.9	523.3	316.1	933.8
401	Great Salt Lake	74,114.4	65.3	237.4	354.3	782.1	915.8	634.2	481.7	350.5	213.3	121.5	68.5	66.7	357.6
222	Grisalva	127,675.5	11,859.4	1,204.2	664.4	610.2	1,646.9	8,622.6	11,809.7	13,230.5	20,505.5	19,300.8	10,998.3	7,877.1	9,027.5
370	Groot-Kei	18,678.3	2.7	6.5	16.9	12.9	8.5	5.9	5.6	6.2	7.9	8.3	6.0	4.4	7.6
341	Groot-Vis	30,441.2	10.7	24.1	23.0	13.9	11.7	9.5	9.9	13.2	21.7	27.3	18.9	18.9	16.9
161	Guadalquivir	56,954.8	677.2	1,306.6	3,139.9	1,977.6	1,055.0	1,003.9	1,110.6	955.3	481.2	185.6	65.2	160.4	1,009.9
157	Guadiana	66,020.0	246.7	715.8	1,619.0	1,022.7	571.6	571.0	703.0	614.9	301.8	105.4	30.3	15.9	543.2

85	Gudena	2,860.9	259.3	115.5	103.3	77.1	41.9	24.7	16.2	10.7	14.8	58.0	124.8	157.3	83.6
379	Han Jiang	30,741.5	429.5	189.6	1,241.1	2,010.5	3,792.4	4,718.0	2,447.5	2,125.6	1,654.9	782.8	466.3	283.9	1,678.5
160	Han-Gang (Han River)	24,771.5	814.3	12.5	1,372.3	1,838.7	1,166.9	1,200.5	4,312.9	3,914.3	2,697.0	1,388.7	1,010.5	540.7	1,689.1
30	Hayes (Trib. Arctic Ocean)	22,992.8	27.0					834.5	225.2	133.1	80.4	48.5	29.3	17.7	116.3
84	Hayes (Trib. Hudson Bay)	105,371.9	339.8	7.0	4.3	2.6	7,011.9	3,778.4	2,054.7	1,111.5	798.0	731.1	339.8	205.3	1,365.4
355	Hong (Red River)	157,656.9	4,779.9	80.7	104.5	254.5	1,566.3	7,439.6	18,447.1	22,644.5	16,383.8	9,602.6	5,433.7	3,149.0	7,490.5
360	Hornaday	14,778.0	12.4	0.0	0.0	0.0	0.0	181.3	145.7	70.2	37.1	22.3	13.5	8.1	40.9
359	Horton	23,926.2	12.4	0.3	0.2	0.1	78.8	314.9	93.6	54.9	33.2	20.0	12.1	7.3	52.3
183	Huai He	174,309.9	1,271.4	1,031.3	2,377.3	4,348.7	4,430.9	4,375.0	4,755.1	3,800.5	3,689.6	1,900.5	1,153.0	823.3	2,829.7
149	Huang He (Yellow River)	988,062.6	2,702.4	608.3	2,320.1	5,166.0	8,177.4	9,102.2	10,309.3	9,805.2	9,930.4	5,872.9	3,052.0	1,802.8	5,737.4
338	Huasco	9,871.6	92.1	16.5	10.1	6.1	3.7	2.4	1.6	1.4	1.6	1.5	0.6	46.0	15.3
137	Hudson	36,892.8	1,000.4	26.3	4,477.0	4,970.6	2,745.5	1,656.2	1,063.5	725.6	742.5	1,044.8	1,764.9	728.3	1,745.5
40	Iijoki	16,163.3	94.2	0.1	0.1	1,326.2	997.0	396.0	232.9	151.1	155.9	299.9	102.3	61.8	318.1
330	Incomati	46,295.7	1,039.9	1,118.5	1,027.4	517.1	276.9	173.5	113.5	83.8	67.7	43.3	129.4	456.6	420.6
6	Indigirka	341,227.8	1,902.5	179.7	108.5	65.6	450.9	16,809.8	14,507.5	6,786.0	3,631.1	2,176.2	1,314.4	793.9	4,060.5
168	Indus	1,139,075.4	11,918.9	9,642.7	18,198.4	21,870.2	18,514.4	18,264.8	32,379.1	40,736.0	31,344.3	19,300.0	10,858.9	6,757.3	19,982.1
199	Irrawaddy	411,516.3	26,908.9	342.4	994.5	4,301.3	10,976.0	59,825.7	98,546.8	111,837.4	93,122.6	59,380.0	30,238.9	17,673.9	42,845.7
135	Ishikari	13,783.3	859.4	2.4	2.4	4,390.2	1,918.3	1,080.5	793.8	776.5	1,178.1	1,434.9	1,481.8	564.3	1,206.9
392	Issyk-Kul	191,032.5	300.5	48.6	2,132.1	3,385.1	4,295.9	3,990.2	3,002.3	1,671.5	1,072.0	585.2	384.5	205.0	1,756.1
163	James	23,528.4	1,600.5	1,239.3	1,248.0	954.0	670.9	429.1	258.7	169.2	122.7	160.2	381.4	857.5	674.3
309	Jequitinhonha	68,548.9	4,207.8	1,648.8	1,430.8	860.3	471.5	297.0	203.6	126.2	74.1	60.4	675.7	3,152.4	1,100.7
41	Joekulsa A Fjoellum	7,311.0	75.0	0.0	0.0	5.7	754.0	592.9	236.8	148.0	147.0	221.4	82.3	49.2	192.7
24	Kalixaelven	17,157.6	78.3	6.6	4.0	388.6	1,237.8	757.3	351.5	203.1	137.7	118.6	57.6	34.8	281.3
80	Kamchatka	54,103.9	689.9	0.0	0.0	0.0	9,603.9	7,199.7	5,399.2	2,604.2	1,918.9	1,546.8	749.0	452.4	2,513.7
253	Kelantan	14,419.9	3,574.8	439.2	342.5	417.5	434.1	481.7	494.1	561.2	1,352.9	2,084.8	2,431.3	2,719.0	1,277.8

continues

DATA TABLE 8 *continued*

								Natural runoff (Mm³/month)							
Basin ID	Basin Name	Area (km²)	Jan	Feb	Mar	Apr	May	Jun	Jul	Aug	Sep	Oct	Nov	Dec	Average
36	Kem	42,080.8	235.3	0.5	0.4	2,642.5	3,902.2	1,352.6	781.6	490.1	429.8	696.2	253.6	153.2	911.5
18	Kemijoki	55,824.7	487.2	3.2	2.0	594.1	8,987.0	2,448.1	1,458.1	934.7	1,082.1	1,366.4	516.4	312.0	1,515.9
1	Khatanga	294,907.5	1,571.7	58.7	35.4	21.4	221.2	25,398.5	12,884.8	6,944.9	4,357.0	2,419.0	1,461.0	882.4	4,688.0
256	Kinabatangan	14,101.7	2,820.1	862.7	675.4	672.8	675.3	1,148.9	787.8	1,142.6	1,468.5	1,388.6	1,249.3	1,909.0	1,233.4
176	Kiso	5,419.1	652.3	218.1	352.4	848.3	836.4	896.9	961.9	750.1	967.8	780.0	608.0	420.8	691.1
158	Kitakami	9,652.4	690.1	172.5	1,357.6	1,082.6	802.0	551.9	581.1	590.1	631.9	775.8	828.3	528.7	716.1
150	Kizilirmak	77,873.6	199.1	836.5	1,201.8	2,397.1	1,578.7	804.4	537.4	393.4	241.0	132.0	83.4	128.5	711.1
139	Klamath	40,040.1	3,010.5	3,574.1	3,546.4	3,046.4	2,001.6	1,051.5	698.7	455.6	276.0	146.5	394.9	1,495.2	1,641.5
28	Kobuk	30,242.4	211.1	31.8	19.2	11.6	2,063.7	1,100.6	619.7	373.5	338.6	159.6	96.4	58.2	423.7
60	Kokemaenjoki	26,615.9	236.7	1.6	1.5	3,662.2	1,092.9	616.1	371.5	225.5	147.2	152.9	501.9	154.8	597.1
13	Kolyma	652,850.5	3,721.1	160.9	97.2	58.8	5,342.4	25,441.2	35,941.4	16,207.8	10,517.4	5,575.9	3,367.8	2,034.2	9,038.8
274	Kouilou	60,000.0	4,007.3	2,400.3	3,950.9	5,306.6	3,068.4	1,426.7	862.7	522.5	316.7	191.7	1,130.5	3,001.1	2,182.1
34	Kovda	10,227.6	30.6	0.0	0.0	0.0	881.9	280.3	152.3	92.5	71.2	78.0	33.2	20.1	136.7
219	Krishna	269,869.0	4,249.9	610.7	1,255.2	1,355.6	1,400.2	3,256.5	16,600.9	15,185.1	11,795.9	6,710.3	4,670.2	3,427.8	5,876.5
131	Kuban	58,935.7	1,008.9	1,275.5	1,664.7	2,123.9	1,943.7	1,594.5	1,393.7	826.2	621.9	471.0	528.0	626.0	1,173.2
145	Kura	182,283.3	559.4	187.2	708.9	3,035.7	4,111.4	2,768.8	1,901.8	1,396.6	913.7	766.2	692.7	413.7	1,454.7
51	Kuskokwim	118,114.0	1,269.5	1.4	0.9	0.5	16,620.3	7,998.3	5,470.1	5,473.3	5,291.7	2,398.6	1,372.5	829.0	3,893.8
55	Kymijoki	33,623.1	186.3	1.0	1.0	4,570.7	1,319.9	758.1	456.3	277.2	216.4	308.4	311.4	122.5	710.7
44	Lagarfljot	3,285.3	112.9	0.0	0.0	22.3	1,190.7	742.9	324.1	231.6	257.6	309.8	128.1	74.0	282.9
356	Lake Chad	2,391,218.9	6,870.6	135.4	144.7	180.2	263.0	1,227.0	7,989.1	36,416.6	27,951.2	14,050.2	7,408.8	4,490.9	8,927.3
395	Lake Mar Chiquita	154,330.1	490.2	594.7	1,026.2	504.2	261.0	162.6	112.7	87.1	80.9	85.5	95.2	137.7	303.2
402	Lake Taymur	138,782.9	723.7	0.0	0.0	0.0	0.0	14,735.9	6,387.8	3,298.4	2,513.9	1,300.5	785.5	474.4	2,518.3
399	Lake Titicaca	107,215.3	7,124.5	5,963.3	4,974.8	2,951.8	1,893.3	1,052.7	658.3	552.1	469.3	688.1	678.7	2,594.8	2,466.8
396	Lake Turkana	181,536.0	2,082.5	9.0	100.4	1,988.3	3,300.7	4,415.2	6,718.5	7,827.8	6,549.2	4,135.2	2,501.6	1,394.5	3,418.6
400	Lake Vattern	11,336.5	109.1	0.5	645.9	535.8	246.0	136.0	81.3	60.0	40.2	82.5	196.1	85.7	184.9

390	Leichhardt	33,399.2	31.9	34.8	11.9	7.2	4.4	2.6	1.6	1.0	0.6	0.4	0.2	0.1	8.1
228	Lempa	18,088.5	888.1	2.9	7.6	10.3	11.7	547.6	1,582.1	1,968.2	3,042.2	2,295.6	976.3	585.2	993.2
7	Lena	2,425,551.1	15,771.8	655.1	396.4	361.9	87,091.9	124,907.8	84,650.4	63,046.2	53,636.1	23,839.3	14,390.4	8,692.2	39,786.6
133	Liao He	194,436.5	678.2	20.4	220.6	1,657.1	2,266.7	2,171.0	2,661.2	3,988.1	2,606.5	1,383.3	782.0	454.6	1,574.1
339	Limari	11,780.3	118.1	101.5	40.8	21.1	12.6	27.9	18.7	17.6	14.6	15.5	11.4	31.8	36.0
320	Limpopo	415,623.1	1,880.1	3,058.9	2,803.2	1,433.2	767.0	501.7	359.2	334.0	330.6	246.8	159.6	308.4	1,015.2
108	Little Mecatina	17,902.9	444.5	0.0	0.0	0.0	6,156.3	2,178.5	1,505.7	1,078.9	1,004.0	1,263.0	482.6	291.5	1,200.4
319	Loa	50,206.4	0.3	0.4	0.3	0.3	0.3	0.3	0.3	0.3	0.3	0.3	0.3	0.3	0.3
125	Loire	115,943.6	5,691.9	3,966.2	3,912.1	3,192.9	2,280.7	1,475.0	937.1	710.2	546.1	804.6	1,839.3	3,262.7	2,384.9
381	Lorentz	4,299.3	493.1	445.0	515.7	475.1	346.5	253.8	280.8	253.8	345.8	199.3	277.0	262.3	345.7
144	Luan He	71,071.5	174.3	47.2	147.9	279.9	408.1	214.3	964.0	1,219.0	725.2	357.8	190.0	115.8	403.6
23	Lule	25,127.6	313.1	1.0	0.6	1,119.1	3,851.3	3,784.3	1,895.4	1,112.2	951.7	683.0	335.9	202.9	1,187.5
371	Lurio	61,172.2	4,604.7	6,185.0	6,089.2	2,522.2	1,435.2	866.8	523.5	316.3	191.1	115.4	69.7	468.0	1,948.9
310	Macarthur	19,673.6	0.4	1.1	55.6	14.5	8.8	5.3	3.2	1.9	1.2	0.7	0.4	0.3	7.8
19	Mackenzie	1,752,001.5	5,637.8	86.9	53.5	9,063.7	79,403.5	79,172.3	46,002.0	24,877.4	15,732.9	10,447.5	5,767.0	3,484.1	23,310.7
224	Mae Klong	28,004.2	1,554.4	20.9	34.2	33.5	1,076.4	3,580.0	5,254.6	5,831.8	5,582.5	3,551.6	1,715.1	1,031.1	2,438.9
235	Magdalena	261,204.9	27,118.0	3,452.3	6,430.2	14,175.3	21,211.1	18,633.1	15,055.6	15,479.8	18,291.2	31,846.5	31,789.4	20,916.3	18,699.9
210	Mahanadi (Mahahadi)	135,061.1	3,751.2	108.6	158.0	151.1	158.2	157.6	5,209.0	21,461.7	14,877.9	6,698.9	4,119.4	2,571.8	4,951.9
205	Mahi	36,237.7	884.2	157.9	246.0	223.2	137.5	37.8	2,640.9	4,426.7	3,152.3	1,389.2	865.8	573.5	1,227.9
307	Majes	18,612.1	893.9	994.8	875.9	436.0	232.5	139.9	84.7	52.7	34.4	31.7	29.3	363.0	347.4
380	Mamberamo	75,416.0	15,446.4	9,453.9	12,610.4	11,467.8	9,723.6	7,842.7	8,385.4	7,964.6	8,697.3	6,547.6	7,173.3	8,630.9	9,495.3
324	Mangoky	43,141.1	1,857.9	2,203.9	1,852.5	898.9	518.2	330.4	220.7	136.7	83.0	50.1	56.6	333.7	711.9
106	Manicouagan (Rivière)	54,205.4	1,252.8	0.3	0.2	524.6	8,666.6	6,924.6	3,691.1	2,983.3	3,220.2	3,479.9	1,358.9	820.8	2,743.6
333	Maputo	30,937.8	924.9	719.0	618.4	324.5	179.7	114.6	75.2	58.1	46.2	33.2	112.3	467.1	306.1
257	Maroni	65,944.9	3,849.5	4,583.6	5,326.3	7,656.9	11,080.5	10,145.5	7,092.8	4,342.7	2,242.1	1,349.5	815.1	565.6	4,920.8
187	Mekong	787,256.9	30,683.5	405.8	584.3	939.8	5,846.3	34,283.6	85,153.9	107,075.0	107,961.6	64,294.1	35,434.1	20,497.8	41,096.7
136	Merrimack	12,645.1	381.1	8.4	3,157.8	1,789.4	1,035.9	634.7	377.8	233.8	212.7	361.4	783.0	252.8	769.1
372	Messalo	24,810.9	941.3	1,818.6	2,197.0	1,098.8	548.0	330.8	199.8	120.7	72.9	44.1	26.6	16.1	617.9
39	Mezen	76,715.3	372.9	4.5	2.7	3,916.9	10,521.7	3,855.7	2,081.0	1,250.6	799.5	867.4	386.3	233.3	2,024.4

continues

DATA TABLE 8 *continued*

Natural runoff (Mm³/month)

Basin ID	Basin Name	Area (km²)	Jan	Feb	Mar	Apr	May	Jun	Jul	Aug	Sep	Oct	Nov	Dec	Average
378	Min Jiang	60,039.7	1,710.0	2,261.2	6,444.7	6,263.4	9,721.2	10,643.9	4,850.2	3,725.5	2,882.6	2,049.4	1,340.9	880.7	4,397.8
267	Mira	13,264.8	2,073.0	1,259.4	1,291.0	1,430.0	2,032.3	1,959.3	1,248.7	1,229.1	1,355.3	1,165.3	1,283.2	912.8	1,436.6
122	Mississippi	3,196,605.4	79,924.0	66,571.6	111,936.2	102,147.6	83,552.9	56,877.9	36,359.7	27,329.8	18,097.4	12,199.9	20,013.0	38,932.9	54,495.2
306	Mitchell (N. Au)	71,725.2	1,080.2	6,217.0	5,876.8	2,616.6	1,435.1	859.9	521.2	317.3	194.5	119.8	72.3	43.0	1,612.8
159	Mogami	6,853.1	969.3	827.6	1,039.4	774.9	556.2	376.7	392.9	356.7	430.5	519.9	740.8	1,023.9	667.4
242	Mono	23,899.0	122.1	0.3	14.5	50.7	126.1	330.7	356.7	319.3	472.0	289.3	132.5	80.1	191.2
116	Moose (Trib. Hudson Bay)	105,615.2	1,000.2	0.6	0.6	11,334.0	9,979.0	4,300.8	2,527.6	1,594.4	2,136.5	3,024.6	1,085.8	656.0	3,136.7
313	Mucuri	16,732.2	1,331.8	412.5	315.0	254.2	167.2	113.9	88.6	50.0	28.9	21.6	212.3	1,046.2	336.8
15	Muonio	37,346.5	143.9	13.1	8.0	213.9	2,005.5	1,110.4	750.1	388.4	294.9	187.3	101.6	61.4	439.9
331	Murray	1,059,507.7	2,868.3	1,379.7	1,501.2	1,110.5	1,279.6	2,153.4	2,511.6	3,164.9	3,371.2	3,337.6	2,314.5	2,000.4	2,249.4
37	Nadym	54,624.7	383.2	0.2	0.1	0.1	4,504.7	7,346.2	2,354.4	1,461.2	1,502.8	687.7	415.4	250.9	1,575.6
167	Naktong	23,325.2	508.2	274.7	755.4	983.6	697.6	959.3	1,987.4	1,879.2	1,673.3	882.4	545.1	344.9	957.6
208	Narmada	95,818.2	2,407.1	430.8	900.5	1,131.7	1,138.1	445.1	7,352.7	11,663.1	8,745.2	3,757.6	2,295.3	1,586.4	3,487.8
70	Narva	58,147.0	710.7	1.2	1.2	6,018.5	1,937.0	1,079.4	639.7	420.0	372.9	731.8	1,459.3	466.4	1,153.2
81	Nass	21,211.7	1,054.0	0.0	810.7	3,441.7	6,020.4	4,634.1	2,047.3	1,515.8	1,839.7	2,607.5	1,420.3	691.1	2,173.5
109	Natashquan (Rivière)	16,948.2	291.0	0.0	0.0	156.2	3,726.8	1,620.1	1,208.4	808.1	691.2	792.3	315.9	190.8	816.7
345	Negro (Argentina)	130,062.1	2,461.3	98.8	347.1	1,009.8	4,025.3	5,859.2	6,210.5	6,075.1	5,030.4	4,368.0	3,162.3	1,740.6	3,365.7
340	Negro (Uruguay)	70,756.4	1,301.6	86.9	492.1	1,499.0	2,274.1	3,311.9	3,222.8	3,290.1	3,379.1	2,823.2	1,553.8	846.5	2,006.8
83	Nelson	1,099,380.3	1,698.4	40.6	35.8	29,311.7	20,864.5	15,143.9	8,074.5	5,014.9	4,579.4	4,110.0	1,799.8	1,090.5	7,647.0
87	Neman	97,299.9	740.2	3.4	2,540.2	8,813.4	3,141.2	1,745.1	1,037.9	641.8	406.1	467.3	1,597.2	486.5	1,801.7
56	Neva	223,309.5	1,195.2	10.6	8.8	32,169.2	9,560.0	5,468.9	3,272.4	1,984.2	1,703.3	2,586.4	1,686.6	773.9	5,034.9

207	Niger	2,117,888.7	21,778.0	99.0	256.2	1,297.8	4,833.0	14,794.5	38,348.0	80,634.7	90,704.8	47,626.3	23,829.5	14,278.6	28,206.7
194	Nile	3,078,088.1	20,724.2	2,310.3	6,660.0	16,300.5	18,779.0	21,757.3	48,384.2	80,541.4	66,624.1	36,444.6	22,470.9	14,322.2	29,609.9
50	Nizhny Vyg (Soroka)	31,334.1	206.5	0.1	0.1	5,451.2	1,622.9	926.7	553.6	334.4	420.4	626.2	224.2	135.5	875.1
20	Noatak	32,319.5	156.5	0.8	0.5	0.3	401.9	1,980.3	1,099.7	624.3	629.7	275.6	166.5	100.6	453.1
48	Northern Dvina (Severnaya Dvina)	323,573.1	1,021.4	35.1	22.2	44,567.2	19,378.9	9,541.8	5,592.3	3,376.1	2,058.6	2,097.6	971.8	587.9	7,437.6
114	Nottaway	118,709.0	2,188.3	0.2	0.2	13,451.3	15,987.7	7,640.1	5,188.1	4,271.3	5,083.6	6,291.3	2,438.0	1,434.9	5,331.3
264	Ntem	33,526.9	2,055.6	8.3	442.8	1,623.3	2,595.7	1,808.1	791.1	470.6	1,558.6	4,430.8	3,192.8	1,428.3	1,700.5
197	Nueces	43,877.9	4.2	8.4	21.4	28.3	35.8	41.5	66.6	49.3	24.4	10.5	5.4	3.8	25.0
67	Nushagak	29,513.6	517.3	0.0	0.0	1,308.4	4,936.8	1,770.2	1,042.6	1,293.2	1,548.5	1,382.8	561.6	339.2	1,225.1
275	Nyanga	12,369.1	1,159.8	691.5	941.1	1,044.4	523.6	267.4	161.5	97.5	58.9	35.6	692.8	805.1	539.9
261	Nyong	34,626.2	1,269.1	0.1	386.2	1,153.1	1,817.1	1,504.5	698.1	677.8	2,434.6	3,389.9	1,699.9	832.3	1,321.9
25	Ob	2,701,040.7	6,930.1	198.1	135.9	80,842.2	148,619.4	68,228.3	36,839.8	23,540.4	18,275.7	14,074.8	6,865.0	4,161.9	34,059.3
365	Ocona	16,063.9	539.4	582.1	517.1	245.1	134.9	80.9	48.8	30.6	19.9	49.2	71.2	235.0	212.9
100	Oder	116,536.3	1,032.5	1,852.6	5,896.3	3,319.4	2,160.7	1,462.2	947.3	682.0	505.5	478.7	627.0	1,046.2	1,667.5
49	Oelfusa	5,678.3	378.5	0.0	323.1	1,726.6	690.3	685.2	565.6	398.0	471.2	658.7	434.6	320.1	554.3
265	Ogooué	222,662.7	22,726.7	7,569.4	15,733.8	20,776.5	19,147.2	7,950.2	4,618.0	2,791.3	2,505.1	8,651.3	23,592.1	17,443.6	12,792.1
357	Okavango	705,055.7	4,075.2	6,488.8	8,619.1	3,971.7	2,041.0	1,233.1	745.9	452.1	274.8	167.1	100.7	882.0	2,421.0
2	Olenek	208,522.0	1,231.0	159.8	96.5	58.3	658.3	16,431.9	5,510.3	3,070.3	1,870.4	1,106.6	668.3	403.7	2,605.5
8	Omoloy	38,871.3	26.9	0.7	0.4	0.2	0.2	426.1	277.1	128.9	73.6	43.7	26.4	15.9	85.0
53	Onega	65,894.0	224.9	2.8	1.8	10,633.8	3,402.5	1,902.0	1,125.9	680.1	436.9	556.4	234.8	141.1	1,611.9
326	Orange	972,388.4	1,857.2	2,095.4	2,246.2	1,402.3	807.0	489.5	342.5	319.8	311.7	362.3	534.8	884.1	971.1
308	Ord	55,686.1	0.0	3.9	1.5	2.7	4.2	5.2	6.4	7.5	8.1	6.6	3.6	0.2	4.2
237	Orinoco	952,173.4	73,559.9	9,908.6	16,110.2	48,197.5	96,502.6	137,370.7	156,922.8	139,130.4	112,036.3	103,445.6	77,389.5	46,830.9	84,783.7
239	Oueme	59,842.6	458.8	1.0	1.1	6.9	240.8	976.7	1,265.1	1,320.9	1,893.3	1,037.9	499.0	301.2	666.9
43	Oulujoki	30,554.5	230.9	1.1	0.8	5,398.3	1,702.8	939.7	562.7	373.1	416.6	710.7	247.3	149.5	894.5
262	Oyapock	27,075.7	4,282.0	4,043.5	4,817.3	6,249.8	6,602.9	5,755.6	3,526.6	2,069.9	1,131.7	683.3	412.7	586.7	3,346.8
405	Ozero Sevan	4,765.3	8.6	0.7	3.8	76.9	106.9	55.9	31.2	21.0	12.9	13.1	13.9	5.9	29.2

continues

DATA TABLE 8 *continued*

Natural runoff (Mm³/month)

Basin ID	Basin Name	Area (km²)	Jan	Feb	Mar	Apr	May	Jun	Jul	Aug	Sep	Oct	Nov	Dec	Average
260	Pahang	28,436.7	5,776.0	1,292.5	1,416.5	1,962.1	1,937.4	1,260.3	863.4	823.5	1,395.6	2,714.8	3,612.5	4,175.9	2,269.2
17	Palyavaam	31,112.8	106.3	0.0	0.0	0.0	0.0	1,878.7	1,036.4	526.4	355.2	191.1	115.4	69.7	356.6
279	Pangani	50,364.8	34.7	22.7	34.9	203.1	493.3	239.9	152.6	87.1	59.8	38.7	28.7	27.5	118.6
212	Panuco	82,929.1	1,540.6	95.9	186.5	197.5	137.9	362.8	2,013.6	2,491.4	6,156.6	3,524.7	1,770.8	1,046.5	1,627.1
221	Papaloapan	39,885.1	1,877.2	12.4	16.0	15.8	11.5	477.3	2,373.0	4,411.2	5,780.4	4,426.4	2,194.2	1,261.7	1,904.8
322	Paraiba Do Sul	58,027.2	7,384.0	4,105.7	3,823.9	2,180.1	1,239.1	759.3	470.0	308.5	264.6	590.0	1,633.4	4,400.3	2,263.2
302	Paraná	2,640,486.1	105,979.7	72,853.1	68,346.4	50,136.2	39,988.6	34,546.3	22,504.9	18,037.0	19,546.7	24,410.8	32,882.5	57,492.3	45,560.4
363	Patacua	24,232.4	1,710.9	118.6	53.8	32.7	19.7	70.9	693.4	892.9	1,438.3	2,176.2	1,802.4	1,293.7	858.6
190	Pearl	22,423.0	1,984.2	2,012.8	2,131.6	1,713.4	1,094.2	598.0	392.6	256.9	162.9	99.8	148.7	752.9	945.7
22	Pechora	312,763.3	2,459.1	31.2	19.2	1,449.0	58,305.9	38,794.9	16,270.2	9,630.5	7,855.7	4,450.1	2,542.8	1,536.1	11,945.4
171	Pee Dee	46,531.3	2,824.5	2,381.4	2,346.1	1,609.2	944.6	589.6	563.3	455.2	460.3	379.9	545.7	1,384.2	1,207.0
226	Penner	54,976.4	568.7	34.4	52.1	45.1	43.8	41.5	158.5	151.2	182.9	360.5	1,017.5	485.3	261.8
129	Penobscot	21,168.9	655.1	0.6	482.1	5,554.7	1,878.0	1,224.9	733.4	451.5	421.6	684.5	1,327.0	429.8	1,153.6
128	Po	73,066.6	4,276.6	2,000.0	3,536.1	5,530.3	6,397.4	4,452.6	2,941.3	2,394.4	2,314.0	2,947.8	3,482.3	2,857.2	3,594.2
33	Ponoy	13,186.0	127.2	0.0	0.0	0.0	1,585.1	438.1	255.8	178.3	268.5	394.1	138.1	83.4	289.1
376	Popigay	48,954.2	84.9	0.8	0.5	0.3	0.2	2,052.3	754.1	410.5	251.2	146.9	88.7	53.6	320.3
156	Potomac	32,380.6	1,355.9	1,163.5	1,705.2	1,374.1	948.5	624.9	370.6	255.1	188.4	213.6	392.6	752.3	778.7
250	Pra	23,479.8	378.6	3.2	102.6	282.2	657.8	1,316.6	691.0	327.9	645.8	1,039.5	483.0	247.6	514.6
31	Pur	111,351.3	740.7	0.3	0.3	0.3	7,186.7	15,161.5	4,615.2	2,795.1	2,896.5	1,331.2	804.1	485.8	3,001.5
286	Purari	32,139.9	7,089.0	4,200.2	5,204.7	5,335.2	4,459.8	3,535.0	2,952.7	3,064.9	3,853.1	3,615.9	3,635.9	4,215.8	4,263.5
375	Pyasina	63,888.8	470.7	4.9	3.1	2.0	1.3	8,582.4	2,828.9	1,874.0	1,803.8	814.2	491.9	297.2	1,431.2
38	Quoich	28,217.6	41.6					1,216.3	349.6	199.5	128.2	74.8	45.2	27.3	173.5
263	Rajang	49,943.5	20,497.2	8,513.0	9,657.7	10,211.2	9,935.6	7,769.6	6,882.1	6,855.9	9,242.6	11,129.0	11,855.9	12,276.9	10,402.2
344	Rapel	15,689.5	1,119.4	178.1	113.7	66.3	510.5	1,373.7	1,356.5	1,185.4	876.8	701.4	412.7	681.5	714.7
110	Rhine	190,522.1	13,179.1	7,657.6	8,094.3	9,602.3	8,013.0	6,055.1	4,779.6	4,183.7	3,971.9	4,514.6	6,546.6	8,363.7	7,080.1
126	Rhone	97,485.2	7,316.8	3,365.1	5,895.8	6,325.3	5,588.7	4,446.6	2,690.2	2,212.7	2,277.1	3,477.1	5,232.4	5,154.7	4,498.5

278	Rio Acarau	14,472.9	22.3	31.8	810.4	1,076.9	646.7	301.3	179.6	110.0	67.9	42.1	26.2	16.4	277.6
266	Rio Araguari	33,771.5	4,727.0	5,377.5	6,935.9	8,461.5	8,118.0	7,123.8	4,272.9	2,486.6	1,376.3	831.0	501.9	398.5	4,217.6
272	Rio Capim	54,888.3	1,537.9	5,606.4	8,799.8	7,926.1	5,924.9	3,782.3	2,616.7	1,569.4	878.6	525.8	317.7	205.5	3,307.6
298	Rio De Contas	56,526.5	297.0	120.1	258.0	556.9	550.1	447.9	457.9	292.5	179.2	110.9	165.1	224.8	305.0
314	Rio Doce	86,085.9	12,563.7	5,298.4	4,242.2	2,461.5	1,301.4	784.3	483.1	302.0	187.0	117.8	2,334.5	9,099.0	3,264.6
271	Rio Gurupi	32,335.3	491.7	2,048.7	4,647.2	4,513.5	3,426.0	2,151.5	1,412.8	764.6	450.8	272.3	164.5	103.1	1,703.9
277	Rio Itapecuru	52,672.0	84.3	1,224.8	3,260.0	3,253.3	1,651.3	883.6	520.8	314.3	190.4	115.4	69.7	42.3	967.5
295	Rio Itapicuru	37,593.4	56.5	2.3	2.1	7.2	99.2	263.0	645.3	362.8	169.7	102.3	62.2	38.4	150.9
337	Rio Jacuí	70,798.0	5,005.1	2,492.1	2,901.2	3,920.9	4,897.0	5,793.0	5,339.7	5,145.1	5,743.1	5,136.4	3,405.9	2,794.2	4,381.1
285	Rio Jaguaribe	72,804.3	64.1	11.7	1,619.3	2,744.3	1,562.6	861.9	515.6	319.8	213.3	142.0	92.4	60.4	683.9
282	Rio Mearim	56,687.0	356.4	2,713.7	5,445.6	4,657.5	2,321.9	1,253.1	745.7	450.8	272.8	165.1	99.9	61.7	1,545.3
296	Rio Paraguacu	54,607.1	274.1	185.7	300.7	705.2	1,494.5	1,165.9	1,466.7	878.0	455.0	267.7	194.3	163.9	629.3
288	Rio Paraíba	18,969.1	37.7	2.0	3.5	89.5	253.8	514.2	396.9	208.3	115.4	73.2	46.6	29.6	147.6
276	Rio Parnaíba	336,584.2	2,039.6	3,746.9	7,333.0	6,991.8	3,124.0	1,723.6	1,047.6	641.5	394.0	242.4	157.1	558.8	2,333.4
280	Rio Pindare	39,112.0	253.4	2,304.7	4,716.9	4,525.5	2,674.7	1,415.4	814.8	488.9	295.4	178.6	108.0	65.3	1,486.8
304	Rio Prado	31,673.7	696.0	283.3	357.3	417.6	243.4	187.0	168.8	106.4	58.7	36.6	137.5	549.1	270.1
329	Rio Ribeira Do Iguape	25,697.5	2,174.8	1,657.5	1,425.9	887.3	735.4	782.1	538.1	442.8	601.1	860.6	773.8	987.3	988.9
294	Rio Vaza-Barris	15,314.2	17.1	1.4	1.7	1.6	61.4	151.1	177.6	102.7	49.8	30.8	18.8	11.8	52.2
166	Roanoke	26,801.0	1,844.9	1,467.9	1,460.0	1,085.8	725.1	448.3	316.0	238.3	184.2	159.1	332.5	853.9	759.7
141	Rogue	14,526.6	1,090.2	1,088.4	909.0	858.5	622.6	302.4	190.5	120.0	73.0	41.5	212.2	514.3	501.9
299	Roper	79,907.5	434.9	1,537.1	1,682.2	599.2	358.4	216.7	131.2	79.7	48.5	29.6	17.7	10.5	428.8
373	Rovuma	151,948.6	9,106.2	15,265.1	18,092.1	9,319.0	4,605.3	2,779.5	1,678.9	1,014.2	612.7	370.2	223.6	516.4	5,298.6
284	Rufiji	204,638.8	1,836.2	3,827.1	7,683.8	8,804.4	4,027.1	2,129.0	1,285.4	779.2	473.6	288.1	176.3	418.8	2,644.1
115	Rupert	16,063.4	311.4	0.0	0.0	794.7	3,201.0	1,189.1	796.1	677.5	758.7	894.3	338.0	204.2	763.8
287	Ruvu	17,541.2	121.7	122.6	361.9	952.4	722.0	287.6	174.0	106.2	64.9	39.6	23.9	73.4	254.2
191	Sabine	25,611.8	1,109.9	1,334.8	1,286.5	1,307.3	936.9	456.1	276.7	165.9	101.9	62.1	86.8	295.2	618.3
148	Sacramento	77,208.9	5,375.8	6,127.3	6,249.1	5,067.8	3,136.9	2,369.0	2,064.4	1,708.0	1,170.4	511.3	343.5	1,439.3	2,963.6
112	Saguenay (Rivière)	91,366.9	1,984.2	1.5	1.5	11,245.1	11,251.7	8,137.7	5,058.6	4,100.4	4,665.6	5,748.5	2,173.4	1,301.6	4,639.2
127	Saint John	55,151.8	1,543.0	1.9	1.9	13,364.2	4,478.5	3,060.9	1,929.0	1,260.9	1,384.3	2,179.7	2,871.4	1,012.4	2,757.4

continues

DATA TABLE 8 *continued*

									Natural runoff (Mm³/month)					
Basin ID	Basin Name	Area (km²)	Jan	Feb	Mar	Apr	May	Jun	Jul	Aug	Sep	Oct	Nov	Dec	Average
153	Sakarya	62,482.7	348.5	1,016.9	1,126.0	888.3	508.1	345.2	273.7	242.4	156.9	68.9	26.5	75.1	423.0
342	Salado	266,263.9	1,011.5	24.7	73.2	787.5	1,228.1	1,234.8	1,105.6	959.3	1,261.2	1,567.9	1,418.0	767.6	953.3
170	Salinas	12,654.6	11.8	34.5	58.5	36.9	32.6	47.2	66.4	68.2	42.9	9.6	2.7	1.6	34.4
354	Salween	258,475.2	8,366.0	95.5	592.3	1,511.4	2,846.6	12,055.4	24,649.1	32,068.1	27,586.6	18,159.4	9,737.1	5,522.3	11,932.5
198	San Antonio	10,952.4	12.3	13.2	14.4	23.4	25.7	19.4	25.1	18.3	9.9	5.7	4.2	8.8	15.0
162	San Joaquin	34,365.6	707.1	829.3	1,285.7	1,335.3	1,136.9	1,098.6	1,267.9	1,193.7	818.4	322.2	74.9	102.3	847.7
232	San Juan	41,659.4	5,223.3	533.6	282.7	261.5	1,036.0	3,952.8	4,778.7	4,793.9	6,119.6	7,350.0	4,839.9	3,921.9	3,591.2
258	San Juan (Columbia-Pacific)	13,898.0	6,064.5	2,203.5	2,616.1	3,456.3	4,124.0	3,912.7	3,904.9	3,971.2	4,103.4	4,529.7	4,529.1	3,905.9	3,943.4
203	San Pedro	29,358.8	192.9	2.6	4.4	6.2	7.1	3.4	137.0	732.2	987.8	369.0	206.3	130.3	231.6
249	Sanaga	134,252.0	4,812.3	3.0	236.5	2,004.2	4,068.5	5,840.0	7,929.0	9,616.8	13,725.9	13,358.0	5,383.1	3,156.1	5,844.5
292	Santa	11,882.5	455.4	563.1	650.9	399.4	210.8	125.2	78.5	57.1	44.0	79.5	128.6	132.8	243.8
352	Santa Cruz	30,599.9	1,652.2	385.0	560.0	1,181.8	1,590.7	1,850.7	1,577.2	2,464.8	3,221.0	3,226.9	1,605.8	1,042.5	1,696.6
175	Santee	40,035.3	804.8	794.4	752.0	496.5	298.4	193.4	174.2	169.8	140.6	124.1	150.6	395.6	374.5
211	Santiago	126,222.3	681.2	116.1	243.6	265.3	168.1	205.1	1,027.8	2,651.8	3,112.7	1,413.6	791.4	492.4	930.8
290	São Francisco	628,629.1	27,716.3	15,251.7	13,981.8	7,511.4	4,656.8	5,028.4	4,947.7	3,343.7	1,665.4	1,268.7	4,929.8	18,730.9	9,086.0
240	Sassandra	68,097.5	2,261.5	1.4	3.2	129.1	309.3	2,250.3	3,322.8	4,638.5	8,322.9	5,416.1	2,603.4	1,487.3	2,562.2
181	Savannah	27,740.4	1,501.0	1,570.4	1,658.6	1,058.3	583.9	360.8	272.1	190.7	187.5	165.1	238.5	572.9	696.6
315	Save	114,957.8	2,203.3	3,356.1	2,440.7	1,065.6	627.7	386.7	242.8	173.8	130.9	85.0	41.4	348.7	925.2
68	Seal	53,439.9	126.4	1.6	1.0	0.6	3,534.7	1,434.9	728.5	432.3	415.7	241.7	130.6	78.9	593.9
179	Sebou	36,201.3	1,058.8	1,081.1	1,248.4	934.9	533.2	346.0	284.6	201.0	141.9	76.4	197.1	563.6	555.6
119	Seine	74,227.9	3,426.9	2,491.0	2,183.7	1,663.9	1,005.1	594.9	402.4	304.5	202.7	234.2	696.1	1,700.9	1,242.2
220	Senegal	436,981.1	1,629.1	10.0	15.2	11.7	12.0	447.8	3,062.5	8,464.3	6,863.4	3,270.9	1,792.3	1,076.7	2,221.3
281	Sepik	81,119.7	15,375.9	9,555.7	12,665.7	12,180.9	9,397.5	7,423.9	6,795.9	6,812.5	7,707.7	8,100.4	8,084.8	9,182.5	9,440.3
89	Severn (Trib. Hudson Bay)	98,590.5	593.2	0.0	0.0	969.8	7,964.8	3,325.3	1,813.8	1,158.6	1,429.9	1,689.2	644.0	388.9	1,664.8

241	Shebelle	805,077.0	1,126.2	49.5	54.4	2,532.2	1,755.0	1,025.3	1,594.0	2,005.1	1,944.7	1,681.9	1,610.5	791.5	1,347.5
165	Shinano, Chikuma	11,158.8	1,100.6	807.7	800.2	1,837.3	1,864.2	1,290.9	1,242.1	1,006.9	1,172.1	1,174.8	1,150.4	1,044.7	1,207.7
215	Sittang	34,265.3	2,254.3	3.3	6.3	6.6	16.9	3,852.3	7,852.6	10,043.8	8,332.3	4,900.8	2,492.4	1,478.4	3,436.7
123	Skagit	7,961.0	951.3	342.8	1,452.8	1,747.8	975.0	491.1	287.0	173.6	108.5	409.0	751.3	600.5	690.9
82	Skeena	42,944.4	1,078.2	0.2	1,041.9	4,491.6	8,819.6	7,160.8	3,082.5	2,010.7	1,937.6	2,313.5	1,518.6	707.1	2,846.9
86	Skjern A	2,817.6	351.5	132.9	121.1	89.0	50.8	28.9	18.4	22.4	70.4	165.2	192.5	228.6	122.6
289	Solo (Benga-wan Solo)	15,146.1	2,855.5	2,415.3	2,433.6	1,722.1	949.3	531.7	332.1	222.4	171.0	106.2	250.0	1,099.8	1,090.8
348	South Esk	10,842.5	186.5	8.9	10.5	32.6	76.3	208.6	392.9	471.5	403.2	358.1	217.0	135.7	208.5
121	Southern Bug	60,121.0	38.8	7.9	1,840.9	967.0	519.4	311.3	205.6	138.9	77.0	43.3	27.5	18.1	349.6
79	Spey	2,942.2	478.1	203.5	174.9	136.5	98.8	58.1	43.8	52.5	81.3	157.4	246.1	292.1	168.6
130	St. Croix	4,638.6	170.5	0.1	0.1	1,441.8	475.5	301.5	171.5	101.8	88.6	166.9	352.7	111.8	281.9
196	St. Johns	22,489.2	379.1	187.4	226.0	126.3	72.7	67.3	331.1	475.2	796.6	731.2	335.3	225.9	329.5
117	St. Lawrence	1,055,021.5	13,835.1	351.1	29,605.6	132,375.7	51,230.9	31,947.4	19,602.9	13,031.9	15,304.9	21,038.1	22,898.9	9,715.3	30,078.1
71	Stikine	51,147.5	1,826.9	603.3	508.0	1,656.9	9,699.8	11,933.0	5,766.5	3,728.6	3,815.5	3,379.2	1,948.6	1,370.5	3,853.1
384	Sungai Kajan	33,171.8	10,769.4	4,092.4	5,488.6	6,619.3	6,870.7	5,795.5	5,104.8	4,917.8	6,431.1	6,999.8	7,819.9	6,829.4	6,478.2
386	Sungai Kapuas	84,902.3	29,491.3	13,356.0	15,227.7	15,516.0	13,513.3	10,291.5	7,393.4	6,608.1	8,936.3	13,775.6	17,293.6	17,622.4	14,085.4
385	Sungai Mahakam	75,822.7	18,606.0	7,876.5	10,134.8	13,756.5	12,600.1	9,668.4	6,831.4	5,649.9	5,951.6	7,402.7	11,152.8	12,616.4	10,187.3
398	Suriname	24,867.7	2,186.8	1,915.0	2,088.7	2,922.6	4,729.6	5,013.8	3,973.9	2,420.7	1,213.0	729.1	440.4	526.6	2,346.7
54	Susitna	49,470.3	1,270.7	2.2	1.4	3,654.1	8,319.0	8,905.7	5,269.7	4,095.0	4,842.3	2,738.5	1,370.5	827.8	3,441.4
143	Susquehanna	69,080.1	2,091.9	1,240.2	8,814.7	5,917.2	3,887.6	2,447.9	1,522.7	1,002.4	867.5	1,337.0	2,499.4	1,658.8	2,774.0
192	Suwannee	26,400.9	1,043.0	1,262.4	1,316.2	797.7	415.7	300.4	613.1	756.9	714.7	385.8	211.7	527.5	695.4
42	Svarta, Skagafiroi	3,429.6	54.9	0.0	0.0	123.6	363.8	392.6	148.0	89.0	76.1	152.2	68.9	36.0	127.9
69	Taku	17,967.6	430.1	0.0	0.0	806.8	2,508.0	2,296.9	1,033.2	783.7	1,027.7	1,263.6	467.0	282.0	908.3
269	Tana	95,715.0	290.5	14.7	38.9	512.9	706.1	333.2	183.6	113.4	69.9	103.6	260.5	305.1	244.4
9	Tana (NO, FI)	14,518.1	71.8	0.0	0.0	0.0	2,851.0	778.3	457.0	276.1	217.4	149.0	77.9	47.1	410.5
247	Tano	15,656.1	321.4	0.2	32.5	186.6	547.1	1,356.6	700.0	337.1	480.9	810.9	421.8	218.4	451.1
214	Tapti	65,096.3	1,243.3	168.4	306.4	365.6	395.0	149.1	3,365.4	5,169.5	5,115.9	2,124.1	1,333.7	886.1	1,718.5

continues

DATA TABLE 8 *continued*

			Natural runoff (Mm³/month)												
Basin ID	Basin Name	Area (km²)	Jan	Feb	Mar	Apr	May	Jun	Jul	Aug	Sep	Oct	Nov	Dec	Average
358	Tarim	1,051,731.4	241.9	77.0	269.6	593.6	1,657.0	2,845.3	3,326.7	2,360.3	1,351.4	540.9	295.2	164.1	1,143.6
27	Taz	152,086.0	989.9	6.0	3.6	2.2	12,922.1	23,163.4	7,321.6	4,370.1	3,193.2	1,738.0	1,049.7	634.0	4,616.2
152	Tejo	70,351.7	2,713.3	2,202.2	3,261.5	2,062.1	1,343.3	834.4	721.8	559.4	292.8	124.7	136.5	1,182.1	1,286.2
174	Tenryu	5,769.0	584.8	291.0	511.3	839.0	748.4	861.4	816.6	724.3	1,026.6	888.4	635.8	438.9	697.2
113	Thames	12,358.9	726.2	447.7	361.1	237.6	136.9	78.4	49.7	32.4	21.8	18.4	131.8	395.2	219.8
45	Thelon	238,839.0	478.4	0.1	0.0	0.0	6,190.9	10,749.7	3,370.2	2,026.8	1,686.6	859.5	519.1	313.5	2,182.9
47	Thjorsa	7,527.1	422.1	33.1	91.6	1,123.0	1,723.6	1,322.4	662.2	584.0	668.7	840.8	478.5	288.8	686.6
62	Thlewiaza	64,399.6	91.9	10.3	6.2	3.7	1,834.1	666.0	339.5	204.2	198.6	94.1	56.8	34.3	295.0
155	Tigris and Euphrates	832,578.6	15,514.6	16,609.9	22,140.0	25,647.8	18,735.0	10,281.7	7,368.3	5,738.1	3,544.0	2,291.3	3,575.7	7,623.3	11,589.1
273	Tocantins	774,718.3	82,937.0	65,826.8	71,926.3	45,721.6	24,110.4	14,567.5	8,930.8	5,585.4	3,923.1	3,979.0	16,536.2	44,543.4	32,382.3
169	Tone	15,739.3	936.2	260.8	691.8	1,057.7	983.0	973.7	1,058.8	1,240.9	1,561.5	1,541.0	979.8	683.0	997.3
225	Tranh (Nr Thu Bon)	9,459.9	2,007.0	45.2	24.6	16.4	71.3	290.8	883.2	1,278.8	1,770.6	2,724.1	2,418.9	1,615.2	1,095.5
102	Trent	9,052.9	691.4	368.2	303.2	213.3	137.6	80.5	56.5	51.0	49.0	67.1	198.0	402.5	218.2
189	Trinity (Texas)	46,168.2	588.6	822.5	766.2	973.4	839.8	365.7	234.6	152.1	98.0	66.6	52.7	255.9	434.7
317	Tsiribihina	61,991.9	9,631.8	9,804.7	9,049.7	4,154.5	2,354.3	1,436.8	893.2	545.1	328.9	198.3	299.2	2,841.4	3,461.5
335	Tugela	30,079.3	758.7	786.8	752.8	372.3	204.5	126.4	86.3	74.3	66.9	64.9	86.0	376.1	313.0
14	Tuloma	26,057.7	94.6	11.6	7.1	4.4	1,732.1	547.3	300.0	180.4	119.2	121.3	55.2	33.5	267.2
91	Tweed	4,770.9	575.8	239.8	225.1	158.2	110.7	72.3	54.4	61.3	91.1	172.0	311.6	366.0	203.2
362	Ulua	26,392.0	1,735.8	77.7	42.4	31.4	19.3	510.5	1,791.6	1,997.4	3,135.9	2,748.5	1,916.8	1,262.0	1,272.4
97	Ural	339,084.2	324.2	5.7	165.2	10,759.3	3,817.4	2,162.0	1,470.8	948.2	527.8	322.1	587.8	214.5	1,775.4
334	Uruguay	265,504.6	15,702.5	5,633.7	8,112.2	13,990.4	16,949.4	19,158.1	16,344.5	15,620.3	18,876.7	20,160.7	12,864.3	9,587.2	14,416.7
383	Uwimbu	29,373.9	8,952.7	5,395.2	6,379.0	5,992.2	6,010.0	5,432.3	5,263.6	5,126.4	5,387.6	4,552.8	4,357.4	5,401.7	5,687.6
61	Vaenern-Goeta	51,791.5	1,198.3	3.2	2,259.8	5,377.8	2,297.5	1,379.5	819.2	707.0	791.8	1,475.7	1,616.9	1,037.2	1,580.3
404	Van Golu	17,736.8	117.9	0.6	127.0	1,118.2	1,378.3	489.2	294.9	184.2	112.7	115.2	230.6	77.6	353.9

ID	Name												
32	Varzuga	8,182.2	47.2	0.0	0.0	0.0	683.9	188.7	110.1	66.6	112.4	140.1	51.2	30.9	119.3
223	Verde	18,342.8	378.7	2.9	6.5	6.9	4.5	10.1	365.4	850.4	1,648.0	893.8	414.1	250.8	402.7
305	Victoria	78,462.4	0.1	20.4	14.0	5.5	3.3	2.0	1.2	0.7	0.4	0.3	0.2	0.1	4.0
64	Volga	1,408,278.9	3,091.7	147.2	747.6	140,132.1	51,088.9	28,581.8	17,230.0	10,747.1	6,887.4	6,486.8	3,165.6	1,902.0	22,517.4
227	Volta	414,004.1	2,522.1	7.7	80.1	291.8	828.4	2,616.6	3,570.8	8,402.5	11,296.3	5,427.0	2,744.7	1,654.9	3,286.9
52	Vuoksi	62,707.4	334.8	1.2	1.2	10,528.9	2,968.3	1,728.3	1,039.2	643.3	560.3	960.4	387.0	220.0	1,614.4
347	Waikato	15,358.7	1,209.9	436.0	382.1	601.3	1,271.4	1,642.5	1,617.1	1,551.8	1,388.8	1,356.2	1,080.3	781.6	1,109.9
103	Weser	43,140.2	3,511.3	2,008.3	1,895.7	1,397.8	902.7	617.5	504.3	484.0	501.8	819.6	1,601.5	2,294.1	1,378.2
76	Western Dvina (Daugava)	89,340.3	772.7	2.1	2.1	9,180.8	2,977.1	1,713.6	1,016.3	619.6	425.3	849.3	1,548.2	507.4	1,634.5
94	Winisk	106,470.3	890.2	2.1	1.3	3,470.5	10,190.2	4,433.5	2,449.4	1,479.2	2,350.9	2,486.7	957.9	578.6	2,440.9
98	Wisla	193,764.0	947.0	37.8	12,812.8	6,810.9	4,190.9	2,577.5	1,781.3	1,243.2	858.0	779.6	1,157.5	847.9	2,837.0
201	Xi Jiang	362,894.3	8,372.9	2,682.6	4,932.3	10,762.2	25,796.8	44,596.4	41,263.7	42,397.2	25,138.3	14,767.9	8,958.4	5,447.9	19,593.0
4	Yana	233,479.4	896.1	72.8	44.0	26.6	159.2	7,535.9	6,820.7	3,807.3	1,832.6	1,106.8	668.5	403.8	1,947.9
177	Yangtze (Chang Jiang)	1,745,094.4	40,746.3	20,673.1	48,340.7	86,583.8	114,954.9	136,756.4	119,291.0	113,382.9	101,024.6	67,829.6	41,278.9	25,519.8	76,365.2
193	Yaqui	76,181.7	15.8	30.2	63.2	72.9	36.0	24.2	49.0	61.5	44.6	29.0	14.8	15.8	38.1
5	Yenisei	2,558,237.3	16,135.2	742.2	452.4	7,752.7	162,714.9	162,047.8	97,190.0	64,240.9	48,091.7	24,440.8	14,455.9	8,735.1	50,583.3
178	Yodo	8,424.2	1,063.4	536.0	654.4	713.4	583.0	743.8	675.6	510.1	768.0	820.6	652.7	632.7	696.1
151	Yongding He	214,406.5	174.7	403.8	1,472.8	2,529.4	2,501.8	1,275.4	1,729.3	2,397.6	1,071.8	442.9	185.1	137.8	1,193.5
16	Yukon	829,632.3	4,850.3	252.5	152.7	943.7	53,166.5	48,766.9	29,776.4	16,607.0	12,906.1	7,648.1	4,212.2	2,544.3	15,152.2
293	Zambezi	1,388,572.2	73,279.9	82,019.0	68,514.8	32,327.8	18,268.1	10,936.3	6,663.6	4,203.7	2,618.2	1,621.8	1,305.4	21,500.8	26,938.3

DATA TABLE 9

Area Equipped for Irrigation Actually Irrigated

Description

The area equipped for agricultural irrigation that is actually irrigated is listed here by major country, in thousands of hectares (1,000 ha), for different five-year periods. The periods are 1988–1992, 1993–1997, 1998–2002, 2003–2007, and 2008–2012. Within each of these periods, data are typically available for only a single year, listed in the column labeled "Actual Year of Data." Thus, for Afghanistan (the first entry), three data points are available, for the years 1990, 2002, and 2011, as noted.

The area actually irrigated is typically less than the area capable of being irrigated (or equipped for irrigation), since actual water availability, climatic conditions, economic factors, and decisions about cropping and fallowing patterns vary annually.

Countries not included on this list either have no irrigation or no reported data—their absence from the list is not an indication of which condition applies.

Limitations

Data are collected by the Food and Agriculture Organization of the United Nations (FAO), usually through country reports and information about crop production. Different methods of collecting these data are used in different regions. All of the data are made available through the FAO's AQUASTAT database, as noted below. For some countries, data are available for only a single year.

Sources

UN FAO. 2013. AQUASTAT database. Rome, Italy: Food and Agriculture Organization of the United Nations. http://www.fao.org (accessed February 6, 2013).

DATA TABLE 9 Area Equipped for Irrigation Actually Irrigated (1,000 ha)

Country	Actual Year of Data	Period 1988–1992	Actual Year of Data	Period 1993–1997	Actual Year of Data	Period 1998–2002	Actual Year of Data	Period 2003–2007	Actual Year of Data	Period 2008–2012
Afghanistan	1990	2,660			2002	1,732			2011	1,896
Albania					2001	453	2007	107	2009	205
Algeria	1992	366								
Angola			1996	35						
Antigua and Barbuda										
Argentina					2002	1,356				
Armenia			1995	173			2006	176		
Australia							2007	43		
Austria			1995	46						
Azerbaijan	1992	1,340			2002	1,352	2004	1,356	2010	1,358
Bahamas										
Bahrain			1994	3						
Bangladesh	1989	2,738								
Barbados										
Belarus							2007	6	2011	31
Belgium										
Belize					2002	9				
Benin			1995	27	1999	128	2007	28	2008	17
Bhutan										
Bolivia (Plurinational State of)										
Bosnia and Herzegovina										
Botswana	1992	1								
Brazil	1989	2,700					2006	4,454		
Brunei Darussalam			1995	1						
Bulgaria			1997	43	2002	31	2007	73	2011	41
Burkina Faso	1992	24								

continues

DATA TABLE 9 continued

Country	Actual Year of Data	Period 1988–1992	Actual Year of Data	Period 1993–1997	Actual Year of Data	Period 1998–2002	Actual Year of Data	Period 2003–2007	Actual Year of Data	Period 2008–2012
Burundi										
Cambodia			1993	242	2001	255	2006	317		
Cameroon										
Canada										
Cape Verde	1988	2	1997	2						
Central African Republic										
Chad	1988	13			2002	26				
Chile							2007	1,094		
China	1992	44,418	1995	45,017			2006	54,219		
Colombia										
Comoros										
Congo										
Costa Rica			1997	103			2006	102		
Côte d'Ivoire			1994	67						
Cuba			1997	738						
Cyprus			1994	33			2007	31		
Czech Republic							2007	20		
Democratic Republic of the Congo			1995	8	2000	7				
Denmark							2007	254		
Djibouti	1989	0			1999	0				
Dominican Republic					1999	270				
Ecuador			1997	613	2000	620				
Egypt	1990	2,585	1993	3,246	2002	3,422				
El Salvador										
Equatorial Guinea										
Eritrea			1993	13						
Estonia			1995	2					2010	0

Ethiopia						
Finland					2007	1,512
France						
Gabon						
Gambia	1991	1	1996	274		
Georgia					2006	235
Germany	1990	933	1994	4	2000	28
Ghana			1997	1,163	2000	1,161	2007	1,280
Greece			1997	130			2003	312
Guatemala					2001	95		
Guinea			1996	23				
Guinea-Bissau								
Guyana	1991	65						
Haiti	1991	55					2006	61
Honduras					2000	67	2007	88
Hungary	1991	47,430			2001	58,130	2005	59,206	2008	62,286
India								
Indonesia	1990	1,935	1993	7,264			2006	6,423
Iran (Islamic Republic of)					1998	1		
Iraq								
Ireland	1990	2,697	1993	2,649	2000	2,453	2006	182
Israel			1997	25			2007	2,666
Italy	1990	3,012	1993	3,128			2006	2,600
Jamaica								
Japan	1992	67					2003	97
Jordan			1994	5			2003	7
Kazakhstan			1994	1,077			2005	1,021	2011	1,021
Kenya	1990	127	1995	138	2000	296	2005	271

continues

DATA TABLE 9 *continued*

Country	Actual Year of Data	Period 1988–1992	Actual Year of Data	Period 1993–1997	Actual Year of Data	Period 1998–2002	Actual Year of Data	Period 2003–2007	Actual Year of Data	Period 2008–2012
Latvia							2007	1		
Lebanon					1998	90				
Lesotho			1994	0	1999	0				
Liberia										
Libya	1990	240			2000	316				
Lithuania							2007	1		
Madagascar	1992	895			2000	550				
Malawi	1992	27								
Mali					2000	176				
Malta	1990	1					2007	3		
Mauritania			1994	27			2004	23		
Mauritius					2002	21				
Mexico	1989	5,150	1997	5,505			2007	5,439		
Mongolia			1993	63						
Morocco					2000	1,407	2004	1,448		
Mozambique			1995	45	2001	40				
Myanmar	1990	1,056	1995	1,582			2004	2,110		
Namibia	1992	6								
Netherlands							2007	202		
New Zealand					2002	384	2007	509		
Nicaragua			1997	51						
Niger					2000	65	2005	66		
Nigeria	1991	172			2000	222	2004	219		
Norway							2007	55		
Oman			1993	62						
Panama			1997	28	2000	27				
Peru			1994	1,109						

Country						
Philippines	1993	1,471				
Poland	1990	631				
Portugal					2007	72
Puerto Rico					2007	422
Qatar	1990	6	1993	8	2007	16
Republic of Moldova	1992	225	1994	210	2004	6
Romania			1994	4,095	2007	32
Russian Federation					2007	173
Rwanda	1992	1,608	1999	1,191	2008	8
Saudi Arabia	1997	69				
Senegal					2011	34
Serbia			2000	111	2003	0
Seychelles			2000	65	2007	39
Slovakia			2000	1,498		
Somalia						
South Africa					2011	1,601
Spain					2009	3,093
Sri Lanka	1995	1,197	2000	800	2006	463
Sudan and South Sudan			1998	51		
Suriname			2002	45		
Swaziland			2000	1,210		
Sweden					2007	54
Switzerland					2010	36
Syrian Arab Republic	1994	719	2002	29	2008	674
Tajikistan					2007	5,060
Thailand	1996	6				
Timor-Leste	1997	3	2000	393		
Togo	1990	2				
Trinidad and Tobago	1991	322				
Tunisia	1994	3,098			2008	5,280
Turkey					2006	4,320

continues

DATA TABLE 9 *continued*

Country	Actual Year of Data	Period 1988–1992	Actual Year of Data	Period 1993–1997	Actual Year of Data	Period 1998–2002	Actual Year of Data	Period 2003–2007	Actual Year of Data	Period 2008–2012
Turkmenistan			1994	1,744			2006	1,991		
Uganda					1998	6			2010	12
Ukraine			1995	1,845	2002	730	2003	731		
United Arab Emirates			1993	54					2010	76
United Kingdom							2007	138		
United Republic of Tanzania										
United States of America			1997	22,779	2002	22,384	2007	22,905	2008	22,229
Uruguay					1998	181				
Uzbekistan			1994	4,202			2005	3,700		
Venezuela (Bolivarian Republic of)			1998	308						
Vietnam	1990	2,100	1994	2,100			2005	4,585	2008	979
Zambia	1992	46			2002	156				
Zimbabwe					1999	124				

DATA TABLE 10

Overseas Development Assistance for Water Supply and Sanitation, by Donating Country, 2004-2011

Description

The annual overseas development assistance, or official development assistance (ODA), for water supply and sanitation is listed here, by donating country (and European Union institutions), for the years 2004 through 2011. Shown are the total amounts committed, in current US dollars in millions.

"ODA" is the term given to funding that flows from governments or multilateral institutions for the purpose of providing aid to countries. This funding is usually provided for the purpose of promoting economic development and welfare and is "concessional in character and conveys a grant element of at least 25 percent." ODA can take a number of forms, including technical assistance, investment projects, debt forgiveness or rescheduling, equity investments, and other assistance. The Organisation for Economic Co-operation and Development's (OECD's) Development Assistance Committee (DAC) identifies eleven subsectors in the "water supply and sanitation" category:

Water resources policy and administrative management

Water resources protection

Water supply and sanitation—large systems

Water supply—large systems

Sanitation—large systems

Basic drinking water supply and basic sanitation

Basic drinking water supply

Basic sanitation

River development

Waste management and disposal

Education and training in water supply and sanitation

Data table 11 provides a breakdown of ODA for water supply and sanitation by these subsectors.

Limitations

ODA does not constitute all the funding that flows to developing countries, such as other public sector or private sector flows; hence, these numbers do not reflect all funding for water projects. In addition, not all ODA described as related to "water supply and sanitation" may flow to purely water projects.

Source

Organisation for Economic Co-operation and Development (OECD). 2013. Query Wizard for International Development Statistics. http://stats.oecd.org/qwids/.

DATA TABLE 10 Overseas Development Assistance for Water Supply and Sanitation, by Donating Country, 2004–2011

Country/Agency	Current (Millions US$)							
	2004	2005	2006	2007	2008	2009	2010	2011
Australia	30.19	34.50	6.43	15.01	13.90	44.68	171.38	230.33
Austria	20.08	16.77	20.40	24.15	36.58	23.16	17.99	26.57
Belgium	24.70	37.79	55.16	47.98	102.96	61.33	48.93	69.84
Canada	79.76	41.09	18.54	24.33	46.96	73.85	19.47	44.99
Denmark	245.43	98.90	144.62	31.66	19.01	164.95	140.81	88.07
Finland	6.05	43.34	44.17	30.59	51.18	54.32	99.35	122.20
France	175.50	114.85	237.42	391.23	359.65	802.77	495.60	323.02
Germany	435.75	382.26	497.14	593.96	906.44	820.47	750.81	1,040.71
Greece	1.39	0.52	1.03	2.78	0.76	2.99	0.12	—
Iceland	—	—	—	—	—	—	—	0.39
Ireland	19.15	16.83	16.91	22.80	28.15	17.47	9.97	10.70
Italy	5.89	69.00	54.50	59.69	163.41	55.94	63.29	13.92
Japan	709.47	2,128.66	1,250.89	1,930.07	1,668.24	2,786.04	1,933.26	1,711.10
Korea	78.31	101.56	80.76	74.52	269.70	70.71	283.15	172.13
Luxembourg	14.38	12.43	10.27	12.95	19.02	22.29	20.30	21.54
Netherlands	146.50	191.69	455.15	359.27	373.08	196.51	69.28	129.53
New Zealand	1.76	2.16	2.77	3.58	3.21	2.58	1.53	5.83
Norway	32.26	42.53	28.02	46.60	44.65	40.66	50.86	17.76
Portugal	2.17	2.48	0.63	1.57	0.32	0.42	0.99	0.63
Spain	78.47	58.02	69.03	121.45	577.07	577.77	308.44	155.11
Sweden	43.77	68.73	64.32	46.72	76.90	75.37	45.31	66.30
Switzerland	31.57	35.77	31.74	34.31	49.28	43.06	49.26	331.51
United Kingdom	29.46	44.26	51.16	104.88	160.66	114.32	156.94	170.45
United States	954.69	1,023.27	817.79	432.14	846.78	461.91	431.34	465.40
EU Institutions	413.17	687.16	726.55	490.50	170.28	528.11	678.11	603.69
Kuwait (KFAED)	—	—	—	—	—	—	43.27	40.22
Multilateral, Total	1,476.04	1,346.75	1,807.02	1,955.88	1,676.51	1,682.27	2,326.53	3,141.53
All Donors Total	5,055.91	6,601.32	6,492.42	6,858.62	7,664.70	8,723.95	8,216.29	9,003.47

DATA TABLE 11

Overseas Development Assistance for Water Supply and Sanitation, by Subsector, 2007-2011 (Total of All Donating Countries)

Description

The allocation of total overseas development assistance (ODA) commitments for water supply and sanitation by all donating countries is provided in this table, from 2007 to 2011, in millions of dollars. The Organisation for Economic Co-operation and Development's (OECD's) Development Assistance Committee (DAC) recently expanded the categories from previous years, and it now identifies the following eleven subsectors in the "water supply and sanitation" category:

- Water resources policy and administrative management
- Water resources protection
- Water supply and sanitation—large systems
- Water supply—large systems
- Sanitation—large systems
- Basic drinking water supply and basic sanitation
- Basic drinking water supply
- Basic sanitation
- River development
- Waste management and disposal
- Education and training in water supply and sanitation

By far the largest expenditure is for large-scale water supply and sanitation projects; basic water supply and sanitation projects generally receive less than half as much funding as large-scale projects. One concern of many analysts is that much ODA is directed to serving wealthier populations, or to improving services to populations that are already at least partly served by existing systems, and hence does not contribute to meeting the water- and sanitation-related Millennium Development Goals.

Limitations

ODA does not constitute all of the funding that flows to developing countries, such as other public sector or private sector flows; hence, these numbers do not reflect all funding for water projects. These data do not readily compare with those published in the previous volume of *The World's Water* as a result of changes in categorizations by the OECD. The development statistics provided by the OECD do not clearly differentiate where there might be double-counting of assistance in multiple categories.

Source

Organisation for Economic Co-operation and Development (OECD). 2013. Query Wizard for International Development Statistics. http://stats.oecd.org/qwids/.

DATA TABLE 11 Overseas Development Assistance for Water Supply and Sanitation, by Subsector, 2007 to 2011 (Total of All Donating Countries)

	Amount Committed in Millions of Dollars				
Water Sector(s)	2007	2008	2009	2010	2011
Water resources policy and administrative management	762.04	927.19	795.70	980.29	842.79
Water resources protection	128.65	247.58	172.47	69.95	193.92
Water supply and sanitation—large systems	3,980.24	4,555.75	4,163.96	2,146.08	1,417.39
Water supply—large systems	—	—	—	1,333.36	1,504.10
Sanitation—large systems	—	—	—	639.25	1,506.55
Basic drinking water supply and basic sanitation	978.62	1,170.90	3,027.27	962.34	1,067.98
Basic drinking water supply	—	—	—	664.46	402.73
Basic sanitation	—	—	—	409.87	160.46
River development	282.18	411.89	219.74	234.88	377.62
Waste management and disposal	414.94	155.04	255.07	270.26	373.46
Education and training in water supply and sanitation	16.17	67.02	16.70	69.35	76.82
Water Supply and Sanitation Total	6,562.82	7,535.37	8,650.92	7,780.09	7,923.82

DATA TABLE 12

Per Capita Water Footprint of National Consumption, by Country, 1996-2005

Description

Conventional estimates of national water use have historically been restricted to statistics on water withdrawals within national boundaries. In recent years, a new approach has been developed and applied to calculate water "footprints," which include water use (typically withdrawals and consumption, estimated separately), the use of "green water" and "grey water," and information about water used in other countries to produce imported products. Quantifying these national water footprints is an evolving field.

The data here include the per capita water footprint for internal and external water consumption by country, measured in cubic meters per person per year (m^3/cap/yr), averaged over the decade 1996–2005. Data on population assumptions are also shown. A distinction is made between "green" and "blue" water. In addition, the authors include the grey water footprint in the estimation of the water footprint in each sector, an estimate of the footprint of farm animal products, and details on the water footprints of national imports and exports.

Freshwater footprints are reported in terms of water volumes consumed (evaporated or incorporated into a product) or polluted per unit of time. A water footprint has three components: green, blue, and grey. The "blue water" footprint refers to use of surface water and groundwater. The "green water" footprint is the volume of rainwater consumed, which is particularly relevant in agricultural production. The "grey water" footprint is a measure of freshwater pollution and is defined as the volume of freshwater required to assimilate a load of pollutants.

Limitations

While the concept of the water footprint is an extremely valuable one, the process for computing footprints continues to evolve, and data sets continue to improve. For example, water withdrawal numbers are often not accurately or consistently reported (see, e.g., the description of data table 2). Estimates of agricultural water use depend on a mix of implicit and explicit assumptions about actual irrigation levels, climatic data for certain crop types, and irrigation technologies. Some consumer categories are not included or accurately estimated. Estimates of the water footprint of imported goods are produced using global averages because of constraints on information of the location of exported goods. The grey water footprint estimates are based on a limited set of pollutants—leaching and runoff of nitrogen fertilizers—and exclude the potential effects of other agricultural chemicals, including pesticides.

See the original source for far more detail on the fundamental assumptions and limitations of the footprint approach.

Source

Mekonnen, M. M., and A. Y. Hoekstra. 2011. National Water Footprint Accounts: The Green, Blue and Grey Water Footprint of Production and Consumption. Value of Water Research Report Series No. 50. Delft, Netherlands: UNESCO-IHE Institute for Water Education. http://www.waterfootprint.org/Reports/Report50-NationalWaterFootprints Vol1.pdf.

DATA TABLE 12 Per Capita Water Footprint of National Consumption, by Country, 1996–2005 (m³/cap/yr)

Water Footprint of National Consumption

Country	Total Population (Thousands)	Internal			External			Total			Total
		Green	Blue	Grey	Green	Blue	Grey	Green	Blue	Grey	
Albania	3,084.9	602.2	106.4	218.9	492.7	78.7	56.4	1,094.9	185.0	275.3	1,555.2
Algeria	30,767.4	634.6	93.8	43.6	700.6	51.2	65.7	1,335.2	145.0	109.4	1,589.5
Angola	14,609.8	798.4	14.2	8.2	108.4	14.7	14.3	906.8	28.9	22.6	958.3
Antigua and Barbuda	77.4	391.2	5.3	44.0	828.0	100.9	198.5	1,219.2	106.2	242.6	1,567.9
Argentina	37,060.0	1,288.3	104.3	152.8	35.0	5.6	20.7	1,323.4	109.9	173.5	1,606.8
Armenia	3,089.5	516.3	134.6	237.1	450.2	41.8	58.8	966.5	176.4	295.8	1,438.8
Australia	19,320.0	1,716.9	194.9	130.7	136.3	21.5	114.3	1,853.3	216.3	245.0	2,314.6
Austria	8,057.3	396.3	16.2	92.9	738.1	82.5	271.5	1,134.4	98.7	364.4	1,597.5
Azerbaijan	8,155.3	510.0	176.6	198.3	310.6	26.0	23.3	820.6	202.6	221.6	1,244.9
Bahamas	306.3	200.8	—	—	1,348.9	167.4	415.8	1,549.7	167.4	415.8	2,132.9
Bangladesh	141,966.7	477.9	57.3	99.8	98.1	22.5	13.0	576.0	79.8	112.8	768.6
Barbados	253.0	553.0	19.9	255.5	941.3	117.3	202.9	1,494.2	137.2	458.4	2,089.9
Belarus	10,030.1	1,091.2	19.3	223.4	215.8	44.2	123.8	1,306.9	63.5	347.2	1,717.6
Belgium	10,401.0	34.5	21.7	148.9	1,181.2	120.5	380.7	1,215.7	142.2	529.6	1,887.5
Belize	254.6	1,154.7	39.2	585.9	120.9	38.5	71.0	1,275.7	77.8	656.9	2,010.4
Benin	6,820.0	1,020.2	1.9	14.5	67.3	20.3	11.9	1,087.5	22.1	26.5	1,136.1
Bermuda	63.0	6.8	2.0	4.8	1,706.9	230.2	1,044.7	1,713.7	232.1	1,049.5	2,995.3
Bolivia	8,408.8	3,063.9	47.4	29.2	295.9	15.3	16.2	3,359.9	62.7	45.3	3,467.9
Bosnia and Herzegovina	3,631.7	659.0	4.7	33.2	408.2	28.8	122.2	1,067.3	33.4	155.5	1,256.1
Botswana	1,725.8	1,202.0	13.9	54.9	561.7	179.8	38.9	1,763.7	193.7	93.8	2,051.2
Brazil	175,307.6	1,644.8	57.7	137.5	159.6	12.7	14.8	1,804.4	70.4	152.3	2,027.1
Brunei Darussalam	336.8	25.5	0.0	0.0	2,796.4	334.1	264.5	2,821.9	334.1	264.5	3,420.6
Bulgaria	7,988.3	1,478.9	35.1	352.7	330.3	47.6	52.2	1,809.2	82.6	405.0	2,296.8
Burkina Faso	11,950.2	1,600.1	16.7	28.5	37.4	14.6	5.7	1,637.5	31.2	34.2	1,702.9

continues

DATA TABLE 12 *continued*

Water Footprint of National Consumption

Country	Total Population (Thousands)	Internal			External			Total			
		Green	Blue	Grey	Green	Blue	Grey	Green	Blue	Grey	Total
Burundi	6,651.7	677.3	15.0	9.8	13.2	1.3	2.0	690.5	16.3	11.8	718.6
Cambodia	12,834.1	955.7	54.1	6.5	47.7	8.3	5.6	1,003.5	62.4	12.1	1,077.9
Cameroon	16,080.9	1,144.8	10.5	21.9	37.5	21.2	9.1	1,182.3	31.7	31.0	1,245.0
Canada	30,889.0	1,345.1	79.7	426.1	254.9	73.8	153.6	1,600.0	153.6	579.7	2,333.3
Cape Verde	442.6	599.3	22.9	4.1	495.1	65.0	57.8	1,094.4	87.9	61.8	1,244.1
Central African Republic	3,772.3	1,147.8	19.3	5.3	14.4	2.9	3.3	1,162.3	22.2	8.6	1,193.1
Chad	8,614.3	1,425.0	20.3	4.2	7.7	2.6	1.9	1,432.7	23.0	6.1	1,461.7
Chile	15,492.1	451.5	150.8	166.5	327.3	20.0	38.5	778.8	170.8	205.0	1,154.6
China	1,277,208.3	614.3	106.3	243.7	86.2	11.2	9.4	700.5	117.5	253.1	1,071.1
Colombia	40,093.6	894.5	48.0	158.8	227.0	18.7	28.0	1,121.5	66.6	186.8	1,374.9
Comoros	710.2	1,107.4	2.9	6.7	102.1	84.9	21.1	1,209.5	87.9	27.8	1,325.2
Congo, Democratic Republic	52,052.6	529.8	2.1	4.0	10.2	3.4	2.7	540.0	5.4	6.6	552.1
Congo, Republic	3,096.2	506.6	7.5	10.6	191.3	47.9	22.3	697.9	55.4	32.9	786.2
Costa Rica	3,962.9	687.7	60.5	277.0	348.5	49.8	66.9	1,036.2	110.3	343.9	1,490.4
Côte d'Ivoire	17,422.3	1,158.5	8.2	31.1	45.5	36.6	15.6	1,204.0	44.8	46.7	1,295.5
Croatia	4,520.8	892.7	1.1	143.0	462.3	36.6	152.1	1,355.0	37.7	295.0	1,687.8
Cuba	11,090.9	1,189.6	93.1	179.6	175.2	12.6	37.0	1,364.9	105.7	216.6	1,687.2
Cyprus	790.9	364.4	229.3	91.0	1,317.9	120.0	262.8	1,682.3	349.3	353.8	2,385.4
Czech Republic	10,228.3	818.6	15.2	203.2	418.2	64.5	130.9	1,236.7	79.7	334.2	1,650.6
Denmark	5,339.5	512.7	13.3	97.1	709.4	82.2	219.9	1,222.1	95.5	317.0	1,634.6
Dominica	68.1	2,169.8	—	23.9	361.8	83.5	77.4	2,531.7	83.5	101.3	2,716.5
Dominican Republic	8,900.7	740.4	105.3	105.9	366.6	25.2	57.9	1,107.0	130.5	163.8	1,401.3
East Timor	868.1	1,748.0	43.0	—	48.1	10.6	13.6	1,796.1	53.6	13.6	1,863.2
Ecuador	12,367.9	1,388.6	136.3	244.8	199.2	11.8	26.0	1,587.9	148.0	270.9	2,006.8

Egypt	70,957.6	150.8	506.6	301.3	333.8	20.1	28.5	484.5	526.7	329.8	1,341.0
El Salvador	5,944.5	578.9	13.6	115.4	249.3	37.4	37.9	828.2	51.0	153.3	1,032.5
Eritrea	3,786.8	533.1	2.7	8.1	494.1	27.1	23.8	1,027.2	29.8	31.9	1,088.9
Estonia	1,372.7	712.5	8.0	110.4	601.9	148.5	138.4	1,314.4	156.5	248.8	1,719.7
Ethiopia	66,511.7	1,107.7	23.3	9.8	20.4	1.8	4.2	1,128.1	25.1	14.0	1,167.2
Fiji Islands	803.6	1,195.0	2.8	51.0	415.6	59.1	43.8	1,610.6	61.9	94.7	1,767.3
Finland	5,181.9	651.0	18.0	79.4	406.2	61.5	197.7	1,057.2	79.5	277.2	1,413.9
France	59,435.5	735.3	48.4	156.4	618.1	86.8	140.6	1,353.4	135.2	297.0	1,785.7
French Polynesia	237.9	907.3	2.6	6.2	630.6	146.9	182.8	1,537.9	149.5	189.0	1,876.4
Gabon	1,245.3	1,020.4	15.0	52.9	271.1	55.2	36.2	1,291.5	70.2	89.1	1,450.8
Gambia	1,324.9	541.9	1.8	12.2	246.4	52.4	32.2	788.2	54.2	44.4	886.9
Georgia	4,719.9	1,219.7	97.0	282.3	587.4	35.3	45.7	1,807.1	132.3	328.0	2,267.5
Germany	82,138.8	311.4	18.0	116.2	741.9	67.2	171.6	1,053.3	85.1	287.8	1,426.3
Ghana	19,784.5	1,091.4	3.3	19.3	57.7	21.0	14.2	1,149.0	24.3	33.5	1,206.9
Greece	10,937.4	890.9	218.6	142.1	761.1	107.4	218.0	1,652.0	326.0	360.1	2,338.1
Grenada	101.5	1,267.2	2.2	—	636.9	94.5	153.3	1,904.1	96.7	153.3	2,154.0
Guatemala	11,412.4	713.0	15.0	68.8	136.1	18.8	31.0	849.0	33.8	99.8	982.7
Guinea	8,463.3	1,492.1	6.6	16.3	52.7	26.4	11.4	1,544.9	33.0	27.7	1,605.6
Guinea-Bissau	1,325.7	1,028.8	65.0	19.7	31.2	40.2	12.7	1,060.0	105.2	32.4	1,197.6
Guyana	759.1	1,046.5	101.0	159.6	165.2	23.4	52.7	1,211.7	124.5	212.2	1,548.4
Haiti	8,719.4	780.9	26.4	6.0	141.0	49.9	25.6	921.9	76.3	31.6	1,029.8
Honduras	6,299.4	913.5	19.4	71.4	123.3	28.0	21.9	1,036.7	47.4	93.3	1,177.5
Hungary	10,199.8	1,547.2	24.5	303.5	369.2	41.2	98.4	1,916.3	65.7	401.8	2,383.9
Iceland	282.2	14.5	31.7	195.9	1,468.9	102.3	295.5	1,483.4	133.9	491.4	2,108.7
India	1,051,289.5	685.5	216.7	159.0	22.4	2.2	3.0	707.9	218.9	162.0	1,088.8
Indonesia	206,706.2	854.8	48.8	107.0	77.8	24.3	10.8	932.7	73.1	117.7	1,123.5
Iran	67,157.7	749.4	568.1	215.6	289.5	20.7	23.2	1,038.9	588.8	238.7	1,866.5
Ireland	3,868.6	278.8	9.7	79.9	660.8	99.0	173.2	939.6	108.7	253.2	1,301.4
Israel	6,134.0	207.4	145.2	72.9	1,583.1	108.1	186.1	1,790.5	253.3	259.0	2,302.7
Italy	57,520.5	629.3	66.7	209.5	1,091.2	126.1	180.3	1,720.5	192.7	389.8	2,302.9

continues

DATA TABLE 12 *continued*

Water Footprint of National Consumption

Country	Total Population (Thousands)	Internal			External			Total		
		Green	Blue	Grey	Green	Blue	Grey	Green	Blue	Grey	Total
Jamaica	2,578.4	838.5	23.8	83.9	585.7	66.2	98.0	1,424.2	90.0	181.9	1,696.2
Japan	126,741.3	176.6	29.3	112.9	832.8	61.7	165.8	1,009.4	91.0	278.7	1,379.0
Jordan	4,956.1	112.5	82.0	43.8	1,126.3	167.3	146.2	1,238.8	249.3	189.9	1,678.0
Kazakhstan	15,160.4	1,710.6	355.5	175.7	81.1	17.9	35.6	1,791.7	373.4	211.3	2,376.4
Kenya	31,935.3	875.1	10.9	24.1	155.8	23.6	11.9	1,031.0	34.4	35.9	1,101.3
Kiribati	85.0	2,341.6	—	—	314.7	148.8	54.1	2,656.3	148.8	54.1	2,859.2
Korea, Democratic People's Rep	22,867.1	542.5	70.1	149.6	91.0	17.0	18.2	633.4	87.1	167.9	888.4
Korea, Republic	46,443.0	252.4	25.9	78.6	1,032.7	85.4	154.2	1,285.2	111.3	232.8	1,629.3
Kuwait	2,251.5	6.6	40.9	170.8	1,412.9	251.0	190.2	1,419.5	291.9	361.0	2,072.4
Kyrgyzstan	4,961.6	495.4	249.7	99.1	593.5	27.9	33.8	1,088.9	277.6	132.9	1,499.4
Laos	5,431.9	891.0	67.8	43.2	27.3	4.3	7.9	918.3	72.1	51.1	1,041.4
Latvia	2,368.7	1,036.6	8.6	150.2	289.9	55.5	256.0	1,326.5	64.1	406.2	1,796.8
Lebanon	3,816.5	303.4	186.6	82.3	1,245.0	148.0	146.2	1,548.4	334.7	228.5	2,111.5
Lesotho	1,890.3	1,533.5	4.2	15.5	40.2	33.0	13.1	1,573.7	37.2	28.7	1,639.5
Liberia	2,790.3	1,064.7	2.6	15.4	64.8	33.2	53.8	1,129.5	35.8	69.3	1,234.6
Libya	5,411.9	301.2	283.6	129.0	981.3	227.7	115.6	1,282.5	511.3	244.5	2,038.3
Lithuania	3,501.4	1,048.4	6.7	58.9	218.3	47.7	135.9	1,266.7	54.4	194.8	1,516.0
Luxembourg	450.5	508.9	7.2	110.2	1,432.2	96.3	359.5	1,941.2	103.5	469.6	2,514.3
Macedonia	2,010.5	602.2	49.6	182.4	382.7	43.0	88.0	984.8	92.6	270.4	1,347.8
Madagascar	15,529.4	1,329.4	148.4	38.7	36.2	16.2	7.2	1,365.6	164.6	45.9	1,576.1
Malawi	12,005.8	804.1	13.8	59.9	48.4	3.1	6.9	852.6	17.0	66.8	936.3
Malaysia	23,458.2	1,240.6	56.5	128.6	503.3	96.3	78.1	1,743.9	152.8	206.7	2,103.3
Maldives	273.5	221.8	5.3	15.6	813.5	164.2	140.6	1,035.3	169.5	156.2	1,361.1
Mali	10,701.0	1,769.6	145.6	54.5	54.8	11.1	8.2	1,824.4	156.7	62.7	2,043.8

Malta	391.0		20.1	88.6	1,600.7	186.6	245.7	1,674.8	206.7	334.3	2,215.8
Mauritania	2,648.1	74.1	84.4	62.1	1,036.8	85.0	34.9	2,298.8	169.4	96.9	2,565.1
Mauritius	1,199.6	1,262.0	26.8	153.6	1,238.9	219.2	150.3	1,610.8	246.0	303.9	2,160.6
Mexico	99,810.4	371.9	101.7	195.3	661.1	88.5	91.4	1,501.1	190.2	286.7	1,978.0
Moldova	4,051.3	840.0	105.9	138.3	55.7	14.8	29.0	1,038.9	120.7	167.2	1,326.8
Mongolia	2,409.3	983.2	40.7	87.0	2,207.8	54.6	68.1	3,524.1	95.3	155.2	3,774.6
Morocco	28,961.9	1,316.3	171.1	69.2	424.2	37.7	41.2	1,405.5	208.8	110.4	1,724.8
Mozambique	18,560.9	981.3	8.3	9.9	61.8	9.3	7.2	1,084.4	17.6	17.2	1,119.2
Myanmar	46,626.5	1,022.6	42.2	27.7	18.0	1.3	3.0	1,142.8	43.5	30.7	1,217.0
Namibia	1,839.1	1,124.8	22.9	53.1	339.6	68.4	28.9	1,509.0	91.3	82.0	1,682.2
Nepal	24,707.0	1,169.4	115.1	20.9	94.9	9.9	15.2	1,039.5	125.0	36.1	1,200.6
Netherlands	15,946.9	944.6	12.7	22.9	1,012.3	116.2	258.0	1,055.9	128.9	280.9	1,465.7
New Caledonia	216.8	43.6	0.2	35.8	1,100.4	128.2	179.6	1,651.0	128.4	215.4	1,994.7
New Zealand	3,905.5	550.6	90.3	104.9	720.0	58.4	137.8	1,197.3	148.7	242.7	1,588.8
Nicaragua	5,125.0	477.3	30.2	53.8	104.6	20.2	20.8	787.2	50.4	74.6	912.1
Niger	11,272.1	682.6	64.5	12.6	59.1	22.6	7.9	3,411.0	87.1	20.5	3,518.7
Nigeria	126,648.8	3,351.9	12.3	19.1	43.9	11.6	8.1	1,191.2	23.9	27.2	1,242.3
Norway	4,501.6	1,147.3	21.6	86.3	678.8	100.9	190.5	1,023.5	122.5	276.8	1,422.8
Occupied Palestinian Territory	3,221.2	344.7	120.4	68.1	30.1	6.7	40.5	819.0	127.1	108.6	1,054.6
Pakistan	149,801.8	788.8	406.6	172.3	191.7	15.2	10.0	727.2	421.8	182.2	1,331.3
Panama	2,979.1	535.5	37.2	162.5	311.5	34.6	70.7	1,058.7	71.8	233.2	1,363.7
Paraguay	5,406.5	747.2	41.9	66.6	26.1	12.2	18.3	1,815.2	54.1	84.9	1,954.1
Peru	26,157.7	1,789.1	135.4	97.2	307.8	16.1	29.2	810.2	151.5	126.3	1,088.0
Philippines	78,472.9	502.4	51.9	145.2	133.0	26.2	27.1	1,127.3	78.1	172.2	1,377.7
Poland	38,408.0	994.4	18.6	262.4	232.6	40.2	73.6	1,010.6	58.8	336.0	1,405.4
Portugal	10,278.4	778.0	161.9	146.1	1,157.9	201.3	141.9	1,854.2	363.2	288.1	2,505.5
Romania	22,089.3	696.3	39.4	217.0	161.5	37.3	50.3	1,344.8	76.7	267.3	1,688.8
Russian Federation	146,082.3	1,183.2	55.9	191.7	174.1	39.1	16.1	1,548.9	95.0	207.8	1,851.6
Rwanda	7,735.9	1,374.8	3.0	6.0	24.4	5.4	7.0	799.6	8.4	13.0	821.0
Saint Kitts and Nevis	46.4	775.2	0.2	1.2	706.7	127.9	202.3	1,300.1	128.1	203.5	1,631.6

continues

DATA TABLE 12 continued

Water Footprint of National Consumption

Country	Total Population (Thousands)	Internal			External			Total			
		Green	Blue	Grey	Green	Blue	Grey	Green	Blue	Grey	Total
Saint Lucia	157.5	56.4	8.0	71.4	1,145.2	115.1	162.9	1,201.6	123.2	234.4	1,559.1
Saint Vincent/Grenadines	108.2	878.8	—	—	594.5	96.9	119.1	1,473.3	96.9	119.1	1,689.3
Samoa	176.1	1,811.4	—	0.5	168.3	64.3	37.9	1,979.7	64.3	38.4	2,082.4
São Tomé and Príncipe	141.5	1,459.9	0.7	—	155.8	59.5	52.0	1,615.8	60.2	52.0	1,728.0
Saudi Arabia	21,114.2	126.3	343.4	157.9	1,004.9	104.1	112.7	1,131.2	447.5	270.6	1,849.3
Senegal	10,055.0	757.9	33.2	16.5	264.7	53.9	24.9	1,022.6	87.1	41.3	1,151.1
Serbia and Montenegro	10,729.5	1,224.5	58.5	928.1	107.5	27.9	43.3	1,332.1	86.3	971.3	2,389.7
Seychelles	80.6	657.7	50.8	126.6	1,010.5	155.1	191.0	1,668.1	205.9	317.6	2,191.6
Sierra Leone	4,408.0	1,132.2	11.5	5.5	248.5	22.9	16.9	1,380.7	34.4	22.5	1,437.6
Slovakia	5,377.2	704.2	21.8	141.3	310.0	51.6	106.4	1,014.3	73.4	247.7	1,335.4
Slovenia	1,986.5	373.8	20.0	351.8	971.8	77.2	217.7	1,345.6	97.2	569.5	2,012.4
Solomon Islands	421.9	530.2	—	—	138.8	22.0	31.8	668.9	22.0	31.8	722.7
South Africa	45,183.8	786.8	85.8	106.3	240.5	11.8	24.1	1,027.3	97.6	130.4	1,255.4
Spain	40,840.6	1,004.9	232.2	167.1	797.3	88.9	170.8	1,802.1	321.2	338.0	2,461.3
Sri Lanka	18,883.5	771.5	107.7	78.4	207.7	54.5	35.9	979.3	162.2	114.2	1,255.7
Sudan	35,224.9	1,428.1	199.4	40.4	56.4	4.8	6.7	1,484.5	204.1	47.1	1,735.7
Suriname	470.8	749.1	122.1	155.4	260.3	18.5	41.6	1,009.4	140.6	196.9	1,346.9
Swaziland	1,073.1	491.4	49.1	28.7	655.7	146.4	26.4	1,147.0	195.5	55.0	1,397.5
Sweden	8,911.5	535.5	18.2	129.5	502.7	61.9	180.1	1,038.1	80.0	309.6	1,427.8
Switzerland	7,232.5	183.6	15.5	71.9	851.4	98.8	307.1	1,035.0	114.4	379.0	1,528.4
Syria	16,859.5	1,187.6	348.4	236.4	270.8	38.7	25.3	1,458.5	387.0	261.7	2,107.2
Tajikistan	6,201.4	647.8	458.6	177.7	1,069.1	15.4	18.6	1,716.8	474.0	196.2	2,387.1
Tanzania	34,719.2	913.4	24.6	18.5	52.9	10.0	6.7	966.3	34.6	25.3	1,026.2
Thailand	62,968.7	939.0	140.1	123.7	148.8	25.1	30.7	1,087.8	165.2	154.4	1,407.4

Togo	5,303.9	845.6	2.9	20.7	94.5	15.2	11.3	940.0	18.0	32.0	990.1
Trinidad and Tobago	1,296.6	83.4	17.6	159.0	1,224.9	92.4	138.7	1,308.3	110.0	297.7	1,716.0
Tunisia	9,482.7	1,299.7	130.9	68.7	504.7	137.3	75.7	1,804.4	268.2	144.4	2,217.0
Turkey	66,848.5	933.8	178.6	182.3	216.0	74.2	57.0	1,149.9	252.8	239.3	1,641.9
Turkmenistan	4,544.7	1,239.2	729.3	90.7	185.2	10.3	18.9	1,424.4	739.6	109.6	2,273.6
Uganda	24,961.9	1,019.2	1.0	7.7	35.9	10.0	5.2	1,055.1	11.0	12.9	1,079.0
UK	59,334.4	243.5	8.6	59.3	672.0	84.1	190.5	915.5	92.7	249.8	1,258.1
Ukraine	48,737.6	1,117.6	46.3	305.9	55.9	9.7	39.2	1,173.6	56.0	345.1	1,574.6
United Arab Emirates	3,329.8	412.7	159.3	188.7	1,508.5	411.3	455.5	1,921.2	570.6	644.2	3,136.0
Uruguay	3,307.1	1,567.4	56.4	70.6	388.7	6.5	43.0	1,956.2	62.9	113.6	2,132.6
USA	288,958.2	1,600.4	190.0	476.6	367.9	48.9	158.7	1,968.3	238.9	635.3	2,842.5
Uzbekistan	24,895.6	509.9	359.7	135.7	255.6	3.6	13.9	765.5	363.3	149.5	1,278.3
Vanuatu	193.8	1,264.0	—	—	45.5	75.0	46.6	1,309.5	75.0	46.6	1,431.2
Venezuela	24,640.2	874.6	64.6	184.5	527.0	23.1	36.6	1,401.6	87.7	221.1	1,710.3
Vietnam	79,188.2	637.7	81.0	271.0	52.4	6.8	9.6	690.0	87.8	280.6	1,058.5
Yemen	18,502.1	71.4	127.8	20.1	551.2	90.4	40.4	622.6	218.2	60.5	901.3
Zambia	10,573.6	754.0	30.8	45.7	76.2	8.0	6.5	830.2	38.9	52.2	921.3
Zimbabwe	12,357.4	958.7	63.3	92.7	81.1	6.8	7.7	1,039.7	70.1	100.4	1,210.2
World	6,154,564.2	796.6	117.8	170.5	218.8	35.5	46.0	1,015.4	153.3	216.5	1,385.2

DATA TABLE 13

Per Capita Water Footprint of National Consumption, by Sector and Country, 1996-2005

Description

Like data table 12, this table includes estimates of water footprints by country, but here they are split into sectoral data on agricultural, industrial, and domestic uses, for "blue," "green," and "grey" water, in cubic meters per person per year (m^3/cap/yr).

Conventional estimates of national water use have historically been restricted to statistics on water withdrawals within national boundaries. In recent years, a new approach has been developed and applied to calculate water "footprints," which include water use (typically withdrawals and consumption, estimated separately), the use of "green water" and "grey water," and information about water used in other countries to produce imported products. Quantifying these national water footprints is an evolving field.

The data here include the water footprints for crop production, industrial production, and domestic water supply, with a distinction made between green and blue water. In addition, the authors include the grey water footprint in the estimation of the water footprint in each sector, an estimate of the footprint of farm animal products, and details on the water footprints of national imports and exports. The data are reported as per capita water consumption, measured in cubic meters per person per year and averaged over the decade 1996–2005. Data on population assumptions are also shown.

Freshwater footprints are reported in terms of water volumes consumed (evaporated or incorporated into a product) or polluted per unit of time. A water footprint has three components: green, blue, and grey. The "blue water" footprint refers to use of surface water and groundwater. The "green water" footprint is the volume of rainwater consumed, which is particularly relevant in agricultural production. The "grey water" footprint is a measure of freshwater pollution and is defined as the volume of freshwater required to assimilate a load of pollutants.

Limitations

While the concept of the water footprint is an extremely valuable one, the process for computing footprints continues to evolve, and data sets continue to improve. For example, water withdrawal numbers are often not accurately or consistently reported (see, e.g., the description of data table 2). Estimates of agricultural water use depend on a mix of implicit and explicit assumptions about actual irrigation levels, climatic data for certain crop types, and irrigation technologies. Some consumer categories are not included or accurately estimated. Estimates of the water footprint of imported goods are produced using global averages because of constraints on information of the location of exported goods. The grey water footprint estimates are based on a limited set of pollutants—leaching and runoff of nitrogen fertilizers—and exclude the potential effects of other agricultural chemicals, including pesticides.

See the original source for far more detail on the fundamental assumptions and limitations of the footprint approach.

Source

Mekonnen, M. M., and A. Y. Hoekstra. 2011. National Water Footprint Accounts: The Green, Blue and Grey Water Footprint of Production and Consumption. Value of Water Research Report Series No. 50. Delft, Netherlands: UNESCO-IHE Institute for Water Education. http://www.waterfootprint.org/Reports/Report50-NationalWaterFootprints-Vol1.pdf.

DATA TABLE 13 Per Capita Water Footprint of National Consumption, by Sector and Country, 1996–2005 (m³/cap/yr)

Country	Population (Thousands)	Water Footprint of Consumption of Agricultural Products						Water Footprint of Consumption of Industrial Products				Water Footprint of Domestic Water Consumption	
		Internal			External			Internal		External			
		Green	Blue	Grey	Green	Blue	Grey	Blue	Grey	Blue	Grey	Blue	Grey
Albania	3,084.9	602.2	88.9	36.5	492.7	77.4	39.5	2.6	48.2	1.3	17.0	14.9	134.2
Algeria	30,767.4	634.6	88.8	9.6	700.6	50.6	59.7	0.7	6.4	0.6	6.1	4.3	27.7
Angola	14,609.8	798.4	13.6	2.4	108.4	14.6	13.1	0.1	0.9	0.1	1.2	0.5	4.9
Antigua and Barbuda	77.4	391.2	1.0	—	828.0	87.8	74.8	0.5	9.2	13.1	123.7	3.9	34.9
Argentina	37,060.0	1,288.3	87.9	45.4	35.0	3.9	3.0	3.1	33.9	1.7	17.7	13.2	73.5
Armenia	3,089.5	516.3	105.6	26.7	450.2	40.7	41.0	1.6	19.4	1.1	17.8	27.3	191.0
Australia	19,320.0	1,716.9	172.0	80.9	136.3	12.0	10.1	4.6	10.4	9.4	104.2	18.2	39.5
Austria	8,057.3	396.3	2.0	52.4	738.1	64.2	81.0	5.1	8.0	18.3	190.5	9.2	32.5
Azerbaijan	8,155.3	510.0	160.2	17.5	310.6	25.1	9.1	10.1	132.1	0.9	14.2	6.4	48.6
Bahamas	306.3	200.8	—	—	1,348.9	140.1	136.7	—	—	27.3	279.2	—	—
Bangladesh	141,966.7	477.9	54.4	72.9	98.1	22.4	11.2	0.1	1.4	0.1	1.8	2.8	25.5
Barbados	253.0	553.0	1.4	24.9	941.3	108.3	120.5	6.6	123.8	9.0	82.4	11.9	106.7
Belarus	10,030.1	1,091.2	7.5	199.3	215.8	37.4	10.8	5.3	4.6	6.8	113.1	6.5	19.5
Belgium	10,401.0	34.5	0.1	10.0	1,181.2	92.6	97.3	14.4	112.1	27.9	283.4	7.1	26.9
Belize	254.6	1,154.7	16.3	246.6	120.9	34.2	25.7	19.1	306.5	4.3	45.3	3.9	32.8
Benin	6,820.0	1,020.2	1.0	5.2	67.3	20.0	8.6	0.2	3.9	0.3	3.3	0.6	5.4
Bermuda	63.0	6.8	2.0	4.8	1,706.9	150.1	128.5	—	—	80.1	916.2	—	—
Bolivia	8,408.8	3,063.9	44.8	7.9	295.9	14.8	11.6	0.5	5.8	0.5	4.6	2.1	15.4
Bosnia and Herzegovina	3,631.7	659.0	4.7	33.2	408.2	24.3	63.0	—	—	4.5	59.3	—	—
Botswana	1,725.8	1,202.0	8.5	—	561.7	179.4	34.3	0.7	13.7	0.4	4.6	4.6	41.2
Brazil	175,307.6	1,644.8	48.5	55.4	159.6	11.8	6.1	2.4	33.5	0.8	8.7	6.9	48.6
Brunei Darussalam	336.8	25.5	0.0	0.0	2,796.4	327.3	169.2	—	—	6.8	95.3	—	—
Bulgaria	7,988.3	1,478.9	7.5	124.7	330.3	45.2	19.7	13.7	139.6	2.3	32.5	13.8	88.5
Burkina Faso	11,950.2	1,600.1	15.8	20.2	37.4	14.4	4.0	0.0	0.5	0.1	1.7	0.9	7.8

Burundi	6,651.7	677.3	14.1	0.8	13.2	1.2	1.3	0.1	2.4	0.1	0.7	0.7	6.6
Cambodia	12,834.1	955.7	53.6	2.1	47.7	8.0	4.2	0.0	0.2	0.2	1.4	0.5	4.2
Cameroon	16,080.9	1,144.8	9.2	8.5	37.5	21.0	6.8	0.2	3.3	0.2	2.2	1.1	10.1
Canada	30,889.0	1,345.1	24.9	172.4	254.9	60.5	30.8	25.7	140.8	13.3	122.8	29.1	112.9
Cape Verde	442.6	599.3	22.5	—	495.1	63.8	46.6	0.0	0.8	1.2	11.2	0.4	3.3
Central African Republic	3,772.3	1,147.8	18.8	0.0	14.4	2.9	3.0	0.0	0.5	0.0	0.2	0.5	4.8
Chad	8,614.3	1,425.0	19.9	0.0	7.7	2.6	1.4	—	—	0.1	0.5	0.5	4.2
Chile	15,492.1	451.5	135.6	122.3	327.3	18.0	16.9	6.0	20.1	2.1	21.6	9.2	24.1
China	1,277,208.3	614.3	98.5	170.0	86.2	10.9	5.3	2.5	31.7	0.3	4.2	5.3	41.9
Colombia	40,093.6	894.5	34.1	30.5	227.0	17.8	19.9	0.4	7.3	0.8	8.1	13.4	121.0
Comoros	710.2	1,107.4	2.2	—	102.1	84.6	18.2	0.0	0.6	0.3	2.9	0.7	6.1
Congo, Democratic Republic	52,052.6	529.8	1.7	0.3	10.2	3.4	2.6	0.0	0.4	0.0	0.1	0.4	3.3
Congo, Republic	3,096.2	506.6	6.5	1.0	191.3	47.7	21.4	0.0	0.3	0.1	0.9	1.0	9.3
Costa Rica	3,962.9	687.7	37.4	43.1	348.5	47.1	42.6	3.1	57.0	2.7	24.3	19.9	176.9
Côte d'Ivoire	17,422.3	1,158.5	6.7	15.0	45.5	36.3	12.1	0.3	4.8	0.3	3.5	1.3	11.4
Croatia	4,520.8	892.7	0.6	136.0	462.3	28.6	49.7	0.5	6.9	8.1	102.4	—	—
Cuba	11,090.9	1,189.6	74.9	41.4	175.2	11.7	26.7	4.2	48.6	0.8	10.3	14.1	89.6
Cyprus	790.9	364.4	220.6	27.6	1,317.9	106.7	96.7	0.2	2.4	13.3	166.1	8.5	61.0
Czech Republic	10,228.3	818.6	1.0	149.2	418.2	56.2	40.8	3.9	15.0	8.3	90.1	10.3	39.1
Denmark	5,339.5	512.7	3.7	75.3	709.4	66.4	58.9	1.9	4.6	15.8	161.0	7.7	17.3
Dominica	68.1	2,169.8	—	23.9	361.8	80.1	45.6	—	—	3.5	31.8	—	—
Dominican Republic	8,900.2	740.4	92.8	—	366.6	23.7	44.2	0.2	4.0	1.5	13.8	12.2	101.9
East Timor	868.1	1,748.0	43.0	—	48.1	10.6	13.6	—	—	—	—	—	—
Ecuador	12,367.9	1,388.6	116.4	40.3	199.2	10.8	16.1	2.7	50.3	1.0	9.9	17.1	154.3
Egypt	70,957.6	150.8	496.7	189.5	333.8	19.6	22.7	2.4	44.6	0.5	5.8	7.5	67.2
El Salvador	5,944.5	578.9	7.1	45.5	249.3	36.1	25.2	1.2	21.4	1.2	12.7	5.4	48.4
Eritrea	3,786.8	533.1	1.9	0.5	494.1	27.0	22.0	0.0	0.2	0.2	1.9	0.8	7.4
Estonia	1,372.7	712.5	0.2	77.8	601.9	139.0	35.0	1.2	4.3	9.5	103.4	6.6	28.2
Ethiopia	66,511.7	1,107.7	22.7	5.0	20.4	1.8	3.1	0.0	0.3	0.1	1.0	0.5	4.5
Fiji Islands	803.6	1,195.0	1.0	29.6	415.6	57.1	18.0	0.5	10.2	2.1	25.8	1.2	11.2

continues

DATA TABLE 13 *continued*

	Population (Thousands)	Water Footprint of Consumption of Agricultural Products						Water Footprint of Consumption of Industrial Products				Water Footprint of Domestic Water Consumption	
		Internal			External			Internal		External			
Country		Green	Blue	Grey	Green	Blue	Grey	Blue	Grey	Blue	Grey	Blue	Grey
Finland	5,181.9	651.0	1.2	17.5	406.2	47.9	40.3	10.2	32.3	13.6	157.4	6.6	29.6
France	59,435.5	735.3	23.1	63.2	618.1	77.0	35.0	14.7	55.9	9.8	105.7	10.6	37.4
French Polynesia	237.9	907.3	2.6	6.2	630.6	135.7	82.7	—	—	11.2	100.2	—	—
Gabon	1,245.3	1,020.4	10.1	6.7	271.1	54.6	30.2	0.1	2.9	0.6	6.1	4.8	43.4
Gambia	1,324.9	541.9	1.2	5.4	246.4	52.2	29.4	0.1	2.0	0.2	2.8	0.5	4.8
Georgia	4,719.9	1,219.7	76.2	42.6	587.4	34.2	29.4	5.6	102.5	1.1	16.3	15.3	137.3
Germany	82,138.8	311.4	1.0	85.9	741.9	56.9	62.3	9.9	10.4	10.3	109.4	7.1	19.9
Ghana	19,784.5	1,091.4	1.9	4.8	57.7	20.6	9.7	0.2	3.8	0.4	4.6	1.2	10.7
Greece	10,937.4	890.9	206.0	88.0	761.1	96.4	78.2	1.0	2.7	11.0	139.9	11.6	51.4
Grenada	101.5	1,267.2	2.2	—	636.9	90.1	113.7	—	—	4.4	39.6	—	—
Guatemala	11,412.4	713.0	13.0	50.3	136.1	17.7	19.8	0.9	10.1	1.1	11.2	1.1	8.4
Guinea	8,463.3	1,492.1	5.1	2.5	52.7	26.3	10.2	0.1	1.4	0.1	1.3	1.4	12.4
Guinea-Bissau	1,325.7	1,028.8	63.1	—	31.2	40.0	11.1	0.2	4.1	0.1	1.6	1.7	15.6
Guyana	759.1	1,046.5	96.6	114.9	165.2	22.1	39.6	0.5	9.1	1.3	13.1	4.0	35.6
Haiti	8,719.4	780.9	25.8	—	141.0	49.8	23.9	0.0	0.8	0.2	1.7	0.6	5.2
Honduras	6,299.4	913.5	18.0	55.6	123.3	27.4	16.1	0.3	5.8	0.7	5.8	1.1	10.0
Hungary	10,199.8	1,547.2	7.2	200.0	369.2	34.6	24.5	10.3	66.2	6.6	73.9	7.0	37.3
Iceland	282.2	14.5	—	5.8	1,468.9	82.3	88.1	14.0	114.3	20.0	207.4	17.7	75.7
India	1,051,289.5	685.5	210.5	91.6	22.4	2.1	1.8	1.2	22.7	0.1	1.2	5.0	44.7
Indonesia	206,706.2	854.8	45.5	77.1	77.8	24.0	7.5	0.1	1.4	0.3	3.3	3.2	28.5
Iran	67,157.7	749.4	558.3	122.9	289.5	20.1	14.9	0.6	9.6	0.6	8.2	9.2	83.1
Ireland	3,868.6	278.8	0.1	37.7	660.8	90.9	65.5	2.9	11.9	8.1	107.7	6.7	30.3
Israel	6,134.0	207.4	134.2	54.3	1,583.1	98.9	89.1	0.6	1.3	9.3	97.0	10.3	17.2
Italy	57,520.5	629.3	43.9	94.2	1,091.2	117.0	82.4	8.7	47.7	9.1	97.8	14.0	67.7

Jamaica	2,578.4	838.5	17.3	16.2	585.7	63.8	75.7	1.0	18.8	2.5	22.3	5.4	48.9
Japan	126,741.3	176.6	11.0	16.6	832.8	54.7	83.0	4.6	26.8	7.0	82.9	13.7	69.5
Jordan	4,956.1	112.5	75.8	10.1	1,126.3	163.8	103.7	0.3	2.3	3.5	42.4	5.9	31.4
Kazakhstan	15,160.4	1,710.6	344.2	16.5	81.1	16.4	13.0	7.4	124.1	1.5	22.6	3.9	35.0
Kenya	31,935.3	875.1	9.3	8.5	155.8	23.3	8.7	0.1	2.4	0.2	3.2	1.5	13.1
Kiribati	85.0	2,341.6	—	—	314.7	148.2	48.0	—	—	0.5	6.2	—	—
Korea, Democratic People's Republic	22,867.1	542.5	58.1	—	91.0	16.7	14.6	4.2	79.2	0.2	3.7	7.8	70.5
Korea, Republic	46,443.0	252.4	9.8	24.3	1,032.7	79.7	91.9	1.9	6.6	5.7	62.4	14.3	47.7
Kuwait	2,251.5	6.6	22.9	6.4	1,412.9	246.0	138.2	0.2	4.3	4.9	51.9	17.8	160.1
Kyrgyzstan	4,961.6	495.4	240.8	14.4	593.5	26.9	16.8	2.4	32.3	1.0	17.0	6.4	52.5
Laos	5,431.9	891.0	64.2	0.2	27.3	3.9	2.4	1.2	21.5	0.3	5.4	2.4	21.5
Latvia	2,368.7	1,036.6	0.2	107.3	289.9	39.6	24.5	1.7	9.8	15.9	231.5	6.8	33.0
Lebanon	3,816.5	303.4	174.8	33.7	1,245.0	142.7	76.7	1.8	11.0	5.4	69.4	10.0	37.7
Lesotho	1,890.3	1,533.5	2.9	0.7	40.2	32.8	11.0	0.3	5.3	0.2	2.1	1.1	9.5
Liberia	2,790.3	1,064.7	1.2	—	64.8	30.5	22.6	0.3	5.8	2.7	31.2	1.1	9.7
Libya	5,411.9	301.2	271.7	16.1	981.3	226.7	105.1	0.6	11.4	1.0	10.5	11.3	101.4
Lithuania	3,501.4	1,048.4	0.3	28.5	218.3	39.2	27.5	0.4	1.8	8.5	108.4	6.0	28.7
Luxembourg	450.5	508.9	0.0	101.1	1,432.2	68.0	87.3	1.6	1.7	28.3	272.2	5.6	7.3
Macedonia	2,010.5	602.2	35.5	42.4	382.7	38.5	30.7	3.4	46.8	4.5	57.3	10.7	93.2
Madagascar	15,529.4	1,329.4	145.1	2.7	36.2	16.0	4.8	0.6	11.7	0.2	2.4	2.7	24.3
Malawi	12,005.8	804.1	12.4	45.0	48.4	3.0	5.2	0.2	3.7	0.1	1.7	1.2	11.2
Malaysia	23,458.2	1,240.6	49.0	57.1	503.3	93.9	56.7	1.0	13.2	2.4	21.4	6.5	58.3
Maldives	273.5	221.8	4.2	8.7	813.5	160.8	89.0	0.0	—	3.4	51.6	1.1	6.9
Mali	10,701.0	1,769.6	139.8	0.0	54.8	10.9	5.7	0.3	4.8	0.2	2.5	5.5	49.6
Malta	391.0	74.1	9.8	7.0	1,600.7	176.4	130.0	0.0	0.6	10.2	115.7	10.2	81.0
Mauritania	2,648.1	1,262.0	78.1	0.0	1,036.8	84.6	30.4	0.6	11.1	0.4	4.5	5.7	51.0
Mauritius	1,199.6	371.9	8.4	3.3	1,238.9	215.3	103.8	0.6	6.9	3.9	46.5	17.8	143.4
Mexico	99,810.4	840.0	86.7	88.1	661.1	84.9	59.6	1.3	16.8	3.6	31.8	13.6	90.4
Moldova	4,051.3	983.2	90.0	29.6	55.7	13.2	4.7	10.5	72.9	1.5	24.3	5.4	35.8

continues

Data Table 13 *continued*

Country	Population (Thousands)	Water Footprint of Consumption of Agricultural Products						Water Footprint of Consumption of Industrial Products				Water Footprint of Domestic Water Consumption	
		Internal			External			Internal		External			
		Green	Blue	Grey	Green	Blue	Grey	Blue	Grey	Blue	Grey	Blue	Grey
Mongolia	2,409.3	1,316.3	35.5	30.2	2,207.8	53.6	52.7	1.4	23.2	1.1	15.4	3.7	33.6
Morocco	28,961.9	981.3	166.4	45.8	424.2	36.7	29.6	0.4	1.6	1.0	11.5	4.2	21.9
Mozambique	18,560.9	1,022.6	7.9	6.2	61.8	9.3	6.6	0.0	0.3	0.1	0.7	0.4	3.4
Myanmar	46,626.5	1,124.8	41.2	17.5	18.0	1.1	1.2	0.1	2.3	0.1	1.8	0.9	7.9
Namibia	1,839.1	1,169.4	18.7	11.8	339.6	67.9	24.0	0.3	5.6	0.5	4.9	4.0	35.7
Nepal	24,707.0	944.6	113.8	8.3	94.9	9.8	12.3	0.1	1.6	0.2	2.9	1.2	10.9
Netherlands	15,946.9	43.6	2.8	15.1	1,012.3	99.5	85.3	6.9	1.2	16.7	172.8	3.1	6.5
New Caledonia	216.8	550.6	0.2	35.8	1,100.4	118.1	79.5	—	—	10.1	100.1	—	—
New Zealand	3,905.5	477.3	62.1	23.0	720.0	49.1	35.3	2.1	8.1	9.3	102.5	26.1	73.8
Nicaragua	5,125.0	682.6	26.2	16.0	104.6	19.4	12.2	0.2	4.4	0.8	8.6	3.7	33.4
Niger	11,272.1	3,351.9	63.7	4.9	59.1	22.6	7.4	0.0	0.4	0.1	0.5	0.8	7.2
Nigeria	126,648.8	1,147.3	10.9	5.5	43.9	11.5	7.4	0.1	1.6	0.1	0.7	1.3	12.0
Norway	4,501.6	344.7	2.0	12.1	678.8	87.7	55.8	8.5	33.5	13.2	134.7	11.1	40.7
Occupied Palestinian Territory	3,221.2	788.8	114.0	—	30.1	6.6	39.7	1.1	20.1	0.1	0.8	5.3	48.0
Pakistan	149,801.8	535.5	403.6	136.7	191.7	15.0	8.1	0.8	15.9	0.1	1.8	2.2	19.6
Panama	2,979.1	747.2	18.2	53.3	311.5	32.1	44.1	0.5	3.7	2.5	26.6	18.5	105.5
Paraguay	5,406.5	1,789.1	39.7	45.9	26.1	11.0	4.4	0.3	5.4	1.2	13.9	1.8	15.3
Peru	26,157.7	502.4	126.1	55.8	307.8	15.4	21.7	2.8	13.9	0.7	7.5	6.4	27.5
Philippines	78,472.9	994.4	42.8	47.4	133.0	25.7	21.7	1.7	30.7	0.4	5.3	7.5	67.1
Poland	38,408.0	778.0	1.7	151.8	232.6	36.2	23.6	11.4	80.2	4.0	50.0	5.5	30.4
Portugal	10,278.4	696.3	147.2	61.0	1,157.9	196.9	95.6	4.2	25.7	4.4	46.4	10.5	59.4
Romania	22,089.3	1,183.2	24.0	58.6	161.5	34.9	17.5	7.7	99.8	2.5	32.8	7.7	58.5
Russian Federation	146,082.3	1,374.8	43.1	49.7	174.1	38.7	10.5	3.6	59.4	0.4	5.6	9.2	82.6

Rwanda	7,735.9	775.2	2.5	0.5	24.4	5.3	6.1	0.1	1.3	0.1	0.9	0.5	4.2
Saint Kitts and Nevis	46.4	593.4	0.2	1.2	706.7	116.7	86.4	—	—	11.2	115.9	—	—
Saint Lucia	157.5	56.4	0.1	—	1,145.2	110.7	122.8	—	—	4.4	40.1	7.9	71.4
Saint Vincent/Grenadines	108.2	878.8	—	—	594.5	94.1	92.5	—	—	2.8	26.7	—	—
Samoa	176.1	1,811.4	—	0.5	168.3	63.2	24.9	—	—	1.2	13.1	—	—
São Tomé and Príncipe	141.5	1,459.9	0.7	—	155.8	59.0	48.0	—	—	0.4	3.9	—	—
Saudi Arabia	21,114.2	126.3	332.4	52.8	1,004.9	101.8	89.7	0.8	14.3	2.3	23.0	10.1	90.8
Senegal	10,055.0	757.9	32.0	4.9	264.7	53.5	19.9	0.3	3.6	0.4	5.0	1.0	7.9
Serbia and Montenegro	10,729.5	1,224.5	10.8	98.4	107.5	25.7	10.7	40.4	764.3	2.2	32.6	7.3	65.4
Seychelles	80.6	657.7	39.0	0.2	1,010.5	146.8	91.1	2.0	37.1	8.3	99.9	9.9	89.3
Sierra Leone	4,408.0	1,132.2	11.0	—	248.5	22.8	15.4	0.1	1.5	0.1	1.5	0.5	4.1
Slovakia	5,377.2	704.2	11.0	75.7	310.0	44.4	32.8	2.8	17.0	7.2	73.6	8.0	48.7
Slovenia	1,986.5	373.8	0.5	205.2	971.8	65.8	93.6	9.4	79.1	11.5	124.2	10.1	67.5
Solomon Islands	421.9	530.2	—	—	138.8	21.5	26.0	—	—	0.4	5.8	—	—
South Africa	45,183.8	786.8	76.7	49.9	240.5	10.7	12.5	0.5	4.1	1.1	11.6	8.6	52.4
Spain	40,840.6	1,004.9	214.5	142.2	797.3	79.3	68.1	6.0	—	9.7	102.7	11.7	25.0
Sri Lanka	18,883.5	771.5	105.6	54.9	207.7	53.8	26.7	0.5	9.2	0.6	9.2	1.6	14.3
Sudan	35,224.9	1,428.1	196.3	10.2	56.4	4.6	4.4	0.3	5.0	0.2	2.3	2.8	25.3
Suriname	470.8	749.1	114.7	79.1	260.3	17.0	27.8	1.0	18.9	1.4	13.8	6.4	57.3
Swaziland	1,073.1	491.4	46.4	—	655.7	146.0	21.5	0.5	8.6	0.4	4.8	2.2	20.1
Sweden	8,911.5	535.5	1.5	85.7	502.7	50.9	53.5	4.4	13.3	11.0	126.6	12.2	30.5
Switzerland	7,232.5	183.6	0.1	45.2	851.4	78.8	82.6	6.9	4.4	20.1	224.6	8.6	22.3
Syria	16,859.5	1,187.6	338.9	148.7	270.8	38.1	17.4	1.0	15.5	0.6	7.9	8.5	72.2
Tajikistan	6,201.4	647.8	447.7	47.8	1,069.1	14.8	9.2	3.8	66.0	0.6	9.4	7.1	63.9
Tanzania	34,719.2	913.4	23.0	4.3	52.9	9.9	5.0	0.0	0.6	0.1	1.7	1.5	13.7
Thailand	62,968.7	939.0	134.5	64.3	148.8	23.5	10.9	1.2	20.3	1.7	19.8	4.3	39.1
Togo	5,303.9	845.6	1.1	5.1	94.5	15.0	8.9	0.0	0.4	0.2	2.4	1.7	15.1
Trinidad and Tobago	1,296.6	83.4	0.1	0.4	1,224.9	89.0	108.5	1.3	16.9	3.4	30.1	16.2	141.7
Tunisia	9,482.7	1,299.7	126.7	42.3	504.7	135.5	57.8	0.3	3.2	1.8	17.9	3.8	23.2
Turkey	66,848.5	933.8	167.9	103.3	216.0	71.2	17.4	2.4	24.6	3.0	39.5	8.3	54.4

continues

DATA TABLE 13 continued

Country	Population (Thousands)	Water Footprint of Consumption of Agricultural Products						Water Footprint of Consumption of Industrial Products					Water Footprint of Domestic Water Consumption	
		Internal			External			Internal		External				
		Green	Blue	Grey	Green	Blue	Grey	Blue	Grey	Blue	Grey	Blue	Grey		
Turkmenistan	4,544.7	1,239.2	719.3	0.1	185.2	9.4	11.6	0.8	7.4	0.9	7.3	9.2	83.2
Uganda	24,961.9	1,019.2	0.4	1.1	35.9	9.8	3.4	0.1	1.9	0.1	1.8	0.5	4.7
UK	59,334.4	243.5	0.9	53.3	672.0	71.9	61.0	4.1	2.0	12.2	129.6	3.5	4.1
Ukraine	48,737.6	1,117.6	28.2	62.0	55.9	7.7	4.4	8.8	159.7	2.0	34.8	9.4	84.2
United Arab Emirates	3,329.8	412.7	140.1	12.3	1,508.5	386.2	185.3	0.6	9.7	25.1	270.2	18.5	166.8
Uruguay	3,307.1	1,567.4	53.4	39.1	388.7	4.0	11.9	0.5	9.7	2.5	31.0	2.4	21.8
USA	288,958.2	1,600.4	136.5	223.5	367.9	37.5	31.7	30.9	164.7	11.3	127.0	22.6	88.4
Uzbekistan	24,895.6	509.9	346.7	0.0	255.6	3.1	5.8	1.9	35.5	0.5	8.1	11.1	100.1
Vanuatu	193.8	1,264.0	—	—	45.5	73.9	33.5	—	—	1.1	13.2	—	—
Venezuela	24,640.2	874.6	48.5	33.2	527.0	22.2	28.7	0.7	12.2	0.9	7.9	15.5	139.2
Vietnam	79,188.2	637.7	68.0	94.2	52.4	6.4	4.2	6.1	113.8	0.4	5.4	7.0	63.0
Yemen	18,502.1	71.4	126.2	5.1	551.2	90.2	38.4	0.1	1.8	0.2	2.0	1.5	13.1
Zambia	10,573.6	754.0	27.8	16.0	76.2	7.9	4.9	0.3	5.1	0.1	1.5	2.7	24.7
Zimbabwe	12,357.4	958.7	57.7	34.5	81.1	6.6	5.5	0.8	15.3	0.2	2.2	4.8	42.9
World	6,154,564.2	796.6	107.1	90.3	218.8	33.2	21.4	3.8	34.3	2.3	24.6	6.8	45.8

DATA TABLE 14

Total Water Footprint of National Consumption, by Country, 1996-2005

Description

The data in this table are similar to those in data table 12 on both internal production and external imports, but on a total (rather than a per capita) basis. Conventional estimates of national water use have historically been restricted to statistics on water withdrawals within national boundaries. In recent years, a new approach has been developed and applied to calculate water "footprints," which include water use (typically withdrawals and consumption, estimated separately), the use of "green water" and "grey water," and information about water used in other countries to produce imported products. Quantifying these national water footprints is an evolving field.

The data here include the total water footprint for internal and external water consumption by country, measured in millions of cubic meters per year (Mm^3/yr) and averaged over the decade 1996–2005. Data on population assumptions are also provided. A distinction is made between green water and blue water. In addition, the authors include the grey water footprint in the estimation of the water footprint in each sector, an estimate of the footprint of farm animal products, and details on the water footprints of national imports and exports. The final column presents the external water footprint as a fraction of the total national water footprint.

Freshwater footprints are reported in terms of water volumes consumed (evaporated or incorporated into a product) or polluted per unit of time. A water footprint has three components: green, blue, and grey. The "blue water" footprint refers to use of surface water and groundwater. The "green water" footprint is the volume of rainwater consumed, which is particularly relevant in agricultural production. The "grey water" footprint is a measure of freshwater pollution and is defined as the volume of freshwater required to assimilate a load of pollutants.

Limitations

While the concept of the water footprint is an extremely valuable one, the process for computing footprints continues to evolve, and data sets continue to improve. For example, water withdrawal numbers are often not accurately or consistently reported (see, e.g., the description of data table 2). Estimates of agricultural water use depend on a mix of implicit and explicit assumptions about actual irrigation levels, climatic data for certain crop types, and irrigation technologies. Some consumer categories are not included or accurately estimated. Estimates of the water footprint of imported goods are produced using global averages because of constraints on information of the location of exported goods. The grey water footprint estimates are based on a limited set of pollutants—leaching and runoff of nitrogen fertilizers—and exclude the potential effects of other agricultural chemicals, including pesticides.

See the original source for more detail on the fundamental assumptions and limitations of the footprint approach.

Source

Mekonnen, M. M., and A. Y. Hoekstra. 2011. National Water Footprint Accounts: The Green, Blue and Grey Water Footprint of Production and Consumption. Value of Water Research Report Series No. 50. Delft, Netherlands: UNESCO-IHE Institute for Water Education. http://www.waterfootprint.org/Reports/Report50-NationalWaterFootprints-Vol1.pdf.

DATA TABLE 14 Total Water Footprint of National Consumption, by Country, 1996–2005 (Mm³/yr)

	Total Water Footprint of National Consumption									Ratio of External/ Total Water Footprint (%)
	Internal (WFcons, nat, int)			External (WFcons, nat, ext)			Total (WFcons, nat)			
Country	Green	Blue	Grey	Green	Blue	Grey	Green	Blue	Grey	
Albania	1,857.8	328.1	675.2	1,519.8	242.7	174.1	3,377.6	570.8	849.3	40.4
Algeria	19,523.7	2,886.1	1,342.2	21,555.6	1,573.9	2,022.6	41,079.4	4,459.9	3,364.9	51.4
Angola	11,663.8	207.7	120.1	1,584.0	215.2	209.6	13,247.8	422.9	329.7	14.3
Antigua and Barbuda	30.3	0.4	3.4	64.1	7.8	15.4	94.4	8.2	18.8	71.9
Argentina	47,746.2	3,864.9	5,661.8	1,298.1	207.5	767.8	49,044.3	4,072.3	6,429.6	3.8
Armenia	1,595.1	415.7	732.5	1,391.0	129.3	181.5	2,986.0	545.0	914.0	38.3
Australia	33,170.9	3,764.5	2,525.7	2,634.0	414.9	2,208.2	35,804.9	4,179.4	4,733.9	11.8
Austria	3,192.9	130.7	748.7	5,947.3	664.6	2,187.5	9,140.2	795.3	2,936.2	68.4
Azerbaijan	4,158.9	1,440.6	1,617.2	2,533.4	212.0	190.2	6,692.3	1,652.6	1,807.4	28.9
Bahamas	61.5	—	—	413.2	51.3	127.4	474.7	51.3	127.4	90.6
Bangladesh	67,849.0	8,130.3	14,168.5	13,923.7	3,196.3	1,848.8	81,772.7	11,326.6	16,017.3	17.4
Barbados	139.9	5.0	64.6	238.1	29.7	51.3	378.0	34.7	116.0	60.4
Belarus	10,944.5	193.6	2,240.4	2,164.1	443.6	1,241.9	13,108.5	637.2	3,482.3	22.3
Belgium	359.2	225.6	1,549.2	12,285.4	1,253.6	3,959.4	12,644.6	1,479.2	5,508.6	89.1
Belize	294.0	10.0	149.2	30.8	9.8	18.1	324.8	19.8	167.3	11.5
Benin	6,957.5	12.6	99.1	458.9	138.4	81.4	7,416.4	151.0	180.4	8.8
Bermuda	0.4	0.1	0.3	107.5	14.5	65.8	108.0	14.6	66.1	99.5
Bolivia	25,764.0	398.7	245.3	2,488.5	128.4	136.0	28,252.5	527.1	381.3	9.4
Bosnia and Herzegovina	2,393.3	17.0	120.6	1,482.6	104.5	444.0	3,876.0	121.4	564.6	44.5
Botswana	2,074.4	23.9	94.8	969.4	310.3	67.1	3,043.8	334.2	161.9	38.0
Brazil	288,345.2	10,119.8	24,109.3	27,980.8	2,222.0	2,596.5	316,326.0	12,341.8	26,705.8	9.2
Brunei Darussalam	8.6	0.0	0.0	941.8	112.5	89.1	950.4	112.5	89.1	99.3
Bulgaria	11,814.2	280.2	2,817.7	2,638.3	379.9	417.2	14,452.5	660.1	3,234.9	18.7
Burkina Faso	19,121.3	199.1	340.2	447.0	174.1	68.5	19,568.3	373.2	408.7	3.4
Burundi	4,505.5	99.6	65.4	87.5	8.6	13.2	4,593.0	108.2	78.6	2.3

continues

343

DATA TABLE 14 continued

	Total Water Footprint of National Consumption										
	Internal (WFcons, nat, int)			External (WFcons, nat, ext)			Total (WFcons, nat)			Ratio of External/ Total Water Footprint (%)
Country	Green	Blue	Grey	Green	Blue	Grey	Green	Blue	Grey	
Cambodia	12,266.0	694.5	84.0	612.6	105.9	71.3	12,878.6	800.4	155.3	5.7
Cameroon	18,409.0	169.0	352.7	603.7	340.9	145.5	19,012.7	509.9	498.2	5.4
Canada	41,548.5	2,463.0	13,161.6	7,874.8	2,280.4	4,746.0	49,423.3	4,743.4	17,907.6	20.7
Cape Verde	265.2	10.1	1.8	219.1	28.8	25.6	484.4	38.9	27.4	49.7
Central African Republic	4,329.9	72.9	20.2	54.5	11.0	12.4	4,384.4	83.9	32.6	1.7
Chad	12,275.2	175.2	36.0	66.2	22.5	16.1	12,341.4	197.7	52.1	0.8
Chile	6,994.0	2,336.0	2,579.5	5,070.5	310.4	597.1	12,064.5	2,646.4	3,176.6	33.4
China	784,627.8	135,713.2	311,238.1	110,077.5	14,298.3	12,048.7	894,705.3	150,011.5	323,286.9	10.0
Colombia	35,863.2	1,922.5	6,365.5	9,100.9	748.7	1,122.2	44,964.1	2,671.2	7,487.7	19.9
Comoros	786.5	2.1	4.8	72.5	60.3	15.0	859.0	62.4	19.8	15.7
Congo, Democratic Republic	27,577.5	108.6	206.1	530.4	175.0	139.6	28,107.9	283.6	345.8	2.9
Congo, Republic	1,568.7	23.3	32.7	592.2	148.2	69.1	2,160.9	171.5	101.8	33.3
Costa Rica	2,725.2	239.8	1,097.8	1,381.0	197.4	265.0	4,106.2	437.2	1,362.8	31.2
Côte d'Ivoire	20,184.3	142.7	542.5	792.2	637.2	271.6	20,976.5	779.9	814.1	7.5
Croatia	4,035.7	5.0	646.4	2,089.9	165.6	687.4	6,125.6	170.6	1,333.8	38.6
Cuba	13,194.0	1,033.0	1,991.4	1,943.7	139.4	410.5	15,137.7	1,172.4	2,401.9	13.3
Cyprus	288.2	181.4	72.0	1,042.3	94.9	207.8	1,330.5	276.3	279.8	71.3
Czech Republic	8,372.5	155.4	2,078.7	4,277.0	660.1	1,339.2	12,649.6	815.5	3,417.9	37.2
Denmark	2,737.5	71.0	518.7	3,787.8	438.8	1,174.1	6,525.3	509.8	1,692.7	61.9
Dominica	147.8	—	1.6	24.6	5.7	5.3	172.4	5.7	6.9	19.2
Dominican Republic	6,589.8	937.2	942.3	3,263.2	224.0	515.7	9,853.0	1,161.3	1,458.0	32.1
East Timor	1,517.4	37.3	—	41.8	9.2	11.8	1,559.2	46.5	11.8	3.9
Ecuador	17,174.5	1,685.4	3,028.1	2,464.1	145.6	321.9	19,638.6	1,831.0	3,350.0	11.8
Egypt	10,697.6	35,944.2	21,381.6	23,682.2	1,426.2	2,023.7	34,379.8	37,370.4	23,405.3	28.5
El Salvador	3,441.3	81.1	685.9	1,482.0	222.1	225.3	4,923.3	303.2	911.2	31.4

Eritrea	2,018.6	10.2	30.7	1,871.0	102.7	90.2	3,889.7	112.9	120.9	50.1
Estonia	978.1	11.0	151.5	826.2	203.8	190.0	1,804.3	214.8	341.5	51.7
Ethiopia	73,675.2	1,546.7	654.5	1,357.8	121.9	276.2	75,033.0	1,668.6	930.7	2.3
Fiji Islands	960.3	2.3	41.0	334.0	47.5	35.2	1,294.3	49.8	76.1	29.3
Finland	3,373.3	93.1	411.6	2,105.0	318.7	1,024.7	5,478.3	411.8	1,436.3	47.1
France	43,704.2	2,879.3	9,295.2	36,738.7	5,156.5	8,358.0	80,443.0	8,035.8	17,653.1	47.3
French Polynesia	215.8	0.6	1.5	150.0	34.9	43.5	365.9	35.6	45.0	51.2
Gabon	1,270.7	18.7	65.8	337.6	68.8	45.1	1,608.3	87.4	111.0	25.0
Gambia	717.9	2.4	16.2	326.4	69.5	42.7	1,044.3	71.9	58.9	37.3
Georgia	5,756.9	457.8	1,332.6	2,772.6	166.7	215.7	8,529.5	624.5	1,548.3	29.5
Germany	25,576.8	1,474.5	9,542.0	60,940.9	5,518.2	14,099.1	86,517.7	6,992.7	23,641.1	68.8
Ghana	21,592.2	65.3	381.7	1,140.8	416.1	281.2	22,733.0	481.4	662.9	7.7
Greece	9,744.3	2,391.0	1,553.7	8,324.7	1,174.3	2,384.6	18,069.0	3,565.2	3,938.4	46.5
Grenada	128.6	0.2	—	64.6	9.6	15.6	193.3	9.8	15.6	41.1
Guatemala	8,136.5	171.7	785.0	1,552.9	214.6	354.0	9,689.4	386.3	1,139.0	18.9
Guinea	12,628.3	55.7	137.7	446.4	223.6	96.8	13,074.7	279.4	234.5	5.6
Guinea-Bissau	1,363.8	86.2	26.2	41.4	53.3	16.8	1,405.2	139.5	43.0	7.0
Guyana	794.4	76.7	121.1	125.4	17.8	40.0	919.8	94.5	161.1	15.6
Haiti	6,808.8	230.0	52.4	1,229.8	435.5	222.9	8,038.5	665.5	275.3	21.0
Honduras	5,754.3	122.3	449.8	776.5	176.4	137.9	6,530.8	298.8	587.7	14.7
Hungary	15,780.8	249.4	3,095.4	3,765.5	420.4	1,003.4	19,546.4	669.8	4,098.8	21.3
Iceland	4.1	8.9	55.3	414.5	28.9	83.4	418.6	37.8	138.7	88.5
India	720,644.0	227,832.7	167,199.4	23,546.0	2,277.7	3,105.3	744,190.0	230,110.4	170,304.7	2.5
Indonesia	176,695.4	10,092.3	22,107.9	16,089.3	5,026.9	2,226.8	192,784.7	15,119.2	24,334.7	10.1
Iran	50,325.3	38,153.7	14,477.1	19,445.0	1,391.1	1,556.1	69,770.3	39,544.8	16,033.2	17.9
Ireland	1,078.6	37.3	309.2	2,556.3	383.0	670.1	3,634.9	420.4	979.4	71.7
Israel	1,272.1	890.4	447.1	9,710.6	663.2	1,141.4	10,982.7	1,553.6	1,588.4	81.5
Italy	36,197.9	3,834.6	12,049.7	62,764.0	7,251.0	10,369.2	98,961.9	11,085.6	22,418.9	60.7
Jamaica	2,162.1	61.3	216.4	1,510.2	170.8	252.8	3,672.3	232.1	469.1	44.2
Japan	22,377.2	3,711.7	14,306.8	105,550.2	7,819.5	21,013.9	127,927.4	11,531.1	35,320.7	76.9

continues

DATA TABLE 14 *continued*

	Total Water Footprint of National Consumption									Ratio of External/Total Water Footprint (%)
	Internal (WFcons, nat, int)			External (WFcons, nat, ext)			Total (WFcons, nat)			
Country	Green	Blue	Grey	Green	Blue	Grey	Green	Blue	Grey	
Jordan	557.6	406.2	217.0	5,582.1	829.2	724.4	6,139.7	1,235.4	941.4	85.8
Kazakhstan	25,933.0	5,389.4	2,663.2	1,229.6	271.5	539.8	27,162.6	5,660.9	3,203.0	5.7
Kenya	27,947.7	346.8	768.4	4,976.8	752.8	378.8	32,924.5	1,099.6	1,147.2	17.4
Kiribati	199.0	—	—	26.8	12.6	4.6	225.8	12.6	4.6	18.1
Korea, Democratic People's Republic	12,405.1	1,603.0	3,421.6	2,080.0	387.7	417.2	14,485.0	1,990.8	3,838.8	14.2
Korea, Republic	11,724.0	1,202.9	3,650.6	47,962.4	3,967.4	7,162.4	59,686.4	5,170.3	10,813.1	78.1
Kuwait	14.9	92.2	384.6	3,181.1	565.0	428.1	3,196.0	657.2	812.8	89.5
Kyrgyzstan	2,458.1	1,238.9	491.9	2,944.6	138.4	167.7	5,402.7	1,377.3	659.6	43.7
Laos	4,839.6	368.2	234.8	148.4	23.3	42.7	4,988.0	391.5	277.6	3.8
Latvia	2,455.4	20.5	355.7	686.6	131.4	606.3	3,142.0	151.9	962.1	33.5
Lebanon	1,158.0	712.2	314.2	4,751.5	565.0	557.8	5,909.5	1,277.2	872.0	72.9
Lesotho	2,898.7	8.0	29.3	76.0	62.3	24.8	2,974.7	70.3	54.2	5.3
Liberia	2,970.9	7.2	43.1	180.7	92.7	150.2	3,151.6	99.9	193.3	12.3
Libya	1,630.0	1,534.6	697.9	5,310.9	1,232.4	625.5	6,940.9	2,767.1	1,323.4	65.0
Lithuania	3,670.9	23.5	206.4	764.3	167.1	475.8	4,435.2	190.6	682.2	26.5
Luxembourg	229.3	3.2	49.6	645.2	43.4	161.9	874.5	46.6	211.6	75.1
Macedonia	1,210.6	99.7	366.7	769.4	86.5	176.9	1,980.0	186.2	543.6	38.1
Madagascar	20,644.5	2,305.2	601.3	562.6	251.1	111.5	21,207.1	2,556.3	712.7	3.8
Malawi	9,654.3	166.1	719.5	581.6	37.5	82.3	10,235.8	203.6	801.8	6.2
Malaysia	29,102.0	1,324.2	3,016.9	11,805.6	2,259.8	1,831.4	40,907.6	3,584.0	4,848.3	32.2
Maldives	60.7	1.5	4.3	222.5	44.9	38.5	283.2	46.4	42.7	82.2
Mali	18,936.7	1,557.7	583.0	586.4	119.1	87.9	19,523.1	1,676.8	670.9	3.6
Malta	29.0	7.9	34.6	625.9	72.9	96.1	654.9	80.8	130.7	91.7
Mauritania	3,341.8	223.4	164.3	2,745.6	225.0	92.4	6,087.4	448.5	256.7	45.1
Mauritius	446.1	32.1	184.3	1,486.1	263.0	180.3	1,932.3	295.1	364.6	74.4

Mexico	83,840.7	10,147.9	19,492.1	65,986.0	8,833.4	9,124.9	149,826.7	18,981.3	28,617.1	42.5
Moldova	3,983.3	429.1	560.3	225.6	59.8	117.3	4,208.9	488.9	677.6	7.5
Mongolia	3,171.4	98.0	209.6	5,319.3	131.6	164.2	8,490.7	229.7	373.8	61.7
Morocco	28,420.2	4,955.5	2,005.4	12,286.8	1,092.0	1,192.7	40,707.0	6,047.5	3,198.1	29.2
Mozambique	18,980.3	153.1	183.8	1,147.6	173.0	134.5	20,128.0	326.1	318.4	7.0
Myanmar	52,447.8	1,968.7	1,291.8	837.9	58.6	140.7	53,285.7	2,027.3	1,432.5	1.8
Namibia	2,150.6	42.2	97.6	624.5	125.7	53.1	2,775.1	168.0	150.7	26.0
Nepal	23,338.5	2,843.7	515.9	2,344.9	245.8	374.8	25,683.4	3,089.5	890.7	10.0
Netherlands	695.2	202.6	365.0	16,142.5	1,852.5	4,115.1	16,837.7	2,055.1	4,480.1	94.6
New Caledonia	119.4	0.0	7.8	238.6	27.8	38.9	357.9	27.8	46.7	70.6
New Zealand	1,864.3	352.8	409.6	2,811.9	228.0	538.3	4,676.2	580.8	947.9	57.7
Nicaragua	3,498.2	154.6	275.5	536.3	103.5	106.6	4,034.5	258.1	382.1	16.0
Niger	37,783.4	727.5	141.7	666.2	254.8	89.5	38,449.6	982.3	231.2	2.5
Nigeria	145,305.7	1,557.3	2,417.6	5,557.1	1,468.0	1,030.0	150,862.8	3,025.3	3,447.6	5.1
Norway	1,551.9	97.2	388.6	3,055.6	454.2	857.5	4,607.5	551.4	1,246.1	68.2
Occupied Palestinian Territory	2,540.9	387.9	219.4	97.1	21.4	130.4	2,638.0	409.3	349.8	7.3
Pakistan	80,221.9	60,913.3	25,806.7	28,718.7	2,274.2	1,494.2	108,940.6	63,187.5	27,300.9	16.3
Panama	2,226.0	110.7	484.1	928.0	103.1	210.7	3,154.0	213.8	694.7	30.6
Paraguay	9,672.6	226.3	360.3	141.1	65.9	98.7	9,813.7	292.2	459.0	2.9
Peru	13,141.8	3,541.9	2,541.3	8,050.4	421.8	763.0	21,192.2	3,963.7	3,304.3	32.4
Philippines	78,032.3	4,074.0	11,392.0	10,433.7	2,054.9	2,123.4	88,466.0	6,128.9	13,515.4	13.5
Poland	29,880.5	713.8	10,078.3	8,934.8	1,544.9	2,827.6	38,815.3	2,258.8	12,905.9	24.7
Portugal	7,156.9	1,664.0	1,502.1	11,901.3	2,069.1	1,458.9	19,058.2	3,733.1	2,960.9	59.9
Romania	26,137.0	869.6	4,793.2	3,567.9	824.9	1,111.8	29,704.8	1,694.5	5,905.0	14.8
Russian Federation	200,827.8	8,160.5	28,000.3	25,437.2	5,711.1	2,353.6	226,265.0	13,871.5	30,353.9	12.4
Rwanda	5,996.7	23.1	46.4	189.1	41.8	53.9	6,185.8	64.9	100.3	4.5
Saint Kitts and Nevis	27.5	0.0	0.1	32.8	5.9	9.4	60.3	5.9	9.4	63.5
Saint Lucia	8.9	1.3	11.3	180.4	18.1	25.7	189.3	19.4	36.9	91.3
Saint Vincent/Grenadines	95.1	—	—	64.3	10.5	12.9	159.4	10.5	12.9	48.0
Samoa	319.0	—	0.1	29.6	11.3	6.7	348.6	11.3	6.8	13.0

continues

347

DATA TABLE 14 *continued*

Total Water Footprint of National Consumption

Country	Internal (WFcons, nat, int)			External (WFcons, nat, ext)			Total (WFcons, nat)			Ratio of External/ Total Water Footprint (%)
	Green	Blue	Grey	Green	Blue	Grey	Green	Blue	Grey	
São Tomé and Príncipe	206.6	0.1	—	22.1	8.4	7.4	228.6	8.5	7.4	15.5
Saudi Arabia	2,667.5	7,250.0	3,334.7	21,217.6	2,198.0	2,378.7	23,885.1	9,448.1	5,713.4	66.1
Senegal	7,620.5	333.8	165.5	2,662.0	542.4	250.2	10,282.4	876.1	415.6	29.8
Serbia and Montenegro	13,138.5	627.2	9,957.7	1,153.9	299.3	464.2	14,292.3	926.5	10,421.8	7.5
Seychelles	53.0	4.1	10.2	81.4	12.5	15.4	134.5	16.6	25.6	61.9
Sierra Leone	4,990.7	50.7	24.4	1,095.2	101.0	74.7	6,085.9	151.7	99.1	20.1
Slovakia	3,786.8	117.1	759.9	1,667.2	277.7	572.1	5,454.0	394.8	1,332.0	35.1
Slovenia	742.6	39.7	698.8	1,930.4	153.5	432.5	2,673.1	193.2	1,131.3	62.9
Solomon Islands	223.7	—	—	58.5	9.3	13.4	282.2	9.3	13.4	26.6
South Africa	35,551.6	3,879.0	4,804.7	10,867.1	532.1	1,089.1	46,418.8	4,411.1	5,893.8	22.0
Spain	41,038.9	9,484.6	6,826.5	32,561.5	3,631.8	6,976.7	73,600.4	13,116.4	13,803.2	42.9
Sri Lanka	14,569.5	2,033.7	1,479.7	3,922.7	1,028.5	677.6	18,492.2	3,062.2	2,157.3	23.7
Sudan	50,305.0	7,022.6	1,424.7	1,985.8	168.4	234.3	52,290.8	7,191.1	1,659.0	3.9
Suriname	352.7	57.5	73.1	122.5	8.7	19.6	475.2	66.2	92.7	23.8
Swaziland	527.3	52.7	30.8	703.6	157.1	28.3	1,230.9	209.8	59.1	59.3
Sweden	4,772.0	161.8	1,154.2	4,479.4	551.5	1,605.3	9,251.4	713.3	2,759.4	52.2
Switzerland	1,328.1	112.5	519.9	6,157.6	714.9	2,221.3	7,485.7	827.3	2,741.2	82.3
Syria	20,023.1	5,873.0	3,984.9	4,566.1	652.3	426.9	24,589.2	6,525.3	4,411.8	15.9
Tajikistan	4,017.0	2,844.1	1,101.7	6,629.8	95.7	115.2	10,646.9	2,939.8	1,216.9	46.2
Tanzania	31,712.9	853.0	644.0	1,836.6	348.4	233.1	33,549.5	1,201.4	877.1	6.8
Thailand	59,129.4	8,821.3	7,791.3	9,367.2	1,582.5	1,931.9	68,496.6	10,403.7	9,723.2	14.5
Togo	4,484.9	15.1	109.7	501.0	80.5	60.1	4,985.9	95.7	169.8	12.2
Trinidad and Tobago	108.2	22.8	206.2	1,588.2	119.8	179.8	1,696.4	142.6	386.0	84.8
Tunisia	12,324.7	1,241.4	651.2	4,785.7	1,302.0	718.2	17,110.5	2,543.4	1,369.4	32.4
Turkey	62,426.0	11,937.3	12,186.7	14,441.1	4,959.3	3,807.2	76,867.2	16,896.7	15,994.0	21.1

Turkmenistan	5,631.8	3,314.7	412.3	841.6	46.7	85.9	6,473.3	3,361.3	498.2	9.4
Uganda	25,440.5	26.2	191.9	897.1	248.5	129.9	26,337.5	274.7	321.8	4.7
UK	14,449.5	508.6	3,519.8	39,873.5	4,989.5	11,304.8	54,323.0	5,498.1	14,824.6	75.2
Ukraine	54,470.5	2,255.0	14,907.4	2,726.0	474.5	1,911.0	57,196.5	2,729.5	16,818.4	6.7
United Arab Emirates	1,374.4	530.3	628.3	5,023.0	1,369.5	1,516.8	6,397.3	1,899.8	2,145.1	75.7
Uruguay	5,183.7	186.4	233.4	1,285.6	21.5	142.1	6,469.3	207.9	375.5	20.5
USA	462,444.5	54,907.2	137,709.8	106,318.7	14,116.8	45,856.7	568,763.2	69,024.0	183,566.5	20.2
Uzbekistan	12,695.0	8,954.8	3,377.1	6,362.5	88.8	345.9	19,057.5	9,043.6	3,723.0	21.4
Vanuatu	245.0	—	—	8.8	14.5	9.0	253.8	14.5	9.0	11.7
Venezuela	21,550.6	1,591.2	4,545.8	12,984.8	568.6	901.0	34,535.4	2,159.8	5,446.9	34.3
Vietnam	50,495.0	6,415.2	21,459.5	4,146.3	537.6	764.1	54,641.2	6,952.8	22,223.6	6.5
Yemen	1,320.7	2,364.8	371.0	10,199.2	1,672.5	747.6	11,520.0	4,037.3	1,118.5	75.7
Zambia	7,972.4	326.2	483.6	805.8	85.0	68.3	8,778.1	411.1	551.8	9.8
Zimbabwe	11,846.5	782.2	1,145.2	1,002.0	84.2	95.3	12,848.5	866.4	1,240.5	7.9
World	4,902,626.4	724,848.3	1,049,238.3	1,346,910.7	218,476.3	282,963.7	6,249,537.1	943,324.6	1,332,202.1	21.7

DATA TABLE 15

Total Water Footprint of National Consumption, by Sector and Country, 1996-2005

Description

The data in this table are similar to those for data table 14 on both internal production and external imports by economic sector. Conventional estimates of national water use have historically been restricted to statistics on water withdrawals within national boundaries. In recent years, a new approach has been developed and applied to calculate water "footprints," which include water use (typically withdrawals and consumption, estimated separately), the use of "green water" and "grey water," and information on water used in other countries to produce imported products. Quantifying these national water footprints is an evolving field.

The data here include the total water footprint for internal and external water consumption by country and sector (agricultural, industrial, domestic), measured in millions of cubic meters per year (Mm^3/yr) and averaged over the decade 1996–2005. Data on population assumptions are also provided. A distinction is made between green water and blue water. In addition, the authors include the grey water footprint in the estimation of the water footprint in each sector, an estimate of the footprint of farm animal products, and details on the water footprints of national imports and exports.

Freshwater footprints are reported in terms of water volumes consumed (evaporated or incorporated into a product) or polluted per unit of time. A water footprint has three components: green, blue, and grey. The "blue water" footprint refers to use of surface water and groundwater. The "green water" footprint is the volume of rainwater consumed, which is particularly relevant in agricultural production. The "grey water" footprint is a measure of freshwater pollution and is defined as the volume of freshwater required to assimilate a load of pollutants.

Limitations

While the concept of the water footprint is an extremely valuable one, the process for computing footprints continues to evolve, and data sets continue to improve. For example, water withdrawal numbers are often not accurately or consistently reported (see, e.g., the description of data table 2). Estimates of agricultural water use depend on a mix

Data Section

of implicit and explicit assumptions about actual irrigation levels, climatic data for certain crop types, and irrigation technologies. Some consumer categories are not included or accurately estimated. Estimates of the water footprint of imported goods are produced using global averages because of constraints on information of the location of exported goods. The grey water footprint estimates are based on a limited set of pollutants—leaching and runoff of nitrogen fertilizers—and exclude the potential effects of other agricultural chemicals, including pesticides.

See the original source for far more detail on the fundamental assumptions and limitations of the footprint approach.

Source

Mekonnen, M. M., and A. Y. Hoekstra. 2011. National Water Footprint Accounts: The Green, Blue and Grey Water Footprint of Production and Consumption. Value of Water Research Report Series No. 50. Delft, Netherlands: UNESCO-IHE Institute for Water Education. http://www.waterfootprint.org/Reports/Report50-NationalWaterFootprints-Vol1.pdf.

DATA TABLE 15 Total Water Footprint of National Consumption, by Sector and Country, 1996–2005 (Mm^3/yr)

	Water Footprint of Consumption of Agricultural Products						Water Footprint of Consumption of Industrial Products				Water Footprint of Domestic Water Consumption	
	Internal (WFcons, nat, int, agr)			External (WFcons, nat, ext, agr)			Internal (WFcons, nat, int, ind)		External (WFcons, nat, ext, ind)			
Country	Green	Blue	Grey	Green	Blue	Grey	Blue	Grey	Blue	Grey	Blue	Grey
Albania	1,857.8	274.1	112.5	1,519.8	238.8	121.7	7.9	148.6	3.9	52.4	46.0	414.0
Algeria	19,523.7	2,731.4	294.4	21,555.6	1,555.6	1,835.6	21.7	196.8	18.3	187.1	133.0	851.1
Angola	11,663.8	198.9	34.9	1,584.0	213.1	192.0	0.9	13.2	2.0	17.6	8.0	72.0
Antigua and Barbuda	30.3	0.1	—	64.1	6.8	5.8	0.0	0.7	1.0	9.6	0.3	2.7
Argentina	47,746.2	3,258.0	1,682.4	1,298.1	146.0	111.2	115.9	1,255.5	61.5	656.5	491.0	2,724.0
Armenia	1,595.1	326.4	82.4	1,391.0	125.8	126.6	5.0	60.0	3.5	54.9	84.3	590.0
Australia	33,170.9	3,323.8	1,562.8	2,634.0	232.4	195.2	88.7	200.5	182.4	2,013.0	352.0	762.5
Austria	3,192.9	16.0	422.4	5,947.3	517.2	652.8	40.8	64.6	147.4	1,534.7	74.0	261.7
Azerbaijan	4,158.9	1,306.1	143.1	2,533.4	205.1	74.2	82.4	1,077.5	6.9	116.1	52.1	396.6
Bahamas	61.5	—	—	413.2	42.9	41.9	—	—	8.4	85.5	—	—
Bangladesh	67,849.0	7,717.2	10,353.4	13,923.7	3,177.6	1,587.5	11.6	200.8	18.6	261.2	401.6	3,614.4
Barbados	139.9	0.4	6.3	238.1	27.4	30.5	1.7	31.3	2.3	20.8	3.0	27.0
Belarus	10,944.5	75.5	1,998.8	2,164.1	375.3	108.0	53.0	46.2	68.3	1,133.9	65.0	195.5
Belgium	359.2	1.2	103.6	12,285.4	963.5	1,012.3	150.1	1,165.5	290.0	2,947.2	74.3	280.0
Belize	294.0	4.1	62.8	30.8	8.7	6.6	4.9	78.0	1.1	11.5	1.0	8.3
Benin	6,957.5	7.1	35.3	458.9	136.5	58.7	1.4	26.8	1.9	22.7	4.1	36.9
Bermuda	0.4	0.1	0.3	107.5	9.5	8.1	—	—	5.0	57.7	—	—
Bolivia	25,764.0	376.8	66.4	2,488.5	124.4	97.3	3.8	49.0	4.0	38.7	18.0	129.9
Bosnia and Herzegovina	2,393.3	17.0	120.6	1,482.6	88.3	228.7	—	—	16.2	215.3	—	—
Botswana	2,074.4	14.7	—	969.4	309.6	59.3	1.3	23.7	0.7	7.9	7.9	71.1
Brazil	288,345.2	8,497.8	9,704.9	27,980.8	2,074.9	1,074.5	420.1	5,878.7	147.1	1,522.0	1,202.0	8,525.7
Brunei Darussalam	8.6	0.0	0.0	941.8	110.2	57.0	—	—	2.3	32.1	—	—
Bulgaria	11,814.2	60.1	995.8	2,638.3	361.3	157.2	109.6	1,115.3	18.6	260.0	110.6	706.6

Burkina Faso	19,121.3	188.4	241.1	447.0	172.3	47.7	0.3	5.5	1.8	20.8	10.4	93.6
Burundi	4,505.5	93.8	5.6	87.5	8.3	8.6	0.8	15.7	0.4	4.6	4.9	44.1
Cambodia	12,266.0	688.1	26.9	612.6	103.3	53.8	0.4	3.1	2.6	17.5	6.0	54.0
Cameroon	18,409.0	148.1	137.4	603.7	337.6	109.9	2.9	53.3	3.3	35.6	18.0	162.0
Canada	41,548.5	769.7	5,326.5	7,874.8	1,868.5	952.3	794.3	4,348.1	411.9	3,793.7	899.0	3,487.0
Cape Verde	265.2	9.9	—	219.1	28.2	20.6	0.0	0.4	0.5	5.0	0.2	1.4
Central African Republic	4,329.9	70.8	0.1	54.5	10.9	11.5	0.1	2.0	0.1	0.9	2.0	18.0
Chad	12,275.2	171.2	0.0	66.2	22.0	11.9	—	—	0.5	4.2	4.0	36.0
Chile	6,994.0	2,100.8	1,895.0	5,070.5	278.1	261.8	93.2	311.5	32.3	335.3	142.0	373.0
China	784,627.8	125,741.5	217,182.4	110,077.5	13,906.1	6,719.4	3,218.7	40,523.2	392.2	5,329.3	6,753.0	53,532.5
Colombia	35,863.2	1,367.9	1,223.2	9,100.9	715.6	798.6	15.7	291.3	33.1	323.6	539.0	4,851.0
Comoros	786.5	1.6	—	72.5	60.1	12.9	0.0	0.4	0.2	2.1	0.5	4.3
Congo, Democratic Republic	27,577.5	88.6	15.5	530.4	174.4	133.3	1.1	19.6	0.6	6.3	19.0	171.0
Congo, Republic	1,568.7	20.0	3.1	592.2	147.8	66.4	0.1	0.8	0.4	2.7	3.2	28.8
Costa Rica	2,725.2	148.3	170.9	1,381.0	186.7	168.8	12.4	226.1	10.6	96.2	79.0	700.9
Côte d'Ivoire	20,184.3	116.2	260.9	792.2	631.8	211.4	4.5	83.6	5.5	60.2	22.0	198.0
Croatia	4,035.7	2.6	615.0	2,089.9	129.1	224.6	2.5	31.4	36.4	462.9	—	—
Cuba	13,194.0	830.5	458.9	1,943.7	130.0	296.2	46.5	539.1	9.4	114.3	156.0	993.4
Cyprus	288.2	174.5	21.8	1,042.3	84.4	76.5	0.1	1.9	10.5	131.3	6.8	48.2
Czech Republic	8,372.5	10.2	1,525.8	4,277.0	574.8	417.3	40.2	153.1	85.3	921.9	105.0	399.8
Denmark	2,737.5	19.6	401.8	3,787.8	354.6	314.6	10.4	24.3	84.2	859.5	41.0	92.5
Dominica	147.8	—	1.6	24.6	5.5	3.1	—	—	0.2	2.2	—	—
Dominican Republic	6,589.8	826.1	—	3,263.2	210.5	393.1	2.1	35.4	13.5	122.6	109.0	906.9
East Timor	1,517.4	37.3	—	41.8	9.2	11.8	—	—	—	—	—	—
Ecuador	17,174.5	1,440.2	498.3	2,464.1	133.9	198.9	33.3	621.8	11.8	123.0	212.0	1,908.0
Egypt	10,697.6	35,246.2	13,446.0	23,682.2	1,394.1	1,609.1	168.0	3,165.7	32.2	414.6	530.0	4,770.0
El Salvador	3,441.3	42.1	270.5	1,482.0	214.7	150.0	7.0	127.4	7.4	75.3	32.0	288.0
Eritrea	2,018.6	7.1	1.9	1,871.0	102.1	83.2	0.0	0.9	0.6	7.1	3.1	27.9

continues

DATA TABLE 15 continued

Country	Water Footprint of Consumption of Agricultural Products						Water Footprint of Consumption of Industrial Products						Water Footprint of Domestic Water Consumption	
	Internal (WFcons, nat, int, agr)			External (WFcons, nat, ext, agr)			Internal (WFcons, nat, int, ind)			External (WFcons, nat, ext, ind)				
	Green	Blue	Grey	Green	Blue	Grey		Blue	Grey		Blue	Grey		Blue	Grey
Estonia	978.1	0.3	106.9	826.2	190.8	48.0		1.7	5.9		13.0	142.0		9.0	38.8
Ethiopia	73,675.2	1,512.4	335.5	1,357.8	116.8	209.2		1.0	19.2		5.1	67.0		33.3	299.7
Fiji Islands	960.3	0.8	23.8	334.0	45.8	14.4		0.4	8.2		1.7	20.7		1.0	9.0
Finland	3,373.3	6.2	90.8	2,105.0	248.4	208.9		52.9	167.3		70.3	815.8		34.0	153.5
France	43,704.2	1,375.0	3,753.5	36,738.7	4,576.9	2,077.9		876.3	3,320.4		579.5	6,280.1		628.0	2,221.3
French Polynesia	215.8	0.6	1.5	150.0	32.3	19.7		—	—		2.7	23.8		—	—
Gabon	1,270.7	12.5	8.3	337.6	68.0	37.6		0.2	3.6		0.7	7.5		6.0	54.0
Gambia	717.9	1.6	7.2	326.4	69.1	39.0		0.1	2.7		0.3	3.7		0.7	6.3
Georgia	5,756.9	359.4	200.9	2,772.6	161.5	138.7		26.4	483.7		5.2	77.0		72.0	648.0
Germany	25,576.8	83.9	7,055.1	60,940.9	4,672.9	5,116.2		809.5	854.9		845.4	8,982.9		581.0	1,632.0
Ghana	21,592.2	37.7	94.8	1,140.8	408.5	191.2		4.1	75.4		7.6	90.0		23.5	211.5
Greece	9,744.3	2,253.0	962.0	8,324.7	1,053.9	854.9		11.0	29.9		120.3	1,529.7		127.0	561.8
Grenada	128.6	0.2	—	64.6	9.1	11.5		—	—		0.4	4.0		—	—
Guatemala	8,136.5	148.6	573.9	1,552.9	201.8	226.0		10.1	114.8		12.9	128.0		13.0	96.3
Guinea	12,628.3	43.0	21.4	446.4	222.7	86.3		0.8	11.7		1.0	10.6		12.0	104.6
Guinea-Bissau	1,363.8	83.6	—	41.4	53.1	14.7		0.3	5.5		0.2	2.1		2.3	20.7
Guyana	794.4	73.3	87.2	125.4	16.8	30.1		0.4	6.9		1.0	9.9		3.0	27.0
Haiti	6,808.8	224.6	—	1,229.8	433.9	208.3		0.4	7.4		1.6	14.6		5.0	45.0
Honduras	5,754.3	113.2	350.1	776.5	172.3	101.2		2.2	36.7		4.1	36.7		7.0	63.0
Hungary	15,780.8	73.7	2,039.6	3,765.5	353.3	249.8		104.7	674.9		67.1	753.6		71.0	380.9
Iceland	4.1	—	1.6	414.5	23.2	24.9		3.9	32.3		5.6	58.5		5.0	21.4
India	720,644.0	221,346.8	96,332.9	23,546.0	2,179.1	1,885.5		1,261.9	23,850.5		98.5	1,219.8		5,224.0	47,016.0
Indonesia	176,695.4	9,413.1	15,941.2	16,089.3	4,969.0	1,553.2		17.2	280.0		57.9	673.6		662.0	5,886.7
Iran	50,325.3	37,496.5	8,253.6	19,445.0	1,347.5	1,002.6		37.1	643.5		43.6	553.4		620.0	5,580.0

Ireland	1,078.6	0.2	145.7	2,556.3	351.6	253.5	11.1	46.1	31.4	416.6	26.0	117.4
Israel	1,272.1	823.4	333.3	9,710.6	606.4	546.4	4.0	8.1	56.8	595.0	63.0	105.7
Italy	36,197.9	2,526.9	5,416.4	62,764.0	6,727.6	4,740.8	500.7	2,741.5	523.4	5,628.4	807.0	3,891.9
Jamaica	2,162.1	44.7	41.8	1,510.2	164.5	195.2	2.6	48.6	6.3	57.5	14.0	126.0
Japan	22,377.2	1,388.0	2,109.0	105,550.2	6,930.0	10,513.4	583.7	3,392.7	889.4	10,500.5	1,740.0	8,805.1
Jordan	557.6	375.7	50.0	5,582.1	812.0	514.0	1.4	11.5	17.2	210.4	29.1	155.5
Kazakhstan	25,933.0	5,218.5	250.7	1,229.6	249.1	197.6	111.9	1,881.6	22.4	342.2	59.0	531.0
Kenya	27,947.7	295.5	273.0	4,976.8	744.8	277.2	4.3	76.5	8.0	101.6	47.0	418.9
Kiribati	199.0	—	—	26.8	12.6	4.1	—	—	0.0	0.5	—	—
Korea, Democratic People's Republic	12,405.1	1,328.4	—	2,080.0	382.1	333.3	95.7	1,810.6	5.7	83.9	179.0	1,611.0
Korea, Republic	11,724.0	454.3	1,130.6	47,962.4	3,700.9	4,266.6	86.6	306.0	266.5	2,895.8	662.0	2,214.1
Kuwait	14.9	51.6	14.4	3,181.1	553.9	311.3	0.5	9.8	11.1	116.9	40.1	360.5
Kyrgyzstan	2,458.1	1,194.9	71.3	2,944.6	133.6	83.5	12.1	160.2	4.8	84.2	32.0	260.3
Laos	4,839.6	348.8	0.8	148.4	21.4	13.2	6.4	117.0	1.8	29.6	13.0	117.0
Latvia	2,455.4	0.5	254.2	686.6	93.9	58.0	4.0	23.3	37.5	548.3	16.0	78.2
Lebanon	1,158.0	667.3	128.7	4,751.5	544.5	292.8	6.9	41.8	20.5	265.1	38.0	143.7
Lesotho	2,898.7	5.4	1.3	76.0	62.0	20.9	0.6	10.0	0.3	4.0	2.0	18.0
Liberia	2,970.9	3.3	—	180.7	85.2	63.2	0.8	16.1	7.5	87.1	3.0	27.0
Libya	1,630.0	1,470.3	87.3	5,310.9	1,226.9	568.8	3.4	61.6	5.5	56.7	61.0	549.0
Lithuania	3,670.9	1.1	99.6	764.3	137.3	96.3	1.4	6.3	29.8	379.4	21.0	100.5
Luxembourg	229.3	0.0	45.6	645.2	30.7	39.3	0.7	0.8	12.7	122.6	2.5	3.3
Macedonia	1,210.6	71.4	85.3	769.4	77.4	61.7	6.8	94.1	9.0	115.1	21.5	187.4
Madagascar	20,644.5	2,253.5	41.5	562.6	247.9	74.4	9.6	181.8	3.2	37.1	42.0	378.0
Malawi	9,654.3	148.7	539.9	581.6	36.0	62.4	2.4	44.6	1.5	19.9	15.0	135.0
Malaysia	29,102.0	1,148.7	1,338.6	11,805.6	2,202.7	1,329.5	23.6	310.3	57.1	502.0	152.0	1,368.0
Maldives	60.7	1.2	2.4	222.5	44.0	24.3	0.0	—	0.9	14.1	0.3	1.9
Mali	18,936.7	1,496.0	0.3	586.4	116.8	60.7	2.7	51.7	2.3	27.2	59.0	531.0
Malta	29.0	3.8	2.7	625.9	69.0	50.8	0.0	0.2	4.0	45.2	4.0	31.7
Mauritania	3,341.8	206.8	0.0	2,745.6	224.0	80.6	1.6	29.3	1.1	11.8	15.0	135.0

continues

DATA TABLE 15 *continued*

	Water Footprint of Consumption of Agricultural Products							Water Footprint of Consumption of Industrial Products					Water Footprint of Domestic Water Consumption	
	Internal (WFcons, nat, int, agr)			External (WFcons, nat, ext, agr)			Internal (WFcons, nat, int, ind)		External (WFcons, nat, ext, ind)		
Country	Green	Blue	Grey	Green	Blue	Grey	Blue	Grey	Blue	Grey	Blue	Grey
Mauritius	446.1	10.0	4.0	1,486.1	258.2	124.5	0.7	8.3	4.7	55.7	21.4	172.1
Mexico	83,840.7	8,654.2	8,796.4	65,986.0	8,475.0	5,950.0	134.6	1,674.0	358.5	3,174.9	1,359.0	9,021.8
Moldova	3,983.3	364.6	119.8	225.6	53.6	18.9	42.5	295.3	6.3	98.4	22.0	145.2
Mongolia	3,171.4	85.6	72.8	5,319.3	129.0	127.0	3.5	55.8	2.6	37.2	9.0	81.0
Morocco	28,420.2	4,819.7	1,326.6	12,286.8	1,062.2	858.5	12.8	45.9	29.8	334.2	123.0	632.9
Mozambique	18,980.3	145.8	115.9	1,147.6	171.8	122.4	0.3	4.9	1.2	12.1	7.0	63.0
Myanmar	52,447.8	1,921.6	815.6	837.9	52.5	57.8	6.1	107.2	6.1	82.9	41.0	369.0
Namibia	2,150.6	34.3	21.7	624.5	124.9	44.1	0.6	10.2	0.9	8.9	7.3	65.7
Nepal	23,338.5	2,811.3	206.1	2,344.9	241.2	302.9	2.3	39.9	4.6	71.9	30.0	270.0
Netherlands	695.2	44.0	241.5	16,142.5	1,586.3	1,359.8	109.7	19.2	266.2	2,755.3	49.0	104.4
New Caledonia	119.4	0.0	7.8	238.6	25.6	17.2	—	—	2.2	21.7	—	—
New Zealand	1,864.3	242.4	89.9	2,811.9	191.6	137.8	8.4	31.6	36.4	400.5	102.0	288.0
Nicaragua	3,498.2	134.4	82.2	536.3	99.5	62.8	1.2	22.3	4.0	43.9	19.0	171.0
Niger	37,783.4	718.2	55.8	666.2	254.3	83.3	0.3	5.0	0.6	6.2	9.0	81.0
Nigeria	145,305.7	1,375.3	699.1	5,557.1	1,459.0	941.5	13.0	197.6	9.0	88.5	169.0	1,521.0
Norway	1,551.9	8.8	54.3	3,055.6	394.7	251.0	38.3	150.9	59.5	606.5	50.0	183.4
Occupied Palestinian Territory	2,540.9	367.3	—	97.1	21.3	127.8	3.4	64.7	0.2	2.6	17.2	154.7
Pakistan	80,221.9	60,459.4	20,480.7	28,718.7	2,252.7	1,220.2	126.9	2,383.0	21.5	274.0	327.0	2,943.0
Panama	2,226.0	54.4	158.8	928.0	95.5	131.4	1.4	10.9	7.6	79.2	55.0	314.4
Paraguay	9,672.6	214.5	248.3	141.1	59.5	23.6	1.8	29.4	6.5	75.1	10.0	82.6
Peru	13,141.8	3,299.5	1,458.4	8,050.4	404.0	567.8	74.5	362.4	17.8	195.1	168.0	720.5
Philippines	78,032.3	3,356.3	3,722.9	10,433.7	2,020.2	1,704.4	132.9	2,405.8	34.7	419.0	584.8	5,263.2
Poland	29,880.5	65.3	5,829.9	8,934.8	1,391.5	906.6	438.5	3,080.7	153.4	1,921.1	210.0	1,167.7

Portugal	7,156.9	1,513.3	627.4	11,901.3	2,023.9	982.3	42.7	264.4	45.3	476.5	108.0	610.2
Romania	26,137.0	529.9	1,295.1	3,567.9	770.2	387.0	170.7	2,205.2	54.7	724.8	169.0	1,292.9
Russian Federation	200,827.8	6,300.2	7,257.7	25,437.2	5,649.8	1,528.4	520.3	8,682.6	61.3	825.2	1,340.0	12,060.0
Rwanda	5,996.7	19.0	4.0	189.1	41.3	47.2	0.5	10.0	0.5	6.7	3.6	32.4
Saint Kitts and Nevis	27.5	0.0	0.1	32.8	5.4	4.0	—	—	0.5	5.4	—	—
Saint Lucia	8.9	0.0	—	180.4	17.4	19.3	—	—	0.7	6.3	1.3	11.3
Saint Vincent/ Grenadines	95.1	—	—	64.3	10.2	10.0	—	—	0.3	2.9	—	—
Samoa	319.0	—	0.1	29.6	11.1	4.4	—	—	0.2	2.3	—	—
São Tomé and Príncipe	206.6	0.1	—	22.1	8.4	6.8	—	—	0.1	0.6	—	—
Saudi Arabia	2,667.5	7,019.4	1,115.7	21,217.6	2,149.9	1,893.1	17.7	302.0	48.1	485.6	213.0	1,917.0
Senegal	7,620.5	321.4	49.5	2,662.0	538.0	200.0	2.5	36.1	4.4	50.1	9.8	79.9
Serbia and Montenegro	13,138.5	116.0	1,055.6	1,153.9	275.5	114.6	433.1	8,200.1	23.8	349.5	78.0	702.0
Seychelles	53.0	3.1	0.0	81.4	11.8	7.3	0.2	3.0	0.7	8.1	0.8	7.2
Sierra Leone	4,990.7	48.3	—	1,095.2	100.4	68.0	0.4	6.4	0.6	6.6	2.0	18.0
Slovakia	3,786.8	59.2	406.8	1,667.2	238.8	176.5	15.1	91.4	38.8	395.5	42.8	261.7
Slovenia	742.6	1.1	407.6	1,930.4	130.7	185.9	18.6	157.1	22.8	246.6	20.1	134.1
Solomon Islands	223.7	—	—	58.5	9.1	11.0	—	—	0.2	2.4	—	—
South Africa	35,551.6	3,466.1	2,253.2	10,867.1	483.9	566.2	22.6	183.2	48.3	522.9	390.4	2,368.3
Spain	41,038.9	8,759.6	5,805.6	32,561.5	3,237.5	2,782.2	246.0	—	394.3	4,194.6	479.0	1,020.9
Sri Lanka	14,569.5	1,993.7	1,036.6	3,922.7	1,016.3	504.2	10.0	173.1	12.3	173.4	30.0	270.0
Sudan	50,305.0	6,914.0	358.3	1,985.8	162.6	155.0	9.6	175.4	5.8	79.3	99.0	891.0
Suriname	352.7	54.0	37.2	122.5	8.0	13.1	0.5	8.9	0.7	6.5	3.0	27.0
Swaziland	527.3	49.8	—	703.6	156.6	23.1	0.5	9.2	0.4	5.2	2.4	21.6
Sweden	4,772.0	13.8	764.1	4,479.4	453.9	477.2	39.1	118.5	97.6	1,128.1	109.0	271.6
Switzerland	1,328.1	0.7	326.8	6,157.6	569.7	597.2	49.8	32.1	145.2	1,624.1	62.0	161.0
Syria	20,023.1	5,713.0	2,506.6	4,566.1	642.1	293.0	17.4	261.5	10.2	133.9	142.6	1,216.9
Tajikistan	4,017.0	2,776.2	296.5	6,629.8	92.0	56.9	23.9	409.3	3.6	58.2	44.0	396.0

continues

DATA TABLE 15   continued

| Country | Water Footprint of Consumption of Agricultural Products ||||||| Water Footprint of Consumption of Industrial Products |||||| Water Footprint of Domestic Water Consumption ||
| | Internal (WFcons, nat, int, agr) ||| External (WFcons, nat, ext, agr) ||| Internal (WFcons, nat, int, ind) || External (WFcons, nat, ext, ind) || | |
	Green	Blue	Grey	Green	Blue	Grey	Blue	Grey	Blue	Grey	Blue	Grey
Tanzania	31,712.9	799.1	147.6	1,836.6	344.1	174.9	1.2	22.1	4.3	58.2	52.7	474.3
Thailand	59,129.4	8,471.4	4,050.3	9,367.2	1,477.0	683.8	76.0	1,275.9	105.5	1,248.1	273.9	2,465.1
Togo	4,484.9	6.1	27.3	501.0	79.4	47.4	0.1	2.3	1.1	12.7	8.9	80.1
Trinidad and Tobago	108.2	0.1	0.5	1,588.2	115.5	140.7	1.7	21.9	4.4	39.1	21.0	183.8
Tunisia	12,324.7	1,201.5	400.8	4,785.7	1,285.0	548.3	3.3	30.2	17.0	169.9	36.5	220.2
Turkey	62,426.0	11,221.3	6,903.6	14,441.1	4,756.7	1,165.3	160.1	1,646.4	202.6	2,642.0	556.0	3,636.7
Turkmenistan	5,631.8	3,269.1	0.7	841.6	42.7	52.8	3.6	33.6	4.0	33.1	42.0	378.0
Uganda	25,440.5	10.7	28.4	897.1	245.5	85.9	2.5	46.5	3.0	44.1	13.0	117.0
UK	14,449.5	56.0	3,160.0	39,873.5	4,264.0	3,617.4	245.6	117.0	725.5	7,687.4	207.0	242.8
Ukraine	54,470.5	1,372.2	3,022.2	2,726.0	375.2	214.9	426.7	7,781.3	99.4	1,696.1	456.0	4,104.0
United Arab Emirates	1,374.4	466.5	40.9	5,023.0	1,285.9	617.1	2.1	32.2	83.6	899.7	61.7	555.3
Uruguay	5,183.7	176.6	129.4	1,285.6	13.1	39.5	1.7	32.0	8.4	102.6	8.0	72.0
USA	462,444.5	39,447.3	64,568.7	106,318.7	10,839.9	9,160.3	8,915.9	47,583.7	3,276.9	36,696.4	6,544.0	25,557.4
Uzbekistan	12,695.0	8,630.3	0.3	6,362.5	76.8	145.3	47.5	883.8	12.1	200.6	277.0	2,493.0
Vanuatu	245.0	—	—	8.8	14.3	6.5	—	—	0.2	2.6	—	—
Venezuela	21,550.6	1,194.1	817.3	12,984.8	547.2	706.6	16.1	299.5	21.4	194.4	381.0	3,429.0
Vietnam	50,495.0	5,381.6	7,459.0	4,146.3	507.5	334.6	479.7	9,014.5	30.1	429.5	554.0	4,986.0
Yemen	1,320.7	2,335.5	94.1	10,199.2	1,669.4	709.8	2.0	34.2	3.1	37.8	27.2	242.6
Zambia	7,972.4	294.1	169.0	805.8	83.6	52.0	3.0	53.6	1.4	16.2	29.0	261.0
Zimbabwe	11,846.5	713.2	426.4	1,002.0	81.7	68.6	10.1	188.7	2.5	26.7	58.9	530.1
World	4,902,626.4	659,210.5	555,954.8	1,346,910.7	204,557.4	131,506.6	23,614.9	211,278.3	13,918.9	151,457.2	42,022.9	282,005.3

DATA TABLES 16A AND 16B

Global Cholera Cases Reported to the World Health Organization, by Country, 1949-2011

Description

Annual cases of cholera are shown here for all reporting countries. Cholera is one of many waterborne diseases that are a consequence of the lack of sanitation services and access to clean drinking water. Some data on cholera cases and deaths are available as early as 1922 from reports to the League of Nations, but consistent reports are available from the World Health Organization only since 1949. The data here are given annually from 1949 to 2011.

In 2011, a total of fifty-eight countries from all continents reported approximately 590,000 cases of cholera to the World Health Organization, of which 32 percent were in Africa and 61.2 percent were in the Americas (defined as North, Central, and South America and the Caribbean). These data reflect a massive upsurge in the Americas from previous years. A particularly severe outbreak started in Haiti at the end of October 2010 and also affected the Dominican Republic.

These data also show the rapid and severe spread of cholera in Latin America during the seventh pandemic, starting with the appearance of cholera in Peru in 1991 and spreading rapidly. That sharp spike in cholera was the first manifestation of the seventh pandemic in the Americas and was the first time that the Americas had seen more than a handful of indigenous cases in over a century.

Limitations

There are serious problems with the reporting of cholera. In particular, the data here represent only the cases identified and reported to the World Health Organization and do not include undiagnosed or unreported cases. The true number of cholera cases is known to be much higher than reported here. Many countries refuse to report cholera cases, or define many severe diarrheal diseases as something other than cholera, often as "acute watery diarrhea." As many as 500,000 to 700,000 cases of acute watery diarrhea occur annually in areas of Central and Southeast Asia and in some African countries.

As a result, the totals shown in the table are likely to seriously underrepresent total cholera cases.

Sources

World Health Organization (WHO). 2011. Number of Reported Cholera Cases. Global Health Observatory. http://www.who.int/gho/epidemic_diseases/cholera/cases_text/en/index.html (accessed June 1, 2013).
———. 2013. Cholera Data Portal. http://apps.who.int/gho/data/node.main.175.

DATA TABLE 16A Global Cholera Cases Reported to the World Health Organization, by Country, 1949–1979

Country	1949	1950	1951	1952	1953	1954	1955	1956	1957	1958	1959	1960	1961	1962	1963	1964	1965	1966	1967	1968	1969	1970	1971	1972	1973	1974	1975	1976	1977	1978	1979
Afghanistan												887					218														
Albania																															
Algeria																							1,332	646	605	738	1,165	286	262	220	2,513
Angola																								1	268	263	934	88	726		
Argentina																															
Armenia																															
Australia																								40					2	1	1
Austria																															
Azerbaijan																															
Bahamas																															
Bahrain																	1							74	37					906	39
Bangladesh	29,809	20,894	21,154	27,631	15,617	16,642	23,699	8,054	15,631	16,915	15,618	2,663	2,524	3,975	3,333	1,123	3,154	664	3,156	7,411	7,419	2,342	1,059	1,969	5,614	4,888	957	10,403	5,576	2,154
Belarus																															
Belgium																															
Belize																															
Benin																						175	2,133	250	3	73	45	146	2	27	
Bhutan																															
Bolivia (Plurinational State of)																															
Botswana																															
Brazil																															
Brunei Darussalam							1										198					24									
Burkina Faso																							1	1,761	1	1,118	632	3			1
Burundi																														8,297	915
Cambodia		10	57	24	11	2		1	6	3						52	187	62	4	8			2,167	362	159	145	66				
Cameroon																									206	83	3	135		3	16

continues

DATA TABLE 16A   continued

Country	1949	1950	1951	1952	1953	1954	1955	1956	1957	1958	1959	1960	1961	1962	1963	1964	1965	1966	1967	1968	1969	1970	1971	1972	1973	1974	1975	1976	1977	1978	1979
Canada																															
Cape Verde																										303	20	219			
Central African Republic																															
Chad																							8,230	9		338					
Chile																															
China	1													384	1				1												85
Colombia																															
Comoros																											2,675	5			
Congo																														51	5
Costa Rica																															
Côte d'Ivoire																						868	668								3
Czech Republic																						4									
Democratic People's Republic of Korea																															
Democratic Republic of the Congo																														3,481	5,515
Denmark																															
Djibouti																						6	440	8				2	2	3	
Dominican Republic																															
Ecuador																															
El Salvador																															
Equatorial Guinea																															
Eritrea																															
Estonia																															
Ethiopia																						850	2,187								
Finland																															
France																						1	3		4	5	9	5			8
Gabon																															5
Gambia																							1	2	6	3	1	2			
Georgia																															
Germany																													1	1	

Country	1	2	3	4	5	6	7	8	9	10	11	12	13	14	15	16	17	18	19	20	21	22	23	24	25	26	27	28	29	30
Ghana																					2,733	13,048	625	677	483	187	102	6,565	1,853	1,783
Greece																														
Guatemala																														
Guinea																					2,000									
Guinea-Bissau																														
Guyana																														
Haiti																														
Honduras																														
Hungary																														
India	625	86,787	101,493	213,225	22,146	23,533	39,829	54,961	66,849	14,852	14,621	47,637	51,082	44,839	53,011	42,677	13,052	13,708	22,587	19,280	15,067	71,386	20,435	41,611	30,903	22,049	14,946	8,376	10,585	5,073
Indonesia											41	1,742	428	535	326		204	700	135	627	5,997	23,555	44,383	52,042	41,474	48,387	41,264	17,112	10,683	18,817
Iran (Islamic Republic of)																5,037	162	221		557	19,663	344	322	55	304	2,966	2,100	10,836	264	1,856
Iraq																	227											133	96	
Ireland																														
Israel																					185	1	11		278	1		2		8
Italy														16	2						8	5				3	6	52	34	11
Japan																					3						152	427		141
Jordan																														
Kazakhstan																														
Kenya																						239	51		402	1,093	1,359	21	673	1,070
Kiribati																												1,307	494	
Kuwait																					4					3	2	13	1	3
Kyrgyzstan																														
Lao People's Democratic Republic																				448										
Lebanon											31										54								30	
Lesotho																														
Liberia																					168	606	947	1,336	512	704	646	512	422	438
Libya																					1,151									
Madagascar																						2								

continues

DATA TABLE 16A continued

Country	1949	1950	1951	1952	1953	1954	1955	1956	1957	1958	1959	1960	1961	1962	1963	1964	1965	1966	1967	1968	1969	1970	1971	1972	1973	1974	1975	1976	1977	1978	1979
Malawi																												19	577	263	
Malaysia						1									349	512	16	5		15	70	106	53	864	369	349	110	246	444	1,635	502
Maldives																														11,336	
Mali																						2,665	4,792	2	219	130					
Marshall Islands																															
Mauritania																							1,139	148	150						
Mauritius																															
Mexico																															
Micronesia (Federated States of)																															
Mongolia																															
Morocco																							56	7				2			
Mozambique		4,212	6,753	347	22	35	42	13	10	7	3	259	1		1	3,019			9	3	205					453	1,018	11	3	18	4,564
Myanmar																922	207	263				911	378	180	386	2,363	2,942	1,519		3,551	874
Namibia																															
Nauru																														38	50
Nepal										2,706							727	2		34	226	391	4	1	7	8	260	185	428	1,662	22
Netherlands																												1	1	4	5
New Zealand																								3							
Nicaragua																															
Niger																						16	9,265	51	121	286					
Nigeria																						15	22,931	1,363	157		38	112	376	197	293
Norway																															
Oman																							9								
Pakistan										366		6,704					21			3,932	717	2	1,185					144	12		
Panama														1,293																	
Papua New Guinea																															
Paraguay																															
Peru																															
Philippines													9,923	13,015	3,979	16,464	4,581	5,261	1,629	2,787	1,541	1,095	3,585	5,601	2,840	1,730	680	1,258	1,363	1,408	1,268
Poland																															

Country																
Portugal					64					2,467	1,066					
Qatar																
Republic of Korea			415	11		1,538	475									
Republic of Moldova																
Romania																
Russian Federation							720				1		1			
Rwanda														838		5
Samoa																
São Tomé and Príncipe																
Saudi Arabia							266	303		91	50	18		30		23
Senegal							265	379						315		103
Serbia and Montenegro (former)											3					
Seychelles																
Sierra Leone							293	211							12	
Singapore			27	24	6	11		1	114	8	10		11		83	10
Slovenia																
Somalia							43	89								
South Africa									1	37			2		2	
Spain								22		5	11					267
Sri Lanka									118	4,559	1,453	728	5	48		46
Sudan																845
Suriname																
Swaziland													2			
Sweden								4		10	1					1
Switzerland													1		1	
Syrian Arab Republic							49	5		505	67		795	2,362	689	
Tajikistan	16	1														
Thailand	10,201	4,368	2,230	960	65	401	148	404	961	844	1,475	1,335	6	383	4,183	1,788
Togo					75		335	16		58	132					6
Tonga																

continues

DATA TABLE 16A *continued*

Country	1949	1950	1951	1952	1953	1954	1955	1956	1957	1958	1959	1960	1961	1962	1963	1964	1965	1966	1967	1968	1969	1970	1971	1972	1973	1974	1975	1976	1977	1978	1979
Tunisia																						27		4	656						
Turkey																													17		
Turkmenistan																						384									
Tuvalu																															
Uganda																							757				3			1,120	940
Ukraine																															
United Arab Emirates																								2							
United Kingdom																						1	3	2	5	3	1	1	2		
United Republic of Tanzania																										10			297	6,608	2,559
United States of America																															
Uzbekistan																															
Venezuela (Bolivarian Republic of)																															
Vietnam		9	26	7	7	3										20,186	1,738	8,353	7,664	1,097	2,573	82	270	146	1,495	139	5	1,068	32	2	365
Yemen																							190	1,064	215	6	1			414	1,953
Zambia																														263	
Zimbabwe																									1	37	615	144			12
Global Total	17	34,665	114,518	123,025	240,927	37,804	40,218	63,542	63,031	95,763	36,138	38,130	61,966	68,727	59,437	95,938	56,609	31,146	24,748	34,164	36,173	63,994	176,058	80,248	108,788	97,635	93,055	68,622	63,276	77,678	57,645

DATA TABLE 16B   Global Cholera Cases Reported to the World Health Organization, by Country, 1980–2011

Country	1980	1981	1982	1983	1984	1985	1986	1987	1988	1989	1990	1991	1992	1993	1994	1995	1996	1997	1998	1999	2000	2001	2002	2003	2004	2005	2006	2007	2008	2009	2010	2011
Afghanistan														37,046	38,735	19,903		8,340	20,000	49,278	8,660	8,998				66			13,152	1,324	2,369	3,733
Albania															626																	
Algeria	614			218	45			1,507	699	393	1,299		69		118												—					
Angola								16,222	15,500	17,601	9,527	8,590	3,608	2,350	3,443	3,295	1,306									—	134,514	36,844	31,533	4,038	1,484	1,810
Argentina															889	188	474	1,274	12													
Armenia																			25													
Australia	2	2	1	4	1	2		1	1				3	5	3	5	2	4	10	8	2	4	10		4	4	6	6			3	6
Austria											2			1	1		1		2	2		—	2			1						
Azerbaijan															18																	
Bahamas																																1
Bahrain																																
Bangladesh								523	571	94	82	8	479	78	562	2,297	418	3,918	2,134	6,880	1,021											
Belarus															3	3																
Belgium	1			1	1	1							1	1	1																	
Belize							243						159	135	6	19	26	4	56	12												
Benin	3	2	3	1	1							7,474	413	10	187	203	6,190	1,592	412	1,710	936	7,886	540	868	1,284	1,498	182		2,955	148	983	755
Bhutan												422	494			25			19													
Bolivia (Plurinational State of)												206	22,260	10,134	2,710	2,293	2,847	3,264	466													
Botswana																										—	—		24	15		
Brazil												2,103	30,317	59,212	49,455	15,915	5,522	5,762	5,142	8,990	1,430	14			42	5						1
Brunei Darussalam			6																	93												3
Burkina Faso					2,191	1,149						537				1,451	425		2,072	186	1,234	954		2		2,100	—		—			20
Burundi	2,039	582	415	512	180	259		474			82	3	479	78	562	2,297	418	3,918	2,134	6,880	2,042	2,006	1,154	864	1,638	2,618	1,772	730	702	710	333	1,072
Cambodia									56	47	5	770	1,229	2,252	3,085	4,190	740	310	2,394	3,422					114					78	588	12
Cameroon	229	243	5	55	392	1,158	165	94	4	4	16	4,026	1,268	648	527	615	5,796	3,418	9,206	3,168	246	518	132	414	16,010	5,694	1,844	20	—	1,608	10,759	22,433

continues

DATA TABLE 16B continued

Country	1980	1981	1982	1983	1984	1985	1986	1987	1988	1989	1990	1991	1992	1993	1994	1995	1996	1997	1998	1999	2000	2001	2002	2003	2004	2005	2006	2007	2008	2009	2010	2011
Canada												2	5	6	2	7	2		4		10	12	8	10	6	14	4	2	3	4	2	9
Cape Verde																12,913	428		266							—	—					
Central African Republic										1								886	44						640							117
Chad												13,915			1,094		7,830	17,602	8,190	434		10,488		110	11,062	180	3,336	—		67	6,395	17,267
Chile													73	32	1		1		8 24													1
China	88								4 7,267	5,628	644	210	583 11,673		34,877	10,337	316	2,354	9,284	9,178	3,692	280		446	488	1,960	322	336	522	170	157	26
Colombia												11,979	15,129	230	996 1,922		4,428	3,016	884	84					2							
Comoros																			14,600	2,360	6,594	452	3,134	112	2	—	—	3,110	4			
Congo																		550	6,444	9,626	18						350	15,570	468	93		
Costa Rica																	19	1														
Côte d'Ivoire			34									604	37	724	1,108	4,993	1,345					11,824	8,376	2,068	210	78	828	16	21	10	32	1,261
Czech Republic																		7					2									
Democratic People's Republic of Korea	170																															
Democratic Republic of the Congo	1,051	2,379	10,328	2,977	162	740	1,059	1,150	295	99	468	4,066	1,949	986	58,057	553	7,886	4,842	69,798	25,422	29,990	11,456	63,316	54,544	15,330	26,860	41,284	56,538	90,450	45,798	13,884	21,700
Denmark											1			2	2	3									2				1			
Djibouti						115								10,055	1,122			4,848	328		3,656							2,174			2,047	127
Dominican Republic																															191	20,851
Ecuador														6,833	1,785	2,160	1,059	130	7,448	180	54	18		50	5							
El Salvador												46,320	31,870	6,573 11,745		2,923	182		16	268	631											
Equatorial Guinea					404	108						947	8,717												118	12,782						
Eritrea			—																							—	—	238	1		—	
Estonia		1																														
Ethiopia		20	18	3	1		37	7							2	1		6	4			4			32	—	108,140	48,242	11,586	63,018	1,682	
Finland																			2							2			1			
France	1	20								1	6	7		5	4	6	8	6	4		10	4	2				4	8	6	1		1
Gabon		7				2																		12	1,258		—		2			
Georgia														8	1	15	7									428		24	1			
Germany	4	4	1				2		2	1	1	1	1	1	5	1		4	10	6	4	2	2	2	6		2	4			4	4

368

Country	1	2	3	4	5	6	7	8	9	10	11	12	13	14	15	16	17	18	19	20	21	22	23	24	25	26	27	28	29	30	31
Ghana	260	943	11,051	14,160	1,015	60				2,937	13,172	228	1,448	2,267	4,698	1,665	758	6,852	18,864	6,662	10,974	7,228	408	814	6,332	6,714	358	3,669	2,588	438	10,628
Greece							1																								
Guatemala											3,674	15,395	30,604	5,282	7,970	1,568	2,526	11,940	4,154	1,224	26	2	1	3,032	7,642	6,484	17,092	1,539	42		3
Guinea						286								31,415	6,506	287		1,762	1,198	1,038	784	122	12	310	50,222	74	306	28,646	5		
Guinea-Bissau						200	6,000							15,296	119	8,397	252					1,684	580								1,187
Guyana												576	66																		
Haiti											17	407	4,013	5,049	4,717	708	200	612	112	16										179,379	340,311
Honduras																				1											
Hungary																															
India	8,344	4,681	4,656	8,542	2,527	5,808	4,208	11,934	17,162	3,583	6,993	6,911	12,084	4,973	3,315	4,396	5,536	14,302	7,678	8,106	8,162	6,910	5,786	9,390	6,310	3,878	5,270	5,360		5,155	
Indonesia	5,541	18,354	10,391	13,832	7,921	4,732	11,915	659	50	155	6,202	25	3,564	47		66	132				1,122				2,676		38	1,007			
Iran (Islamic Republic of)	1,599	6,034	427	270	531	1,208	20	295	486	178	1,892	97	1,347	15	2,177		2,212	540	2,738	690	210	236	192	188	2,266			72			1,187
Iraq									5,222						119																
Ireland											877	97	280	838	820	1		106	3,970	1,064	1,120	1,436	374	70		9,394		1,850			
Israel		1						2														1									
Italy									1						12	1	4									2	1				
Japan	23	19	16	35	55	36	26	34	33	73	90	46	89	91	321	39	178	120	80	68	22	54	32	132	86	68	—		12		
Jordan	870									2																					
Kazakhstan														3	16		8			2											
Kenya	2,808	2,424	3,498	1,049	14	1,352	839	255	918			3,388	880	1,547		482	34,400	44,864	22,078	2,314	2,002	582		1,740	1,632	1,740	2,412	6,182	22,850	74	
Kiribati																												1			
Kuwait		8				113	38		133				1				1														
Kyrgyzstan														4																	
Lao People's Democratic Republic													5,521	9,660	1,365		720										338	402	237		
Lebanon													344	3																	
Lesotho																									1	—	—				
Liberia	2,690	1,582	670	183	492	355	59	33	68	132				764	3,420	8,922	182	4,246	430	730	2,124	2,230	69,480	5,572	7,646	9,858	6,126	2,472	2,140	1,546	1,146
Libya									28					22									10								
Madagascar																—				19,946	58,610	14,438	54			—	—				

continues

DATA TABLE 16B continued

Country	1980	1981	1982	1983	1984	1985	1986	1987	1988	1989	1990	1991	1992	1993	1994	1995	1996	1997	1998	1999	2000	2001	2002	2003	2004	2005	2006	2007	2008	2009	2010	2011
Malawi		261		487					6	8,351	13,457	8,088	298	25,193	107	1	1	260	3,490	53,016	4,782	4,790	65,236	5,472	1,350	2,210	8,296	950	1,662	11,502	1,155	120
Malaysia	97	469	516	2,195	67	52	55	1,168	1,324	393	2,071	506	474	995	534	2,209	1,486	778	2,608	1,070	248	1,114			32		474	—		374	443	586
Maldives																																
Mali					1,795	3,759	1,916	352								2,191	5,723	12		12	3,770	134	36	2,910	5,678	2,356	14		153			2,220
Marshall Islands																					300											
Mauritania					166	259	3,734	1,578	575	700							4,534	924					160	68		8,264	50	3				46
Mauritius																										—	—					
Mexico												2,690	8,162	10,712	4,059	16,430	1,088	4,712	142		10	2							2		1	1
Micronesia (Federated States of)											34						1				6,904	28				—	—					
Mongolia																										—						
Morocco											4,573				6																	
Mozambique	1,212	1,753	2,301	10,745	521	3	1			371	4,152	7,847	30,802	19,803	692		—	17,478	85,344	88,658	35,298	17,588	48,750	27,516	40,160	4,452	12,612	5,244	18,174	39,358	7,430	1,279
Myanmar	1,018	28		989						597		924	826	1,758	421	1,296													45			16
Namibia																											370	28	6,992	159		
Nauru	1	24																														
Nepal		7			3,788					166		3,238	798	31	32	157	274	490	3,490	2,372	450	162	472	584	4,356	1,106	2,464	528	1,944	164	1,790	12
Netherlands		2	248	178	1,667	30	91	1,290	137	1,078	3			3	1	9	3	4	8	4	2		2	2	2	8	6	48	5		1,154	2,324
New Zealand								2			5				2	2	2		2	2		2				1		1				
Nicaragua												1	3,067	6,631	7,821	8,825	2,813	2,566	2,874	1,104	12											
Niger	138	107							1		23,888	30,648				264	3,957	518					10,858	3,866	6,372	8,954	4,056	3,322	10,820	27,382	44,456	23,377
Nigeria												59,478	8,687	4,160	2,859	1,059	12,374	2,644	6,928	52,716	5,598	4,624										
Norway								1		1							1		4							2	2	1				
Oman																						14	1									
Pakistan		4												12,092																	164	527
Panama												1,178	2,416	42																		
Papua New Guinea														3			4									—	—			3,914	8,997	1,535
Paraguay																									1,066					5		
Peru	836	864	930									322,562	242,237	72,404	23,887	22,397	4,518	6,966	83,434	3,092	1,868	988	16									
Philippines													345	708	3,340	847	1,402	1,210	1,458	660	550	348				278	132	—			33	120
Poland		1				10									1											1						

Country	C1	C2	C3	C4	C5	C6	C7	C8	C9	C10	C11	C12	C13	C14	C15	C16	C17	C18	C19	C20	C21	C22	C23	C24	C25	C26	C27
Portugal																											
Qatar																							2	2			
Republic of Korea	145						113	6	5	34	74			20				276	2		20		2	2		1	3
Republic of Moldova								3	1	8	240																
Romania						270	226	3	15	80	118					106											
Russian Federation	30	24 *201*				49	78	6	23	1,048	9	4	22	20	16		314	106				2	2				3
Rwanda		54	161	21	226	101	70	1	679	503	568	10	3	106	548	6,440	654	2,470	24	1,172	178	2,906	46	67			
Samoa	1																										
São Tomé and Príncipe							3														3,932	1,852	180	119			
Saudi Arabia	2	13				804																					
Senegal			428		728 *2,988*	476 *3,150*	390					3,332		742					38	2,454		730	7,968	2,566		8	3	5
Serbia and Montenegro (former)		2									4	2															
Seychelles																					—	—	356	—			
Sierra Leone					8,957 *15,980*	557						9,709 *10,285*				4,192 *1,726*				1,026		12	5,120	4,438		62	
Singapore	18	34	31	14	40	27	27	63	19	39	26	34	17	24	41	14	19	38	62	22	20	16		11	2		2
Slovenia									3	11									1		2						
Somalia	859	4,180 *11,968*	1,182 *2,742*	10,199 *15,980*	120	37			10	1	78	4			28,334 *35,514*	8,834	14,992	5,134 *212,318*	5,550	22,040	8,980	83,296		2,562	416 *3,510*		77,636
South Africa			4,715									9,255 *10,274*				40	136		20,008	7,802	5,534	7,006		7,814		10,520
Spain	4	2	2	4		2					1	6	1						2		2		4	4	5		
Sri Lanka	104	574	309	86			154			70	121	1			860	3,072	216			9							
Sudan	17			4,457																4			61,324	27,462	34,482	13,681	
Suriname								12																			
Swaziland		238	538						2,281							2		282		64	2,150		36	2		19	
Sweden							1			1	1					14		11,224	268		2,150	—					1
Switzerland		2			1			1	2	2					4			2				1	1				
Syrian Arab Republic																											
Tajikistan	4,331								165	10																	
Thailand	39	638 *1,497*	645	899	213	6,353	3,418		2,281	19	47	3,487	2		14	282		676	514	768	2,160	2,640	70	2,856	872	630 *1,974*	279
Togo				1			2,396	753	19	47	65	146		84	6,434	1,334		5,102				2,318	130	794	436	72	4

continues

Data Table 16b *continued*

Country	1980	1981	1982	1983	1984	1985	1986	1987	1988	1989	1990	1991	1992	1993	1994	1995	1996	1997	1998	1999	2000	2001	2002	2003	2004	2005	2006	2007	2008	2009	2010	2011
Tonga																						145										
Tunisia																																
Turkey																																
Turkmenistan																		110				1,122	30									
Tuvalu											42		293		1																	
Uganda	1,539		190					140				279	5,072		704	538	291	5,220	99,028	10,358	5,614	494	4,548	8,754	6,760	9,848	10,388	552	7,452	2,190	2,341	
Ukraine															813	525				4			3					—				33
United Arab Emirates												75																				
United Kingdom	6	12	1	4	5	4	11	2	10	1	6	8	5	13	18	10	13	12	36		66			18	26		98	64	16	32	8	32
United Republic of Tanzania	5,196	4,241	4,071	1,816	2,600	1,984	1,231	1,892	5,267	2,150	2,230	5,676	18,526	792	2,240	2,957	1,464	80,498	28,976	24,532	9,274	2,600	23,840	1,420	20,638	5,890	28,594	3,218	5,822	15,400	4,469	942
United States of America								6	9		7	26	103	19	47	19	3	8	34	12	8	8	4	4	10	24	16	14	10	20	15	42
Uzbekistan																1																
Venezuela (Bolivarian Republic of)												15	2,842	409			269	5,102	626	752	141											49
Vietnam	978	157	57	392	22	502	525	216	343	143	358	52	4,260	3,361	5,776	6,088	566	8	26	338						—		3,892	1,706	942	606	3
Yemen	720																													110	300	31,789
Zambia	57	14	1,403	233						44	3,717	13,154	11,659	6,766			2,172	72	342	27,022	9,008	6,218	678	2,098	24,298	3,006	10,720	4,572	4,122	9,424	6,794	330
Zimbabwe						144	2						2,048	5,385					1,990	11,274	3,350	1,300	6,250	2,018	238	1,032	1,578	130	120,110	136,306	951	1,220
Global Total	42,776	51,633	64,925	65,251	29,324	45,279	52,738	56,105	53,920	54,108	74,764	595,334	492,964	380,714	385,408	214,802	144,727	298,829	596,982	526,065	285,783	372,077	284,513	223,137	202,744	263,745	473,544	357,351	430,991	417,778	317,528	589,104

DATA TABLES 17A AND 17B

Global Cholera Deaths Reported to the World Health Organization, by Country, 1949-2011

Description

Annual deaths from cholera are shown here for all reporting countries. Cholera is one of many waterborne diseases that are a consequence of the lack of sanitation services and access to clean drinking water. Some data on cholera cases and deaths are available as early as 1922 from reports to the League of Nations, but consistent reports are available from the World Health Organization only since 1949. The data here are given annually from 1949 to 2011.

Since 2006, between 7,500 and 15,000 people have died each year from cholera, not counting unreported or misreported data. Particularly hard hit in 2011 were Haiti and Somalia.

In 2011, a total of fifty-eight countries from all continents reported approximately 590,000 cases of cholera to the World Health Organization, of which 32 percent were in Africa and 61.2 percent were in the Americas (defined as North, Central, and South America and the Caribbean). These data reflect a massive upsurge in the Americas from previous years. A particularly severe outbreak started in Haiti at the end of October 2010 and also affected the Dominican Republic.

These data also show the rapid and severe spread of cholera in Latin America during the seventh pandemic, starting with the appearance of cholera in Peru in 1991 and spreading rapidly. That sharp spike in cholera was the first manifestation of the seventh pandemic in the Americas and was the first time that the Americas had seen more than a handful of indigenous cases in over a century.

Limitations

There are serious problems with the reporting of cholera. In particular, the data here include only the deaths identified and reported to the World Health Organization and do not include undiagnosed or unreported cases. The true number of cholera deaths is known to be much higher than reported here. Many countries refuse to report cholera as such, or define many severe diarrheal diseases as something other than cholera, often as "acute watery diarrhea." As many as 500,000 to 700,000 cases and an unknown number of deaths from acute watery diarrhea occur annually in vast areas of Central and Southeast Asia and in some African countries.

As a result, the totals shown in the table are likely to seriously underrepresent total cholera deaths.

Sources

World Health Organization (WHO). 2011. Number of Reported Cholera Cases. Global Health Observatory. http://www.who.int/gho/epidemic_diseases/cholera/cases_text/en/index.html (accessed June 1, 2013).

———. 2013. Cholera Data Portal. http://apps.who.int/gho/data/node.main.175.

DATA TABLE 17A Global Cholera Deaths Reported to the World Health Organization, by Country, 1948–1979

Country	1949	1950	1951	1952	1953	1954	1955	1956	1957	1958	1959	1960	1961	1962	1963	1964	1965	1966	1967	1968	1969	1970	1971	1972	1973	1974	1975	1976	1977	1978	1979
Afghanistan												199					55														
Albania																															
Algeria																							110	71	74	50	118	30	9	13	94
Andorra																															
Angola																								24	9	34	4		31		
Argentina																															
Armenia																															
Australia																					0			0					0	0	0
Austria																															
Azerbaijan																															
Bahrain																		0						4	0					2	0
Bangladesh	12,947	12,372	12,884	16,904	9,443	9,802	15,310	5,134	10,119	11,056	6,272	1,703	1,304	1,248	2,419	683	1,234	369	614	1,556	1,889	386	201	369	173	117	62	354	81	21
Belarus																															
Belgium																															
Belize																						37	274	29	2	4	1	14	0	0	
Benin																															
Bhutan									0																						
Bolivia (Plurinational State of)																															
Botswana																															
Brazil																															
Brunei Darussalam							0									6					0										
Burkina Faso																						1	495	0	249	66	3			0	
Burundi																														248	31
Cambodia	1	40	10	8	2		1	1	0					13	27		5	0	1					14	1	0				
Cameroon																						336	27	16	16	0	27		0	4

continues

DATA TABLE 17A *continued*

Country	1949	1950	1951	1952	1953	1954	1955	1956	1957	1958	1959	1960	1961	1962	1963	1964	1965	1966	1967	1968	1969	1970	1971	1972	1973	1974	1975	1976	1977	1978	1979
Canada																															
Cape Verde																										8	3	17			
Central African Republic																															
Chad																							2,411	3		87					
Chile																															
China														24	0				0												2
Colombia	1																														
Comoros																											238	1			
Congo																														14	0
Costa Rica																															
Côte d'Ivoire																						39	9								
Czech Republic																						1									0
Democratic People's Republic of Korea																															
Democratic Republic of the Congo																														183	511
Denmark																															
Djibouti																						3	81	3				0	0	2	
Dominican Republic																															
Ecuador																															
El Salvador																															
Equatorial Guinea																															
Eritrea																															
Estonia																															
Ethiopia																						72	124								0
Finland																						0	0	0	0	0	0	0	0		
France																															
Gabon																												0			2
Gambia																															
Georgia																						0	0	0	0	0	0	0	0		
Germany																															

Country																								
Ghana																								113
Greece																								
Guatemala																								
Guinea																								
Guinea-Bissau																								
Guyana																								
Haiti																								
Honduras																								
Hungary																				180	353	120	113	
India	396	41,543	58,252		19,368	7,662	17,206	49,715	49,818	7,696		26,946	27,141	16,357	17,035	12,664	2,799		3,420	10,478	2,845	5,406	2,289	2,330	662	289	246	113
Indonesia												474	80	83	69	36	139		1,257	3,607	6,956	3,058	3,920	3,715	2,665	780	507	769
Iran (Islamic Republic of)										22								262									56
Iraq																	20									1	0	
Israel																			10	0	0			0		0		0
Italy															0	1						23						0
Japan																								0	0	1	1	0
Jordan																			0						0		1	0
Kazakhstan																												
Kenya																				38	0		20	74	59	0	17	2
Kiribati																										21	0	0
Kuwait																			0					1	0	0	0	0
Kyrgyzstan																												
Lao People's Democratic Republic				18															21									
Lebanon																			1							0		
Lesotho																												0
Liberia																			27	106	28	27	12	16	10	18	9	6
Libya																			28									
Madagascar																				0		20						
Malawi																									0	0	0	24

continues

DATA TABLE 17A continued

Country	1949	1950	1951	1952	1953	1954	1955	1956	1957	1958	1959	1960	1961	1962	1963	1964	1965	1966	1967	1968	1969	1970	1971	1972	1973	1974	1975	1976	1977	1978	1979
Malaysia						0									24	143	1	0		3		3	1	11	17	0	8	4	12	64	10
Maldives																														220	
Mali																						201	1,321	0	32	21					
Marshall Islands																															
Mauritania																								53	24	22					
Mauritius																															
Mexico																															
Micronesia (Federated States of)																															
Mongolia																															
Morocco																							0	0				0			
Mozambique																									87	138	3	2	3		547
Myanmar		2,785	4,685	245	9	19	12	6	3	3	2	200	0	0	783	171	8	17	0	0	34	76	42	7	24	203	172	173		372	75
Namibia																															
Nauru																														0	0
Nepal										384							97	0		17	6	1	0	0	0	0	0	1	4	10	0
Netherlands																												1	0	0	0
New Zealand																								1							
Nicaragua																															
Niger																						8	2,344	12	16	75					
Nigeria																						4	2,945	96	27		1	25	10	20	4
Norway																															
Oman																							8								
Pakistan										86		1,107					1			55	17	0	43						1		
Panama																															
Papua New Guinea														464																	
Paraguay																															
Peru													1,405	1,682	433	1,518	448	273	52	89	106	38	109	78	447	33	15	160	180	122	408
Philippines																															
Poland																															
Portugal																							4			48	8				

Qatar																						
Republic of Korea												38		1				137	23			
Republic of Moldova																						8
Romania																				0		
Russian Federation																		0	0		0	
Rwanda																					48	0
Samoa																						
São Tomé and Príncipe																						
Saudi Arabia																14		0	48	0	1	0
Senegal																60	37				8	3
Serbia and Montenegro (former)																			1			
Seychelles																				0		
Sierra Leone																74	13				0	
Singapore												2	3		0	4		3			0	0
Slovenia																						
Somalia																10	10					
South Africa																				0		0
Spain																			0	0		
Sri Lanka																			13	333	67	16	2	0
Sudan																							176
Suriname																						
Swaziland																						
Sweden																	0				0	
Switzerland																					0	0
Syrian Arab Republic																0	0	25		6	11	66	30
Tajikistan																							
Thailand	2	0							1,426	613			156	30	4	21	6	21	40	33	64	88	0	5	123	44
Togo																3	26	0			8		12	6
Tonga																							
Tunisia																0		0	16		0		

continues

DATA TABLE 17A *continued*

Country	1949	1950	1951	1952	1953	1954	1955	1956	1957	1958	1959	1960	1961	1962	1963	1964	1965	1966	1967	1968	1969	1970	1971	1972	1973	1974	1975	1976	1977	1978	1979
Turkey																						52							0		
Turkmenistan																															
Tuvalu																															
Uganda																							30				0			110	60
Ukraine																															
United Arab Emirates																								0							
United Kingdom																						0	0	0	0	0	0	0	0		
United Republic of Tanzania																										6			26	733	305
United States of America																															
Uzbekistan																															
Venezuela (Bolivarian Republic of)																															
Vietnam		4	21	6	7	3										872	29	167	71	8	26	1	1	1	18	2	0	0	0	0	28
Yemen																							26	140	23	1	0			12	45
Zambia																														36	11
Zimbabwe																										0	33	9			
Total	3	16,133	58,661	71,397	141,173	28,835	17,476	32,523	54,853	61,836	19,367	7,800	30,528	30,695	19,137	22,289	13,996	4,572	637	830	2,209	7,533	26,132	10,658	10,081	7,649	7,079	3,950	2,177	3,354	3,478

DATA TABLE 17B Global Cholera Deaths Reported to the World Health Organization, by Country, 1980–2011

Country	1980	1981	1982	1983	1984	1985	1986	1987	1988	1989	1990	1991	1992	1993	1994	1995	1996	1997	1998	1999	2000	2001	2002	2003	2004	2005	2006	2007	2008	2009	2010	2011
Afghanistan														931	118	624		250	0	304	396	228		14		0	0		44	22	20	44
Albania															25																	
Algeria	18			9	0			0	30	0	20		0		4											0	0					
Andorra														0																		
Angola								1,403	1,542	925	809	582	184		187	248	42									0	5,444	1,026	486	176	60	110
Argentina													15	34	15	1	5	24														
Armenia																			0													
Australia	0	0	0	0	0	0		0	0		0		0	0	0	0	0	0		0						0	0	0			0	0
Austria		0									0				0		0			0							0					
Azerbaijan															0																	
Bahrain																																
Bangladesh								23	43	2	4	0	29	0	41	61	0	190	52	126	32											
Belarus															0	0																
Belgium	0	0		0	0									0	0	1	0	0	2	0						0	0					
Belize													4	3	1	1	0	32	2	50	22	142	26	22	18	22	2	2	10	2	16	4
Benin	0	0	0	0	0		13					259	17	2	20	12	203															
Bhutan												19	6			0			0													
Bolivia (Plurinational State of)												12	395	254	46	54	68	36	14													
Botswana																																
Brazil												33	363	644	452	85	33	74	54	166	34	0				0	0	0	2	4		0
Brunei Darussalam			0																	0												
Burkina Faso					117	156						61					58		104	14	28	14				32		0	0			2
Burundi	28	11	9	35	10	15							29	0	41	113		190	52	126	32	40	16	36	28	36	14	4	8	0	2	1
Cambodia								55	0	2	0	97	120	104	211	123	20	0	132	260					2					0	2	
Cameroon	23	17	3	18	60	103	12	6	0	2	0	491	66	19	61	21	485	360	632	70	58	14	16	72	274	220	70	2	0	178	1,314	783

continues

DATA TABLE 17B *continued*

Country	1980	1981	1982	1983	1984	1985	1986	1987	1988	1989	1990	1991	1992	1993	1994	1995	1996	1997	1998	1999	2000	2001	2002	2003	2004	2005	2006	2007	2008	2009	2010	2011
Canada	0								0	0				1	0	0	0	0	0		0		0		0	0	0	0	0	0	0	
Cape Verde															12	240	3		2							0	0		0	0		
Central African Republic																		150	6						96							15
Chad												1,344			43		448	886	288	36		452		14	544	28	142	0		6	350	458
Chile	0											2	1	0	0		0	0	4													0
China									89	59	5	0	1	94	319	88	4	58	0	0	6	0					4			0	0	
Colombia												207	158	0	14	35	70	64	14	0												
Comoros																			136	84	182	8	92	0	0	0	0	58	0			
Congo																		120	44	40	4					0	20	266	8			
Costa Rica													0	0	0	0	1	0														
Côte d'Ivoire			0									116	7	42	55	184	22					610	286	100	18	12	30	2	2	4	0	24
Czech Republic	0																															
Democratic People's Republic of Korea																																
Democratic Republic of the Congo	81	800	693	245	3	30	125	218	32	6	61	294	59	175	4,181	22	638	652	5,852	1,566	1,882	780	3,958	1,978	456	488	852	1,200	1,096	474	364	584
Denmark											0			0	0	0									0				0			
Djibouti						7								99	15			100	6		64							54			38	1
Dominican Republic																															0	336
Ecuador												697	208	72	16	23	12	6	74	0	2	0			0							
El Salvador												34	45	14	38	2	2	0	0	0	4											
Equatorial Guinea					31	1																			2	66						
Eritrea																									0	0	0	0	0			
Estonia														0														18	0			
Ethiopia			0																						0	0	1,150	544	46	868	42	
Finland			0													0	0	0	0							0			0			
France	0			0	0		0	0			1	0					0	0	0		0						0	0	0	0		0
Gabon		0																											0	0		
Gambia						0											0									0	0	0	0			
Georgia														0										0		26		2	0		2	
Germany	0		0				0		0	0	0	0	0	0	0	0	0	0		0					0		0	0	0			0

Country																														
Ghana	21	62	819	92	0				60	409	23	38	54	93	70	24	216	520	148	320	130	8	12	102	214	36	50	18	6	105
Greece		0										0																		
Guatemala										50	207	306	36	95	14	0	114	0	12	0	0	2	234	214	438	622	64	0	0	0
Guinea					58								671	565	17		106	56	64	44	22	4	6	798	0	16	450	0		
Guinea-Bissau					50	68						285			84	1,810	4			12										
Guyana											8	2																		
Haiti																													7,980	2,869
Honduras										0	17	102	102	77	14	2	26	6	0	0		4		42	22	6	2	0		
Hungary																	0													18
India	307	150	213	309	27	155	70	240	380	93	83	149	55	32	5	34	32	22	12	40	12	20	4	14	12	6	6	2		
Indonesia	150	463	535	346	59	85	391	18		0		0	25	0	0		0		12					38			54			
Iran (Islamic Republic of)	32	108	0	0	5	9	1	4	15	85	3	32	4	5	59	18	2	42	2				0	2	22	0	8			12
Iraq													6						6											
Israel		0						0	0			0	0	3	0	2	2	60	8			0	0	0	0	48	22	0	0	0
Italy													0	9	0		18													
Japan	0	0	0	0	0	0	0	0	0		1	0	0	1	0	0	0	0	0	12		0	0	0	0	0	0	0		
Jordan		4							0	0			0	0	0	0														
Kazakhstan																											0			
Kenya	42	108	91	53	0	102	45	12		32		100	255	28	39	14	1,110	2,474	700	156	110	20		30	42	22	134	226	528	126	2
Kiribati																														
Kuwait		0			0	0	0			0			0																	
Kyrgyzstan														0																
Lao People's Democratic Republic													255	606	174	33												6	0	8
Lebanon												23															0			
Lesotho																									0		0			
Liberia	55	33	8	12	8	3	0	1	1	0	40		17	17	126	169	0	34	0	20	0	0	76	8	36	34	14	6	4	0
Libya															2															
Madagascar			49														148	1,098	3,344	826	1,822									
Malawi		0							0	16	481	245	8	524	11	0	0	30	148	1,296	112	84	1,822	68	8	22	110	10	52	250	34	4

continues

Data Table 17b continued

Country	1980	1981	1982	1983	1984	1985	1986	1987	1988	1989	1990	1991	1992	1993	1994	1995	1996	1997	1998	1999	2000	2001	2002	2003	2004	2005	2006	2007	2008	2009	2010	2011
Malaysia	7	14	17	38	1	2	2	18	32	14	38	6	8	13	0	27	0	8	38	0	0	22			0		4	0			2	12	10
Maldives																																95
Mali					406	838	151	84								231	761	6		6	102	18	4	238	410	152	0		10			
Marshall Islands																					12											
Mauritania					15	20	113	101	38	27							148	14					0	16		140	0	0				3
Mauritius																										0	0					
Mexico												34	99	193	56	137	5	2	0	0	0	0							0		0	0
Micronesia (Federated States of)											1						0				40	0				0						
Mongolia																	12									0						
Morocco											137				0																	
Mozambique	85	132	199	447	11	0	0			12	194	328	727	507	27		0	518	2,706	2,388	476	204	684	204	220	48	58	44	204	310	234	5
Myanmar	50	3		29						11		39	50	52	4	6													2			
Namibia																											10	0	76	6		
Nauru																																
Nepal	0	0								2	495	873	15	0	0	0	1	0	34									0	0	0	18	
Netherlands				0					0		0			0	0	0	0	0		0					0	0	0					
New Zealand								0		0	0				0	0				0						0	0	0				
Nicaragua												0	46	220	134	164	107	72	58	14	0											
Niger		0			308					22		367			83	15	206	26		170	76	2	24	22	114	110	162	4	140		132	60
Nigeria	25	18	18	2	185	0	23	111	10	61		7,654	686	266	313	87	1,193	268	374	4,170	86	238	308	174	370	348	254	96	494	862	3,424	742
Norway		0						0		0						0			0							0	0	0				
Oman																																
Pakistan		0																													0	219
Panama												29	49	4																		
Papua New Guinea														206																90	190	2
Paraguay														0			0									0				0		
Peru						1						2,909	801	575	199	171	21	58	1,136	12	12	0										
Philippines	61	51	31										0	10	27	4	14	0	34	0	12				6	4	2	0			4	3
Poland												0			0											0						
Portugal		0													0	0																

Country	C1	C2	C3	C4	C5	C6	C7	C8	C9	C10	C11	C12	C13	C14	C15	C16	C17	C18	C19	C20	C21	C22	C23	C24	C25	C26	C27	C28	C29	C30	C31	C32	C33
Qatar	4																												0	0			
Republic of Korea												4		0										0									
Republic of Moldova																0	5																
Romania									1			9	0			0	3																
Russian Federation	0	0			0							0	0		0	2	0	1	0										0	0			
Rwanda	4	1	2	5	13		0	7	0	6	3			35	35	34	2	15	0	0	10	32	148	98	16	4	0		0	0	42	42	0	0
Samoa	0																																
São Tomé and Príncipe									64				9	1														66	18	0	8		
Saudi Arabia	0	0					3																										
Senegal	0							63	288	1									160	765	22							20	916	20	48	40	0	0
Serbia and Montenegro (former)	0																	0														0	0
Seychelles																												0		0	2	0	
Sierra Leone								669	30									623	447			114	10					84	0	198	168	2	
Singapore	0	0		8	1	0	0							2	0	0	1	0				4	0		0	0		2				0	
Slovenia																																0	
Somalia					1,589	1,945													1,206				1,386		1,126	318	112	52			2,364	32	6	206	1,130
South Africa	16	52	224	63	20	4	1		0			0	0	0			2	0				2		4	136	464	64	90	70	56	0	0	44	114
Spain	0	0	0									0	0		0	0	0	0	0			0	0		0							0	0
Sri Lanka	4	57	21	2					11			2	3	0		0					24	98	10	0		2							
Sudan	2				239																		0							2,022	1,000	236	104
Suriname											1																						
Swaziland	17	13										30										0	0		0	32	214	4	26	32	0	0	0	0
Sweden				0		0	0					0	0	0		0	0	0					0	0					0	0			
Switzerland	0				0					0		0	0	0			0	0											0				
Syrian Arab Republic																																	
Tajikistan														5	0	5	0																
Thailand	85	2	16	41	19	14	1	36	41			81	53	2	5		5	4		17	14	442	62	46	260	28	56	66	30	0	14	6	4	30	4
Togo										0																				50	2	6	2	6	0
Tonga																								4									

continues

DATA TABLE 17B continued

Country	1980	1981	1982	1983	1984	1985	1986	1987	1988	1989	1990	1991	1992	1993	1994	1995	1996	1997	1998	1999	2000	2001	2002	2003	2004	2005	2006	2007	2008	2009	2010	2011
Tunisia																																
Turkey																																
Turkmenistan																		0				12	4									
Tuvalu											0		8																			0
Uganda	181		48					50				28	104		39	66	40	376	4,128	456	344	8	266	258	182	196	212	6	240	46	156	
Ukraine												0			20	10				0								0				0
United Arab Emirates																																
United Kingdom		0	0	0	0	1	0	0	0	0	0	0	0	0	0	0	0	0	0						0		0	0	0	0	0	0
United Republic of Tanzania	587	330	324	127	338	250	215	232	580	232	81	572	2,173	74	155	96	36	4,462	1,216	1,168	312	104	594	64	544	188	508	130	176	226	118	11
United States of America								0	0		0	0	0	1	0	0	0	0	0	0	0				0	0	0	0	0	0	0	0
Uzbekistan															0																	
Venezuela (Bolivarian Republic of)													2	68	10		9	118	14	16	0											0
Vietnam	49	8	0	0	0	9	8	5	7	0	4	1	110	9	68	45	2	0	2	0						0		0	0	2	0	
Yemen	20																													6	8	134
Zambia	17	0	131	19						44	315	1,091	913	426			42	0	0	1,070	606	330	24	58	746	14	366	70	78	160	124	7
Zimbabwe						6							105	332					88	770	446	28	384	70	18	52	122	8	5,856	5,412	42	2
Total	1,954	2,441	3,821	2,668	1,830	4,114	3,966	3,009	2,855	1,711	2,802	19,302	8,214	6,761	10,750	5,045	6,418	12,742	21,664	18,442	10,538	5,794	9,128	3,788	4,690	4,544	12,600	8,066	10,286	9,886	15,086	7,781

DATA TABLE 18A

Perceived Satisfaction with Water Quality in Sub-Saharan Africa

Description

In 2012, the Gallup Organization released the results of a survey on perceived satisfaction with water quality taken in cities and other areas of sub-Saharan Africa (and in other regions; see data table 18B). The question asked was, "In the city or area where you live, are you satisfied or dissatisfied with the quality of water?"

This table summarizes the results of that survey, which may be the first of its kind. The complete survey was done worldwide. The results shown here are only for sub-Saharan Africa. Perceived water quality is generally lowest in this region, the Middle East, and countries that were formerly in the Soviet Union. The summary provided by Gallup notes, "Seven of the 10 countries where residents are least satisfied are in sub-Saharan Africa, where residents sometimes have to walk miles for water, and waterborne and water-related diseases such as cholera are common."

Limitations

Like any poll, this one has limitations. As described by Gallup, the results presented here are based on telephone and face-to-face interviews with 1,000 adults, aged fifteen and older, conducted in more than 140 countries in 2011. The maximum margin of sampling error is described as ranging from a low of ±2 percentage points to a high of ±5.1 percentage points. In addition to sampling error, one challenge with polls such as this involves the wording and interpretation of the questions asked. For example, what does "satisfied" mean?

Moreover, perceived levels of satisfaction say nothing about actual water quality. Many people may perceive problems where none exist and, conversely, may not perceive problems when major contamination exists.

Source

Gallup. 2012. Air Quality Rated Better Than Water Quality Worldwide: Residents of Sub-Saharan Africa Rate Their Water the Worst. Report by A. Pugliese and J. Ray. May 14. http://www.gallup.com/poll/154646/Air-Quality-Rated-Better-Water-Quality-Worldwide.aspx (accessed June 3, 2013).

DATA TABLE 18A Perceived Satisfaction with Water Quality in Sub-Saharan Africa

Country or City	Satisfied (%)	Dissatisfied (%)
Mauritius	88	12
Rwanda	70	30
Zimbabwe	64	36
South Africa	62	38
Mozambique	62	38
Botswana	62	38
Malawi	62	38
Ghana	61	39
Kenya	61	39
Somaliland region	61	39
Uganda	61	39
Benin	59	41
Swaziland	58	42
Niger	57	43
Senegal	56	44
Burundi	56	44
Madagascar	53	47
Djibouti	52	48
Angola	51	49
Mali	50	50
Zambia	49	51
Guinea	48	52
Cameroon	48	52
Mauritania	47	53
Comoros	47	53
Liberia	47	53
Conga Brazzaville	47	53
Burkina Faso	46	54
Chad	46	54
Nigeria	43	57
Gabon	42	58
Central African Republic	41	59
Togo	41	59
Lesotho	39	61
Tanzania	36	64
Sudan	35	65
Congo Kinshasa	31	69
Sierra Leone	30	70

DATA TABLE 18B

Regional Assessment of Satisfaction with Water (and Air) Quality

Description

In 2012, the Gallup Organization released the results of a survey on perceived satisfaction with water quality taken in cities and other areas around the world. The question asked was, "In the city or area where you live, are you satisfied or dissatisfied with the quality of [air or water]?"

This table summarizes the results of that survey for different regions of the world. In general, with the exception of Europe, air quality in all regions was thought to be better than water quality. Ratings of water quality are highest in the richer countries of North America, much of Europe, and developed regions in Asia. Perceived water quality is generally lowest in sub-Saharan Africa (see data table 18A), the Middle East, and countries of the former Soviet Union. The summary provided by Gallup notes, "Seven of the 10 countries where residents are least satisfied are in sub-Saharan Africa, where residents sometimes have to walk miles for water, and waterborne and water-related diseases such as cholera are common."

Limitations

Like any poll, this one has limitations. As described by Gallup, the results presented here are based on telephone and face-to-face interviews with 1,000 adults, aged fifteen and older, conducted in more than 140 countries in 2011. The maximum margin of sampling error is described as ranging from a low of ±2 percentage points to a high of ±5.1 percentage points. In addition to sampling error, one challenge with polls such as this involves the wording and interpretation of the questions asked. For example, what does "satisfied" mean?

Moreover, perceived levels of satisfaction say nothing about actual water quality. Many people may perceive problems where none exist and, conversely, may not perceive problems when major contamination exists.

Source

Gallup. 2012. Air Quality Rated Better Than Water Quality Worldwide: Residents of Sub-Saharan Africa Rate Their Water the Worst. Report by A. Pugliese and J. Ray. May 14. http://www.gallup.com/poll/154646/Air-Quality-Rated-Better-Water-Quality-Worldwide.aspx (accessed May 28, 2013).

DATA TABLE 18B Regional Assessment of Satisfaction with Water (and Air) Quality

Region	Satisfied with Air Quality (%)	Satisfied with Water Quality (%)
Global	75	68
Sub-Saharan Africa	78	51
Americas	78	73
Asia	84	78
Europe	75	79
Former Soviet Union	68	61
Middle East and North Africa	68	63

DATA TABLE 18C

Countries Most and Least Satisfied with Water Quality

Description

In 2012, the Gallup Organization released the results of a survey on perceived satisfaction with water quality taken in cities and other areas around the world. The question asked was, "In the city or area where you live, are you satisfied or dissatisfied with the quality of [air or water]?"

This table presents the countries "most" and "least" satisfied with water quality. Ratings of water quality are highest in the richer countries of the Americas, much of Europe, and parts of developed regions in Asia. Perceived water quality is generally lowest in sub-Saharan Africa (see data table 18A), the Middle East, and parts of the former Soviet Union.

Limitations

Like any poll, this one has limitations. As described by Gallup, the results presented here are based on telephone and face-to-face interviews with 1,000 adults, aged fifteen and older, conducted in more than 140 countries in 2011. The maximum margin of sampling error is described as ranging from a low of ±2 percentage points to a high of ±5.1 percentage points. In addition to sampling error, one challenge with polls such as this involves the wording and interpretation of the questions asked. For example, what does "satisfied" mean?

Moreover, perceived levels of satisfaction say nothing about actual water quality. Many people may perceive problems where none exist and, conversely, may not perceive problems when major contamination exists.

Source

Gallup. 2012. Air Quality Rated Better Than Water Quality Worldwide: Residents of Sub-Saharan Africa Rate Their Water the Worst. Report by A. Pugliese and J. Ray. May 14. http://www.gallup.com/poll/154646/Air-Quality-Rated-Better-Water-Quality-Worldwide.aspx (accessed May 28, 2013).

DATA TABLE 18c   Countries Most and Least Satisfied with Water Quality

Country	Most Satisfied with Water Quality %
United Kingdom	97
Germany	96
Sweden	96
Denmark	96
Singapore	95
Finland	94
Austria	94
Netherlands	93
Australia	92
Luxembourg	91
Uruguay	91

Country	Least Satisfied with Water Quality %
Congo (Kinshasha)	24
Sierra Leone	30
Iraq	32
Haiti	32
Lesotho	39
Central African Republic	41
Togo	41
Tanzania	41
Ukraine	42
Gabon	42

Water Units, Data Conversions, and Constants

Water experts, managers, scientists, and educators work with a bewildering array of different units and data. These vary with the field of work: engineers may use different water units than hydrologists; urban water agencies may use different units than reservoir operators; academics may use different units than water managers. But they also vary with regions: water agencies in England may use different units than water agencies in France or Africa; hydrologists in the eastern United States often use different units than hydrologists in the western United States. And they vary over time: today's water agency in California may sell water by the acre-foot, but its predecessor a century ago may have sold miner's inches or some other now arcane measure.

These differences are of more than academic interest. Unless a common "language" is used, or a dictionary of translations is available, errors can be made or misunderstandings can ensue. In some disciplines, unit errors can be more than embarrassing; they can be expensive, or deadly. In September 1999, the $125 million Mars Climate Orbiter spacecraft was sent crashing into the face of Mars instead of into its proper safe orbit above the surface because one of the computer programs controlling a portion of the navigational analysis used English units incompatible with the metric units used in all the other systems. The failure to translate English units into metric units was described in the findings of the preliminary investigation as the principal cause of mission failure.

This table is a comprehensive list of water units, data conversions, and constants related to water volumes, flows, pressures, and much more. Most of these units and conversions were compiled by Kent Anderson and initially published in P. H. Gleick, 1993, *Water in Crisis: A Guide to the World's Fresh Water Resources*, Oxford University Press, New York.

Water Units, Data Conversions, and Constants

Prefix (Metric)	Abbreviation	Multiple	Prefix (Metric)	Abbreviation	Multiple
deka-	da	10	deci-	d	0.1
hecto-	h	100	centi-	c	0.01
kilo-	k	1000	milli-	m	0.001
mega-	M	10^6	micro-	μ	10^{-6}
giga-	G	10^9	nano-	n	10^{-9}
tera-	T	10^{12}	pico-	P	10^{-12}
peta-	P	10^{15}	femto-	f	10^{-15}
exa-	E	10^{18}	atto-	a	10^{-18}

LENGTH (L)

1 micron (μ)	$= 1 \times 10^{-3}$ mm	**10 hectometers**	= 1 kilometer
	$= 1 \times 10^{-6}$ m	**1 mil**	= 0.0254 mm
	$= 3.3937 \times 10^{-5}$ in		$= 1 \times 10^{-3}$ in
1 millimeter (mm)	= 0.1 cm	**1 inch (in)**	= 25.4 mm
	$= 1 \times 10^{-3}$ m		= 2.54 cm
	= 0.03937 in		= 0.08333 ft
1 centimeter (cm)	= 10 mm		= 0.0278 yd
	= 0.01 m	**1 foot (ft)**	= 30.48 cm
	$= 1 \times 10^{-5}$ km		= 0.3048 m
	= 0.3937 in		$= 3.048 \times 10^{-4}$ km
	= 0.03281 ft		= 12 in
	= 0.01094 yd		= 0.3333 yd
1 meter (m)	= 1000 mm		$= 1.89 \times 10^{-4}$ mi
	= 100 cm	**1 yard (yd)**	= 91.44 cm
	$= 1 \times 10^{-3}$ km		= 0.9144 m
	= 39.37 in		$= 9.144 \times 10^{-4}$ km
	= 3.281 ft		= 36 in
	= 1.094 yd		= 3 ft
	$= 6.21 \times 10^{-4}$ mi		$= 5.68 \times 10^{-4}$ mi
1 kilometer (km)	$= 1 \times 10^5$ cm	**1 mile (mi)**	= 1609.3 m
	= 1000 m		= 1.609 km
	= 3280.8 ft		= 5280 ft
	= 1093.6 yd		= 1760 yd
	= 0.621 mi	**1 fathom (nautical)**	= 6 ft
10 millimeters	= 1 centimeter	**1 league (nautical)**	= 5.556 km
10 centimeters	= 1 decimeter		= 3 nautical miles
10 decimeters (dm)	= 1 meter	**1 league (land)**	= 4.828 km
			= 5280 yd
10 meters	= 1 dekameter		= 3 mi
10 dekameters (dam)	= 1 hectometer	**1 international nautical mile**	= 1.852 km
			= 6076.1 ft
			= 1.151 mi

Water Units, Data Conversions, and Constants *(continued)*

AREA (L^2)

1 square centimeter (cm^2)	$= 1 \times 10^{-4} m^2$	**1 square foot (ft^2)**	$= 929.0\ cm^2$
	$= 0.1550\ in^2$		$= 0.0929\ m^2$
	$= 1.076 \times 10^{-3}\ ft^2$		$= 144\ in^2$
	$= 1.196 \times 10^{-4}\ yd^2$		$= 0.1111\ yd^2$
1 square meter (m^2)	$= 1 \times 10^{-4}$ hectare		$= 2.296 \times 10^{-5}$ acre
	$= 1 \times 10^{-6}\ km^2$		$= 3.587 \times 10^{-8}\ mi^2$
	$= 1$ centare (French)	**1 square yard (yd^2)**	$= 0.8361\ m^2$
	$= 0.01$ are		$= 8.361 \times 10^{-5}$ hectare
	$= 1550.0\ in^2$		$= 1296\ in^2$
	$= 10.76\ ft^2$		$= 9\ ft^2$
	$= 1.196\ yd^2$		$= 2.066 \times 10^{-4}$ acres
	$= 2.471 \times 10^{-4}$ acre		$= 3.228 \times 10^{-7}\ mi^2$
1 are	$= 100\ m^2$	**1 acre**	$= 4046.9\ m^2$
1 hectare (ha)	$= 1 \times 10^4\ m^2$		$= 0.40469$ ha
	$= 100$ are		$= 4.0469 \times 10^{-3}\ km^2$
	$= 0.01\ km^2$		$= 43{,}560\ ft^2$
	$= 1.076 \times 10^5\ ft^2$		$= 4840\ yd^2$
	$= 1.196 \times 10^4\ yd^2$		$= 1.5625 \times 10^{-3}\ mi^2$
	$= 2.471$ acres	**1 square mile (mi^2)**	$= 2.590 \times 10^6\ m^2$
	$= 3.861 \times 10^{-3}\ mi^2$		$= 259.0$ hectares
1 square kilometer (km^2)	$= 1 \times 10^6\ m^2$		$= 2.590\ km^2$
	$= 100$ hectares		$= 2.788 \times 10^7\ ft^2$
	$= 1.076 \times 10^7\ ft^2$		$= 3.098 \times 10^6\ yd^2$
	$= 1.196 \times 10^6\ yd^2$		$= 640$ acres
	$= 247.1$ acres		$= 1$ section (of land)
	$= 0.3861\ mi^2$	**1 feddan (Egyptian)**	$= 4200\ m^2$
1 square inch (in^2)	$= 6.452\ cm^2$		$= 0.42$ ha
	$= 6.452 \times 10^{-4}\ m^2$		$= 1.038$ acres
	$= 6.944 \times 10^{-3}\ ft^2$		
	$= 7.716 \times 10^{-4}\ yd^2$		

(continues)

Water Units, Data Conversions, and Constants (continued)

VOLUME (L³)

1 cubic centimeter (cm³)	= 1 × 10⁻³ liter	**1 cubic foot (ft³)**	= 2.832 × 10⁴ cm³
	= 1 × 10⁻⁶ m³		= 28.32 liters
	= 0.06102 in³		= 0.02832 m³
	= 2.642 × 10⁻⁴ gal		= 1728 in³
	= 3.531 × 10⁻³ ft³		= 7.481 gal
1 liter (l)	= 1000 cm³		= 0.03704 yd³
	= 1 × 10⁻³ m³	**1 cubic yard (yd³)**	= 0.7646 m³
	= 61.02 in³		= 6.198 × 10⁻⁴ acre-ft
	= 0.2642 gal		
	= 0.03531 ft³		= 46656 in³
1 cubic meter (m³)	= 1 × 10⁶ cm³		= 27 ft³
	= 1000 liter	**1 acre-foot**	= 1233.48 m³
	= 1 × 10⁻⁹ km³	**(acre-ft or AF)**	= 3.259 × 10⁵ gal
	= 264.2 gal		= 43560 ft³
	= 35.31 ft³	**1 Imperial gallon**	= 4.546 liters
	= 6.29 bbl		= 277.4 in³
	= 1.3078 yd³		= 1.201 gal
	= 8.107 × 10⁻⁴ acre-ft		= 0.16055 ft³
		1 cfs-day	= 1.98 acre-feet
1 cubic decameter (dam³)	= 1000 m³		= 0.0372 in-mi²
	= 1 × 10⁶ liter	**1 inch-mi²**	= 1.738 × 10⁷ gal
	= 1 × 10⁻⁶ km³		= 2.323 × 10⁶ ft³
	= 2.642 × 10⁵ gal		= 53.3 acre-ft
	= 3.531 × 10⁴ ft³		= 26.9 cfs-days
	= 1.3078 × 10³ yd³	**1 barrel (of oil) (bbl)**	= 159 liter
	= 0.8107 acre-ft		= 0.159 m³
1 cubic hectometer (ha³)	= 1 × 10⁶ m³		= 42 gal
	= 1 × 10³ dam³		= 5.6 ft³
	= 1 × 10⁹ liter	**1 million gallons**	= 3.069 acre-ft
	= 2.642 × 10⁸ gal	**1 pint (pt)**	= 0.473 liter
	= 3.531 × 10⁷ ft³		= 28.875 in³
	= 1.3078 × 10⁶ yd³		= 0.5 qt
	= 810.7 acre-ft		= 16 fluid ounces
1 cubic kilometer (km³)	= 1 × 10¹² liter		= 32 tablespoons
	= 1 × 10⁹ m³		= 96 teaspoons
	= 1 × 10⁶ dam³	**1 quart (qt)**	= 0.946 liter
	= 1000 ha³		= 57.75 in³
	= 8.107 × 10⁵ acre-ft		= 2 pt
			= 0.25 gal
	= 0.24 mi³	**1 morgen-foot (S. Africa)**	= 2610.7 m³
1 cubic inch (in³)	= 16.39 cm³		
	= 0.01639 liter	**1 board-foot**	= 2359.8 cm³
	= 4.329 × 10⁻³ gal		= 144 in³
	= 5.787 × 10⁻⁴ ft²		= 0.0833 ft³
1 gallon (gal)	= 3.785 liters	**1 cord**	= 128 ft³
	= 3.785 × 10⁻³ m³		= 0.453 m³
	= 231 in³		
	= 0.1337 ft³		
	= 4.951 × 10⁻³ yd³		

Water Units, Data Conversions, and Constants *(continued)*

VOLUME/AREA (L³/L²)

1 inch of rain	= 5.610 gal/yd²	**1 box of rain**	= 3,154.0 lesh
	= 2.715 × 10⁴ gal/acre		

MASS (M)

1 gram (g or gm)	= 0.001 kg	**1 ounce (oz)**	= 28.35 g
	= 15.43 gr		= 437.5 gr
	= 0.03527 oz		= 0.0625 lb
	= 2.205 × 10⁻³ lb	**1 pound (lb)**	= 453.6 g
1 kilogram (kg)	= 1000 g		= 0.45359237 kg
	= 0.001 tonne		= 7000 gr
	= 35.27 oz		= 16 oz
	= 2.205 lb	**1 short ton (ton)**	= 907.2 kg
1 hectogram (hg)	= 100 gm		= 0.9072 tonne
	= 0.1 kg		= 2000 lb
1 metric ton (tonne or te or MT)	= 1000 kg	**1 long ton**	= 1016.0 kg
	= 2204.6 lb		= 1.016 tonne
	= 1.102 ton	**1 long ton**	= 2240 lb
	= 0.9842 long ton		= 1.12 ton
1 dalton (atomic mass unit)	= 1.6604 × 10⁻²⁴ g	**1 stone (British)**	= 6.35 kg
			= 14 lb
1 grain (gr)	= 2.286 × 10⁻³ oz		
	= 1.429 × 10⁻⁴ lb		

TIME (T)

1 second (s or sec)	= 0.01667 min	**1 day (d)**	= 24 hr
	= 2.7778 × 10⁻⁴ hr		= 86400 s
1 minute (min)	= 60 s	**1 year (yr or y)**	= 365 d
	= 0.01667 hr		= 8760 hr
1 hour (hr or h)	= 60 min		= 3.15 × 10⁷ s
	= 3600 s		

DENSITY (M/L³)

1 kilogram per cubic meter (kg/m³)	= 10⁻³ g/cm³	**1 metric ton per cubic meter (te/m³)**	= 1.0 specific gravity
	= 0.062 lb/ft³		= density of H₂O at 4°C
1 gram per cubic centimeter (g/cm³)	= 1000 kg/m³		= 8.35 lb/gal
	= 62.43 lb/ft³	**1 pound per cubic foot (lb/ft³)**	= 16.02 kg/m³

(continues)

Water Units, Data Conversions, and Constants (continued)

VELOCITY (L/T)

1 meter per second (m/s)	= 3.6 km/hr = 2.237 mph = 3.28 ft/s	**1 foot per second (ft/s)**	= 0.68 mph = 0.3048 m/s
1 kilometer per hour (km/h or kph)	= 0.62 mph = 0.278 m/s	**velocity of light in vacuum (c)** **1 knot**	= 2.9979×10^8 m/s = 186,000 mi/s = 1.852 km/h = 1 nautical mile/hour
1 mile per hour (mph or mi/h)	= 1.609 km/h = 0.45 m/s = 1.47 ft/s		= 1.151 mph = 1.688 ft/s

VELOCITY OF SOUND IN WATER AND SEAWATER
(assuming atmospheric pressure and sea water salinity of 35,000 ppm)

Temp, °C	Pure water, (meters/sec)	Sea water, (meters/sec)
0	1,400	1,445
10	1,445	1,485
20	1,480	1,520
30	1,505	1,545

FLOW RATE (L³/T)

1 liter per second (1/sec)	= 0.001 m³/sec = 86.4 m³/day = 15.9 gpm = 0.0228 mgd = 0.0353 cfs = 0.0700 AF/day	**1 cubic decameters per day (dam³/day)**	= 11.57 1/sec = 1.157×10^{-2} m³/sec = 1000 m³/day = 1.83×10^6 gpm = 0.264 mgd
1 cubic meter per second (m³/sec)	= 1000 1/sec = 8.64×10^4 m³/day = 1.59×10^4 gpm = 22.8 mgd = 35.3 cfs = 70.0 AF/day	**1 gallon per minute (gpm)**	= 0.409 cfs = 0.811 AF/day = 0.0631 1/sec = 6.31×10^{-5} m³/sec = 1.44×10^{-3} mgd = 2.23×10^{-3} cfs = 4.42×10^{-3} AF/day
1 cubic meter per day (m³/day)	= 0.01157 1/sec = 1.157×10^{-5} m³/sec = 0.183 gpm = 2.64×10^{-4} mgd = 4.09×10^{-4} cfs = 8.11×10^{-4} AF/day	**1 million gallons per day (mgd)**	= 43.8 1/sec = 0.0438 m³/sec = 3785 m³/day = 694 gpm = 1.55 cfs = 3.07 AF/day

Water Units, Data Conversions, and Constants *(continued)*

FLOW RATE (L^3/T) (continued)

1 cubic foot per second (cfs)	= 28.3 l/sec = 0.0283 m³/sec = 2447 m³/day = 449 gpm = 0.646 mgd = 1.98 AF/day	**1 miner's inch**	= 0.02 cfs (in Idaho, Kansas, Nebraska, New Mexico, North Dakota, South Dakota, and Utah) = 0.026 cfs (in Colorado) = 0.028 cfs (in British Columbia)
1 acre-foot per day (AF/day)	= 14.3 l/sec = 0.0143 m³/sec = 1233.48 m³/day = 226 gpm = 0.326 mgd = 0.504 cfs	**1 weir** **1 quinaria (ancient Rome)**	= 0.02 garcia = 0.47–0.48 l/sec
1 miner's inch	= 0.025 cfs (in Arizona, California, Montana, and Oregon: flow of water through 1 in² aperture under 6-inch head)		

ACCELERATION (L/T^2)

standard acceleration of gravity	= 9.8 m/s² = 32 ft/s²

FORCE (ML/T^2 = Mass × Acceleration)

1 newton (N)	= kg-m/s² = 10⁵ dynes = 0.1020 kg force = 0.2248 lb force	**1 dyne** **1 pound force**	= g·cm/s² = 10⁻⁵ N = lb mass × acceleration of gravity = 4.448 N

(continues)

Water Units, Data Conversions, and Constants *(continued)*

PRESSURE (M/L^2 = Force/Area)		**1 kilogram per sq. centimeter (kg/cm^2)**	= 14.22 lb/in^2
1 pascal (Pa)	= N/m^2		
1 bar	= 1×10^5 Pa		
	= 1×10^6 $dyne/cm^2$	**1 inch of water at 62°F**	= 0.0361 lb/in^2
	= 1019.7 g/cm^2		= 5.196 lb/ft^3
	= 10.197 te/m^2		= 0.0735 inch of mercury at 62°F
	= 0.9869 atmosphere	**1 foot of water at 62°F**	= 0.433 lb/in^2
	= 14.50 lb/in^2		= 62.36 lb/ft^2
	= 1000 millibars		= 0.833 inch of mercury at 62°F
1 atmosphere (atm)	= standard pressure		= 2.950×10^{-2} atmosphere
	= 760 mm of mercury at 0°C	**1 pound per sq. inch (psi or lb/in^2)**	= 2.309 feet of water at 62°F
	= 1013.25 millibars		= 2.036 inches of mercury at 32°F
	= 1033 g/cm^2		
	= 1.033 kg/cm^2		= 0.06804 atmosphere
	= 14.7 lb/in^2		
	= 2116 lb/ft^2		= 0.07031 kg/cm^2
	= 33.95 feet of water at 62°F	**1 inch of mercury at 32°F**	= 0.4192 lb/in^2
	= 29.92 inches of mercury at 32°F		= 1.133 feet of water at 32°F
TEMPERATURE			
degrees Celsius or Centigrade (°C)	= (°F–32) × 5/9	**degrees Fahrenheit (°F)**	= 32 + (°C × 1.8)
	= K–273.16		= 32 + ((°K–273.16) × 1.8)
Kelvins (K)	= 273.16 + °C		
	= 273.16 + ((°F– 32) × 5/9)		

Water Units, Data Conversions, and Constants *(continued)*

ENERGY (ML^2/T^2 = Force × Distance)

1 joule (J)	= 10^7 ergs	**1 kilowatt-hour**	= 3.6×10^6 J
	= N·m	**(kWh)**	= 3412 Btu
	= W·s		= 859.1 kcal
	= kg·m^2/s^2	**1 quad**	= 10^{15} Btu
	= 0.239 calories		= 1.055×10^{18} J
	= 9.48×10^{-4} Btu		= 293×10^9 kWh
1 calorie (cal)	= 4.184 J		= 0.001 Q
	= 3.97×10^{-3} Btu		= 33.45 GWy
	(raises 1 g H$_2$O	**1 Q**	= 1000 quads
	1°C)		≈ 10^{21} J
1 British thermal	= 1055 J	**1 foot-pound (ft-lb)**	= 1.356 J
unit (Btu)	= 252 cal (raises		= 0.324 cal
	1 lb H$_2$O 1°F)	**1 therm**	= 10^5 Btu
	= 2.93×10^{-4} kWh	**1 electron-volt (eV)**	= 1.602×10^{-19} J
1 erg	= 10^{-7} J	**1 kiloton of TNT**	= 4.2×10^{12} J
	= g·cm^2/s^2	**1 10^6 te oil equiv.**	= 7.33×10^6 bbl oil
	= dyne·cm	**(Mtoe)**	= 45×10^{15} J
1 kilocalorie (kcal)	= 1000 cal		= 0.0425 quad
	= 1 Calorie (food)		

POWER (ML^2/T^3 = rate of flow of energy)

1 watt (W)	= J/s	**1 horsepower**	= 0.178 kcal/s
	= 3600 J/hr	**(H.P. or hp)**	= 6535 kWh/yr
	= 3.412 Btu/hr		= 33,000 ft-lb/min
1 TW	= 10^{12} W		= 550 ft-lb/sec
	= 31.5×10^{18} J		= 8760 H.P.-hr/yr
	= 30 quad/yr	**H.P. input**	= 1.34 × kW input
1 kilowatt (kW)	= 1000 W		to motor
	= 1.341 horsepower		= horsepower
	= 0.239 kcal/s		input to motor
	= 3412 Btu/hr	**Water H.P.**	= H.P. required to
10^6 bbl (oil)/day	≈ 2 quads/yr		lift water at a
(Mb/d)	≈ 70 GW		definite rate to
1 quad/yr	= 33.45 GW		a given distance
	≈ 0.5 Mb/d		assuming 100%
1 horsepower	= 745.7 W		efficiency
(H.P. or hp)	= 0.7457 kW		= gpm × total head
			(in feet)/3960

(continues)

Water Units, Data Conversions, and Constants *(continued)*

EXPRESSIONS OF HARDNESS[a]

1 grain per gallon	= 1 grain $CaCO_3$ per U.S. gallon	**1 French degree**	= 1 part $CaCO_3$ per 100,000 parts water
1 part per million	= 1 part $CaCO_3$ per 1,000,000 parts water	**1 German degree**	= 1 part CaO per 100,000 parts water
1 English, or Clark, degree	= 1 grain $CaCO_3$ per Imperial gallon		

CONVERSIONS OF HARDNESS

1 grain per U.S. gallon	= 17.1 ppm, as $CaCO_3$	**1 French degree**	= 10 ppm, as $CaCO_3$
1 English degree	= 14.3 ppm, as $CaCO_3$	**1 German degree**	= 17.9 ppm, as $CaCO_3$

WEIGHT OF WATER

1 cubic inch	= 0.0361 lb	**1 imperial gallon**	= 10.0 lb
1 cubic foot	= 62.4 lb	**1 cubic meter**	= 1 tonne
1 gallon	= 8.34 lb		

DENSITY OF WATER[a]

Temperature		Density
°C	°F	gm/cm³
0	32	0.99987
1.667	35	0.99996
4.000	39.2	1.00000
4.444	40	0.99999
10.000	50	0.99975
15.556	60	0.99907
21.111	70	0.99802
26.667	80	0.99669
32.222	90	0.99510
37.778	100	0.99318
48.889	120	0.98870
60.000	140	0.98338
71.111	160	0.97729
82.222	180	0.97056
93.333	200	0.96333
100.000	212	0.95865

Note: Density of Sea Water: approximately 1.025 gm/cm³ at 15°C.

[a]*Source:* van der Leeden, F., Troise, F. L., and Todd, D. K., 1990. *The Water Encyclopedia*, 2d edition. Lewis Publishers, Inc., Chelsea, Michigan.

Comprehensive Table of Contents

Volume 1

The World's Water 1998-1999: The Biennial Report on Freshwater Resources

Foreword by Anne H. and Paul R. Erhlich   ix

Acknowledgments   xi

Introduction   1

ONE   **The Changing Water Paradigm**   5
  Twentieth-Century Water-Resources Development   6
  The Changing Nature of Demand   10
  Economics of Major Water Projects   16
  Meeting Water Demands in the Next Century   18
  Summary: New Thinking, New Actions   32
  References   33

TWO   **Water and Human Health**   39
  Water Supply and Sanitation: Falling Behind   39
  Basic Human Needs for Water   42
  Water-Related Diseases   47
  Update on Dracunculiasis (Guinea Worm)   50
  Update on Cholera   56
  Summary   63
  References   64

THREE   **The Status of Large Dams: The End of an Era?**   69
  Environmental and Social Impacts of Large Dams   75
  New Developments in the Dam Debate   80
  The Three Gorges Project, Yangtze River, China   84
  The Lesotho Highlands Project, Senqu River Basin, Lesotho   93
  References   101

FOUR   **Conflict and Cooperation Over Fresh Water**   105
  Conflicts Over Shared Water Resources   107
  Reducing the Risk of Water-Related Conflict   113
  The Israel-Jordan Peace Treaty of 1994   115
  The Ganges-Brahmaputra Rivers: Conflict and Agreement   118
  Water Disputes in Southern Africa   119

　　　　　Summary   124
　　　　　Appendix A
　　　　　　Chronology of Conflict Over Waters in the Legends,
　　　　　　Myths, and History of the Ancient Middle East   125
　　　　　Appendix B
　　　　　　Chronology of Conflict Over Water:
　　　　　　1500 to the Present   128
　　　　　References   132

FIVE　　Climate Change and Water Resources:
　　　　What Does the Future Hold?   137
　　　　　What Do We Know?   138
　　　　　Hydrologic Effects of Climate Change   139
　　　　　Societal Impacts of Changes in Water Resources   144
　　　　　Is the Hydrologic System Showing Signs of Change?   145
　　　　　Recommendations and Conclusions   148
　　　　　References   150

SIX　　　New Water Laws, New Water Institutions   155
　　　　　Water Law and Policy in New South Africa: A Move Toward
　　　　　　Equity   156
　　　　　The Global Water Partnership   165
　　　　　The World Water Council   172
　　　　　The World Commission on Dams   175
　　　　　References   180

SEVEN　Moving Toward a Sustainable Vision for the Earth's
　　　　Fresh Water   183
　　　　　Introduction   183
　　　　　A Vision for 2050: Sustaining Our Waters   185

WATER BRIEFS

　　　　　The Best and Worst of Science: Small Comets and the New
　　　　　　Debate Over the Origin of Water on Earth   193
　　　　　Water Bag Technology   200
　　　　　Treaty Between the Government of the Republic of India and
　　　　　　the Government of the People's Republic of Bangladesh
　　　　　　on Sharing of the Ganga/Ganges Waters at Farakka   206
　　　　　United Nations Conventions on the Law of the Non-
　　　　　　Navigational Uses of International Watercourses   210
　　　　　Water-Related Web Sites   231

DATA SECTION

Table 1　　Total Renewable Freshwater Supply by Country   235
Table 2　　Freshwater Withdrawal by Country and Sector   241

Comprehensive Table of Contents

Table 3	Summary of Estimated Water Use in the United States, 1900 to 1995 245
Table 4	Total and Urban Population by Country, 1975 to 1995 246
Table 5	Percentage of Population with Access to Safe Drinking Water by Country, 1970 to 1994 251
Table 6	Percentage of Population with Access to Sanitation by Country, 1970 to 1994 256
Table 7	Access to Safe Drinking Water in Developing Countries by Region, 1980 to 1994 261
Table 8	Access to Sanitation in Developing Countries by Region, 1980 to 1994 263
Table 9	Reported Cholera Cases and Deaths by Region, 1950 to 1997 265
Table 10	Reported Cholera Cases and Deaths by Country, 1996 and 1997 268
Table 11	Reported Cholera Cases and Deaths in the Americas, 1991 to 1997 271
Table 12	Reported Cases of Dracunculiasis by Country, 1972 to 1996 272
Table 13	Waterborne Disease Outbreaks in the United States by Type of Water Supply System, 1971 to 1994 274
Table 14	Hydroelectric Capacity and Production by Country, 1996 276
Table 15	Populations Displaced as a Consequence of Dam Construction, 1930 to 1996 281
Table 16	Desalination Capacity by Country (January 1, 1996) 288
Table 17	Desalination Capacity by Process (January 1, 1996) 290
Table 18	Threatened Reptiles, Amphibians, and Freshwater Fish, 1997 291
Table 19	Irrigated Area by Country and Region, 1961 to 1994 297

Index   303

Volume 2
The World's Water 2000-2001: The Biennial Report on Freshwater Resources

Foreword by Timothy E. Wirth xiii

Acknowledgments xv

Introduction xvii

ONE **The Human Right to Water** 1
    Is There a Human Right to Water? 2
    Existing Human Rights Laws, Covenants, and Declarations 4
    Defining and Meeting a Human Right to Water 9
    Conclusions 15
    References 15

TWO **How Much Water Is There and Whose Is It? The World's Stocks and Flows of Water and International River Basins** 19
    How Much Water Is There? The Basic Hydrologic Cycle 20
    International River Basins: A New Assessment 27
    The Geopolitics of International River Basins 35
    Summary 36
    References 37

THREE **Pictures of the Future: A Review of Global Water Resources Projections** 39
    Data Constraints 40
    Forty Years of Water Scenarios and Projections 42
    Analysis and Conclusions 58
    References 59

FOUR **Water for Food: How Much Will Be Needed?** 63
    Feeding the World Today 63
    Feeding the World in the Future: Pieces of the Puzzle 65
    How Much Water Will Be Needed to Grow Food? 78
    Conclusions 88
    Referencess 89

FIVE **Desalination: Straw into Gold or Gold into Water** 93
    History of Desalination and Current Status 94
    Desalination Technologies 98
    Other Aspects of Desalination 106
    The Tampa Bay Desalination Plant 108
    Summary 109
    References 110

Comprehensive Table of Contents

SIX   The Removal of Dams: A New Dimension to an Old Debate   113
   Economics of Dam Removal   118
   Dam Removal Case Studies: Some Completed Removals   120
   Some Proposed Dam Removals or Decommissionings   126
   Conclusion   134
   References   134

SEVEN   Water Reclamation and Reuse: Waste Not, Want Not   137
   Wastewater Uses   139
   Direct and Indirect Potable Water Reuse   151
   Health Issues   152
   Wastewater Reuse in Namibia   155
   Wastewater Reclamation and Reuse in Japan   158
   Wastewater Costs   159
   Summary   159
   References   161

WATER BRIEFS

   Arsenic in the Groundwater of Bangladesh and West Bengal, India   165
   Fog Collection as a Source of Water   175
   Environment and Security: Water Conflict Chronology—Version 2000   182
   Water-Related Web Sites   192

DATA SECTION

Table 1   Total Renewable Freshwater Supply, by Country   197
Table 2   Freshwater Withdrawal, by Country and Sector   203
Table 3   World Population, Year 0 to A.D. 2050   212
Table 4   Population, by Continent, 1750 to 2050   213
Table 5   Renewable Water Resources and Water Availability, by Continent   215
Table 6   Dynamics of Water Resources, Selected Countries, 1921 to 1985   218
Table 7   International River Basins of the World   219
Table 8   Fraction of a Country's Area in International River Basins   239
Table 9   International River Basins, by Country   247
Table 10   Irrigated Area, by Country and Region, 1961 to 1997   255
Table 11   Irrigated Area, by Continent, 1961 to 1997   264
Table 12   Human-Induced Soil Degradation, by Type and Cause, Late 1980s   266
Table 13   Continental Distribution of Human-Induced Salinization   268
Table 14   Salinization, by Country, Late 1980s   269
Table 15   Total Number of Reservoirs, by Continent and Volume   270

Table 16	Number of Reservoirs Larger than 0.1 km³, by Continent, Time Series 271
Table 17	Volume of Reservoirs Larger than 0.1 km³, by Continent, Time Series 273
Table 18	Dams Removed or Decommissioned in the United States, 1912 to Present 275
Table 19	Desalination Capacity, by Country, January 1999 287
Table 20	Total Desalination Capacity, by Process, June 1999 289
Table 21	Desalination Capacity, by Source of Water, June 1999 290
Table 22	Number of Threatened Species, by Country/Area, by Group, 1997 291
Table 23	Countries with the Largest Number of Fish Species 298
Table 24	Countries with the Largest Number of Fish Species per Unit Area 299
Table 25	Water Units, Data Conversions, and Constants 300

Index   311

Comprehensive Table of Contents

Volume 3
The World's Water 2002-2003: The Biennial Report on Freshwater Resources

Foreword by Amory B. Lovins xiii

Acknowledgments xv

Introduction xvii

ONE   The Soft Path for Water   1
by Gary Wolff and Peter H. Gleick

A Better Way   3
Dominance of the Hard Path in the Twentieth Century   7
Myths about the Soft Path   9
One Dimension of the Soft Path: Efficiency of Use   16
Moving Forward on the Soft Path   25
Conclusions   30
References   30

TWO   Globalization and International Trade of Water   33
by Peter H. Gleick, Gary Wolff, Elizabeth L. Chalecki, and Rachel Reyes

The Nature and Economics of Water, Background and Definitions   34
Water Managed as Both a Social and Economic Good   38
The Globalization of Water: International Trade   41
The Current Trade in Water   42
The Rules: International Trading Regimes   47
References   54

THREE   The Privatization of Water and Water Systems   57
by Peter H. Gleick, Gary Wolff, Elizabeth L. Chalecki, and Rachel Reyes

Drivers of Water Privatization   58
History of Privatization   59
The Players   61
Forms of Privatization   63
Risks of Privatization: Can and Will They Be Managed?   67
Principles and Standards for Privatization   79
Conclusions   82
References   83

FOUR　Measuring Water Well-Being: Water Indicators and Indices　87
　　by Peter H. Gleick, Elizabeth L. Chalecki, and Arlene Wong

　　Quality-of-Life Indicators and Why We Develop Them　88
　　Limitations to Indicators and Indices　92
　　Examples of Single-Factor or Weighted Water Measures　96
　　Multifactor Indicators　101
　　Conclusions　111
　　References　112

FIVE　Pacific Island Developing Country Water Resources and Climate Change　113
　　by William C.G. Burns

　　PIDCs and Freshwater Sources　113
　　Climate Change and PIDC Freshwater Resources　119
　　Potential Impacts of Climate Change on PIDC Freshwater Resources　124
　　Recommendations and Conclusions　125
　　References　127

SIX　Managing Across Boundaries: The Case of the Colorado River Delta　133
　　by Michael Cohen

　　The Colorado River　134
　　The Colorado River Delta　139
　　Conclusions　144
　　References　145

SEVEN　The World Commission on Dams Report: What Next?　149
　　by Katherine Kao Cushing

　　The WCD Organization　149
　　Findings and Recommendations　151
　　Strategic Priorities, Criteria, and Guidelines　153
　　Reaction to the WCD Report　155
　　References　172

WATER BRIEFS

　　The Texts of the Ministerial Declarations from The Hague March 2000) and Bonn (December 2001)　173
　　The Southeastern Anatolia (GAP) Project and Archaeology　181
　　by Amar S. Mann

Comprehensive Table of Contents

Water Conflict Chronology 194
by Peter S. Gleick

Water and Space 209
by Elizabeth L. Chalecki

Water-Related Web Sites 225

DATA SECTION 237

Table 1 Total Renewable Freshwater Supply, by Country (2002 Update) 237
Table 2 Fresh Water Withdrawals, by Country and Sector (2002 Update) 243
Table 3 Access to Safe Drinking Water by Country, 1970 to 2000 252
Table 4 Access to Sanitation by Country, 1970 to 2000 261
Table 5 Access to Water Supply and Sanitation by Region, 1990 and 2000 270
Table 6 Reported Cases of Dracunculiasis by Country, 1972 to 2000 273
Table 7 Reported Cases of Dracunculiasis Cases, Eradication Progress, 2000 276
Table 8 National Standards for Arsenic in Drinking Water 278
Table 9 United States National Primary Drinking Water Regulations 280
Table 10 Irrigated Area, by Region, 1961 to 1999 289
Table 11 Irrigated Area, Developed and Developing Countries, 1960 to 1999 290
Table 12 Number of Dams, by Continent and Country 291
Table 13 Number of Dams, by Country 296
Table 14 Regional Statistics on Large Dams 300
Table 15 Commissioning of Large Dams in the 20th Century, by Decade 301
Table 16 Water System Rate Structures 303
Table 17 Water Prices for Various Households 304
Table 18 Unaccounted-for Water 305
Table 19 United States Population and Water Withdrawals, 1900 to 1995 308
Table 20 United States GNP and Water Withdrawals, 1900 to 1996 310
Table 21 Hong Kong GDP and Water Withdrawls, 1952 to 2000 313
Table 22 China GDP and Water Withdrawls, 1952 to 2000 316

Water Units, Data Conversions, and Constants 318

Index 329

Volume 4
The World's Water 2004–2005: The Biennial Report on Freshwater Resources

Foreword by Margaret Catley-Carlson xiii

Introduction xv

ONE   The Millennium Development Goals for Water: Crucial Objectives, Inadequate Commitments   1
by Peter H. Gleick

- Setting Water and Sanitation Goals   2
- Commitments to Achieving the MDGs for Water   2
- Consequences: Water-Related Diseases   7
- Measures of Illness from Water-Related Diseases   9
- Scenarios of Future Deaths from Water-Related Diseases   10
- Conclusions   14
- References   14

TWO   The Myth and Reality of Bottled Water   17
by Peter H. Gleick

- Bottled Water Use History and Trends   18
- The Price and Cost of Bottled Water   22
- The Flavor and Taste of Water   23
- Bottled Water Quality   25
- Regulating Bottled Water   26
- Comparison of U.S. Standards for Bottled Water and Tap Water   36
- Other Concerns Associated with Bottled Water   37
- Conclusions   41
- References   42

THREE   Water Privatization Principles and Practices   45
by Meena Palaniappan, Peter H. Gleick, Catherine Hunt, Veena Srinivasan

- Update on Privatization   46
- Principles and Standards for Water   47
- Can the Principles Be Met?   48
- Conclusions   73
- References   74

FOUR   Groundwater: The Challenge of Monitoring and
        Management 79
        by Marcus Moench

   Conceptual Foundations 80
   Challenges in Assessment 80
   Extraction and Use 81
   Groundwater in Agriculture 88
   The Analytical Dilemma 90
   A Way Forward: Simple Data as a Catalyst for Effective
       Management 97
   References 98

FIVE   Urban Water Conservation: A Case Study of Residential
        Water Use in California 101
        by Peter H. Gleick, Dana Haasz, Gary Wolff

   The Debate over California's Water 102
   Defining Water "Conservation" and "Efficiency" 103
   Current Urban Water Use in California 105
   A Word About Agricultural Water Use 107
   Economics of Water Savings 107
   Data and Information Gaps 108
   Indoor Residential Water Use 109
   Indoor Residential Water Conservation: Methods and
       Assumptions 112
   Indoor Residential Summary 118
   Outdoor Residential Water Use 118
   Current Outdoor Residential Water Use 119
   Existing Outdoor Conservation Efforts and Approaches 120
       Outdoor Residential Water Conservation: Methods and
       Assumptions 121
   Residential Outdoor Water Use Summary 125
   Conclusions 126
   Abbreviations and Acronyms 126
   References 127

SIX   Urban Water Conservation: A Case Study of Commercial and
        Industrial Water Use in California 131
        *by Peter H. Gleick, Veena Srinivasan, Christine Henges-Jeck,
        Gary Wolff*

   Background to CII Water Use 132
   Current California Water Use in the CII Sectors 133
   Estimated CII Water Use in California in 2000 138

Data Challenges   139
The Potential for CII Water Conservation and Efficiency
   Improvements: Methods and Assumptions   140
Methods for Estimating CII Water Use and Conservation
   Potential   143
Calculation of Conservation Potential   145
Data Constraints and Conclusions   148
Recommendations for CII Water Conservation   150
Conclusions   153
References   154

SEVEN Climate Change and California Water Resources   157
by Michael Kiparsky and Peter H. Gleick

The State of the Science   158
Climate Change and Impacts on Managed Water-Resource
   Systems   172
Moving From Climate Science to Water Policy   175
Conclusions   183
References   184

WATER BRIEFS

One 3rd World Water Forum in Kyoto: Disappointment and
     Possibility   189
  by Nicholas L. Cain

  Ministerial Declaration of the 3rd World Water Forum:
     Message from the Lake Biwa and Yodo River Basin   198

  NGO Statement of the 3rd World Water Forum   202

Two The Human Right to Water: Two Steps Forward, One Step
     Back   204
  by Peter H. Gleick

  Substantive Issues Arising in the Implementation of the International Covenant on Economic, Social, and Cultural Rights. United Nations General Comment No. 15 (2002)   213

Three The Water and Climate Bibliography   228
  by Peter H. Gleick and Michael Kiparksy

Four Environment and Security: Water Conflict Chronology
     Version 2004–2005   234
  by Peter H. Gleick

  Water Conflict Chronology   236
  by Peter H. Gleick

DATA SECTION

Data Table 1 Total Renewable Freshwater Supply by Country
              (2004 Update)   257
Data Table 2 Freshwater Withdrawals by Country and Sector
              (2004 Update)   263

Comprehensive Table of Contents   417

Data Table 3   Deaths and DALYs from Selected Water-Related Diseases   272
Data Table 4   Official Development Assistance Indicators   274
Data Table 5   Aid to Water Supply and Sanitation by Donor, 1996 to 2001   278
Data Table 6   Bottled Water Consumption by Country, 1997 to 2002   280
Data Table 7   Global Bottled Water Consumption by Region, 1997 to 2002   283
Data Table 8   Bottled Water Consumption, Share by Region, 1997 to 2002   285
Data Table 9   Per-capita Bottled Water Consumption by Region, 1997 to 2002   286
Data Table 10   United States Bottled Water Sales, 1991 to 2001   288
Data Table 11   Types of Packaging Used for Bottled Water in Various Countries, 1999   289
Data Table 12   Irrigated Area by Region, 1961 to 2001   291
Data Table 13   Irrigated Area, Developed and Developing Countries, 1961 to 2001   293
Data Table 14   Global Production and Yields of Major Cereal Crops, 1961 to 2002   295
Data Table 15   Global Reported Flood Deaths, 1900 to 2002   298
Data Table 16   United States Flood Damage by Fiscal Year, 1926 to 2001   301
Data Table 17   Total Outbreaks of Drinking Water-Related Disease, United States, 1973 to 2000   304
Data Table 18
    18.a Extinction Rate Estimates for Continental North American Fauna (percent loss per decade)   309
    18.b Imperiled Species for North American Fauna   309
Data Table 19   Proportion of Species at Risk, United States   311
Data Table 20   United States Population and Water Withdrawals 1900 to 2000   313
Data Table 21   United States Economic Productivity of Water, 1900 to 2000   317

WATER UNITS, DATA CONVERSIONS, AND CONSTANTS   321

COMPREHENSIVE TABLE OF CONTENTS   331

Volume 1:   The World's Water 1998–1999: The Biennial Report on Freshwater Resources   331
Volume 2:   The World's Water 2000–2001: The Biennial Report on Freshwater Resources   334
Volume 3:   The World's Water 2002–2003: The Biennial Report on Freshwater Resources   337

COMPREHENSIVE INDEX   341

Volume 5
The World's Water 2006-2007: The Biennial Report on Freshwater Resources

Foreword by Jon Lane xiii

Introduction xv

ONE   Water and Terrorism   1
by Peter H. Gleick

Introduction   1
The Worry   2
Defining Terrorism   3
History of Water-Related Conflict   5
Vulnerability of Water and Water Systems   15
Responding to the Threat of Water-Related Terrorism   22
Water Security Policy in the United States   25
Conclusion   25

TWO   Going with the Flow: Preserving and Restoring Instream Water Allocations   29
by David Katz

Environmental Flow: Concepts and Applications   30
Legal Frameworks for Securing Environmental Flow   34
The Science of Determining Environmental Flow Allocations   38
The Economics and Finance of Environmental Flow Allocations   40
Making It Work: Policy Implementation   43
Conclusion   45

THREE   With a Grain of Salt: An Update on Seawater Desalination   51
by Peter H. Gleick, Heather Cooley, Gary Wolff

Introduction   51
Background to Desalination   52
History of Desalination   54
Desalination Technologies   54
Current Status of Desalination   55
Advantages and Disadvantages of Desalination   66
Environmental Effects of Desalination   76
Desalination and Climate Change   80
Public Transparency   81

Comprehensive Table of Contents

       Summary   82
       Desalination Conclusions and Recommendations   83

FOUR   Floods and Droughts   91
by Heather Cooley

Introduction   91
Droughts   92
Floods   104
The Future of Droughts and Floods   112
Conclusion   113

FIVE   Environmental Justice and Water   117
by Meena Palaniappan, Emily Lee, Andrea Samulon

Introduction   117
A Brief History of Environmental Justice in the United States   119
Environmental Justice in the International Water Context   123
Environmental Justice and International Water Issues   124
Recommendations   137
Conclusion   141

SIX   Water Risks that Face Business and Industry   145
by Peter H. Gleick, Jason Morrison

Introduction   145
Water Risks for Business   146
Some New Water Trends: Looking Ahead   150
Managing Water Risks   153
An Overview of the "Water Industry"   158
Conclusion   163

WATER BRIEFS

One   Bottled Water: An Update   169
by Peter H. Gleick

Two   Water on Mars   175
by Peter H. Gleick

    Introduction   175
    Background   175
    Martian History of Water   178
    Future Mars Missions   180

Three   Time to Rethink Large International Water Meetings   182
by Peter H. Gleick

    Introduction   182
    Background and History   182

	Outcomes of International Water Meetings 183
	Ministerial Statements 184
	Conclusions and Recommendations 185
	4th World Water Forum Ministerial Declaration 186
Four	Environment and Security: Water Conflict Chronology Version, 2006–2007 189
	by Peter H. Gleick
Five	The Soft Path in Verse 219
	by Gary Wolff

DATA SECTION

Data Table 1 Total Renewable Freshwater Supply by Country 221

Data Table 2 Freshwater Withdrawal by Country and Sector 228

Data Table 3 Access to Safe Drinking Water by Country, 1970 to 2002 237

Data Table 4 Access to Sanitation by Country, 1970 to 2002 247

Data Table 5 Access to Water Supply and Sanitation by Region, 1990 and 2002 256

Data Table 6 Annual Average ODA for Water, by Country, 1990 to 2004 (Total and Per Capita) 262

Data Table 7 Twenty Largest Recipients of ODA for Water, 1990 to 2004 268

Data Table 8 Twenty Largest Per Capita Recipients of ODA for Water, 1990 to 2004 270

Data Table 9 Investment in Water and Sewerage Projects with Private Participation, by Region, in Middle- and Low-Income Countries, 1990 to 2004 273

Data Table 10 Bottled Water Consumption by Country, 1997 to 2004 276

Data Table 11 Global Bottled Water Consumption, by Region, 1997 to 2004 280

Data Table 12 Per Capita Bottled Water Consumption by Region, 1997 to 2004 282

Data Table 13 Per Capita Bottled Water Consumption, by Country, 1999 to 2004 284

Data Table 14 Global Cholera Cases and Deaths Reported to the World Health Organization, 1970 to 2004 287

Data Table 15 Reported Cases of Dracunculiasis by Country, 1972 to 2005 293

Data Table 16 Irrigated Area, by Region, 1961 to 2003 298

Data Table 17 Irrigated Area, Developed and Developing Countries, 1961 to 2003 301

Data Table 18 The U.S. Water Industry Revenue (2003) and Growth (2004–2006) 303

Data Table 19 Pesticide Occurrence in Streams, Groundwater, Fish, and Sediment in the United States 305

Data Table 20 Global Desalination Capacity and Plants— January 1, 2005 308

Data Table 21 100 Largest Desalination Plants Planned, in Construction, or in Operation—January 1, 2005   310

Data Table 22 Installed Desalination Capacity by Year, Number of Plants, and Total Capacity, 1945 to 2004   314

WATER UNITS, DATA CONVERSIONS, AND CONSTANTS   319

COMPREHENSIVE TABLE OF CONTENTS   329

Volume 1: The World's Water 1998–1999: The Biennial Report on Freshwater Resources   329

Volume 2: The World's Water 2000–2001: The Biennial Report on Freshwater Resources   332

Volume 3: The World's Water 2002–2003: The Biennial Report on Freshwater Resources   335

Volume 4: The World's Water 2004–2005: The Biennial Report on Freshwater Resources   338

COMPREHENSIVE INDEX   343

Volume 6
The World's Water 2008–2009: The Biennial Report on Freshwater Resources

Foreword by Malin Falkenmark xi

Acknowledgments xiii

Introduction xv

ONE   Peak Water   1
Meena Palaniappan and Peter H. Gleick
Concept of Peak Oil   2
Comparison of Water and Oil   3
Utility of the Term "Peak Water"   9
A New Water Paradigm: The Soft Path for Water   12
Conclusion   14
References   15

TWO   Business Reporting on Water   17
Mari Morikawa, Jason Morrison, and Peter H. Gleick
Corporate Reporting: A Brief History   17
Qualitative Information: Water Management Policies, Strategies, and Activities   22
Water Reporting Trends by Sector   32
Conclusions and Recommendations   36
References   38

THREE   Water Management in a Changing Climate   39
Heather Cooley
The Climate Is Already Changing   39
Projected Impacts of Rising Greenhouse Gas Concentrations   40
Climate Change and Water Resources   43
Vulnerability to Climate Change   44
Adaptation   45
Conclusion   53
References   54

FOUR   Millennium Development Goals: Charting Progress and the Way Forward   57

Meena Palaniappan

Millennium Development Goals   57
Measuring Progress: Methods and Definitions   60
Progress on the Water and Sanitation MDGs   62
A Closer Look at Water and Sanitation Disparities   70
Meeting the MDGs: The Way Forward   73
Conclusion   77
References   78

FIVE   China and Water   79

Peter H. Gleick

The Problems   80
Water-Related Environmental Disasters in China   82
Water Availability and Quantity   83
Groundwater Overdraft   85
Floods and Droughts   86
Climate Change and Water in China   87
Water and Chinese Politics   88
Growing Regional Conflicts Over Water   90
Moving Toward Solutions   91
Improving Public Participation   96
Conclusion   97
References   97

SIX   Urban Water-Use Efficiencies: Lessons from United States Cities   101

Heather Cooley and Peter H. Gleick

Use of Water in Urban Areas   101
Projecting and Planning for Future Water Demand   102
Per-Capita Demand   104
Water Conservation and Efficiency Efforts   106
Comparison of Water Conservation Programs   110
Rate Structures   112
Conclusion   120
References   120

WATER BRIEFS

One   Tampa Bay Desalination Plant: An Update   123
Heather Cooley

Two   Past and Future of the Salton Sea   127
Michael J. Cohen

Background   129

	Restoration 131
	Conclusion 137
Three	Three Gorges Dam Project, Yangtze River, China 139
	Peter H. Gleick
	Introduction 139
	The Project 140
	Major Environmental, Economic, Social, and Political Issues 141
	Conclusion 148
Four	Water Conflict Chronology 151
	Peter H. Gleick

DATA SECTION

Data Table 1 Total Renewable Freshwater Supply, by Country   195
Data Table 2 Freshwater Withdrawal by Country and Sector   202
Data Table 3 Access to Safe Drinking Water by Country,
 1970 to 2004   211
Data Table 4 Access to Sanitation by Country, 1970 to 2004   221
Data Table 5 MDG Progress on Access to Safe Drinking Water
 by Region   230
Data Table 6 MDG Progress on Access to Sanitation by Region   233
Data Table 7 United States Dams and Dam Safety Data, 2006   236
Data Table 8 Dams Removed or Decommissioned in the United States,
 1912 to Present   239
Data Table 9 Dams Removed or Decommissioned in the United States,
 1912 to Present, by Year and State   265
Data Table 10 United States Dams by Primary Purposes   270
Data Table 11 United States Dams by Owner   272
Data Table 12 African Dams: Number and Total Reservoir Capacity
 by Country   274
 African Dams: The 30 Highest   276
Data Table 13 Under-5 Mortality Rate by Cause and Country, 2000   279
Data Table 14 International River Basins of Africa, Asia, Europe,
 North America, and South America   289
Data Table 15 OECD Water Tariffs   312
Data Table 16 Non-OECD Water Tariffs   320
Data Table 17 Fraction of Arable Land that Is Irrigated, by Country   324
Data Table 18 Area Equipped for Irrigation, by Country   329
Data Table 19 Water Content of Things   335
Data Table 20 Top Environmental Concerns of the American Public:
 Selected Years, 1997–2008   339

WATER UNITS, DATA CONVERSIONS, AND CONSTANTS   343

COMPREHENSIVE TABLE OF CONTENTS   353

Volume 1: The World's Water 1998–1999: The Biennial Report on
 Freshwater Resources   353

Volume 2: The World's Water 2000–2001: The Biennial Report on
          Freshwater Resources   356
Volume 3: The World's Water 2002–2003: The Biennial Report on
          Freshwater Resources   359
Volume 4: The World's Water 2004–2005: The Biennial Report on
          Freshwater Resources   362
Volume 5: The World's Water 2006–2007: The Biennial Report on
          Freshwater Resources   366
Volume 6: The World's Water 2008–2009: The Biennial Report on
          Freshwater Resources   370

COMPREHENSIVE INDEX   373

Volume 7
The World's Water Volume 7: The Biennial Report on Freshwater Resources

Foreword by Robert Glennon xi

Introduction xiii

ONE   Climate Change and Transboundary Waters   1
Heather Cooley, Juliet Christian-Smith, Peter H. Gleick, Lucy Allen, and Michael Cohen

Transboundary Rivers and Aquifers   2
Managing Transboundary Basins   4
Transboundary Water Management Policies and Climate Change   7
Case Studies   10
Conclusions and Recommendations   18

TWO   Corporate Water Management   23
Peter Schulte, Jason Morrison, and Peter H. Gleick

Global Trends that Affect Businesses   24
Water-Related Business Risks   25
Key Factors that Determine Extent and Type of Risk   28
Risk and Impact Assessment   30
Strategies for Improved Corporate Water Management   34
Conclusions: A Framework for Action   41

THREE   Water Quality   45
Meena Palaniappan, Peter H. Gleick, Lucy Allen, Michael J. Cohen, Juliet Christian-Smith, and Courtney Smith

Current Water-Quality Challenges   46
Consequences of Poor Water Quality   54
Moving to Solutions and Actions   65
Mechanisms to Achieve Solutions   66
Conclusion   67

FOUR   Fossil Fuels and Water Quality   73
Lucy Allen, Michael J. Cohen, David Abelson, and Bart Miller

Fossil-Fuel Production and Associated Water Use   74
Fossil Fuels and Water Quality: Direct Impacts   76
Impacts on Freshwater Ecosystems   84
Impacts on Human Communities   87
Conclusion   92

Comprehensive Table of Contents   427

FIVE   Australia's Millennium Drought: Impacts and Responses   97
Matthew Heberger

Water Resources of Australia   98
Impacts of the Millennium Drought   102
Responses to Drought   106
Conclusion   121

SIX   China Dams   127
Peter H. Gleick

Dams in China   129
Dams on Chinese International Rivers   130
Exporting Chinese Dams   133
Growing Internal Concern Over Chinese Dams   135
International Principles Governing Dam Projects: The World Commission on Dams   136
Conclusions   140

SEVEN   U.S. Water Policy Reform   143
Juliet Christian-Smith, Peter H. Gleick, and Heather Cooley

Background   143
International Water Reform Efforts   144
Common Themes and "Soft Path" Solutions   149
A 21st Century U.S. Water Policy   150
Conclusions   154

WATER BRIEFS   157

One   Bottled Water and Energy   157
Peter H. Gleick and Heather Cooley

Energy to Produce Bottled Water   158
Summary of Energy Uses   163
Conclusions   163

Two   The Great Lakes Water Agreements   165
Peter Schulte

History of Shared Water Resource Management   165
Conclusion   169

Three   Water in the Movies   171
Peter H. Gleick

Popular Movies/Films   171
Water Documentaries   174
Short Water Videos and Films   174

Four   Water Conflict Chronology   175
Peter H. Gleick and Matthew Heberger

DATA SECTION

Data Table 1 Total Renewable Freshwater Supply, by Country 215
Data Table 2 Freshwater Withdrawal by Country and Sector 221
Data Table 3 Access to Safe Drinking Water by Country, 1970–2008 230
Data Table 4 Access to Sanitation by Country, 1970–2008 241
Data Table 5 MDG Progress on Access to Safe Drinking Water by Region (proportion of population using an improved water source) 251
Data Table 6 MDG Progress on Access to Sanitation by Region (proportion of population using an improved sanitation facility) 254
Data Table 7 Under-5 Mortality Rate by Cause and Country, 2008 257
Data Table 8 Infant Mortality Rate by Country (per 1,000 live births) 264
Data Table 9 Death and DALYs from Selected Water-Related Diseases, 2000 and 2004 270
Data Table 10 Overseas Development Assistance for Water Supply and Sanitation, by Donating Country 273
Data Table 11 Overseas Development Assistance for Water Supply and Sanitation, by Subsector (total of all donating countries) 275
Data Table 12 Organic Water Pollutant (BOD) Emissions by Country (% from various industries), 2005 278
Data Table 13 Top Environmental Concerns of the American Public: Selected Years, 1997–2010 (% who worry "a great deal") 282
Data Table 14 Top Environmental Concerns Around the World 285
Data Table 15 Satisfaction With Local Water Quality, by Country, 2006–2007 289
Data Table 16 Extinct (or Extinct in the Wild) Freshwater Animal Species 292
Data Table 17 U.S. Federal Water-Related Agency Budgets 303
Data Table 18 Overseas Dams With Chinese Financiers, Developers, or Builders (as of August 2010) 308
Data Table 19 Per-Capita Bottled Water Consumption by Top Countries, 1999–2010 (liters per person per year) 339

WATER UNITS, DATA CONVERSIONS, AND CONSTANTS 341

COMPREHENSIVE TABLE OF CONTENTS 351

Volume 1: The World's Water 1998–1999: The Biennial Report on Freshwater Resources 351
Volume 2: The World's Water 2000–2001: The Biennial Report on Freshwater Resources 354
Volume 3: The World's Water 2002–2003: The Biennial Report on Freshwater Resources 357
Volume 4: The World's Water 2004–2005: The Biennial Report on Freshwater Resources 360

Volume 5: The World's Water 2006–2007: The Biennial Report on Freshwater Resources   364

Volume 6: The World's Water 2008–2009: The Biennial Report on Freshwater Resources   368

Volume 7: The World's Water Volume 7: The Biennial Report on Freshwater Resources   372

COMPREHENSIVE INDEX   377

Volume 8
The World's Water Volume 8: The Biennial Report on Freshwater Resources

Foreword by Ismail Serageldin   xi

Introduction   xiii

ONE   Global Water Governance in the Twenty-First Century   1

Heather Cooley, Newsha Ajami, Mai-Lan Ha, Veena Srinivasan, Jason Morrison, Kristina Donnelly, and Juliet Christian-Smith

Global Water Challenges   2
The Emergence of Global Water Governance   6
Conclusions   15

TWO   Shared Risks and Interests: The Case for Private Sector Engagement in Water Policy and Management   19

Peter Schulte, Stuart Orr, and Jason Morrison

The Business Case for Investing in Sustainable Water Management   21
Utilizing Corporate Resources While Ensuring Public Interest Outcomes and Preventing Policy Capture   28
Moving Forward: Unlocking Mutually Beneficial Corporate Action on Water   31

THREE   Sustainable Water Jobs   35

Eli Moore, Heather Cooley, Juliet Christian-Smith, and Kristina Donnelly

Water Challenges in Today's Economy   36
Job Quality and Growth in Sustainable Water Occupations   54
Conclusions   57
Recommendations   58

FOUR   Hydraulic Fracturing and Water Resources: What Do We Know and Need to Know?   63

Heather Cooley and Kristina Donnelly

Overview of Hydraulic Fracturing   64
Concerns Associated with Hydraulic Fracturing Operations   65
Water Challenges   67
Conclusions   77

FIVE Water Footprint   83
Julian Fulton, Heather Cooley, and Peter H. Gleick

The Water Footprint Concept   83
Water, Carbon, and Ecological Footprints and Nexus Thinking   85
Water Footprint Findings   87
Conclusion   90

SIX Key Issues for Seawater Desalination in California: Cost and Financing   93
Heather Cooley and Newsha Ajami

How Much Does Seawater Desalination Cost?   93
Desalination Projects and Risk   101
Case Studies   106
Conclusions   117

SEVEN Zombie Water Projects   123
Peter H. Gleick, Matthew Heberger, and Kristina Donnelly

The North American Water and Power Alliance—NAWAPA   124
The Reber Plan   131
Alaskan Water Shipments   133
Las Vegas Valley Pipeline Project   136
Diverting the Missouri River to the West   140
Conclusions   144

WATER BRIEF

One   The Syrian Conflict and the Role of Water   147
   Peter H. Gleick

Two   The Red Sea–Dead Sea Project Update   153
   Kristina Donnelly

Three   Water and Conflict: Events, Trends, and Analysis (2011–2012)   159
   Peter H. Gleick and Matthew Heberger

Four   Water Conflict Chronology   173
   Peter H. Gleick and Matthew Heberger

DATA SECTION

Data Table 1:   Total Renewable Freshwater Supply by Country (2013 Update)   221

Data Table 2:   Freshwater Withdrawal by Country and Sector (2013 Update)   227

Data Table 3A:	Access to Improved Drinking Water by Country, 1970–2008	236
Data Table 3B:	Access to Improved Drinking Water by Country, 2011 Update	247
Data Table 4A:	Access to Improved Sanitation by Country, 1970–2008	252
Data Table 4B:	Access to Improved Sanitation by Country, 2011 Update	263
Data Table 5:	MDG Progress on Access to Safe Drinking Water by Region	268
Data Table 6:	MDG Progress on Access to Sanitation by Region	271
Data Table 7:	Monthly Natural Runoff for the World's Major River Basins, by Flow Volume	274
Data Table 8:	Monthly Natural Runoff for the World's Major River Basins, by Basin Name	292
Data Table 9:	Area Equipped for Irrigation Actually Irrigated	310
Data Table 10:	Overseas Development Assistance for Water Supply and Sanitation, by Donating Country, 2004–2011	317
Data Table 11:	Overseas Development Assistance for Water Supply and Sanitation, by Subsector, 2007–2011	320
Data Table 12:	Per Capita Water Footprint of National Consumption, by Country, 1996–2005	323
Data Table 13:	Per Capita Water Footprint of National Consumption, by Sector and Country, 1996–2005	332
Data Table 14:	Total Water Footprint of National Consumption, by Country, 1996–2005	341
Data Table 15:	Total Water Footprint of National Consumption, by Sector and Country, 1996–2005	350
Data Table 16A:	Global Cholera Cases Reported to the World Health Organization, by Country, 1949–1979	359
Data Table 16B:	Global Cholera Cases Reported to the World Health Organization, by Country, 1980–2011	367
Data Table 17A:	Global Cholera Deaths Reported to the World Health Organization, by Country, 1949–1979	373
Data Table 17B:	Global Cholera Deaths Reported to the World Health Organization, by Country, 1980–2011	381
Data Table 18A:	Perceived Satisfaction with Water Quality in Sub-Saharan Africa	387
Data Table 18B:	Regional Assessment of Satisfaction with Water (and Air) Quality	389
Data Table 18C:	Countries Most and Least Satisfied with Water Quality	392

WATER UNITS, DATA CONVERSIONS, AND CONSTANTS 395

COMPREHENSIVE TABLE OF CONTENTS 405

Volume 1: The World's Water 1998–1999: The Biennial Report on Freshwater Resources 405

Volume 2: The World's Water 2000–2001: The Biennial Report on Freshwater Resources 408

Volume 3: The World's Water 2002–2003: The Biennial Report on Freshwater Resources 411

Volume 4: The World's Water 2004–2005: The Biennial Report on Freshwater Resources 414

Volume 5: The World's Water 2006–2007: The Biennial Report on Freshwater Resources 418

Volume 6: The World's Water 2008–2009: The Biennial Report on Freshwater Resources 422

Volume 7: The World's Water, Volume 7: The Biennial Report on Freshwater Resources 426

Volume 8: The World's Water, Volume 8: The Biennial Report on Freshwater Resources 430

COMPREHENSIVE INDEX 435

Cumulative Index
The World's Water, Volumes 1-8

KEY (book volume in boldface numerals)
- **1:** The World's Water 1998-1999: The Biennial Report on Freshwater Resources
- **2:** The World's Water 2000-2001: The Biennial Report on Freshwater Resources
- **3:** The World's Water 2002-2003: The Biennial Report on Freshwater Resources
- **4:** The World's Water 2004-2005: The Biennial Report on Freshwater Resources
- **5:** The World's Water 2006-2007: The Biennial Report on Freshwater Resources
- **6:** The World's Water 2008-2009: The Biennial Report on Freshwater Resources
- **7:** The World's Water Volume 7: The Biennial Report on Freshwater Resources

A
ABB, **1:**85, **7:**133
Abi-Eshuh, **1:**69, **5:**5
Abou Ali ibn Sina, **1:**51
Abu-Zeid, Mahmoud, **1:**174, **4:**193
Acceleration, measuring, **2:**306, **3:**324, **4:**331, **7:**347
Access to water. *See* Conflict/cooperation concerning freshwater; Drinking water, access; Environmental flow; Human right to water; Renewable freshwater supply; Sanitation services; Stocks and flows of freshwater; Withdrawals, water
Acidification:
    acid rain, **7:**87
    and fossil-fuel mining/processing, **7:**73–74, 86, 87
    mine drainage, **7:**52, 65
    overview, **7:**47
Adams, Dennis, **1:**196
Adaptation to climate change. *See* Climate change, adaptation
Adaptive capacity, **4:**236, **8:**5, 13
Adaptive management and environmental flows, **5:**45
Adriatic Sea, **3:**47
Afghanistan, **8:**163
Africa:
    aquifers, transboundary, **7:**3
    bottled water, **4:**288, 289, 291, **5:**281, 283
    cholera, **1:**57, 59, 61–63, 266, 269, **3:**2, **5:**289, 290
    climate change, **1:**148
    conflict/cooperation concerning freshwater, **1:**119–24, **5:**7, 9
    conflicts concerning freshwater, 2011-2012, **8:**167–68
    costs of poor water quality, **7:**63
    dams, **3:**292, **5:**151, **6:**274–78
    desalination, **2:**94, 97
    dracunculiasis, **1:**52–55, 272, **3:**274, **5:**295, 296
    drinking water, **1:**252, 262, **3:**254–55, **5:**240, 241, **6:**65
        access to, **2:**217, **6:**58, 214–15, **7:**24, 38, 41, 233–35
        progress on access to, by region, **7:**253, **8:**270
    droughts, **5:**93, 100
    economic development derailed, **1:**42
    environmental flow, **5:**32
    fog collection as a source of water, **2:**175
    Global Water Partnership, **1:**169
    groundwater, **4:**85–86
    human needs, basic, **1:**47
    human right to water, **4:**211
    hydroelectric production, **1:**71, 277–78
    irrigation, **1:**298–99, **2:**80, 85, 256–57, 265, **3:**289, **4:**296, **5:**299, **6:**324–26, 330–31

Millennium Development Goals, **4:**7
mortality rate
    childhood, **6:**279
    under-5, **6:**279, **7:**259–63
Northern
    air quality, satisfaction, 2012, **8:**391
    climate change and drought, **8:**149
    conflicts concerning freshwater, 2011-2012, **8:**163–64
    drinking water, progress on access to, **7:**253, **8:**270
    sanitation, progress on access to, **7:**256, **8:**273
    water quality, satisfaction, 2012, **8:**391
    water quality, satisfaction by country, **7:**291
population data/issues, **1:**247, **2:**214
reclaimed water, **1:**28, **2:**139
renewable freshwater supply, **1:**237–38, **2:**199–200, 217, **3:**239–40, **4:**263–64, **5:**223, 224
    2011 update, **7:**217–18
    2013 update, **8:**223–24
reservoirs, **2:**270, 272, 274
river basins in, **6:**289–96
rivers, transboundary, **7:**3, **8:**164
salinization, **2:**268
sanitation services, **1:**257, 264, **3:**263–65, 271, **5:**249–50, 259, **6:**67
    access to, **7:**243–45
    progress on access to, **7:**246, **8:**273
sub-Saharan, **6:**65–66, 76–77
    air quality, satisfaction, 2012, **8:**391
    drinking water, access, **7:**24
    drinking water, progress on access to, **7:**253, **8:**270
    sanitation, progress on access to, **7:**256, **8:**273
    schistosomiasis, **7:**58
    water quality, satisfaction
        2012, **8:**391
    water quality, satisfaction by country, **7:**290
        2012, **8:**387–88
threatened/at risk species, **1:**292–93
well-being, measuring water scarcity and, **3:**96
withdrawals, water, **1:**242, **2:**205–7, **3:**245–47, **4:**269–70, **5:**230–31, **6:**204
    by country and sector, **7:**223–25, **8:**229–31
African Development Bank (AfDB), **1:**95–96, 173, **3:**162–63
Agreements, international:
    amendment and review process, **7:**9
    General Agreement on Tariffs and Trade (GATT), **3:**47–52
    general principles, **7:**4–5

Agreements, international (*continued*)
:   Great Lakes–St. Lawrence River Basin, **7:**165–69
    joint institutions, role of, **7:**9–10
    North American Free Trade Agreement (NAFTA), **3:**47–48, 51–54, **8:**130
    state/provincial rights within, **7:**167–68, 169
    transboundary waters, **7:**2–7, 9, 165–69, **8:**14–15
    See also Law/legal instruments/regulatory bodies; United Nations

Agriculture:
:   best management practices, **8:**53–54
    California, **4:**89, **8:**89–90
    carbon footprint, **8:**85
    cereal production, **2:**64, **4:**299–301, **7:**109–10
    conflict/cooperation concerning freshwater, **1:**111
    cropping intensity, **2:**76
    crops as a military target, **8:**165
    crop yields and food needs for current/future populations, **2:**74–76
    data problems, **3:**93
    droughts, **5:**92, 94, 98, 103
        impacts in Australia, **7:**102–5
        impacts in Syria, **8:**147–51
        management in Australia, **7:**97, 107–10
    floods, **5:**109, **8:**128
    groundwater, **4:**83, 87, 88–90
    harvesting technology, **2:**77
    irrigation
        arable land, **6:**324–28
        area equipped for, actually irrigated, by country, **6:**329–34, **8:**310–16
        basin, **2:**82
        border, **2:**82
        business/industry, water risks, **5:**162, **7:**23, 26
        and climate change, **4:**174–75, **7:**16
        conflict/cooperation concerning freshwater, **1:**110, **7:**110–11
        conflicts concerning freshwater, **8:**148, 164
        by continent, **2:**264–65
        by country and region, **1:**297–301, **2:**255–63, **4:**295–96, **5:**298–300, **6:**324–28, 329–34
        by crop type, **2:**78–80
        developing countries, **1:**24, **3:**290, **5:**301–2
        drip, **1:**23–24, **2:**82, 84, **7:**111, **8:**50
        Edwards Aquifer, **3:**74
        furrow, **2:**82, 86
        government involvement, **1:**8
        hard path for meeting water-related needs, **3:**2
        how much water is needed, **2:**81–87
        infrastructure for, **1:**6, **8:**50–53
        North American Water and Power Alliance, **1:**74, **8:**124–31
        projections, review of global water resources, **2:**45
        reclaimed water, **2:**139, 142, 145–46, **7:**54, **8:**74
        Southeastern Anatolia Project, **3:**182
        sprinkler, **2:**82, **8:**50
        surface, **2:**82, 84, 84
        total irrigated areas, **4:**297–98
        water-energy-food nexus, **8:**5–6, 85
        water quality, **2:**87
        water rights, **7:**111, 112, 113
        water-use efficiency, **3:**4, 19–20, **4:**107, **7:**109–10, 146, **8:**50–54
    land availability/quality, **2:**70–71, 73–74
        Africa, **8:**167–68
        impact of the North American Water and Power Alliance, proposed, **8:**127, 128
        Syria, **8:**148–49
    pricing, water, **1:**117, **7:**111–12
    projections, review of global water resources, **2:**45–46
    reclaimed water, **1:**28, 29
    by region, **3:**289
    runoff (*See* Runoff, agricultural)
    subsidies, **1:**24–25, 117, **7:**108, 109, 111, 152
    sustainable, **1:**187–88, **8:**10
    water assessments, **2:**46–49, 54–58
    water footprint, **8:**86, 87 (*See also* Water footprint, per-capita, of national consumption, by sector and country)
    and water quality, **7:**49–50, 62, 64, **8:**52 (*See also* Runoff, agricultural)
    well-being, measuring water scarcity and, **3:**99
    World Water Forum (2003), **4:**203
    See also Food needs for current/future populations

Aguas Argentinas, **3:**78, **4:**47
Aguas de Barcelona, **3:**63
Aguas del Aconquija, **3:**70
AIDS, **6:**58, **7:**259–63
Air quality:
:   certified emission credits, **6:**51
    dust storms, **7:**105–6
    fossil-fuel combustion, **7:**77, 80, 84, 86
    methane contamination, **8:**69
    satisfaction, by region, 2012, **8:**389–91

AkzoNobel, **6:**23
Alaska, Sitka, **8:**135–36
Alaskan water shipments, proposed, **8:**133, 135–36
Albania, **1:**71, **3:**47, **7:**309
Albright, Madeleine K., **1:**106
Algae, **5:**79, **6:**83, 95
Algeria, **7:**75, 309
Alkalinization, **7:**52
Alliance for Water Efficiency, **8:**39
Alliance for Water Stewardship, **7:**26–27
Al-Qaida, **5:**15
Alstom, **7:**133
Altamonte Springs (FL), **2:**146
American Association for the Advancement of Science, **1:**149, **4:**176
American Convention on Human Rights (1969), **2:**4, 8
American Fisheries Society, **2:**113, 133
American Geophysical Union (AGU), **1:**197–98
American Rivers and Trout Unlimited (AR & TU), **2:**118, 123
American Society of Civil Engineers, **5:**24
American Water/Pridesa, **5:**62
American Water Works Association (AWWA), **2:**41, **3:**59–61, **4:**176, **5:**24
Americas:
:   air quality, satisfaction, 2012, **8:**391
    cholera, **8:**359, 373
    water quality, satisfaction, 2012, **8:**391
    See also Caribbean; Central America; North America; South America

Amoebiasis, **1:**48
Amount of water. *See* Stocks and flows of freshwater
Amphibians, **1:**291–96
:   effects of endocrine disruptors, **7:**49
    extinct or extinct in the wild species, **7:**56–57, 296–97

Anatolia region. *See* Southeastern Anatolia Project
Angola, **1:**119, 121, **2:**175, **3:**49, **5:**9
:   dams with Chinese financiers/developers/builders, **7:**309

Anheuser-Busch, **5:**150, **6:**24
Ankara University, **3:**183
Antiochus I, **3:**184–85
Apartheid, **1:**158–59
Appleton, Albert, **4:**52–53
Aquaculture, **2:**79
Aquafina, **4:**21
Aquarius Water Trading and Transportation, Ltd., **1:**201–2, 204, **8:**135
Aquifers, **6:**10

aquifer storage and recovery (ASR), **8**:49
climate change management issues, **7**:2–3
Coca-Cola recharging, in India, **8**:21–22, 23
contamination by fossil-fuel production, **7**:51, 76
percent of global freshwater in, **7**:3
transboundary, **7**:3
Aquifers, specific:
Disi, **8**:157
Edwards, **3**:74–75, **4**:60–61
Ogallala, **3**:50, **8**:128, 141
Aral Sea, **1**:24, **3**:3, 39–41, 77, **7**:52, 131
Archaeological sites, **3**:183–89, 191. *See also* Southeastern Anatolia Project
Area, measuring, **2**:302, **3**:320, **4**:327, **7**:345
Argentina, **3**:13, 60, 70, **4**:40, 47
arsenic in groundwater, **7**:59
Arizona, **3**:20, 138, **8**:28
Army Corps of Engineers, U.S. (ACoE), **1**:7–8, **2**:132, **3**:137, **7**:305
Reber Plan, proposed, **8**:132–33
response to drought, Missouri River and Mississippi River policy, **8**:143–44
Arrowhead bottled water, **4**:21, **7**:161
Arsenic, **2**:165–73, **3**:278–79, **4**:87, **5**:20, **6**:61, 81
in fly ash, **7**:84
from fossil-fuel production, **7**:76, 79
health effects, **7**:59
Artemis Society, **3**:218
Ascariasis, **7**:272
Ascension Island, **2**:175
Asia:
agriculture, **4**:88–90 (*See also* Asia, irrigation)
air quality, satisfaction, 2012, **8**:391
aquifers, transboundary, **7**:3
bottled water, **4**:18, 41, 291, **5**:163, 281, 283
Central (*See* Caucasus and Central Asia)
cholera, **1**:56, 58, 61, 266, 269–70, **3**:2, **5**:289, 290
climate change, **1**:147
conflict/cooperation concerning freshwater, **1**:111
dams, **3**:294–95
dracunculiasis, **1**:272–73, **3**:274–75, **5**:295, 296
drinking water, **1**:253–54, 262, **3**:257–59, **5**:243–45, **6**:66
access to, **7**:24, 237–38
progress on access to, by region, **7**:253, **8**:270
Eastern
progress on access to drinking water, **7**:24, 253, **8**:270
progress on access to sanitation, **8**:273
environmental flow, **5**:32–33
floods, **5**:108
food needs for current/future populations, **2**:75, 79
Global Water Partnership, **1**:169
groundwater, **4**:84, 88–90, 96
human needs, basic, **1**:47
hydroelectric production, **1**:71, 278–79
irrigation, **1**:299–300, **2**:80, 86, 259–61, 265, **3**:289, **4**:296, **5**:299, **6**:326–27, 331–32
population data/issues, **1**:248–49, **2**:214
pricing, water, **1**:24, **3**:69
privatization, **3**:61
renewable freshwater supply, **1**:238–39, **2**:217, **3**:240–41, **4**:264–65, **5**:225, 226
2011 update, **7**:218–19
2013 update, **8**:224–25
reservoirs, **2**:270, 272, 274
river basins, **2**:30, **6**:289, 296–301
rivers, transboundary, **7**:3
runoff, **2**:23
salinization, **2**:268
sanitation services, **1**:258–59, 264, **3**:266–68, 271, **5**:252–54, 258–61
access to, **7**:247–48
progress on access to, **7**:256, **8**:273
South-Eastern, **7**:262–63
progress on access to drinking water, **7**:24, 253, **8**:270
progress on access to sanitation, **7**:256, **8**:273
Southern
progress on access to drinking water, **7**:24, 253, **8**:270
progress on access to sanitation, **7**:256, **8**:273
water conflicts, **8**:162–63
supply systems, ancient, **1**:40
threatened/at risk species, **1**:293–94
water access, **2**:24, 217
water quality
satisfaction, 2012, **8**:391
satisfaction by country, **7**:290
well-being, measuring water scarcity and, **3**:96
Western
progress on access to drinking water, **7**:24, 253, **8**:270
progress on access to sanitation, **7**:256, **8**:273
water conflicts, 2011-2012, 162–63
withdrawals, water, **1**:243–44, **2**:208–9, **3**:248–49, **4**:272–73, **5**:233–34
by country and sector, **7**:226–28, **8**:232–33
Asian Development Bank (ADB), **1**:17, 173, **3**:118, 163, 169, **4**:7, **5**:125
reaction to World Commission on Dams report, **7**:138–39
Asmal, Kader, **1**:160, **3**:169
Assessments:
AQUASTAT database, **4**:81–82, **7**:215, 221, **8**:221, 310
Colorado River Severe Sustained Drought study (CRSSD), **4**:166
Comprehensive Assessment of the Freshwater Resources of the World, **2**:10, **3**:90
Dow Jones Indexes, **3**:167
Global Burden of Disease assessment, **7**:270
The High Efficiency Laundry Metering and Marketing Analysis project (THELMA), **4**:115
Human Development Index, **2**:165
Human Development Report, **4**:7, **6**:74, **7**:61
Human Poverty Index, **3**:87, 89, 90, 109–11, **5**:125
hydrologic cycle and accurate quantifications, **4**:92–96
International Journal on Hydropower and Dams, **1**:70
International Rice Research Institute (IRRI), **2**:75, 76
intl. river basins, **2**:27–35
measurements, **2**:300–309, **3**:318–27, **4**:325–34
Millennium Ecosystem Assessment, **7**:63
National Assessment on the Potential Consequences of Climate Variability and Change, **4**:176
Palmer Drought Severity Index, **5**:93
Standard Industrial Classification (SIC), **4**:132
Standard Precipitation Index, **5**:93
Stockholm Water Symposiums (1995/1997), **1**:165, 170
Third Assessment Report, **3**:121–23
water footprint sustainability assessment, **8**:86
Water-Global Assessment and Prognosis (WaterGAP), **2**:56
Water in Crisis: A Guide to the World's Fresh Water Resources, **2**:300, **3**:318
World Health Reports, **4**:7
World Resources Reports, **3**:88
World Water Development Report, **8**:6
See also Data issues/problems; Groundwater, monitoring/management problems; Projections, review of global water resources; Stocks and flows of freshwater; Well-being, measuring water scarcity and

Association of Southeast Asian Nations (ASEAN), **1:**169
Assyrians, **3:**184
Atlanta (GA), **3:**62, **4:**46–47
　description, **6:**103
　per-capita water demand, **6:**104–6
　population growth, **6:**103
　precipitation, **6:**104
　temperature, **6:**104
　wastewater rate structure, **6:**118–19
　water conservation, **6:**108–9, 110–12
　water rate structures, **6:**115–16
　water-use efficiency, **6:**108–9
Atlantic Salmon Federation, **2:**123
Atmosphere, harvesting water from the, **2:**175–81. *See also* Outer space, search for water in
Austin (TX), **1:**22
Australia:
　agriculture
　　impact of drought, **7:**102–5
　　irrigation, **2:**85, **7:**97
　　production, 1960-2009, **7:**103
　bottled water, **4:**26
　conflict/cooperation concerning freshwater, **5:**10
　desalination, **5:**69, **7:**114–15, **8:**96, 97, 98, 102, 103, 118
　drought
　　historical background, **7:**98, 99
　　impacts, **7:**102–6
　　management, **7:**106–21, 146–47
　　overview, **7:**97, 98–99, 101, 121
　environmental flow, **5:**33, 35, 42
　fossil-fuel production, **7:**75
　globalization and intl. trade of water, **3:**45, 46
　legislation and policy
　　Millennium Development Goals, **4:**7
　　Water Efficiency Labelling and Standards Act, **7:**28, 118–19
　　water policy reform, **7:**146–47
　privatization, **3:**60, 61
　reservoirs, **2:**270, 272, 274
　terrorism, **5:**16
　water resources, **2:**24, 217, **7:**98–99
　water-use efficiency, **8:**38
Austria, **3:**47
Availability, water, **2:**24–27, 215–17. *See also* Conflict/cooperation concerning freshwater; Drinking water, access; Environmental flow; Human right to water; Renewable freshwater supply; Sanitation services; Stocks and flows of freshwater; Withdrawals, water
Azov Sea, **1:**77

B
Babbitt, Bruce, **2:**124–25, 128
Babylon, ancient, **1:**109, 110
Bag technology, water, **1:**200–205, **8:**135
Baker, James, **1:**106
Bakersfield (CA), **1:**29
Balfour Beatty, **3:**166
Balkan Endemic Neuropathy (BEN), **7:**89
Bangladesh:
　agriculture, **4:**88
　Arsenic Mitigation/Water Supply Project, **2:**172
　conflict/cooperation concerning freshwater, **1:**107, 109, 118–19, 206–9
　dams with Chinese financiers/developers/builders, **7:**309
　drinking water, **3:**2
　floods, **5:**106
　groundwater, **4:**88
　　arsenic in, **2:**165–73, **6:**61, **7:**55, 59, **8:**3
　Rural Advancement Committee, **2:**168

Banks, Harvey, **1:**9
Barlow, Nadine, **3:**215
Basic water requirement (BWR), **1:**44–46, **2:**10–13, **3:**101–3
Bass, **2:**123
Bath Iron Works, **2:**124
Bayer, **6:**28
Beard, Dan, **2:**129
Bechtel, **3:**63, 70
Belgium, **5:**10
Belize, **7:**309
Benin, **1:**55, **7:**309, 336
Benzene, **6:**83, **8:**72
Best available technology (BAT), **3:**18, 22–23, **4:**104, **5:**157, **6:**27
Best management practices (BMP), **8:**53–54
Best practicable technology (BPT), **3:**18, **4:**104
Beverage Marketing Corporation (BMC), **4:**18, **7:**157
BHIP Billiton, **6:**23
Biodiversity, **7:**56–57
Biofuel, **7:**25, 74, **8:**6, 167
Biological oxygen demand (BOD), **6:**28, **7:**278–81
Biologic attacks, vulnerability to, **5:**16–22
Bioretention and bioinfiltration, **8:**42
Bioswales, **8:**42
Birds:
　effects of drought, **7:**101
　effects of endocrine disruptors, **7:**49
　effects of fossil-fuel extraction, **7:**87, 91
　extinct or extinct in the wild species, **7:**297–98
　Lesotho Highlands project, **1:**98
　Siberian crane, **1:**90
　Yuma clapper rail, **3:**142
Birth, premature, **7:**259–63
Birth asphyxia, **7:**259–63
Birth defects, **7:**58, 59, 259–63
Bivalves. *See* Invertebrates, clams; Invertebrates, mussels
Black Sea, **1:**77
"Blue" water, in water footprint concept, **8:**86, 89–90, 323, 332, 341, 350
BMW, **6:**27
Bolivia, **3:**68, 69–72, **4:**54, 56–57
　conflicts concerning freshwater, 2011-2012, **8:**165–66
Bonneville Power Administration, **1:**69, **8:**129
Books, **2:**129, **5:**14
Boron, **5:**75, **7:**76
Bosnia, **1:**71
Botswana, **1:**119, 122–24, **7:**309
Bottled water:
　bottle and packaging, **4:**293–94, **7:**158–59
　brands, leading, **4:**21–22, **7:**161–62
　business/industry
　　company assessments, **6:**23
　　standards/rules, **4:**34–35
　　water risks that face, **5:**163
　consumption
　　by country, **4:**284–86, **5:**276–79, 284–86
　　increase in, **7:**157
　　per-capita by country, **7:**339–40
　　per-capita by region, **4:**290–91, **5:**171, 282–83
　　by region, **4:**287–88, **5:**169–70, 280–81
　　share by region, **4:**289
　　U.S., **4:**288–91, **5:**170, 281, 283, **7:**157, 158
　developing countries, **3:**44, 45
　energy considerations
　　bottle manufacture, **7:**158–59, 164
　　to clean, fill, seal, and label bottles, **7:**160–61, 164
　　cooling process, **7:**162–63, 164
　　equivalent barrels of oil used, **7:**164
　　transport, **7:**161–62, 163, 164
　　water processing, **7:**159–60, 164

environmental issues, **4:**41
flavor and taste, **4:**23–24
history and trends, **4:**18–22, **7:**157
hydrogeological assessments of sites for, **6:**23
intl. standards, **4:**35–36
labeling, **4:**28–31, **7:**159, 160–61, 164
overview, **4:**xvi, 17, **7:**88
price and cost, **4:**22–23
recalls, **4:**37–40, **5:**171–74
sales
    global, **3:**43, **5:**169
    and imports in, U.S., **3:**44, 343, **4:**292, **7:**157, 158
    to the poor, **4:**40–41
standards/regulations, **4:**26–27, 34–37, **5:**171, 174
    vs. tap water, **4:**36–37
summary/conclusions, **4:**41
U.S. federal regulations
    adulteration, food, **4:**32–33
    enforcement/regulatory action, **4:**34
    good manufacturing practices, **4:**32
    identity standards, **4:**27–31
    sampling/testing/FDA inspections, **4:**33–34
    water quality, **4:**31–32
U.S., consumption, **4:**288–91, **5:**170, 281, 283, **7:**157, 158
U.S., sales and imports in, **3:**44, 343, **4:**292, **7:**157, 158
water quality, **4:**17, 25–26, 31–32, 37–40
    treatment processes, **7:**159–60, 161
water sources for, **7:**159
Boundaries:
    managing across, **3:**133–34, **7:**1
    transboundary waters
        agreements, limitations and recommendations, **8:**14–15
        and climate change, **7:**1–20
        Snake Valley basin, **8:**139
    See also International river basins
Brazil:
    bottled water, **4:**40, **5:**170
    business/industry, water risks, **5:**149, 151–52, **7:**26, 89
    cholera, **1:**59
    conflict/cooperation concerning freshwater, **1:**107
    conflicts concerning freshwater, 2011-2012, **8:**165
    dams, **1:**16, **5:**134
    drought, **7:**26, **8:**22–23, 165
    energy production, **7:**26, **8:**22–23
        hydroelectric, **1:**71, **7:**129
    environmental concerns, top, **7:**287
    environmental flow, **5:**34
    human needs, basic, **1:**46
    monitoring and privatization, **3:**76–77
    privatization, **3:**76–78
    runoff, **2:**23
    sanitation services, **3:**6
    Three Gorges Dam, **1:**89
Brine, as a contaminant, **2:**107, **5:**77–80, **8:**156
British Columbia Hydro International (BC Hydro), **1:**85, 88
British Geological Survey (BGS), **2:**169
British Medical Association, **4:**63–64
Brownfield redevelopment, **8:**46
Bruce Banks Sails, **1:**202
Bruvold, William, **4:**24
Buildings:
    construction site runoff, volume, **8:**76
    green roofs, **8:**42
Burkina Faso, **1:**55, **4:**211, **8:**167
Burma (Myanmar):
    dams with Chinese financiers/developers/builders, **7:**130, 132, 133, 310–15
    Mekong River Basin, **7:**14
Burns, William, **3:**xiv

Burundi, **1:**62, **7:**11, 315–16
Business for Social Responsibility, **5:**156
Business/industry, water risks:
    assessment of, **6:**23, **7:**25–34
    China, **5:**147, 149, 160, 165
    climate change, **5:**152
    costs of poor water quality on production, **7:**64
    developing countries, **5:**150
    energy and water links, **5:**150, 151–52
    India, **5:**146, 147, 165, **8:**21–22
    management
        best available technology, **5:**157
        best management practices (BMP), **8:**53–54
        "beyond the fence line," **8:**19–20, 28
        companies, review of specific, **5:**153–55, **7:**26–27
        continuous improvement, commit to, **5:**158
        efficiency, improving water-use, **5:**161–62, **8:**20, 28
        five motivations, **7:**23
        global trends that affect, **7:**24–25
        hydrological/social/economic/political factors, **5:**153, 156, **7:**24–30
        partnerships, form strategic, **5:**158
        performance, measure and report, **5:**157–58
        public disclosure (*See* Corporate reporting)
        risks factored into decisions, **5:**157, **6:**27–28, **8:**31
        stakeholder issues, **5:**145, **6:**24, **7:**26–27, 40–41, **8:**20, 26–27
        strategies for policies/goals/targets, **5:**156–57, **7:**34–42, **8:**20–21, 31
        supply chain
            availability/reliability of, **5:**146, **7:**26
            company reporting of, **6:**24, **7:**27
            overview, **5:**145–46, **7:**28–29
            performance evaluation, **7:**35–37
            privatization, public opposition to, **5:**152–53
            public's role in water policy, **5:**150–51
            summary/conclusions, **5:**163–65
            vulnerability, **5:**149, **7:**26, **8:**23
            water quality, **5:**146–49, **8:**23
            working collaboratively with, **5:**156, **6:**24
        sustainable, **7:**23, **8:**19–22, 30, 31
    water accounting tools, **7:**32–34, 146
    water industry
        bottled water (*See* Bottled water, business/industry)
        desalination, **5:**161 (*See also* Desalination, plants)
        disinfection/purification of drinking water, **5:**159
        distribution, infrastructure for, **5:**160–61
        high-quality water, processes requiring, **5:**147–48, 160
        irrigation, **5:**162
        overview, **5:**158–59
        revenue and growth in U.S., **5:**303–4
        utilities, water, **5:**162–63, **7:**28
        wastewater treatment, **5:**153, 159–60
    organic contaminants by industry, **7:**279–81
    overview, **7:**23, 25–34, 50–51, **8:**19–31
    public perception, **7:**26–27
    water footprint, **7:**30–34, **8:**84, 87 (*See also* Water footprint, per capita, of national consumption, by sector and country)
Business sector, **3:**22–24, 169–70, **8:**27. *See also* Bottled water, business/industry; Companies; Corporate reporting; Privatization; Water conservation, California commercial/industrial water use
Bussi, Antonio, **3:**70
Byzantines, **3:**184

C
Cabot Oil & Gas Corp., **8:**70
Cadbury, **7:**37
Calgon, **3:**61
California:
    agriculture, **4:**89, **8:**89–90
    Bakersfield, **1:**29
    bottled water, **4:**24
    business/industry, water risks, **5:**158, **6:**162
    California Central Valley Project, **4:**173, **8:**131
    California Regional Assessment Group, **4:**176
    California State Water Project, **8:**131
    Carlsbad, **8:**93, 95, 98, 112–19
    climate change (*See* Climate change, California)
    conflict/cooperation concerning freshwater, **1:**109
    dams, **1:**75, **2:**120–23
    desalination, **1:**30, 32, **5:**51, 52, 63–69, 71, 73, 74
        Carlsbad plant case study, **8:**93, 95, 98, 112–19
        overview, **8:**93
        proposed and recently constructed plants, **8:**98
        Sand City plant, **8:**93
    droughts, **8:**89, 102
    East Bay Municipal Utilities District, **1:**29
    economics of water projects, **1:**16
    environmental justice, **5:**122–23
    floods, **5:**111
    fog collection as a source of water, **2:**175
    food needs for current/future populations, **2:**87
    groundwater, **4:**89
    import of water, **8:**89, 131
    industrial water use, **1:**20–21
    Irvine Ranch Water District, **4:**124–25
    Kesterson National Wildlife Refuge, **7:**47–48
    Los Angeles
        Chinatown political intrigue, **8:**138
        Los Angeles Aqueduct, **8:**131
        sustainable water jobs, **8:**39, 44, 52
    Metropolitan Water District of Southern California, **1:**22
    Monterey County, **2:**151
    North American Water and Power Alliance, **8:**124–31
    Orange County, **2:**152
    Pomona, **2:**138
    privatization, **3:**73
    projections, review of global water resources, **2:**43
    reclaimed water
        agriculture, **1:**29, **2:**142–46
        drinking water, **2:**152
        first state to attempt, **2:**137–38
        groundwater recharge, **2:**151
        health issues, **2:**154–55
        Irvine Ranch Water District, **2:**147
        Kelly Farm marsh, **2:**149
        San Jose/Santa Clara Wastewater Pollution Control Plant, **2:**149–50
        uses of, **2:**141–45
        West Basin Municipal Water District, **2:**148–49
    San Diego County, **8:**99–100
    San Francisco Bay, **3:**77, **4:**169, 183, **5:**73, **8:**132
    San Francisco Bay Project (Reber Plan), **8:**131–34
    Santa Barbara, **5:**63–64, **8:**102–3, 119
    Santa Rosa, **2:**145–46
    soft path for meeting water-related needs, **3:**20–22, 24–25
    subsidies, **1:**24–25
    toilets, energy-efficient, **1:**22, **8:**39
    twentieth-century water-resources development, **1:**9
    Visalia, **1:**29
    water conservation (*See under* Water conservation)
    water footprint, **8:**88–90, 91
    water-use efficiency, **1:**19, **8:**38
    Western Canal Water District, **2:**121
    See also Legislation, California
California-American Water Company (Cal AM), **5:**74
Cambodia, **7:**14, 59, 130, 316–17
Camdessus, Michael, **4:**195–96
Cameroon, **1:**55, **5:**32, **7:**317–18
Campylobacter jejuni, **7:**57
Canada:
    adaptation in, **6:**46
    bottled water, **4:**25, 26, 39, 288, 289, 291, **5:**281, 283
    Canadian International Development Agency, **2:**14
    cholera, **1:**266, 270
    climate change, **1:**147, 148
    conflict/cooperation concerning freshwater, **5:**6, 8
    dams, **1:**75, **3:**293
    data problems, **3:**93
    dracunculiasis, **1:**52
    drinking water, **1:**253, **3:**256, **5:**242
    environmental concerns, top, **7:**287, 288
    environmental flow, **5:**34
    Export Development Corporation, **6:**141
    fog collection as a source of water, **2:**179
    fossil-fuel production
        data, **7:**75
        natural gas, **7:**79
        tar sands, **7:**78, 79, 87, 88, 91–92
    General Agreement on Tariffs and Trade, **3:**50
    Great Lakes Basin, intl. agreements, **7:**165–69, **8:**126
    groundwater, **4:**86
    hydroelectric production, **1:**71, 74, 278, **7:**129, **8:**131
    intl. river basin, **2:**33
    irrigation, **1:**299, **2:**265, **3:**289, **4:**296, **5:**299
    James Bay Project, Quebec, **8:**131
    mortality rate, under-5, **7:**260
    North American Free Trade Agreement, **3:**47–48, 51–54, **8:**130
    North American Water and Power Alliance, **8:**124–31
    population data/issues, **1:**247, **2:**214
    renewable freshwater supply, **1:**238, **2:**200, 217, **3:**240, **4:**264
        response to North America Water and Power Alliance, **8:**129
        2011 update, **7:**218
        2013 update, **8:**224
    reservoirs, **2:**270, 272, 274
    Rocky Mountain Trench, proposed reservoir, **8:**125–26
    runoff, **2:**23
    salinization, **2:**268
    sanitation services, **1:**258, **3:**265, 272, **7:**245
    threatened/at risk species, **1:**293
    transboundary waters, **7:**3, **8:**124–31
    water availability, **2:**217
    water transfer issues, **8:**129–30
    withdrawals, water, **1:**242, **2:**207, **3:**247, **4:**271
        2011 update, **7:**225
        2013 update, **8:**231
    World Water Council, **1:**172
Canary Islands, **3:**46
Cancer, **6:**81, **7:**58, 59
Cap and trade market and environmental flows, **5:**42
Cape Verde Islands, **2:**175
Carbon dioxide, **1:**138, 139, **3:**120, 215, **4:**160, 164
    and natural gas production, **7:**80
Carbon Disclosure Project (CDP), **7:**41, **8:**22, 24
Carbon footprint, **8:**85, 86
Caribbean:
    dams, **3:**293–94
    drinking water, **1:**253, 262, **3:**255–57, **5:**241, 242
        access to, **7:**24, 235–36
        progress on access to, **7:**253, **8:**270

groundwater, **4:**86
hydroelectric production, **1:**278
irrigation, **6:**334
population data/issues, **1:**247–48, **2:**214
sanitation, **1:**258, **3:**265–66, 271, **5:**250, 251, 259
   access to, **7:**245–46
     progress on access to, **7:**256, **8:**273
threatened/at risk species, **1:**293
water quality, satisfaction by country, **7:**290–91
Caribbean National Forest, **5:**34
Carlsbad (CA), **8:**93, 95, 98, 112–19
Caspian Sea, **1:**77
Catley-Carlson, Margaret, **4:**xii–xiv
Caucasus and Central Asia:
  conflicts concerning freshwater, 2011-2012, **8:**166
  progress on access to drinking water, **8:**270
  progress on access to sanitation, **8:**273
Cellatex, **5:**15
Census of Agriculture, **8:**50
Centers for Disease Control and Prevention (CDC), **1:**52, 55, 57, **7:**305
Central African Republic, **7:**318
Central America:
  cholera, **1:**266, 270, 271
  dams, **3:**165, 293–94
  drinking water, **1:**253, **3:**255, 256, **5:**241–42, **7:**235–36
  environmental flow, **5:**34
  groundwater, **4:**86
  hydroelectric production, **1:**278
  irrigation, **1:**299, **2:**265, **4:**296, **6:**328, 334
  mortality rate, under-5, **7:**260–61
  population data, total/urban, **1:**247–48
  renewable freshwater supply, **1:**238, **2:**200, 217, **3:**240, **4:**264, **5:**224
    2011 update, **7:**218
    2013 update, **8:**224
  reservoirs, **2:**270, 272, 274
  rivers and aquifers, transboundary, **7:**3
  salinization, **2:**268
  sanitation services, **1:**258, **3:**265–66, **5:**250–51, **7:**245–46
  threatened/at risk species, **1:**293
  water availability, **2:**217
  withdrawals, water, **1:**242–43, **2:**207, **3:**247, **4:**271, **5:**231–32
    by country and sector, **7:**225, **8:**231
  See also Latin America
Centre for Ecology and Hydrology, **3:**110–11
Centro de Investigaciones Sociales Alternativas, **2:**179
CEO Water Mandate, **7:**34, **8:**9, 24
Cereal production, **2:**64, **4:**299–301
Certified emission credits (CECs), **6:**51
Chad, **1:**55, **7:**264
Chakraborti, Dipankar, **2:**167
Chalecki, Elizabeth L., **3:**xiv
Chanute (KS), **2:**152
Chemical attacks, vulnerability to, **5:**16–22
Chemical oxygen demand, **6:**28
Chiang Kai-shek, **5:**5
Childhood mortality:
  by cause, **6:**279–88, **7:**257–63
  by country, **6:**279–88, **7:**257–63, 264–69
  infant, by country, **7:**264–69
  limitations in data and reporting, **7:**257–58, 264
  under-5, **6:**279–88, **7:**257–63
  from water-related disease, **6:**58, **7:**57–58, 61–62
Children:
  infant brands of bottled water, **4:**28
  responsibility for water collection, **7:**61
  See also Birth *listings*
Chile:
  arsenic in groundwater, **7:**59

cholera, **1:**59
environmental concerns, top, **7:**287
environmental flow, **5:**34, 37
fog collection as a source of water, **2:**177–78
General Agreement on Tariffs and Trade, **3:**49
privatization, **3:**60, 66, 78
Silala/Siloli River transboundary dispute, **8:**165–66
subsidies, **4:**57–58
China:
agriculture, **2:**86, **4:**88, 90
algae outbreaks, **6:**83
Beijing, **6:**90
benzene contamination, **6:**83
bottled water, **4:**21, 40, **5:**170
business/industry, water risks, **5:**147, 149, 160, 165
business/industry water use, **5:**125, **6:**81, **7:**27–28
canal system, **8:**144
cancer rates in, **6:**81
climate change in, **6:**87–88
dams, **1:**69, 70, 77, 78, 81, **5:**15–16, 133, 134, **6:**91
  construction overseas, **7:**133–35, 308–38
  hydroelectric production, **1:**71, **6:**92, **7:**129, 130
  internal concern over, **7:**135–36
  on intl. rivers, **7:**130–33
  overview, **7:**127–28, 140
  and the World Commission on Dams, **3:**170–71, **7:**128–29, 136–40
  See also under Dams, specific; Three Gorges Dam
desalination, **6:**93
diarrhea-related illness in, **6:**85
diseases, water-related, **6:**85
drinking water
  shortage of, **6:**86
  standards for, **6:**94
droughts, **5:**97, **6:**86–87, **7:**131–32
economic growth in, **6:**79
economics of water projects, **1:**16, **6:**95–96
environment
  grassroots efforts, **6:**80
  pollution in, **6:**79, 81–82 (*See also* China, water, pollution of)
  protections for, **6:**94–95
  top concerns, **7:**287, 288
  water-related disasters, **6:**82–83
environmental flow, **5:**32
Environmental Impact Assessment law, **6:**96
floods, **5:**106, **6:**84–87
food needs for current/future populations, **2:**74
foreign investment in water markets, **6:**92
fossil-fuel production, **7:**75, 82, **8:**63
glaciers in, **6:**87–88
globalization and intl. trade of water, **3:**46
Great Wall of China, **6:**86
groundwater, **2:**87, **3:**2, 50, **4:**79, 82, 83, 88, 90, 96–97, **5:**125, **6:**85–86
  arsenic in, **6:**81, **7:**59
  fluoride in, **6:**81
Guangdong Province, **6:**85
human needs, basic, **1:**46
Jilin Province, **6:**83
natural gas production, **8:**63
nongovernmental organizations in, **6:**89–90, 96
North China Plains, **6:**85–86, 90
politics, **6:**88–90
population, **6:**79
privatization, **3:**59, 60
protests in, **6:**97
provinces, **6:**82
public-private partnerships in, **6:**92–93
Qinghai-Tibetan Plateau, **6:**88
rivers, **6:**79, 81–82, 84, 88, 90, **7:**55 (*See also under* Rivers, specific)

China (*continued*)
    sanitation services, **5:**124, 147
    South-to-North Water Transfer Project, **6:**91
    State Environmental Protection Administration (SEPA), **6:**80–81, 94
    wastewater treatment plants in, **6:**92
    water
        availability of, **6:**83–85
        average domestic use of, **1:**46
        basic requirement, **2:**13
        centralized management of, **6:**89
        effects of climate change, **6:**87–88
        efficiency improvements, **6:**93–94
        expanding the supply of, **6:**91–93
        industrial use, **5:**125, **6:**81
        politics affected by, **6:**88–90
        pollution of, **6:**79, 97, **7:**64, **8:**24–25
        pricing of, **1:**25
        public participation efforts, **6:**96–97
        quality of, **6:**80–82, **8:**3
        quantity of, **6:**83–85
        regional conflicts over, **6:**90–91
        shortage of, **6:**86
        surface, **6:**81
        sustainable management of, **6:**97
        water use per unit of GDP, **6:**93
    water laws in, **6:**88–89
    wetlands in, **6:**86, 88
    withdrawals, water, **3:**316–17, **6:**85–86, **7:**27–28
    Xiluodu hydropower station, **6:**92
China International Capital Corporation (CICC), **6:**142
Chitale, Madhav, **1:**174
Chlorination, **1:**47, 60, **4:**39, **5:**159
Cholera, **1:**48, 56–63, 265–71, **3:**2, **5:**287–92
    cases reported to the WHO, by country
        1949-1979, **8:**361–66
        1980-2011, **8:**367–72
    deaths reported to the WHO, by country
        1949-1979, **8:**375–80
        1980-2011, **8:**381–86
    epidemic diarrheal diseases caused by, **7:**57
    limitations in data and reporting, **8:**359, 373–74
    See also *Vibrio*; *Vibrio cholerae*
Cincinnati Enquirer, **5:**171
CIPM Yangtze Joint Venture, **1:**85
Clementine spacecraft, **1:**197, **3:**212–13
Climate change:
    adaptation
        Adaptation Fund, **6:**51
        Adaptation Policy Framework, **6:**48
        assessments, **6:**47–49
        capacity for, **4:**236, **8:**5
        community participation in, **6:**49
        costs of, **6:**50–51, 53
        definition of, **6:**45
        demand-side options, **6:**46
        economic cost of, **6:**50–51
        equity issues for, **6:**51–52
        funding for, **6:**52
        general circulation models, **6:**47
        Interagency Climate Change Adaptation Task Force, **7:**153
        mainstreaming of, **6:**47
        national adaptation programs of action, **6:**48–49
        options for, **6:**45–46
        Oxfam Adaptation Financing Index, **6:**52
        participation in, **6:**49
        supply-side options, **6:**46
    air temperature increase, predicted, **7:**53
    Bibliography, The Water & Climate, **4:**xvii, 232–37
    business/industry, water risks, **5:**152
    California (*See* Climate change, California)
    changes occurring yet?, **1:**145–48
    China, **6:**87–88
    Colorado River Basin, **1:**142, 144, **4:**165–67, **7:**16–18
    costs of, **6:**50–51
    desalination, **5:**80–81
    developing countries' vulnerability to, **6:**45, 51, 54
    droughts (*See* Droughts)
    ecological effects, **4:**171–72
    environmental flow, **5:**45
    environmental justice, **5:**136–37
    floods (*See* Floods)
    food needs for current/future populations, **2:**87–88
    groundwater affected by, **4:**170, **6:**43
    hydrologic cycle, **1:**139–43, **5:**117
    hydrologic extremes, **6:**43–44
    hydrologic impacts of, **6:**48
    impacts of fossil-fuel extraction/processing, **7:**84–85
    and interbasin agreements, **8:**15
    IPCC (*See* Law/legal instruments/regulatory bodies, Intergovernmental Panel on Climate Change)
    Meking River Basin, **7:**14–15
    Nile River Basin, **7:**12–13
    overview, **1:**137–39, **7:**1, **8:**5
    precipitation, **1:**140–41, 146–47, **4:**159, 166, **6:**40, 87
    recommendations and conclusions, **1:**148–50
    reliability, water-supply, **5:**74
    renewability of resources affected by, **6:**9
    societal impacts, **1:**144–45
    summary of, **6:**53–54
    surface water effects, **6:**43, **7:**53–54
    sustainable vision for the Earth's freshwater, **1:**191
    and Syrian water conflicts, **8:**149–51
    and the Three Gorges Dam, **6:**146–47
    transboundary water management issues, **7:**1–2, 7–20
    vulnerability to, **6:**44–45, 51, 54
    water demand and, **6:**44, **7:**150
    water policy reform which addresses, **7:**153
    water quality and, **4:**167–68, **6:**44, **7:**53–54
    water resources affected by, **6:**43–44
    See also Greenhouse effect; Greenhouse gases
Climate change, California, **1:**144–45
    overview, **4:**xvii, 157–58
    policy
        economics/pricing/markets, **4:**180–81
        information gathering/reducing uncertainty, **4:**182–83
        infrastructure, existing, **4:**175–77
        institutions/institutional behaviors, new, **4:**181–82
        monitoring, hydrologic and environmental, **4:**183
        moving from science to demand management/conservation/efficiency, **4:**179–80
        new supply options, **4:**178–79
        overview, **4:**175
        planning and assessment, **4:**178
        reports recommending integration of science/water policy, **4:**176
    science
        evaporation and transpiration, **4:**159–60
        groundwater, **4:**170
        lake levels and conditions, **4:**168–69
        overview, **4:**158–59, 172–73
        precipitation, **4:**159
        sea level, **4:**169–70
        snowpack, **4:**160–61
        soil moisture, **4:**167
        the state of the ecosystems, **4:**171–72
        storms/extreme events and variability, **4:**161–63

Index 443

temperature, **4:**159
water quality, **4:**167–68
summary/conclusions, **4:**183–84
systems, managed water resource
agriculture, **4:**174–75
hydropower and thermal power generation, **4:**173–74
infrastructure, water supply, **4:**173
Clinton, Bill, **2:**127, 134
Clothes, water footprint of production, **8:**84, 86
Clothes washing. *See* Laundry
CNN, Water Conflict Chronology cited by, **8:**173
Coal:
energy content, **7:**75
extraction and processing, **7:**73, 74, 82–84, 90
production data, **7:**75, 82
transport, **7:**83–84
water consumed and energy production, **7:**25
Coastal zones:
consequences of poor water quality, **7:**55–56
development and desalination, **5:**80
erosion, **7:**46
floods, **5:**104, 108
legislation, **5:**80
Coca-Cola, **4:**21, 38, **5:**146, 163, **6:**31–32
beverage transport, **7:**161
bottle development, **7:**158–59
environmental justice, **5:**127
and groundwater depletion in India, **8:**25
recharging aquifers in India, **8:**21–22, 23
water consumption and reputation, **7:**26
Cogeneration systems and desalination, **2:**107
Colombia, **1:**59, **3:**60, **5:**12, **7:**318, **8:**25–26
Colorado, **4:**95–96, **7:**52
Missouri River diversion project, **8:**140–44
Colorado River:
climate change, **1:**142, 144, **4:**165–67, **7:**16–18
conflict/cooperation concerning freshwater, **1:**109, 111, **7:**6–7, **8:**91
dams, **7:**52 (*See also* Dams, specific, Glen Canyon; Dams, specific, Hoover)
delta characteristics, **3:**139–43, **6:**130
fisheries, **1:**77
hydrology, **3:**135–37, **7:**15, **8:**127
institutional control of, **3:**134
intl. agreements, **7:**6, 8–9, 15–16, **8:**91
Las Vegas Valley Pipeline Project, **5:**74, **8:**136–40
legal framework ("Law of the River"), **3:**137–39, **7:**15, 17
restoration opportunities, **3:**143–44
salinity, **7:**6–7
Salton Sea inflows, **6:**129, 132
summary/conclusions, **3:**144–45
vegetation, **3:**134, 139–42
water transfer, proposed, **8:**140–41
wildlife, **1:**77, **3:**134
Columbia River Alliance, **2:**133
Comets (small) and origins of water on Earth, **1:**193–98, **3:**209–10, 219–20
Commercial sector, **3:**22–24, 169–70. *See also* Business/industry, water risks; Privatization; Water conservation, California commercial/industrial water use
Commissions. *See* International *listings;* Law/legal instruments/regulatory bodies; United Nations; World *listings*
Commodification, **3:**35
Commonwealth Development Corporation, **1:**96
Commonwealth of Independent States. *See also* Caucasus and Central Asia
water quality, satisfaction by country, **7:**290–91
Communities:
agricultural, loss of farms due to drought, **7:**104–5

consequences of poor water quality, **7:**61–62
empowerment through education and outreach, **8:**13–14
engagement (*See* Public participation)
impacts of fossil-fuel extraction/processing, **7:**87–88, 91–92
risks of privatization, **3:**68, 79
sustainable-water job production, **8:**59
Community structure. *See* Biodiversity; Extinct species; Introduced/invasive species; Threatened/at risk species
Companies:
bottled water (*See* Bottled water, business/industry)
continuous improvement commitment by, **6:**32
current water use, **6:**22
decision making, **6:**27–28
environmental management system, **6:**32
funding by, and environmental/lending standards, **8:**10
history, **6:**17–18
public relations and social responsibility, **8:**23, 29–30
recommendations for, **6:**37–38, **8:**31
small and medium enterprises (SME), **8:**26
stakeholder issues, **5:**145, **6:**24, **7:**26–27, **8:**20, 26–27
strategic partnerships, **6:**31–32, **7:**40, **8:**27–31
supply chain involvement, **6:**24
sustainable practices, **7:**23, **8:**19–22, 27–28, 30, 31
sustainable water jobs, **8:**35–59
water management, **6:**20
water performance data published by, **6:**28
water-policy statement, **6:**25–26
See also Business/industry, water risks; Corporate reporting
Company, **2:**126
Concession models, privatization and, **3:**66–67
Conferences/meetings, international. *See* International *listings;* Law/legal instruments/regulatory bodies; United Nations; World *listings*
Conflict/cooperation concerning freshwater:
dispute-resolution procedures, **7:**7, 168, **8:**161
droughts, **5:**99, **7:**8–9, **8:**149–51
economic/social development context, **7:**176, **8:**173
environmental deficiencies and resource scarcities, **1:**105–6, **8:**161
geopolitics and intl./transboundary waters, **2:**35–36, **7:**1, **8:**14–15, 147–51, 164
historical (*See* Water Conflict Chronology)
inequities in water distribution/use/development, **1:**111–13, **8:**2
instrument/tool of conflict
definition, **8:**173
natural gas access as a, **8:**166
water as a, **1:**108–10, **7:**1, 176, **8:**163, 173
military target
crops as, **8:**165
definition, **8:**173
water as a, **1:**110–11, **7:**176, **8:**149, 163, 164, 165, 167, 173
Non-Navigational Uses of International Watercourses (*See* Convention of the Law of the Non-Navigational Uses of International Watercourses)
number per year, 1931-2012, **8:**161
overview, **1:**107, **3:**2–3, **5:**189–90, **7:**1
2011-2012 events, trends, and analysis, **8:**159–69
privatization, **3:**xviii, 70–71, 79, **4:**54, 67
reducing the risk of conflict, **1:**113–15
security analysis, shift in intl., **1:**105, **8:**161

Conflict/cooperation concerning freshwater (*continued*)
    subnational/intrastate *vs.* international/interstate, **8:**159, 161–62, 168–69
    summary/conclusions, **1:**124, **8:**168–69
    sustainable vision, **1:**190
    Syria, **8:**147–51
    water "wars" *vs.* water-related violence, **8:**159, 161
    See also Terrorism; *specific countries*
Congo, Democratic Republic of the, **5:**9
    dams with Chinese financiers/developers/builders, **7:**133, 318
    infant mortality rate, **7:**266
    Nile River Basin, **7:**10, 11
Congo, Republic of the:
    dams with Chinese financiers/developers/builders, **7:**133, 333
    infant mortality rate, **7:**264, 265
    location, **7:**11
Conoco-Phillips, **6:**23
Conservation. *See* Environmental flow; Land conservation; Soft path for water; Sustainable vision for the Earth's freshwater; Twenty-first century water-resources development; Water conservation; Water-use efficiency
Construction:
    building site runoff, volume, **8:**76
    major water projects, delivery methods, **8:**104–5, 107–10
Consumption/consumptive use, **1:**12, 13, **3:**103, **6:**7, 117. *See also* Projections, review of global water resources; Water footprint; Water use, consumptive/nonconsumptive; Withdrawals, water
Contracts:
    for desalination plants, 104–12, 116–19
    privatization and, **4:**65–67
Convention of the Law of the Non-Navigational Uses of International Watercourses (1997), **1:**107, 114, 124, 210–30, **2:**10, 36, 41–42, **3:**191, **5:**35
    China's vote against, **7:**129
    critical importance of, **7:**19
    overview, **7:**4–5, **8:**14
Conventions, international legal/law. *See* International *listings;* Law/legal instruments/regulatory bodies; United Nations; World *listings*
Cook Islands, **3:**118
Cooling and water use, **4:**135, **8:**86
Copper, **7:**59–60, **8:**166
Corporate issues. *See also* Bottled water, business/industry; Business/industry, water risks; Companies; Corporate reporting; Privatization; Water conservation, California commercial/industrial water use
Corporate reporting:
    bottled water, **6:**23
    continuous improvement, **6:**32
    current water use, **6:**22
    history, **6:**17–18
    non-financial reports, **6:**18–20
    recommendations for, **6:**37–38
    stakeholder consultations and engagement, **6:**24
    strategic partnerships, **6:**31–32
    summary of, **6:**36–37
    supply chain involvement, **6:**24
    sustainability reports, **6:**18, 20
    water management, **6:**20, **7:**40–41
    water performance (*See* Water performance reporting)
    water-policy statement, **6:**25–26
    water risk-assessment programs, **6:**23–24
Costa Rica, **5:**34

Costs. *See* Economy/economic issues
Côte d'Ivoire, **1:**55, **4:**65–67, **7:**319
Councils. *See* Law/legal instruments/regulatory bodies
Court decisions and conflict/cooperation concerning freshwater, **1:**109, 120. *See also* Law/legal instruments/regulatory bodies; Legislation
Covenant, the Sword, and the Arm of the Lord (CSA), **5:**21–22
Covenants. *See* Law/legal instruments/regulatory bodies; United Nations
Crane, Siberian, **1:**90
Crayfish, **7:**56
Crime:
    corruption, **8:**12–13, 76, 148
    tax fraud, **8:**131
    vandalism, **8:**76
Critical Trends, **3:**88
Croatia, **1:**71
Crocodiles, **7:**56
Crowdsourcing, **8:**12
Cryptosporidium, **1:**48, **2:**157, **4:**52, **5:**2, 159, **7:**47
Cucapá people, **3:**139
Cultural importance of water, **3:**40
Curaçao, **2:**94–95
Current good manufacturing practice (CGMP), **4:**27
CW Leonis, **3:**219–20
Cyanide, **5:**20
Cyberterrorism, **5:**16, **6:**152, **7:**176
Cyprus, **1:**202–4, **2:**108, **8:**135
Cyrus the Great, **1:**109

D
Dams:
    Africa, **3:**292, **5:**151, **6:**274–78
    business/industry, water risks, **5:**151
    Central Asia, **8:**166
    China (*See* China, dams)
    by continent and country, **3:**291–99
    debate, new developments in the, **1:**80–83
    economic issues, **1:**16, **2:**117–19, 122–24, 127, 129–30
        development and construction by China, **7:**133–35
        private-sector funding, **1:**82
    environmental flow, **5:**32–34
    environmental impacts
        impact statements, **7:**134–35
        overview, **7:**130
        threatened/at risk species due to, **3:**3, **7:**154, **8:**128–29
    environmental/social impacts, **1:**15, 75–80, 83
    and floods, **5:**106, **6:**144, **8:**128
    Gabcikovo-Nagymaros project, **1:**109, 120
    grandiose water-transfer plans, **1:**74–75
    Korean peninsula, **1:**109–10
    large
        historical background, **1:**69
        North American Water and Power Alliance, proposed, **8:**127
        San Francisco Bay, proposed, **8:**132
        total worldwide, **7:**127
        U.S. begins construction of, **1:**69–70
    opposition to, **1:**80–82, **7:**127, 132, 134–36
    by owner, **6:**272–73
    power generation (*See* Hydroelectric production)
    primary purposes of, **6:**270–71
    removal (*See* Dams, removing/decommissioning)
    runoff, humanity appropriating half of the world's, **5:**29
    safety data, **6:**236–38
    social impacts
        displaced people, **1:**77–80, 85, 90, 97–98, 281–87, **5:**134, 151

Index
445

Sudan, **7:**133, 134
U.S., **7:**154
environmental justice, **5:**133–36
schistosomiasis outbreaks, **1:**49, **7:**58
specific sites (*See* Dams, specific)
Syria, strategic attacks on, **8:**149
by the Tennessee Valley Authority, **7:**153
terrorism risks, **5:**15–16
twentieth-century water-resources development, **1:**6
World Water Forum (2003), **4:**193
See also World Commission on Dams
Dams, removing/decommissioning:
  case studies, completed removals
    Edwards, **2:**xix, 123–25
    Maisons-Rouges and Saint-Etienne-du-Vigan, **2:**125
    Newport No. 11, **2:**125–26
    Quaker Neck, **2:**126
    Sacramento River valley, **2:**120–23
  economics, **2:**118–19, **8:**46
  employment, **8:**46, 48
  hydroelectric production, **1:**83, **2:**114–15
  1912 to present, **2:**275–86, **6:**239–69
  overview, **2:**xix, 113–14
  proposed
    Elwha and Glines Canyon, **2:**127–28
    Glen Canyon, **2:**128–31
    Pacific Northwest, **2:**131–34, **8:**129
    Peterson, **2:**128
    Savage Rapids, **2:**128
    Scotts Peak, **2:**126–27
  purpose for being built no longer valid, **2:**117
  renewal of federal hydropower licenses, **2:**114–15
  safety issues, **2:**130
  by state, **6:**265–69
  states taking action, **2:**117–18
  summary/conclusions, **2:**134
  twentieth century by decade, **3:**301–2
  by year, **6:**265–69
Dams, specific:
  American Falls, **7:**153
  Aswan, **7:**12
  Ataturk, **1:**110, **3:**182, 184, 185
  Auburn, **1:**16
  Bakun, **1:**16
  Balbina, **5:**134
  Banqiao, **5:**15–16
  Batman, **3:**187
  Belinga, **7:**134–35
  Belo Monte, **8:**165
  Birecik, **3:**185, 186–87
  Bonneville, **1:**69, **8:**129
  Chixoy, **3:**13
  Cizre, **3:**189–90
  Condit, **2:**119
  Edwards, **1:**83, **2:**xix, 119, 123–25
  Elwha, **1:**83
  Farakka Barrage, **1:**118–19
  Fort Peck, **1:**69
  Fort Randall, **1:**70
  Garrison, **1:**70, **5:**123
  Gezhou, **6:**142
  Gibe, **7:**134
  Glen Canyon, **1:**75–76, **2:**128–31
  Glines Canyon, **1:**83
  Gorges, **5:**133
  Grand Coulee, **1:**69
  Grand Ethiopian Renaissance (Hidase; Millennium), **8:**164
  Hetch Hetchy, **1:**15, 80–81
  Hoa Binh, **5:**134
  Hoover, **1:**69, **3:**137, **8:**127
  Ice Harbor, **2:**131–34
  Ilisu, **3:**187–89, 191
  Imperial, **7:**7
  Itaipu, **1:**16
  Kabini, **8:**163
  Kalabagh, **1:**16
  Karakaya, **3:**182, 184
  Kariba, **5:**134
  Katse, **1:**93, 95
  Keban, **3:**184
  Kenzua, **7:**153
  Koyna, **1:**77
  Krishna Raja Sagar, **8:**163
  Laguna, **3:**136
  Little Goose, **2:**131–34
  Lower Granite, **2:**131–34
  Lower Monumental, **2:**131–34
  Machalgho, **8:**163
  Manitowoc Rapids, **2:**119
  Merowe, **7:**133–34
  Morelos, **3:**138, 142, **7:**7
  Myitsone, **7:**132–33
  Nam Theun I, **1:**16
  Nujiang, **7:**135
  Nurek, **1:**70, **8:**127
  Oahe, **1:**70
  Pak Mun, **5:**134
  Peterson, **2:**128
  Pubugou, **7:**136
  Quaker Neck, **2:**126
  Rogun, **8:**166
  Sadd el-Kafara, **1:**69
  St. Francis, **5:**16
  Salling, **2:**119
  Sandstone, **2:**119
  Sapta Koshi High, **1:**16
  Sardar Sarovar, **5:**133
  Savage Rapids, **2:**119, 128
  Sennâr, **7:**58
  Shasta, **1:**69, **5:**123, **7:**153
  Shimantan, **5:**15–16
  Snake River, **2:**131–33
  Tabqa (al-Thawrah), **8:**149
  Ta Bu, **5:**134
  Tantangara, **7:**114
  Three Gorges Dam (*See* Three Gorges Dam)
  Tiger Leaping Gorge, **6:**89, **7:**135
  Tishrin, **8:**149
  Welch, **2:**118
  Woolen Mills, **2:**117–19
  Wullar, **8:**163
  Xiaowan, **7:**131
  Yacyreta, **3:**13
  Yangliuhu, **7:**135
  Yangtze River, **7:**46, 129
  See also China, dams, construction overseas; Lesotho Highlands project; Southeastern Anatolia Project
Dasani bottled water, **4:**21
Data issues/problems:
  centralized data portal, **8:**12
  climate change, **4:**183
  collection, recommendations, **8:**12
  conversions/units/constants, **2:**300–309, **3:**318–27, **4:**325–34, **5:**319–28, **7:**75, 341–50, **8:**395–404
  global water footprint, **8:**87, 88
  global water resources, projections, **2:**40–42
  groundwater, **4:**97–98
  mobile connectivity, **8:**12
  need for data and water policy reform, **7:**152
  open access to information, **4:**70–73, **8:**11
  polls, **7:**283, 286, **8:**387, 389, 392

Data issues/problems (*continued*)
  sustainable water jobs, **8:**58
  urban commercial/industrial water use in California, **4:**139–40, 148–50, 152–53
  urban residential water use in California, **4:**108–9
  well-being, measuring water scarcity and, **3:**93
da Vinci, Leonardo, **1:**109
Dead Sea:
  mineral extraction from evaporation ponds, **8:**155, 156
  overview, **8:**153, 154–55
  water transfer from the Red Sea to, **8:**144, 153–57
Deaths. *See* Mortality
Decision making:
  joint, for transboundary waters, **7:**13, 167, **8:**14–15
  open/democratic, **4:**70–73, **8:**11
  precautionary principle, **7:**167
  water risks factored into, **6:**27–28
Declarations. *See* Law/legal instruments/regulatory bodies; United Nations
Deer Park, **4:**21
Deforestation, **8:**26
Deltas, river, **3:**xix
  Colorado River, **3:**139–43, **6:**130
  Sacramento River, **8:**133
Demand management, **4:**179–80, **7:**153–54. *See also* Water-use efficiency; Withdrawals, water
  with desalination projects, **8:**102, 106, 115–16, 119
Demographic health surveys (DHS), **6:**61
Demographics of sustainable water job employees, **8:**56
Dengue fever, **3:**2
Denmark, **1:**52
Density, measuring, **2:**304, **3:**327, **4:**329, 334, **7:**345, 350
Desalination:
  advantages and disadvantages, **5:**66–76, **8:**108–9
  Australia, **5:**69, **7:**114–15, **8:**96, 97, 98, 102, 103, 118
  business/industry, water risks, **5:**161
  California (*See* California, desalination)
  capacity by country/process/source of water, **1:**131, 288–90, **2:**287–90, **5:**58–60
  capacity statistics, **5:**56–57, 59–60, 308–17
  Carlsbad (CA), **8:**93, 95, 98, 112–19
  case studies, **8:**106–17
  China, **6:**93
  climate change, **5:**80–81, **7:**16
  concentrate disposal, **2:**107
  economic issues (*See* Economy/economic issues, desalination)
  energy use/reuse, **2:**107, **5:**69–71, 75–76
  environmental effects of, **5:**76–80
  global status of, **5:**55–58
  health, water quality and, **5:**74–75
  history and current status, **2:**94–98, **5:**54
  intakes, water, impingement/entrainment, **5:**76–77
  Israel, **5:**51, 69, 71, 72, **8:**157
  Nauru, **3:**118
  oversight process, regulatory and, **5:**81–82
  overview, **1:**29–30, **2:**93–94, **5:**51–53
  plants
    capacity of actual/planned, **5:**308–17
    Carlsbad (CA), **8:**93, 95, 98, 112–19
    costs, **8:**93–101, 117–18
    project delivery methods, **8:**104–5, 107–10
    projects and risks, **8:**101–6, 118–19
    project structure, **8:**119
    proposed and recently constructed, **8:**98, 103, 155–56
    Sand City (CA), **8:**93
    Tampa Bay (FL), **2:**108–9, **5:**61–63, **6:**123–25, **8:**98, 103, 106–12, 118
  processes, **5:**54–55
    freezing, **2:**104
    ion-exchange methods, **2:**104
    membrane
      costs, **8:**94
      electrodialysis, **2:**101–2
      overview, **2:**101
      reverse osmosis (*See* Reverse osmosis)
    membrane distillation, **2:**104–5
    overview, **2:**103–4
    solar and wind-driven systems, **2:**105–6
    thermal
      multiple-effect distillation, **2:**99–100
      multistage flash distillation, **2:**96, 100
      overview, **2:**98–99
      vapor compression distillation, **2:**100–101
  production statistics, monthly, **8:**107
  Red Sea, proposed, **8:**154, 155–56
  reliability value of, **5:**73–74
  salt concentrations of different waters, **2:**94, **5:**53
  source of water/process, capacity by, **5:**56–57, 59–60
  summary/conclusions/recommendations, **2:**109–10, **5:**82–86, **8:**117–18
  Tampa Bay (FL), **2:**108–9, **5:**61–63, **6:**123–25, **8:**98, 103, 106–12, 118
  U.S., **5:**58–63
Desertification, Australia, **7:**105–6
Deutsche Morgan Grenfell, **1:**96
Developing countries:
  agriculture, **1:**24
    irrigation, **3:**290, **4:**297–98, **5:**301–2
  bottled water, **4:**40–41
  business/industry, water risks, **5:**150
  cholera, **1:**56
  climate change vulnerability of, **6:**44–45, 51, 54
  dams, **1:**82
  diseases, water-related, **5:**117
  dracunculiasis, **1:**51
  drinking water, **1:**261–62
  economic development derailed, **1:**42
  education and expertise in water quality, **7:**66
  efficiency, improving water-use, **1:**19
  food needs for current/future populations, **2:**69
  industrial water use, **1:**21
  knowledge and technology transfer, **8:**10–11
  Pacific Islands developing countries (PIDC), **5:**136
  pollution sources, **8:**3
  population increases and lack of basic water services, **3:**2
  privatization, **3:**79, **4:**46
  sanitation services, **1:**263–64
  toilets, energy-efficient, **1:**22
  unaccounted for water, **4:**59
  See also Environmental justice; *specific countries*
Development:
  assistance (*See* Overseas Development Assistance)
  economic/social, as context for water conflicts, **7:**176, **8:**173 (*See also* Water Conflict Chronology)
  the right to, **2:**8–9
  technology (*See* Technology development; Technology transfer)
Development Assistance Committee (DAC). *See* Overseas Development Assistance
Diageo's Water of Life, **7:**38
Diarrhea, **1:**48, **4:**8, 11, **6:**58, 75, 85
  disability adjusted life year (DALY), **7:**272
  morbidity, **7:**58
  mortality, **7:**57–58, 272
    childhood, under-5, **7:**259–63
Dioxins, **7:**48, 60
Diptheria, **7:**272
Disability-adjusted life year (DALY), **4:**9, 276–77, **7:**57, 270–72

Diseases, water-related, **1:**186–87
    amoebiasis, **1:**48
    ascariasis, **7:**272
    Balkan Endemic Neuropathy, **7:**89
    Campylobacter jejuni, **7:**57
    in China, **6:**85
    cholera (*See* Cholera)
    dams, removing/decommissioning, **2:**130
    death, **6:**58, 73
        and disability-adjusted life year from, **4:**276–77, **7:**57, 270–72
        limitations in data and reporting, **7:**270–71
        projected, **4:**9–10, 12–13
    dengue fever, **3:**2
    diarrhea (*See* Diarrhea)
    diptheria, **7:**272
    dysentery, **1:**42, **4:**64
    emerging diseases/pathogens, **2:**155, **7:**49
    encephalitis, **7:**47
    environmental justice, **5:**128–29
    failure, **3:**2, **5:**117
    fecal coliform bacteria, **7:**52–53 (*See also* Fecal contamination)
    Guinea worm (*See* Dracunculiasis)
    hepatitis, **7:**58
    hookworm, **7:**61, 272
    malaria, **1:**49–50, **6:**58, **7:**259–63
    meningitis, **7:**47
    outbreaks in U.S., **4:**308–12
    overview, **1:**47–50, 274–75, **7:**47, 57–58
    poliomyelitis, **7:**272
    roundworm, **7:**61–62
    schistosomiasis, **1:**48, 49, **7:**58, 272
    Shigella, **7:**57
    trachoma, **1:**48, **7:**272
    trichuriasis, **1:**48, **7:**272
    trypanosomiasis, **7:**272
    typhoid, **1:**48, **7:**57, 58
    waterborne, **1:**47–49, 274–75, **4:**8, **7:**57–58
    waterborne *vs.* water-based, **7:**57–58
    whipworm, **7:**61
    See also Millennium Development Goals
Dishwashers, **4:**109, 116, **6:**106
Displaced people, **8:**148, 164, 167. *See also* Dams, social impacts, displaced people
Dolphins, **1:**77, 90, **3:**49, 50, **7:**56
Dow Jones Indexes, **3:**167
Downstream users, **5:**37. *See also* Human right to water
Dracunculiasis (Guinea worm), **1:**39, 48–56, 272–73, **3:**273–77, **5:**293–97
    host zooplankton, **7:**58
    overview, **7:**47
Drinking water, access:
    collection distance, **7:**231, **8:**237
    collection the responsibility of children/women, **7:**61, 89, **8:**168
    in conflict areas (*See* Conflict/cooperation concerning freshwater, military target, water as a)
    corporate efforts to improve, **7:**38, 41
    costs of, **6:**73
    by country, **1:**251–55, **3:**252–60, **5:**237–46, **6:**211–20
        urban and rural, 1970-2008, **7:**230–40, **8:**236–46
        urban and rural, 2011, **8:**236–38, 247–51
    deaths due to lack, **8:**164
    defining terms, **4:**28, **7:**230–31, **8:**236–37
    developing countries, **1:**261–62, **7:**62
    disinfection and purification, **5:**159
    fluoride, **4:**87
    funding of, **6:**73, **8:**9
    "improved," use of term, **7:**230–31, 251–52, **8:**236–37, 239
    infrastructure for tap water, **8:**35–36
    intl. organizations, recommendations by, **2:**10–11
    limitations in data and reporting, **6:**61–62, **7:**231, 251–52, **8:**237–38, 268–69
    Overseas Development Assistance, **7:**273–77
        by donating country 2004-2011, **8:**317–19
        by subsector 2007-2011, **8:**320–22
    reclaimed water, **2:**151–52
    by region, MDG progress on, **6:**65–67, 230–32, **7:**24, 251–53, **8:**268–70
    rural areas, **6:**70–71 (*See also* Drinking water, access, by country)
    shortages, **6:**44, 53, 86, **8:**1, 164, 167
    statistics regarding, **6:**44, **7:**24, 61, **8:**1
    twentieth-century water-resources development, **3:**2
    urban areas, **6:**70–71, **7:**97 (*See also* Drinking water, access, by country)
    violence during long walks to water source, **8:**168
    well-being, measuring water scarcity and, **3:**96–98
    World Health Organization, **4:**208
    World Water Forum (2003), **4:**202
    See also Health, water issues; Human right to water; Millennium Development Goals; Soft path for water; Water quality; Well-being, measuring water scarcity and; *specific contaminants*
Droughts, **1:**142, 143, **4:**163, 203–4, **6:**44
    agricultural effects, **5:**92, 94, 98, 103, **7:**102–5, **8:**148
    Atlanta (GA), **6:**108
    Australia, **7:**97–121, 146–47
    beginning of, determination, **5:**93
    Brazil, **7:**26, **8:**22–23, 165
    California, **8:**89, 102
    causes, **5:**95–96, **7:**8, 98, 100
    China, **5:**97, **6:**86–87, **7:**131–32
    Colorado River basin, **8:**138
    defining terms, **5:**92, **7:**99–100
    disturbances promoting ecosystem diversity, **5:**91
    due to climate change, **5:**112–13, **6:**44, **7:**8–9, 54, 101–2, **8:**149–51
    ecological effects, **5:**92, **7:**54, 100–101
    economy/economic issues, **5:**91–92, 98, 103
    effects of, **5:**95–99
    fires, **5:**98, 102, **7:**105
    forecasting, **7:**109, 146
    future of, **5:**112–13
    management
        agricultural, **7:**107–14
        crisis management, **5:**99, 111, **7:**8–9, 97, 100, 106–7
        impact and vulnerability assessment, **5:**100–101
        mitigation and response, **5:**101–3
        monitoring and early warning, **5:**99–100, **7:**20
        national policy development, **7:**106–7, 112
        public participation, **7:**116–17
        risk management, **5:**99, **7:**109
        water market/water trading, **7:**110, 111–14, 146
    National Drought Mitigation Center, **5:**94
    overview, **5:**91–92
    short-lived or persistent, **5:**93
    summary/conclusions, **5:**113–14
    Syria, **8:**147–51
    transboundary agreements, **7:**8–9, 12–13, 15
    urban areas, **5:**98, **7:**114–21
    U.S., **5:**93, **6:**44, **8:**130
DuPont, **1:**52
Dust Bowl (U.S.), **5:**93
Dutch Water Line strategy, **5:**5
Dynamics, water, **2:**218
Dysentery, **1:**42, **4:**64

E

Early Warning Monitoring to Detect Hazardous Events in Water Supplies, **5:**2, 20
Earth, origins of water on, **1:**93–98, **3:**209–12, **6:**5
Earthquakes. *See* Seismic activity
Earth Water, **5:**163
East Bay Municipal Utilities District (EBMUD), **5:**73
East Timor. *See* Timor Leste
Economy/economic issues:
    access to water, **5:**125, **7:**67
    bag technology, water, **1:**200, 204–5
    bottled water, **4:**17, 22–23
    budgets, U.S. federal agency water-related, **7:**303–7
    cholera, **1:**60
    climate change, **6:**50–51
    Colorado River, **3:**143
    conflict/cooperation concerning freshwater, **5:**15
    cost effectiveness, **4:**105, **8:**38
    cost of water (*See* Pricing, water)
    costs of water in San Diego County, **8:**100
    credit rating for project, **8:**117
    dams, **1:**82, **2:**117–19, 122–24, 127, 129–30, 132
    desalination, **1:**30, **2:**95, 105–9, **5:**62–63, 66, 68–73
        cost comparisons, **8:**99–101, 118
        cost estimates, **8:**95–99, 110–11, 112–13, 118
        cost terminology, **8:**94–95
        overview, **8:**93–94, 117
    developing countries, **1:**42, **3:**127 (*See also* Overseas Development Assistance)
    disadvantages of corporate sustainability, **8:**30
    disincentives for corporate sustainability, **8:**30
    droughts, **5:**91–92, 98, 103
    economic development, **7:**176, **8:**27
    economic good, treating water as an, **3:**xviii, 33–34, 37–38, 58, **4:**45
    economies of scale, and hard path for meeting water-related needs, **3:**8
    efficiency, economic, **4:**104, 105, **8:**100
    environmental flow, **5:**32, 40–43
    financial assistance, **6:**74
    fishing, **2:**117
    floods, **5:**91–92, 108, 109
    Global Water Partnership, **1:**171
    human needs, basic, **1:**46–47
    human right to water, **2:**13–14, **8:**29
    incentives and rebates for water conservation, **6:**110–12, **7:**119, 121
    incentives for corporate sustainability, **8:**23, 31
    industrial water use, **1:**21, **7:**67
    infrastructure projects, funding, **8:**105
    intl. water meetings, **5:**183
    investment issues, **1:**6–7, **5:**273–75, **8:**12–13, 23
    jobs, sustainable water, **8:**35–59
    Lesotho Highlands project, **1:**95–97, 99
    Millennium Development Goals, **4:**6–7, **7:**64
    overruns, water-supply project, **3:**13
    poor water quality, **7:**63–65
    privatization, **3:**77, **4:**53–60
    productivity of water, U.S., **4:**321–24
    property values and green space, **8:**42
    recession, U.S., **8:**130
    reclaimed water, **2:**156, 159, **7:**121
    recycled water, **8:**11
    revenue/growth of the water industry, **5:**303–4
    sanitation services, **5:**273–75
    soft path for meeting water-related needs, **3:**5, 6–7, 12–15, 23–25, **7:**150
    Southeastern Anatolia Project, **3:**182, 187, 190–91
    subsidies
        agriculture, **1:**24–25, 117, **7:**108, 109, 152
        desalination, **5:**69
        engineering projects, large-scale, **1:**8
    government and intl. organizations, **1:**17, **7:**67, 108, 109, 111, 152, **8:**35
    privatization, **3:**70–72, **4:**50, 53–60
    twenty-first century water-resources development, **1:**24–25
    supply-side solutions, **1:**6
    tariffs, water, **6:**312–23
    terrorism, **5:**20
    Three Gorges Dam, **1:**86–89
    transfer of water, **8:**136, 140, 142
    transparency and accountability, **8:**12–14, 112
    twentieth-century water-resources development, end of, **1:**16–17
    urban commercial/industrial water use in California, **4:**140–43, 151–52
    urban residential water use in California, **4:**107–8
    wages for sustainable water jobs, **8:**56
    water policy reform, **7:**152–53
    water scarcity effects on, **6:**45
    water use *vs.* Gross Domestic Product, **8:**142
    withdrawals, water, **3:**310–17
    World Commission on Dams, **3:**158, 162–64, 167–69
    World Water Forum (2003), **4:**195–96
    See also Business/industry, water risks; Environmental justice; Globalization and international trade of water; Privatization; World Bank
Ecosystems:
    bioproductivity, **8:**86
    classification of impacts, three-tier, **7:**52
    climate change, **4:**171–72
    community structure (*See* Biodiversity; Extinct species; Introduced/invasive species; Threatened/at risk species)
    costs of poor water quality, **7:**63
    ecological footprint, **8:**85, 86
    environmental flow, **5:**30–31
    impacts of dams and water withdrawals, **3:**3, **8:**37
    impacts of desalination, **8:**94
    impacts of fossil-fuel extraction/processing, **7:**84–87
    impacts of water quality degradation factors, **4:**168, **7:**46–49, 52, 54–57
    privatization, **4:**51–53, **7:**146
    reclaimed water, **2:**149–50
    reserve, and nature's right to water, **7:**145
    restoration/protection, **2:**149–50, **7:**66, 146, **8:**44–48
    World Water Forum (2003), **4:**203
    See also Environmental flow; Environmental issues; Fish; Soft path for water; Sustainable vision for the Earth's freshwater
Ecuador, **1:**59, 71, **2:**178–79, **3:**49, **5:**124
    dams with Chinese financiers/developers/builders, **7:**319
    fossil-fuel production, **7:**88
Education:
    effects of inadequate sanitation on, **6:**58, **7:**61
    empowerment through, **8:**13–14
    Save Water and Energy Education Program, **4:**114
    and sustainable water jobs, **8:**55, 57, 58, 59
    in water quality, **7:**66
    in water supply and sanitation, **7:**277
    in water supply and sanitation, Overseas Development Assistance
        by donating country 2004-2011, **8:**317–19
        by subsector 2007-2011, **8:**320–22
Edwards Manufacturing Company, **2:**124

Index 449

Eels, **2:**123
Efficiency (*See* Water-use efficiency)
Egypt, **1:**118, **2:**26, 33, **4:**18, 40, **5:**163
    conflicts concerning freshwater, 2011-2012, **8:**164
    decline in infant mortality rate, **7:**264
    Nile River Basin, **7:**11, **8:**164
    Suez Canal, **8:**153
    tourism, **8:**153, 164
Eighteen District Towns project, **2:**167
El Niño/Southern Oscillation (ENSO), **1:**139, 143, 147, **3:**119, **4:**162, **5:**95, 106
    and the Millennium Drought in Australia, **7:**98
Employees, demographics of sustainable water jobs, **8:**56
Employment, sustainable water jobs, **8:**35–59
Encana Oil and Gas Inc., **8:**71, 73
Encephalitis, **7:**47
Endocrine disruptors, **7:**49, 59, 60
End use of water as a social concern, **3:**7, 8–9
Energy issues:
    bottled water, **7:**157–64
    business/industry, water risks, **5:**150, 151–52
    desalination, **2:**107, **5:**69–71, 75–76, **8:**95
    droughts, **5:**98, **8:**22–23
    energy efficiency, **3:**xiii, **7:**84, 162
    measuring energy, **2:**308, **3:**326, **4:**333, **7:**75, 349
    power generation
        by fossil fuel type, **7:**75
            renewable technologies (*See* Hydroelectric production; Nuclear power; Solar power; Wind power)
            retirement of aging power plants, **7:**84–85
            thermal pollution from, **7:**85
            water consumption, **7:**25, 26, 31, 73, 74–75
            water footprint, **8:**5–6, 84
    tap water, **7:**163
    transfer of water, **8:**142
    transportation method, **7:**162
    unit conversions, **7:**75
    U.S. Department of Energy, **1:**23, **4:**115
    water-energy-food nexus, **8:**5–6
    water treatment, **7:**60
Engineering projects, large-scale. *See* Dams; Transfers, water; Twentieth-century water-resources development; Zombie water projects
England:
    cholera, **1:**56
    desalination, **2:**94
    droughts, **5:**92
    eutrophication, **7:**63
    human right to water, **4:**212
    Lesotho Highlands project, **1:**96
    Office of Water Services, **4:**64–65
    privatization, **3:**58, 60, 61, 78, **4:**62–65
    sanitation services, **5:**129–31
    Southeastern Anatolia Project, **3:**191
    World Commission on Dams, **3:**162, 170
    See also United Kingdom
Enron, **3:**63
Environment, **2:**182
Environmental flow:
    characteristics of hydrologic regimes, **5:**31
    economics/finance, **5:**40–43
    General Accounting Office, **5:**119
    legal framework, **5:**34–37
    policy implementation, **5:**43–45
    projects in practice, **5:**32–34
    science of determining, **5:**38–40
    summary/conclusions, **5:**45–46
    water quality link, **5:**40
    World Commission on Dams' recommendations, **5:**30

Environmental Impact Assessment (EIA), **6:**96, **7:**31–32
Environmental issues:
    bottled water, **4:**41
    change, global, **1:**1
    cleanup (*See* Habitat restoration; Remediation)
    contaminants in water (*See* Wastewater; Water quality; *specific contaminants*)
    as context for business/industry water risk, **7:**29, **8:**24
    dams/reservoirs, **1:**15, 75–80, 83, 91, **8:**128
    desalination, **5:**76–80
    ecological impacts (*See* Ecosystems)
    environmental flow (*See* Environmental flow)
    environmental justice (*See* Environmental justice)
    green infrastructure (low-impact development), **8:**39, 41
    green jobs, **8:**35–59
    Lesotho Highlands project, **1:**98
    nature's right to water, **7:**145
    reclaimed water, **2:**149–50
    shrimp/tuna and turtle/dolphin disputes, **3:**49, 50
    stormwater, pollution by (*See* Stormwater runoff)
    sustainable vision for the Earth's freshwater, **1:**188–90
    Three Gorges Dam, **1:**89–9, **6:**142
    top concerns around the world, **7:**285–88
    top concerns of U.S. public, **7:**282–84
    twentieth-century water-resources development, end of, **1:**12, 15–16
    U.S., **6:**339–41, **7:**282–84
Environmental justice:
    climate change, **5:**136–37
    Coca-Cola, **5:**127
    dams, **5:**133–36
    discrimination, environmental, **5:**118–19
    economic issues, **2:**13–14, **8:**29
    environmentalism of the poor, **5:**123–24
    Environmental Justice Coalition for Water, **5:**122–23
    good governance, **5:**138–41
    history of movement in U.S., **5:**119–20, 122–23
    intl. context, **5:**118
    overview, **5:**117–18
    principles of, **5:**120–22
    privatization, **5:**131–33
    right to water (*See* Human right to water)
    sanitation services, **5:**127–31
    summary/conclusions, **5:**141–42
    water access, **5:**124–25, 127, **6:**44, **8:**29
    water quality, **5:**127–29
    women and water, **5:**126, 134, **7:**61, 89, **8:**168
Environmental management system, **6:**32
Eritrea, **5:**95, **7:**11
Ethanol, **7:**74
Ethical issues:
    land ethic of Leopold, **8:**129
    public relations and social responsibility, **8:**23, 29–30
Ethiopia, **1:**55, **4:**211, **5:**95, 97
    conflicts concerning freshwater, 2011-2012, **8:**164
    dams
        with Chinese financiers/developers/builders, **7:**133, 134, 319–20
        Grand Ethiopian Renaissance (Hidase; Millennium), **8:**164
    drinking water, access to, **6:**71–73, **7:**41
    Nile River Basin, **7:**11, **8:**164
    sanitation, **6:**71–73
Ethos Water, **5:**163
Europa Orbiter, **3:**218

Europe:
    air quality, satisfaction, 2012, **8**:391
    aquifers, transboundary, **7**:3
    bottled water, **4**:22, 25, 288, 289, 291, **5**:163, 281, 283
        per-capita consumption, **7**:340
    cholera, **1**:56, 267, 270, **5**:289, 290
    dams, **3**:165, 292–93
    drinking water, **5**:159, 245–46, **7**:239–40
    Eco-Management and Audit Scheme, **6**:18
    environmental flow, **5**:33
    European Convention on Human Rights (1950), **2**:4, 8
    European Union Development Fund, **1**:96
    European Union Water Framework Directive (2000), **7**:25, 147–49
    food needs for current/future populations, **2**:69
    globalization and intl. trade of water, **3**:45, 47
    Global Water Partnership, **1**:169
    groundwater, **4**:84–85
    human right to water, **4**:209, 214
    hydroelectric production, **1**:71, 279–80
    irrigation, **1**:301, **2**:261–62, 265, **3**:289, **6**:332–33
    Lesotho Highlands project, **1**:96
    mortality rate, under-5, **7**:261–62
    population data/issues, **1**:249–50, **2**:214
    privatization, **3**:58, 60, 61
    renewable freshwater supply, **1**:239–40, **2**:201–2, 217, **3**:241–42, **4**:265–66, **5**:225–26
        2011 update, **7**:219–20
        2013 update, **8**:225–26
    reservoirs, **2**:270, 274
    river basins, **6**:289, 301–06
        transboundary, **2**:29–31, **7**:3
    salinization, **2**:268
    sanitation services, **3**:272, **5**:255, **7**:249–50
    threatened/at risk species, **1**:294–95, **7**:56
    water availability, **2**:217
    waterborne diseases, **1**:48
    water quality, satisfaction
        by country, **7**:290–91
        2012, **8**:391
    withdrawals, water, **1**:244, **2**:209–11, **3**:249–50, **4**:273–75, **5**:235–36
        by country and sector, **7**:228–29, **8**:233–35
Evaporation of water into atmosphere, **1**:141, **2**:20, 22, 83, **4**:159–60, **6**:43, 107
    reduction in drought management, **7**:110
    reduction with aquifer recharge, **8**:49
Evaporation ponds, potash extraction from, **8**:155
Everglades, **8**:48
Evian bottled water, **7**:162
Excreta. *See* Fecal contamination
Ex-Im Bank, **1**:88–89, **3**:163
Export credit agencies (ECAs), **3**:191
Extinct species:
    freshwater animal species, **7**:292–302
    rates for fauna from continental North America, **4**:313–14
    See also Threatened/at risk species

F
Faucets, **4**:117, **6**:106, **7**:28, **8**:38
Fecal contamination, **1**:47–48, **7**:52–53, 57–58
Fertilizer. *See also* Runoff, agricultural
Field flooding, **2**:82, 86
Fiji, **3**:45, 46, 118, **7**:320
Fiji Spring Water, **7**:162
Films:
    and portrayals of terrorism, **5**:14
    water in, **7**:171–74
Filtration, water, **1**:47, **5**:55, **7**:159–60, 161
Finn, Kathy, **4**:69–70

Fire, and drought, **5**:98, 102, **7**:105
First Peoples. *See* Indigenous populations
Fish:
    aquaculture, **2**:79
    bass, **2**:123
    carp, **2**:118
    climate change, **4**:168
    Colorado River, **3**:142
    dams, removing/decommissioning, **2**:118, 123, 126, 128, 131–33
    dams/reservoirs affecting, **1**:77, 83, 90, 98, **2**:117, **6**:142–43
        North American Water and Power Alliance, **8**:128–29
        percent of North American species threatened, **7**:154
    desert pupfish, **3**:142
    droughts, **5**:98, 102
    eels, **2**:123
    endocrine disruptors, **7**:49
    extinct or extinct in the wild species, **7**:294–96
    floods, **5**:109
    food needs for current/future populations, **2**:79
    fossil-fuel extraction affecting, **7**:87, 91
    herring, **2**:123
    impacts of desalination, **8**:94
    largest number of species, countries with, **2**:298–99
    pesticides, **5**:305–7
    Sacramento River, **2**:120–21, **8**:128
    salmon (*See* Salmon)
    San Francisco Bay, **8**:132
    sturgeon, **1**:77, 90, **2**:123
    threatened/at risk species, **1**:291–96, **3**:3, 39–40, 142
        due to dams, **7**:154, **8**:37, 128–29
        due to mining drainage, **7**:52
        percent endangered, **8**:3
        percent of U.S. species, **7**:56, **8**:37
    tuna, **3**:49, 50
Floods, **1**:142–43, **4**:162–63, 203–4, 302–7, **5**:104–9
    Australia, **7**:97
    causes, **5**:106, **7**:9
    China, **6**:84–87, 144
    control, **5**:110–12
    and dams, **5**:106, **6**:144, **8**:128
    Dead Sea, **8**:155
    definition, **5**:104–5
    disturbances promoting ecosystem health, **5**:91, **7**:134
    economy/economic issues, **5**:91–92
    effects of, **5**:106–10
    flash, **5**:104
    frequency, calculation, **5**:105
    future of, **5**:112–13
    Johnstown Flood of 1889, **5**:16
    management, **5**:110–12
    Mekong River Basin, **7**:14–15
    overview, **5**:91–92
    summary/conclusions, **5**:113–14
    transboundary agreements, **7**:9
Florida:
    Altamonte Springs, **2**:146
    reclaimed water, **2**:146–47
    St. Petersburg, **2**:146–47
    Tampa Bay (*See under* Desalination, plants)
Flow-limited resources, **6**:6
Flow rates, **2**:305–6, **3**:104, 324, **4**:168, 172, 331. *See also* Hydrologic cycle; Stocks and flows of freshwater
    measuring, **7**:346–47
Fluoride, **4**:87, **6**:81
Fog collection as a source of water, **2**:175–81
Food. *See also* Agriculture, water footprint

Index 451

access, impacts of fossil-fuel extraction/
    processing, **7:**91
adulteration, **4:**32–33
BOD emissions by country, **7:**279–81
diets, regional, **2:**64–66
fish, **2:**79
genetically-modified, **7:**110
grains (*See* Grains)
meat consumption, **2:**68–69, 72, 79–80
water footprint to produce, **8:**83–84
Food needs for current/future populations:
    agricultural land (*See* Agriculture, land availability/
        quality)
    climate change, **2:**87–88
    cropping intensity, **2:**76
    crop yields, **2:**74–76
    eaten by humans, fraction of crop production,
        2:76–77
    inequalities in food distribution/consumption,
        2:64, 67–70
    kind of food will people eat, what, **2:**68–70
    need and want to eat, how much food will people,
        2:67–68
    overview, **2:**65
    people to feed, how many, **2:**66–67
    production may be unable to keep pace with
        future needs, **2:**64
    progress in feeding Earth's population, **2:**63–64
    summary/conclusions, **2:**88
    water-energy-food nexus, **8:**5–6
    water needed to grow food (*See* Agriculture,
        irrigation)
Footprint. *See* Carbon footprint; Ecosystems,
    ecological footprint; Water footprint
Force, measuring, **2:**306, **3:**324, **4:**331, **7:**347
Foreign Affairs, **3:**xiii
Forestry, urban, **8:**42
Fossil fuels. *See* Petroleum and fossil fuels
Fossil groundwater, **6:**9–10
France:
    conflict/cooperation concerning freshwater, **5:**15
    dams
        Three Gorges Dam, **1:**89
        World Commission on Dams, **3:**159
    dracunculiasis, **1:**52
    environmental concerns, top, **7:**287
    globalization and intl. trade of water, **3:**45
    Global Water Partnership, **1:**171
    human right to water, **4:**209
    Lesotho Highlands project, **1:**96
    privatization, **3:**60, 61
Frank, Louis A., **1:**194–98, **3:**209–10
French Polynesia, **3:**118
Freshwater:
    percent of global in aquifers, **7:**3
    percent of global in the Great Lakes, **7:**165
    See also Drinking water; Lakes; Renewable
        freshwater supply; Rivers; Streams;
        Surface water; Withdrawals, water
Freshwater Action Network, **8:**13
Fuels. *See* Biofuel; Petroleum and fossil fuels
Furans, **7:**48, 60
Fusion Energy Foundation, **8:**130, 131
Future, the. *See* Projections, review of global water
    resources; Soft path for water; Sustainable
    vision for the Earth's freshwater; Twenty-first
    century water-resources development

G
Gabon, **7:**133, 134–35, 321
Galileo spacecraft, **3:**217, 218
Gallup Organization, surveys. *See* Air quality,
    satisfaction; Water quality, satisfaction
Gambia, **4:**211, **7:**321
Gap, Inc., **5:**156, **8:**22
Gardens, **1:**23, **4:**122–23, **7:**115, 116, **8:**42
Garrison Diversion Project, **8:**140
Gases. *See* Greenhouse gases; Natural gas
Gaza Strip, **8:**154
Gaziantep Museum, **3:**186, 187
GEC Alsthom, **1:**85
GE Infrastructure, **5:**159–61
General circulation models (GCMs), **3:**121–23, **4:**158,
    159, 162, 167, **6:**41–42, 47
General Electric, **1:**85, **7:**133
Geophysical Research Letters, **1:**194
Georgia (country), **7:**321
Georgia (state). *See* Atlanta
Germany:
    conflict/cooperation concerning freshwater, **5:**5, 7
    dams
        Three Gorges Dam, **1:**88, 89
        World Commission on Dams, **3:**159, 170
    environmental concerns, top, **7:**287
    fossil-fuel production, **7:**75
    intl. river basin, **2:**29
    Lesotho Highlands project, **1:**96
    privatization, **3:**61
    terrorism, **5:**20
Ghana, **1:**46, 52, 54, **7:**321–22
Giardia, **1:**48, **2:**157, **4:**52, **7:**47, 57
Glen Canyon Institute, **2:**130
Glennon, Robert, **7:**xiii–xiv
Global action networks (GAN), **8:**9
Global Environmental Facility (GEF), **6:**51–52
Global Environmental Management Initiative (GEMI),
    7:33, 34
Global Environmental Outlook, **3:**88
Global Environment Monitoring System/Water
    Programme (GEMS/Water), **7:**50, 53
Globalization and international trade of water:
    Alaskan water shipments, proposed, **8:**133, 135–36
    business/industry, water risks, **5:**151
    defining terms
        commodification, **3:**35
        economic good, **3:**37–38
        globalization, **3:**34–35
        private/public goods, **3:**34
        privatization, **3:**35
        social good, **3:**36–37
        "virtual water," **8:**4, 83, 91
    General Agreement on Tariffs and Trade, **3:**48–51
    governance issues, **8:**6–16
    North American Free Trade Agreement, **3:**47–48,
        51–54, **8:**130
    overview, **3:**33–34, 41–42, **8:**4
    raw or value-added resource, **3:**42–47
    rules, intl. trading regimes, **3:**47–48
    social and economic good, water managed as
        both, **3:**38–40
    water footprint issues, **8:**84, 86, 88, 89, 90, 91
    World Water Forum (2003), **4:**192, 193
Global Reporting Initiative (GRI), **5:**158, **6:**18
    G3 Guidelines, **7:**41
    Sustainability Reporting Guidelines, **6:**28–29
    Water Protocol, **6:**28, 36
Global Runoff Data Centre, **8:**274, 292
Global Water Partnership (GWP), **1:**165–72, 175, 176,
    5:183, **6:**73
The Goddess of the Gorges, **1:**84
Goh Chok Tong, **1:**110
Goodland, Robert, **1:**77
Good manufacturing practice (GMP), **4:**32, 33
Good manufacturing practices, current (CGMP), **4:**27
Gorbachev, Mikhail, **1:**106
Gorton, Slade, **2:**134

Government/politics:
    business/industry water risk and, **7**:30
    desalination, **8**:101
    droughts, **5**:92, 94–95, 103, **7**:106–7, **8**:147–51
    environmental justice, **5**:138–41
    governance issues, **8**:6–16
    human right to water, **2**:3, **5**:137–38, **8**:29
    infrastructure damage due to political violence, **8**:164–65
    irrigation, **1**:8
    military targets (*See under* Conflict/cooperation concerning freshwater; Terrorism)
    military tools (*See under* Conflict/cooperation concerning freshwater)
    North American Water and Power Alliance issues, **8**:124–25, 131
    "policy capture," **8**:28–31
    policy reform, **7**:144, **8**:59
    privatization, **3**:68, **4**:60–73
    Red Sea-Dead Sea projects, **8**:154, 156
    subnational/intrastate *vs.* international, **8**:159, 161–62, 168–69
    subsidies, **1**:17, **7**:67, 108, 109, 111, 152, **8**:35
    twentieth-century water-resources development, **1**:7–8, 17
    World Commission on Dams report, **3**:170–71, **7**:139–40
    See also Climate change, California, policy; Conflict/cooperation concerning freshwater; Human right to water; Law/legal instruments/regulatory bodies; Legislation; Stocks and flows of freshwater; *specific countries*
Grains:
    cereal, **2**:64, **4**:299–301, **7**:109–10
    production, **2**:64, 299–301
    rice, **2**:74–79
    wheat, **2**:75, **4**:89
Grand Banks, **1**:77
Grand Canyon, **1**:15, **2**:138, 146
Granite State Artesian, **4**:39
Grants Pass Irrigation District (GPID), **2**:128
Great Lakes. *See under* Lakes, specific
Greece:
    ancient water systems, **2**:137
    bag technology, water, **1**:202, 204, 205, **8**:135
    conflict/cooperation concerning freshwater, **5**:5
    hydroelectric production, **1**:71
    supply systems, ancient, **1**:40
Greenhouse effect, **1**:137, 138, **3**:126, **4**:171, **6**:39. *See also* Climate change *listings;* Greenhouse gases
Greenhouse gases, **6**:40–43, 53, **7**:80, 84, **8**:85
Greenroof & Greenwall Projects Database, **8**:42
"Green" water, in water footprint concept, **8**:85–86, 90, 323, 332, 341, 350
Greywater (reclaimed water), **7**:120, **8**:38, 48–49
Gross national product (GNP) and water withdrawals, **3**:310–17
Groundwater:
    arsenic in, **2**:165–73, **3**:278–79, **4**:87, **6**:61, **7**:59
    as "blue water" in water footprint concept, **8**:86, 89–90, 323, 332, 341, 350
    climate change, **4**:170, **6**:43
    consequences of poor water quality, **4**:83, 87, **7**:55
    contamination by fossil-fuel production, **7**:51, 76, 79, **8**:66, 68, 69–71
    data problems, **3**:93
    desalination of brackish, **8**:114
    food needs for current/future populations, **2**:87
    fossil, **6**:9–10
    General Agreement on Tariffs and Trade, **3**:49–50
    hard path for meeting water-related needs, **3**:2
    monitoring/management problems
        agriculture, **4**:88–90
        analytical dilemma, **4**:90–97
        challenges in assessments, **4**:80–81
        conceptual foundations of assessments, **4**:80
        data and effective management, **4**:97–98
        extraction and use, **4**:81–88, **8**:2
        overview, **4**:79
    overextraction, **5**:125, 128, **7**:54, 55
        India, **8**:25
        Las Vegas (NV) area, **8**:137
        overview, **8**:2
        Syria, **8**:148–49
    Pacific Island developing countries, **3**:116–18
    pesticides, **5**:307
    privatization, **3**:77, **4**:60–61
    public ownership rights and privatization, **3**:74
    recharge, **8**:49
    reclaimed water, **2**:150–51, **8**:49
    reliability, water-supply, **5**:74, **7**:55
    and sinkhole formation, **8**:155
    stocks and flows of freshwater, **2**:20
    water quality, **4**:83, 87, **8**:114
    well-being, measuring water scarcity and, **3**:104
    See also Aquifers
Groupe DANONE, **7**:41
Guatemala, **3**:13, **8**:166
Guidelines for Drinking-Water Quality (WHO), **4**:26–27, 31
Guinea, **3**:76, **7**:322
Guinea worm. *See* Dracunculiasis
Gulf of California, **3**:141, 142
Gulf of Mexico, **1**:77, **7**:73
Guyana, **7**:323
Gwembe Tonga people, **5**:134

H
Habitat degradation/loss:
    from drought, **5**:98, 103, 109
    from fossil-fuel extraction/processing, **7**:85, 87
    wetlands, **7**:56, 63, **8**:4
Habitat restoration:
    Dead Sea, **8**:155–56
    ecosystems, **2**:149–50, **7**:66, 146, **8**:44–48
    rivers, **2**:xix, 127, **3**:143–44, **8**:45–46, 48
    Salton Sea, **6**:131–37, **8**:144
    streams, **8**:45–46
Habitat simulation, as environmental flow methodology, **5**:39
Haiti, **1**:46, **8**:359, 373
Halogen Occultation Experiment (HALOE), **1**:196
Hamidi, Ahmed Z., **1**:110
Harcourt, Mike, **1**:88
Hardness, measuring, **2**:309, **3**:327, **4**:334, **7**:350
Hard path for meeting water-related needs, **3**:xviii, 2, **6**:13–14. *See also* Soft path for water; Twentieth-century water-resources development
Harran, **3**:185
Harvard School of Public Health, **4**:9
    Global Burden of Disease assessment, **7**:270
Hazardous waste landfills, **5**:119, 124
Health:
    and high concentrations of metals, **7**:59–60
    and high concentrations of nutrients, **7**:58–59
    hunger and malnutrition, **2**:70, **6**:58, **7**:57–58
    maternal, **6**:58
    and persistent organic pollutants, **7**:48, 60
    sanitation issues (*See* Sanitation services)
    toxins (*See* Water quality, contaminants)
    water issues
        arsenic (*See* Groundwater, arsenic in)
        costs of poor water quality, **7**:63–64

desalination, **5:**74–75
diseases (*See* Diseases, water-related)
droughts, **5:**102
floods, **5:**109
fluoride, **4:**87
human needs for water, basic, **1:**42–47
privatization, **4:**47
reclaimed water, **2:**152–56
summary/conclusions, **1:**63–64
The Heat Is On (Gelbspan), **5:**136
Helmut Kaiser, **5:**161
Hepatitis, **7:**58
Herodotus, **1:**109
Herring, **2:**123
Historic flow as environmental flow methodology, **5:**39
Hittites, **3:**184
HIV, **6:**58, **7:**259–63
Hoecker, James, **2:**124
Holistic approaches to environmental flow methodology, **5:**39
Holland. *See* Netherlands
Holmberg, Johan, **1:**166, 175
Honduras, **1:**71, **4:**54–55
Hong Kong, **3:**46, 313–15
Hookworm, **7:**61, 272
Hoppa, Gregory, **3:**217
Human Development Report, **4:**7, **6:**74, **7:**61
Human rights and international law, **2:**4–9
Human right to adequate sanitation, **8:**271
Human right to water, **2:**1–2
    barriers to, **4:**212–13
    defining terms, **2:**9–13
    economic issues, **2:**13–14, **8:**29
    environmental flow, **5:**37
    and environmental justice, **5:**137–38
    failure to meet, consequences of the, **2:**14–15
    is there a right?, **2:**2–3
    laws/covenants/declarations, **2:**4–9, **7:**36, 251, **8:**268
    legal obligations, translating rights into, **2:**3, 13–14
    overview, **4:**207–8
    Prior Appropriation Doctrine, **5:**37
    progress toward acknowledging, **4:**208–11
    summary/conclusions, **2:**15
    why bother?, **4:**214
    See also Environmental justice; Law/legal instruments/regulatory bodies, International Covenant on Economic, Social, and Cultural Rights
Hungary, **1:**109, 120
Hunger. *See* Health, hunger and malnutrition
Hurricane Katrina, **5:**24, 110
Hydraulic geometry as environmental flow methodology, **5:**39
Hydroelectric production:
    California, **4:**173–74
    capacity, countries with largest installed, **1:**72, 276–80, **7:**129
    Central Asia, **8:**166
    China, **6:**92
    Colorado River, **4:**165
    dams, removing/decommissioning, **1:**83
    Dead Sea, proposed, **8:**154, 156
    electricity generation data, **7:**73–74, 130
    Glen Canyon Dam, **2:**129–30
    grandiose water-transfer schemes, **1:**74–75
    James Bay Project, Quebec, **8:**131
    North American Water and Power Alliance, **8:**128
    percentage of electricity generated with hydropower, **1:**73–74
    by region, **1:**70–71
    Snake River, **2:**132–33, **8:**129
Southeastern Anatolia Project, **3:**182
Syria, **8:**149
Three Gorges Dam, **1:**84, **6:**140
transboundary water agreements, **7:**6, 132
water consumption (water footprint), **7:**25, **8:**84
well-being, measuring water scarcity and, **3:**103
Hydrogen sulfide, **7:**76, 80
Hydrologic cycle:
    climate change, **1:**139–43, **4:**183, **5:**117
    desalination, **2:**95, **5:**52
    droughts, **5:**94
    quantifications, accurate, **4:**92–96
    stocks and flows of freshwater, **2:**20–27
    See also Environmental flow
Hydrologic extremes, **6:**43–44
Hydro-Quebec International, **1:**85

I

Iceland, **1:**71
Idaho Rivers United, **2:**133
Identity standards and bottled water, **4:**27–31
Illinois, Cache River basin, **8:**46, 48
India:
    agriculture, **4:**88, 89, **8:**10
        irrigation, **2:**85, 86
    basic water requirement, **2:**13
    bottled water, **4:**22, 25, 40
    business/industry, water risks that face, **5:**146, 147, 165
    Chipko movement, **5:**124
    cholera, **1:**61
    conflict/cooperation concerning freshwater, **1:**107, 109, 118–19, 206–9, **5:**13, 15
        Cauvery River Basin, **7:**3
        PepsiCo and Coca-Cola, **7:**26
    conflicts concerning freshwater 2011-2012, **8:**163
    dams, **1:**70, 78, 81, **5:**133
        displaced people due to, **1:**78
        Kabini, **8:**163
        Krishna Raja Sagar, **8:**163
        World Commission on Dams, **3:**159, 170, **7:**139
        Wullar, and conflict with Pakistan, **8:**163
    Dhaka Community Hospital, **2:**170
    dracunculiasis, **1:**53, 55
    economics of water projects, **1:**16, 17
    environmental concerns, top, **7:**287, 288
    environmental justice, **5:**124, 127
    floods, **5:**106
    fossil-fuel production, **7:**75, 88, 89
    groundwater, **3:**2, 50, **4:**82, 83, 88–90, 92–95, **5:**125, 128
        arsenic in, **2:**165–73, **4:**87, **7:**59
        and Coca-Cola, **8:**21–22, 23, 25
        overextraction, **7:**55, **8:**25
    human right to water, **4:**211
    hydroelectric production, **7:**5, 129
    industrial water use, **1:**21
    intl. river basin, **2:**27, **8:**163
    New Delhi, **8:**163
    renewable water availability in, **6:**83
    sanitation services, **5:**128
    Tamil Nadu state, **8:**163
    water quality, **8:**3
    water use, domestic, **1:**46
Indian Ocean Dipole, **7:**98
Indicators/indices, water-related, **3:**87. *See also* Well-being, measuring water scarcity and
Indigenous populations, **5:**123, 124, **7:**91–92. *See also* Environmental justice
    Africa, **8:**167
    Brazil, **8:**165
    Canada, **8:**128

Indonesia:
    bottled water, **5:**170
    cholera, **1:**58
    climate change, **1:**147
    conflicts concerning freshwater, **8:**168
    dams with Chinese financiers/developers/builders, **7:**323
    fossil-fuel production, **7:**75
    General Agreement on Tariffs and Trade, **3:**49
    human needs, basic, **1:**46
    pricing, water, **1:**25, **3:**69
Industrial sculptures, **5:**219, 220
Industrial water treatment, **5:**160
Industrial water use, **1:**20–21, **5:**124–25. *See also* Business/industry, water risks; Projections, review of global water resources; Water conservation, California commercial/industrial water use; Water footprint, per capita, of national consumption, by sector and country
Infrared Space Observatory, **3:**220
Insects:
    extinct or extinct in the wild species, **7:**293–94
    stone fly, **7:**56
    as vectors for water-related diseases, **1:**49–50, **4:**8–9
Institute of Marine Aerodynamics, **1:**202
Integrated water planning, **1:**17, **3:**21, **8:**59. *See also* Global Water Partnership
Integrated water resource management (IWRM), **8:**27
Intel, **8:**28
Intensity, water, **3:**17–19
Inter-American Development Bank, **3:**163
Interferometry, **3:**221
International alliances/conferences/meetings, time to rethink large, **5:**182–85. *See also* Law/legal instruments/regulatory bodies
International Association of Hydrological Sciences (IAHS), **5:**183
International Bottled Water Association (IBWA), **4:**26, 34, **5:**174
International Council of Bottled Water Association (ICBWA), **4:**26
International Drinking Water Supply and Sanitation Decade (1981-90), **3:**37
International Food Policy Research Institute (IFPRI), **2:**64
International Freshwater Conference in Bonn (2001), **3:**xviii
International Hydrological Program (IHP), **5:**183
International Law Association (ILA), **5:**35, **7:**4
International Law Commission, **1:**107
International Maize and Wheat Improvement Center, **2:**75
International river basins:
    Africa, **6:**289–96, **7:**11–13, **8:**164
    Asia, **2:**30, **6:**289, 296–301
    assessments, **2:**27–35, **7:**2–3
    Central Asia, **8:**166
    climate change and management issues, **7:**2–10, **8:**15
    by country, **2:**247–54, **6:**289–311
    Europe, **6:**289, 301–06
    fraction of a country's area in, **2:**239–46
    geopolitics, **2:**35–36, **8:**15
    North America, **6:**289, 306–08, **7:**165–69 (*See also* Colorado River)
    runoff (*See* Rivers, runoff)
    South America, **6:**289, 308–11
    use of water footprint data for management, **8:**91
    of the world, **2:**219–38
International Rivers Network, **7:**308
International trade. *See* Globalization and international trade of water

International Union for Conservation of Nature (IUCN), **7:**292–302. *See also* World Conservation Union
International Water Association (IWA), **5:**182
International Water Ltd., **3:**70
International Water Management Institute (IWMI), **3:**197, **4:**88, 108, **8:**2
International Water Resources Association (IWRA), **1:**172, **5:**183, **7:**19
Internet, **1:**231–34, **2:**192–96, **3:**225–35. *See also* Websites, water-related
    increased conflict reporting due to, **8:**159
    WikiLeaks and conflict transparency, **8:**164
Introduced/invasive species, **7:**48
Invertebrates:
    clams, **3:**142–43
    crayfish, **7:**56
    effects of acid rain, **7:**87
    extinct or extinct in the wild species, **7:**292, 298–302
    mercury in, **7:**59
    mussels, **7:**56
    shrimp, **3:**49, 50, 141, 142
Iran, **1:**58, **5:**8
    conflict concerning freshwater, **8:**163
    dams with Chinese financiers/developers/builders, **7:**323
    fossil-fuel production, **7:**75, 79
Iraq, **1:**59, 110–11, 118, **5:**13, 15–16
Irrigation. *See* Agriculture, irrigation; Gardens; Lawns
ISO 14001, **6:**32
Israel:
    conflict/cooperation concerning freshwater, **1:**107, 109, 110–11, 115–16, **5:**6, 7, 10, 14–15
    conflicts concerning freshwater, 2011-2012, **8:**165
    desalination, **5:**51, 69, 71, 72, **8:**157
    drip irrigation, **1:**23
    environmental flow, **5:**33
    globalization and intl. trade of water, **3:**45
    reclaimed water, **1:**25, 29, **2:**138, 142
    Red Sea-Dead Sea projects, **8:**144, 153–57
    terrorism, **5:**21
    tourism, **8:**155
    well-being, measuring water scarcity and, **3:**98
Italy, **3:**47, 61, **5:**11

J
Japan:
    conflict/cooperation concerning freshwater, **5:**5
    dracunculiasis, **1:**52, 53
    environmental flow, **5:**42
    industrial water use, **1:**20
    reclaimed water, **2:**139, 140, 158–59
    soft path for meeting water-related needs, **3:**23
    World Commission on Dams, **3:**159
Jarboe, James F., **5:**4
Jefferson, Thomas, **2:**94
Jerusalem Post, **5:**71
Jobs. *See* Employees; Employment
Joint Monitoring Programme (WHO), **6:**60, 73, **7:**230, 241, **8:**236, 252
Jolly, Richard, **2:**3, **4:**196
Jordan, **1:**107, 109, 115–16, **2:**33, **5:**12, 33
    Disi Water Conveyance Project, **8:**157
    Red Sea-Dead Sea projects, **8:**144, 153–57
JPMorgan, **7:**27, 134
Jupiter, search for water on moons, **3:**217–18

K
Kansas:
    Chanute, **2:**152
    Missouri River diversion project, **8:**140–44
Kantor, Mickey, **3:**51–52

Kazakhstan, **7:**75, 323
Kennebec Hydro Developers, **2:**124
Kennedy, John F., **2:**95
Kenya:
    conflicts concerning freshwater, 2011-2012, **8:**167
    dams
        with Chinese financiers/developers/builders, **7:**323–24
        effects of Lake Turkana dam, **7:**134
    dracunculiasis, **1:**53, 55
    droughts, **5:**92, 99
    environmental concerns, top, **7:**287
    fog collection as a source of water, **2:**175
    food needs for current/future populations, **2:**76–77
    Lake Naivasha, **8:**26, 27
    Nairobi, **8:**167
    Nile River Basin, **7:**11 (*See also* Rivers, specific, Nile)
    public-private strategies for shared risk, **8:**26, 27
    sanitation services, **1:**42
Kerogen, **7:**79
Khan, Akhtar H., **1:**39, **4:**71
King, Angus, **2:**123
Kiribati, **3:**118
Kirin, **6:**27
Kitchens and CII water use, **4:**135, 137
Knowledge transfer, **8:**10–11
Kokh, Peter, **3:**218
Korea, **7:**274
Korean peninsula, **1:**53, 109–10
Korean War, **1:**110
Kosovo, **5:**9
Kruger National Park, **1:**120–23, **5:**32
Kurdish Workers' Party (PKK), **5:**22
Kuwait, **1:**111, **2:**94, 97, **4:**18, **5:**69, 160, 163
    fossil-fuel production, **7:**75
Kyoto Protocol, **6:**51
Kyrgyzstan, **7:**324, **8:**166

L
Labeling and bottled water, **4:**28–31, **7:**159, 160–61, 164
Lagash-Umma border dispute, **5:**5
Lakes, **4:**168–69, **7:**55
Lakes, specific:
    Cahuilla, **6:**129
    Chad, **1:**111, 148
    Chapala, **3:**77
    Dead Sea, **8:**153–57
    Great Lakes, **1:**111, **3:**50, **7:**165–69, **8:**126, 128
    Kostonjärvi, **5:**33
    Mead, **3:**137, 140, **7:**8–9, 17, **8:**137, 138
    Mono, **5:**37, 41
    Naivasha, **8:**26, 27
    Oulujärvi, **5:**33
    Powell, **1:**76, **2:**129, 130, **7:**8–9, 17
    Taihu, **6:**95
    Turkana, **7:**134
Land, agricultural. *See* Agriculture, land availability/quality
Land conservation, **8:**42
Landfills, hazardous waste, **5:**119, 124
Landscape design, **1:**23, **4:**122–23, 135, 137–38, **6:**107
Land-use management and floods, **5:**111–12
La Niña, **3:**119, **5:**95, **7:**98
Lao People's Democratic Republic, **7:**14, 324–27
Laos, **1:**16, 71, **7:**130
La Paz-El Alto, **3:**68, 69–72
Laser leveling, agriculture and, **3:**20
Las Vegas (NV):
    description, **6:**103
    Las Vegas Pipeline Project, **8:**136–40
    per-capita water demand, **6:**104–6
    population growth, **6:**103
    precipitation, **6:**104
    temperature, **6:**104
    wastewater rate structure, **6:**118–19
    water conservation, **6:**107–8, 110–12
    water rate structures, **6:**115–17
    and water supply reliability, **5:**74
    water-use efficiency, **6:**107–8
Latin America:
    bottled water, **4:**40
    cholera, **1:**56, 57, 59–61, **3:**2, **8:**359, 373
    climate change, **1:**147
    conflicts concerning freshwater, 2011-2012, **8:**165–66
    dams, **1:**77, 81
    drinking water, **1:**262, **6:**65
        access to, **7:**24, 253, **8:**270
    human needs, basic, **1:**47
    hydroelectric production, **1:**71
    irrigation, **2:**86
    population, **2:**214
    sanitation, **1:**264, **3:**271, **5:**259
        progress on access to, **7:**256, **8:**273
    water quality, satisfaction by country, **7:**290–91
    See also Central America; South America
Laundry:
    emerging technologies, **5:**219, 220
    The High Efficiency Laundry Metering and Marketing Analysis project (THELMA), **4:**115
    laundry water and CII water use, **4:**138
    washing machines, **1:**23, **4:**114–16, **5:**219, 220
Lavelin International, **1:**85
Law/legal instruments/regulatory bodies:
    Agenda 21, **5:**34
    Agreement on Technical Barriers to Trade (TBT), **4:**35
    Agreement on the Application of Sanitary and Phytosanitary Measures (SPS), **4:**35
    Appalachian Regional Commission, **7:**307
    Beijing Platform of Action, **2:**8
    Berlin Conference Report (2004), **5:**35
    Berlin Rules (2004), **7:**5
    Bonn Declaration (2001), **3:**173, 178–80
    Boundary Waters Treaty (1909), **7:**167, 168
    Budapest Treaty, **7:**5
    Bureau of Government Research (BGR), **4:**69–70
    Cairo Programme of Action, **2:**8
    California Bay-Delta Authority, **4:**181
    California Coastal Commission (CCC), **5:**2
    California Department of Water Resources, **1:**9, 29, **3:**11, **4:**170, 232
    California Energy Commission, **4:**157, 176, 232, **5:**76, 232
    Climate and Water Panel, **1:**149
    Climate Change and Water Intra-governmental Panel, **7:**153
    Code of Federal Regulations (CFR), **4:**31–32
    Codex Alimentarius Commission (CAC), **4:**26, 35–36
    Colorado River, **3:**137–39
    Consortium for Energy Efficiency (CEE), **4:**115
    Consultative Group on International Agricultural Research (CGIAR), **3:**90
    Convention of the Rights of the Child (1989), **2:**4, 9, **4:**209
    Convention on Biological Diversity (CBD), **3:**166, **5:**34
    Copenhagen Declaration, **2:**8
    Corporate Industrial Water Management Group, **5:**157
    Declaration on the Right to Development (1986), **2:**4

Law/legal instruments/regulatory bodies (*continued*)
   Dublin Conference (1992), **1**:24, 165–66, 169, **3**:37, 58, 101, **5**:34
   Earth Summit (1992), **3**:38, 88, 101, **5**:137
   Emergency Management and Emergency Preparedness Office, **5**:24
   environmental flow, **5**:34–37
   Environmental Modification Convention (1977), **1**:114, **5**:4
   European Convention on Human Rights (1950), **2**:4, 8
   European Union Water Framework Directive (2000), **7**:25, 147–49
   Federal Bureau of Investigation (FBI), **5**:24
   Federal Emergency Management Agency (FEMA), **5**:24, 96–97
   Federal Energy Regulatory Commission (FERC), **1**:83, **2**:123–24, 126, **5**:36
   Federal Maritime Commission, **7**:307
   First National People of Color Environmental Leadership Summit (1991), **5**:120
   Food and Agricultural Organization (*See under* United Nations)
   Ganges Water Agreement (1977), **1**:119
   General Agreement on Tariffs and Trade, **3**:47–52
   Geneva Conventions, **1**:114, **5**:4
   Global Water Partnership (GWP), **1**:165–72, 175, 176, **5**:183
   Great Lakes–St. Lawrence River Basin Sustainable Water Resources Agreement (2005), **7**:165, 167–68
   Great Lakes–St. Lawrence River Basin Water Resources Compact (2008), **7**:165, 167–69
   Great Lakes–St. Lawrence River Basin Water Resources Council, **7**:168
   groundwater, **4**:95–96
   Hague Declaration (2000), **3**:173–77, **4**:2, **5**:139, 140
   Harmon Doctrine, **7**:4
   Helsinki Rules (1966), **1**:114, **7**:4
   human rights and intl. law, **2**:4–9
   India-Bangladesh, **1**:107, 119, 206–9
   Indus Water Treaty, **8**:163
   Interagency Climate Change Adaptation Task Force, **7**:153
   intergovernmental, limitations and recommendations, **8**:7–8
   Intergovernmental Panel on Climate Change (IPCC), **1**:137, 138, 140, 145, 149, **3**:120–23, **5**:81, 136, **6**:39–40, 45
     Fourth Assessment Report, **7**:7
     2008 report, **8**:5
   International Boundary and Water Commission, **7**:17, 307
   International Commission on Irrigation and Drainage (ICID), **5**:183
   International Commission on Large Dams (ICOLD), **1**:70
   International Court of Justice, **7**:5, 7
   International Covenant on Economic, Social, and Cultural Rights, **2**:4, **4**:208
     actors other than states, obligations of, **4**:231
     Article 2 (1), **2**:6–7
     Article 11, **2**:7
     Article 12, **2**:7
     Declaration on the Right to Development, **2**:9
     implementation at national level, **4**:228–30
     introduction, **4**:216–18
     normative content of the right of water, **4**:218–21
     special topics of broad application, **4**:220–21
     states parties' obligations, **4**:222–26
     violations, **4**:226–28
   International Joint Commission, **3**:50, **7**:167, 168, 307
   intl. law, role of, **1**:114–15
   intl. waters, **5**:182–85, **7**:9–10, 165–69
   Israel-Jordan Peace Treaty (1994), **1**:107, 115–16
   Joint Declaration to Enhance Cooperation in the Colorado River Delta, **3**:144
   Kyoto Protocol, **5**:137
   Kyoto Third World Water Forum (2003), **5**:183
   Mar del Plata Conference (1977), **1**:40, 42, **2**:8, 10, 47, **4**:209, **5**:183, 185
   Massachusetts Water Resources Authority, **3**:20
   Mekong River Basin Agreement (1995), **7**:13
   Mekong River Commission (MRC), **1**:82, **3**:165, **5**:35, **7**:13, 15
   Minute 306, **3**:144–45
   Multilateral Working Group on Water Resources, **8**:154
   National Rainwater and Graywater Initiative, **7**:120
   National Water Commission (Australia), **7**:107, 113, 118
   National Water Commission (U.S.), **7**:152
   Natural Resources Council of Maine, **2**:123
   Nile Basin Initiative, **7**:12
   Nile River Basin Commission, **7**:12
   Nile Waters Treaty (1959), **7**:11–13
   Non-Navigational Uses of International Watercourses (*See* Convention of the Law of the Non-Navigational Uses of International Watercourses)
   North American Free Trade Agreement (NAFTA), **3**:47–48, 51–54, **8**:130
   North American Water and Power Alliance (NAWAPA), **1**:74, **8**:124–31
   OECD (*See* Organisation for Economic Cooperation and Development)
   Okavango River Basin Commission (OKACOM), **1**:122, 124
   Organisation for African Unity (OAU), **1**:120
   overview, **1**:155, **7**:66–67
   Ramsar Convention, **3**:166, **5**:34, **7**:110
   Russian Federation Water Code, **7**:149
   Secretariat, Global Water Partnership's, **1**:170–71
   Snake River Dam Removal Economics Working Group, **2**:132
   South African Department of Water Affairs and Forestry, **1**:96
   South Asian Association for Regional Cooperation (SAARC), **1**:118
   Southern Africa Development Community (SADC), **1**:156–58, 169, **3**:165
   Southern Nevada Water Authority, **8**:136–40, 141
   Southwest Florida Water Management District (SWFWMD), **2**:108, **5**:62
   Stockholm Convention on Persistent Organic Pollutants, **7**:60
   Surface Transportation Board, **7**:307
   Surface Water Treatment Rule (SWTR), **4**:52
   Swedish International Development Agency (SIDA), **1**:165, 170, 171, **2**:14, **3**:162
   Third World Centre for Water Management, **5**:139
   Upper Occoquan Sewage Authority, **2**:152
   U.S. Agency for International Development (USAID), **1**:44, **2**:10, 14, **6**:51, **7**:306, **8**:7
   U.S. Bureau of Land Management (BLM), **8**:138
   U.S. Bureau of Reclamation (BoR), **1**:7, 69, 88, **2**:128, **3**:137, **4**:10, **6**:135
     Colorado River Basin policy, **7**:17, **8**:140
     Mekong River Basin policy, **7**:13
     North American Water and Power Alliance, **8**:128
     Reber Plan, **8**:133
     water policy reform, **7**:152
   U.S. Congress, **7**:307
   U.S. Department of Agriculture, **7**:305

U.S. Department of Commerce, **7:**305
U.S. Department of Defense, **7:**305
U.S. Department of Energy, **1:**23, **4:**115
U.S. Department of Health and Human Services, **5:**24
U.S. Department of Homeland Security, **5:**23, **7:**305
U.S. Department of Housing and Urban Development, **4:**117, **7:**305
U.S. Department of Interior, **2:**123, 127, **7:**305–6, **8:**69
U.S. Department of Justice, **7:**306
U.S. Department of Labor, **7:**306
U.S. Department of State, **7:**306
U.S. Department of the President, **7:**306
U.S. Department of Transportation, **7:**306
U.S. Environmental Protection Agency (EPA), **2:**123, 152, **4:**37, 52, **5:**23, 24
   infrastructure reports, **8:**36, 39
   reclaimed water reuse, **8:**49
   water-related budget, **7:**307
   water withdrawals for natural gas extraction, **8:**67
   well monitoring, **8:**73
U.S. Fish and Wildlife Service, **2:**126, 128, **8:**46
U.S. Food and Drug Administration (FDA), **4:**26, 37, 40, **5:**171
   water-related budget, **7:**305
U.S. Mexico Treaty on the Utilization of the Colorado and Tijuana Rivers, **3:**138
U.S. National Park Service, **2:**117, 127
U.S. National Primary Drinking Water Regulation (NPDWR), **3:**280–88
Vienna Declaration, **2:**8
Water Aid and Water for People, **2:**14
Water Environment Federation (WEF), **3:**78, **4:**62, **5:**182–83, **8:**10
Water Law Review Conference (1996), **1:**161
Water Sentinel Initiative, **5:**23
Water Supply and Sanitation Collaborative Council (WSSCC), **2:**3, 13–14, **5:**138, **7:**230, 241, **8:**236, 252
See also See also American *listings;* Bottled water, U.S. federal regulations; International *listings;* Environmental justice; National *listings;* World *listings*
Lawns, **1:**23, **4:**122–23, **7:**115, 116
Law of Conservation of Energy, **6:**7
Lead, **7:**59
League of Nations, **8:**359, 373
Leak rates, **4:**109, 117–18
Leases and environmental flows, **5:**42
Leasing contracts, **3:**66, **5:**42
Least Developed Countries Fund (LDCF), **6:**51–52
Lebanon, **1:**115
Lechwe, Kafue, **5:**32
Lecornu, Jacques, **1:**174
Legislation:
   Australia. Water Efficiency Labelling and Standards Act, **7:**28, 118–19
   California
      Central Valley Project Improvement Act, **5:**36, **7:**152
      Coastal Act, **5:**80
      Water Conservation in Landscaping Act of 1990, **4:**120–21
   Israel. Water Law of 1959, **5:**35
   Japan. River Law of 1997, **5:**35
   South Africa
      Act 54 of 1956, **1:**93, 160
      Apartheid Equal Rights Amendment (ERA), **1:**158–59
      National Water Act of 1998, **7:**145
      National Water Law of 1998, **5:**35, 37
   Switzerland. Water Protection Act of 1991, **5:**35
   U.S.
      American Recovery and Reinvestment Act of 2009, **8:**35
      Bioterrorism Act of 2002, **5:**23
      Clean Air Act, **4:**68
      Clean Water Act of 1972, **1:**15, **4:**68, **5:**36, **7:**24, 144, **8:**77
      Electric Consumers Protection Act, **5:**36
      Elwha River Ecosystem and Fisheries Restoration Act of 1992, **2:**127
      Endangered Species Act of 1973, **1:**15, **5:**37
      Federal Energy Policy Act of 1992, **1:**21, **7:**28
      Federal Food, Drug, and Cosmetic Act, **4:**27
      Federal Power Act, **5:**36
      Federal Reclamation Act of 1902, **1:**8
      Federal Wild and Scenic Rivers Act of 1968, **1:**15, **8:**129
      Flood Control Act of 1936, **1:**16
      National Environmental Policy Act of 1969/1970, **2:**130, **5:**36, **8:**129
      National Water Commission Act, **7:**143–44
      National Wild and Scenic Rivers Act of 1997, **2:**120, **5:**36
      Nutrition Labeling and Education Act of 1990, **4:**28
      Public Health, Security, and Bioterrorism Preparedness and Response Act of 2002, **5:**23
      Safe Drinking Water Act of 1974, **1:**15, **3:**280, **4:**27
      Saline Water Conversion Act of 1952, **2:**94, 95
      Secure Water Act of 2009, **7:**152
      Water Desalination Act, **2:**95
      Water Resources Development Act, **7:**166, 167
Le Moigne, Guy, **1:**172, 174
Lesotho, **3:**159–60
Lesotho Highlands project:
   chronology of events, **1:**100
   components of, **1:**93, 95
   displaced people, **1:**97–98
   economic issues, **1:**16
   financing the, **1:**95–97
   impacts of the, **1:**97–99
   Kingdom of Lesotho, geographical characteristics of, **1:**93, 94
   Lesotho Highlands Development Authority, **1:**96, 98
   management team, **1:**93
   opposition to, **1:**81, 98, 99
   update, project, **1:**99–101
Levees and flood management, **5:**111
Levi Strauss, **5:**156
Li Bai, **1:**84
Liberia, **1:**63, **7:**264
Libya, **8:**164
Licenses for hydropower dams, **2:**114–15, 123–24
Life cycle assessment (LCA), **7:**32–34, 158
Linnaeus, **1:**51
Living with Water (Netherlands), **6:**49
Lovins, Amory B., **3:**xiii–xiv
Low-energy precision application (LEPA), **1:**23
Lunar Prospector spacecraft, **1:**197, **3:**213

M
Macedonia, **1:**71, **5:**10
Machiavelli, **1:**109
Madagascar, **7:**327
Madres del Este de Los Angeles Santa Isabel, **8:**39
Malaria, **1:**49–50, **6:**58, **7:**259–63
Malawi, **4:**22

Malaysia:
- conflict/cooperation concerning freshwater, **1:**110
- dams with Chinese financiers/developers/builders, **7:**327–28
- data, strict access to water, **2:**41
- disputes with Singapore, **1:**22
- economics of water projects, **1:**16
- floods, **5:**106
- globalization and intl. trade of water, **3:**46
- hydroelectric production, **1:**71
- prices, water, **3:**69
- privatization, **3:**61

Maldives, **5:**106, **7:**264
Mali, **1:**52–53, 55, **7:**328, **8:**167
Mallorca, **3:**45, 46
Malnutrition. *See* Health, hunger and malnutrition
Mammals:
- dolphins, **1:**77, 90, **3:**49, 50, **7:**56
- extinct or extinct in the wild species, **7:**297–98

Manila Water Company, **4:**46
Mao Tse-tung, **1:**85
Mariner 4, **5:**175, 177
Marion Pepsi-Cola Bottling Co., **4:**38
Mars, water on:
- exploration, **3:**214–17
- future Mars missions, **5:**180
- history, **5:**178–80
- instrumental analyses, **5:**178
- missions to Mars, **5:**175–78
- overview, **5:**175
- visual evidence of, **5:**177–78

Mars Climate Orbiter, **2:**300
Mars Express, **5:**178
Mars Global Surveyor (MGS), **3:**214, **5:**177
Marshall Islands, **3:**118, **5:**136
Mars Odyssey, **3:**xx
Mars Orbital Camera (MOC), **3:**215, **5:**178
Mars Reconnaissance Orbiter, **5:**180
Maryland, Montgomery County, **8:**44
Mass, measuring, **2:**304, **7:**345
Mauritania, **1:**55, **4:**82, **8:**167
Maximum available savings (MAS), **3:**18, **4:**105
Maximum cost-effective savings (MCES), **3:**18, 24, **4:**105
Maximum practical savings (MPS), **3:**18, 24, **4:**105
Maytag Corporation, **1:**23
McDonald's, **6:**24
McKernan, John, **2:**123
McPhee, John, **2:**113
Measurements, water, **2:**25, 300–309, **3:**318–27, **4:**325–34. *See also* Assessments; Well-being, measuring water scarcity and
Media. *See* Books; Films
Mediterranean Region:
- climate change and drought, **8:**150
- Eastern, mortality rate, under-5, **7:**261

Mediterranean Sea, **1:**75, 77, **8:**153–57
Medusa Corporation, **1:**204
Meetings/conferences, international. *See* International *listings*; Law/legal instruments/regulatory bodies; United Nations; World *listings*
Meningitis, **7:**47
Merck, **6:**27
Mercury:
- as a contaminant from energy production, **7:**50, 52
- and fossil fuel production, **7:**82, 83, 84, 88, 91
- health impacts, **7:**59
- in measurement of pressure, **7:**348
- terrorism, water contamination with, **7:**194

Metals, as contaminants, **4:**168
- ecological effects, **7:**47–48
- fossil-fuel production, **7:**76, 79, 86, **8:**72
- health effects, **7:**59 60
- industrial wastewater, **7:**50
- mining, **7:**52
- road runoff, **7:**53

Meteorites, water-bearing, **3:**210–12, 216
Methane, **1:**138, 139, **7:**80, 81–82, 89
- coal bed methane, **8:**63
- contamination of drinking water, **8:**69, 70–71, 73

Methemoglobinemia (blue-baby syndrome), **7:**58
Mexico:
- bottled water, **4:**40, **5:**170
- Colorado River, **3:**134, 137, 138, 141, 144–45
  - intl. agreements, **7:**6, 8–9, 15–16
- environmental concerns, top, **7:**287, 288
- environmental flow, **5:**37
- fossil-fuel production, **7:**75
- groundwater, **4:**82, 83, 96, **5:**125
  - arsenic in, **7:**59
- hydroelectric production, **1:**71
- irrigation, **3:**289, **7:**16
- monitoring efforts, **3:**76–77
- North American Free Trade Agreement, **3:**47–48, 51–54, **8:**130
- North American Water and Power Alliance, **8:**124–31
- privatization, **3:**60
- sanitation services, **3:**272
- surface water, effects of climate change, **6:**43
- water-use efficiency, improving, **1:**19

Michigan, Watervliet dam removal, **8:**48
Middle East:
- air quality, satisfaction, 2012, **8:**391
- bottled water, **4:**288, 289, 291, **5:**281, 283
- climate change and drought, **8:**149–51
- conflict/cooperation concerning freshwater, **1:**107–18, **2:**182, **5:**15
- conflicts concerning freshwater
  - effects of Syrian conflict, **8:**147–51, 154
  - 2011-2012, **8:**163–64
- desalination, **2:**94, 97, **5:**54, 55, 57, 58, 68–69
- dracunculiasis, **1:**272–73, **3:**274–75, **5:**295, 296
- environmental flow, **5:**33
- groundwater, **4:**82, 85, **5:**125
- irrigation, **2:**87
- reclaimed water, **1:**28, **2:**139
- water quality, satisfaction
  - by country, **7:**290–91
  - 2012, **8:**391
- *See also* Mediterranean Region, Eastern; Southeastern Anatolia Project; *specific countries*

Military targets. *See under* Conflict/cooperation concerning freshwater; Terrorism
Military tools. *See under* Conflict/cooperation concerning freshwater; Terrorism
Millennium Development Goals (MDGs), **6:**57–78
- commitments to achieving the goals, **4:**2, 6–7
- creation of, **6:**57
- diseases, water-related
  - classes of, four, **4:**8–9
  - future deaths from, **4:**10, 12–13
  - measures of illness/death, **4:**9–11
  - mortality from, **4:**9–10, **6:**58, 73
  - overview, **4:**7–8
- drinking water, access to
  - baseline conditions, **6:**62
  - description of, **6:**58
  - Ethiopia, **6:**71–73
  - goals, **6:**211, **7:**230, **8:**236
  - limitations in data and reporting, **6:**61–62, **7:**231, 251–52, **8:**237–38, 268–69
  - need for, **6:**63
  - population growth effects on, **6:**63

Index 459

progress by region, **6:**65–67, 230–32, **7:**251–53, **8:**268–70
progress on, **6:**62–70
targets for, **4:**2–5, **6:**62
economic return on meeting, **7:**64
funding of, **6:**73
future of, **6:**73–75
overview, **4:**xv, 1, **6:**57, **8:**3–4
progress measurements, **6:**60–62
projections for meeting, **4:**13–14
sanitation
    baseline conditions, **6:**62
    description of, **6:**58
    Ethiopia, **6:**71–73
    limitations in data and reporting, **6:**61–62, **7:**242, 254–55, **8:**253, 271–72
    need for, **6:**63
    population growth effects on, **6:**63
    prioritizing of, **6:**75
    progress by region, **3:**270–72, **5:**256–61, **6:**67–70, 233–35, **7:**254–56, **8:**271–73
    targets for, **4:**2–5, **6:**62
    in urban areas, **6:**70–71
summary/conclusions, **4:**14, **6:**77
targets for, **6:**57–58, 233
technology improvements, **6:**60
within-country disparities, **6:**77
Minas Conga, **8:**166
Mineral extraction, from evaporation ponds, **8:**155, 156
Mineral water, **4:**29
Mining:
    copper, **8:**166
    fossil fuels, **6:**33, **7:**73–74, 83, 85–86, 89
    gold, **8:**166
    processes, **7:**73–74
    protests and violence, Latin America, **8:**166
    wastewater, **8:**74
    water footprint, **7:**31, **8:**84
    See also Petroleum and fossil fuels
Ministerial statements/declarations at global water conferences, **5:**184–85. *See also* Law/legal instruments/regulatory bodies
Minoan civilization, **1:**40, **2:**137
Missouri River diversion project, **8:**140–44
Mohamad, Mahathir, **1:**110
Mokaba, Peter, **1:**123
Moldavia, **5:**8
Mongolia, **7:**328
Monitoring:
    drought, joint intl., **5:**99–100, **7:**20
    environmental, of desalination plants, **8:**94
    and privatization, **3:**75–77, 81–82, **4:**59–60, 62–65 (*See also* Groundwater, monitoring/management problems)
    recommendations, **8:**12
    water quality, technology, **8:**49
The Monkey Wrench Gang (Abbey), **2:**129, **5:**14
Monterey County (CA), **2:**151
Moon, search for water on the, **3:**212–14
Morocco, **7:**328
Mortality:
    childhood (*See* Childhood mortality)
    from cholera (*See* Cholera, deaths reported to the WHO, by country)
    due to lack of water (*See* Drinking water, access, deaths due to lack)
    from water-related disease (*See* Diseases, water-related, death)
Mothers of East Los Angeles, **1:**22
Mount Pelion, **4:**39
Movies. *See* Films
Mozambique, **1:**63, 119–21, **5:**7, **7:**328–29

Mueller, Robert, **5:**4
Muir, John, **1:**80–81
Municipal water, **1:**29, **4:**29, **5:**73, **6:**101
    infrastructure projects, project delivery, **8:**104
    pipeline from desalination plant to, **8:**115
Myanmar. *See* Burma (Myanmar)

N
Nalco, **3:**61
Namibia:
    conflict/cooperation concerning freshwater, **1:**119, 122–24
    fog collection as a source of water, **2:**175
    Lesotho Highlands project, **1:**98–99
    reclaimed water, **1:**28, **2:**152, 156–58
Narmada Project, **1:**17
National Academy of Sciences, **1:**28, **2:**155, 166
National adaptation programs of action (NAPAs), **6:**48–49
National Aeronautics and Space Administration (NASA), **5:**96, 178. *See also* Outer space, search for water in
National Arsenic Mitigation Information Centre, **2:**172
National Council of Women of Canada, **3:**78, **4:**62
National Drought Policy Commission, **5:**95
National Environmental Protection Agency of China, **1:**92
National Fish and Wildlife Foundation, **2:**124
National Geographic, **3:**89
National Institute of Preventative and Social Medicine, **2:**167
National Marine Fisheries Service, **2:**128, 132
National Oceanic and Atmospheric Administration (NOAA):
    Earth System Research Laboratory, **8:**151
National Pollutant Discharge Elimination System (NPDES), **8:**77
National Radio Astronomy Observatory, **3:**221
Native Americans. *See* Indigenous populations
Natural gas:
    access as a military goal, **8:**166
    coal bed methane, **8:**63
    consumption in the U.S., **7:**80
    energy content, **7:**75
    extraction/processing, **7:**74, 75, 79–80, 81, **8:**63–78
    hydraulic fracturing, **8:**63–78
    overview, **8:**64–65, 77–78
    production by country, **7:**75
    shale gas, **8:**63, 67, 68
    tight gas, **8:**63
    unconventional reservoirs, **7:**80–82, **8:**63
    water consumption (water footprint), **7:**25
Natural Springs, **4:**38
The Nature Conservancy, **8:**26
Nature's right to water, **7:**145
Nauru, **3:**45, 46, 118
NAWAPA Foundation, **8:**124
Needs, basic water, **1:**185–86, **2:**10–13, **4:**49–51. *See also* Drinking water, access; Health, water issues; Human right to water; Sanitation services; Well-being, measuring water scarcity and
Negev desert, **4:**89
Nepal:
    arsenic in groundwater, **4:**87
    bottled water, **4:**22–23
    conflict/cooperation concerning freshwater, **5:**11
    dams, **1:**16, 71, 81
        with Chinese financiers/developers/builders, **7:**329–30
        World Commission on Dams, **3:**160
    hydropower potential, **7:**6
    irrigation, **2:**86

Nestle, **4:**21, 40, 41, **5:**163
    bottled water, **7:**158, 161–62
Netherlands, **5:**5
    agriculture and external water resources, **8:**88
    arsenic in groundwater, **2:**167
    climate change, **4:**158, 232
    dracunculiasis, **1:**52
    fossil-fuel production, **7:**75
    Global Water Partnership, **1:**169
    Living with Water strategy, **6:**49
    Millennium Development Goals, **4:**7
    open access to information, **4:**72–73
    public-private partnerships, **3:**66
Neufeld, David, **1:**197
Nevada. *See also* Las Vegas
    North American Water and Power Alliance, **8:**124–31
    Snake Valley, **8:**139
The New Economy of Water (Gleick), **4:**47
New Hampshire, **2:**118
New Mexico, North American Water and Power Alliance, **8:**126
New Orleans (LA), **4:**67–70
New Orleans City Business, **4:**69
New Orleans Times Picayune, **4:**69
Newton Valley Water, **4:**38
New York (NY), **4:**51–53
New Yorker, **2:**113
New York Times, **2:**113, **5:**20
New Zealand:
    bottled water, **4:**26
    climate change, **5:**136
    environmental flow, **5:**33
    globalization and intl. trade of water, **3:**45, 46
    privatization, **3:**61
    reservoirs, **2:**270, 272, 274
Nexus thinking, **8:**5–6, 85–86
Niger, **7:**264, 331
Nigeria, **1:**52, 55, **5:**41
    dams with Chinese financiers/developers/builders, **7:**331
    environmental concerns, top, **7:**287
    fossil-fuel production, **7:**74, 77, 91
Nike, **5:**156, **6:**24, 27
Nitrogen compounds, effects on human health, **7:**58–59. *See also* Nutrients; Runoff, agricultural
Nitrous oxide, **1:**138, 139
Nongovernmental organizations (NGOs), **1:**81, **3:**157, **4:**198, 205–6, **6:**31, 52, 74, 89–90, 96. *See also specific organizations*
    corporate partnerships with, **7:**40
    expanding role and recommendations for, **8:**8–9
Nonrenewable resources, **6:**6–7, 15
Nordic Water Supply Company, **1:**202–5, **8:**135
North Africa. *See* Africa, Northern
North America:
    irrigation, **6:**328, 334
    river basins, **6:**289, 306–08
    transboundary waters, **7:**3, 165–69, **8:**124–31, 139
North American Water and Power Alliance (NAWAPA), **1:**74, **8:**124–31
North Dakota, **8:**140
Northstar Asset Management, **7:**36
Norway, **1:**52, 89, **3:**160
    fossil-fuel production, **7:**75
    hydroelectric production, **7:**129
Nuclear power:
    Dead Sea project, proposed, **8:**154
    water consumption and energy generation, **7:**25
Nutrients:
    cycling/loading, **1:**77, **4:**172, **5:**128
    effects of high concentrations on human health, **7:**58–59
    enrichment/eutrophication, **7:**46, 49–50, 58–59, 63

O

Oak Ridge Laboratory, **1:**23, **4:**115
Oberti Olives, **3:**22–23
Occupational Information Network (O*NET), **8:**55
Oceania:
    bottled water, **4:**288, 289, 291, **5:**281, 283
    cholera, **1:**267, 270, **5:**289, 290
    dams, **3:**295
    drinking water, **1:**254–55, 262, **3:**259–60, **5:**245, **6:**65
        access to, **7:**24, 239
        progress on access to, **7:**253, **8:**270
    groundwater, **4:**86
    hydroelectric production, **1:**71, 280
    irrigation, **1:**301, **2:**263, 265, **3:**289, **4:**296, **5:**299, **6:**328, 334
    population data/issues, **1:**250, **2:**214
    privatization, **3:**61
    renewable freshwater supply, **1:**240, **2:**202, 217, **3:**242, **4:**266, **5:**227
        2011 update, **7:**220
        2013 update, **8:**226
    salinization, **2:**268
    sanitation, **1:**259–60, 264, **3:**268–69, 272, **5:**254–55, 261
        progress on access to, **7:**249, **8:**273
    threatened/at risk species, **1:**295–96
    water access, **2:**24, 217
    withdrawals, water, **1:**244, **2:**211, **3:**251, **4:**275, **5:**236
        by country and sector, **7:**229, **8:**235
    See also Pacific Island developing countries
Ogoni people, **5:**124
Ohio, Euclid Creek dam removal, **8:**46, 48
Oil:
    extraction and refining (*See* Petroleum and fossil fuels)
    oil production by country, **7:**75
    peak (*See* Peak oil)
    spills, **7:**73, 74, 77, 87, 90
    substitutes for, **6:**8–9, 12
    transport of, **6:**14–15, **7:**74, 91
    vs. water, **6:**3–9, 14
Oil shale, **7:**75, 78–79
Olivero, John, **1:**196
Oman, the Sultanate of, **2:**179–80, **7:**331
Ontario Hydro, **1:**88
Orange County (CA), **2:**152
Order of the Rising Sun, **5:**20
Oregon, **2:**128, **8:**129
Organic contaminants:
    bioaccumulation and bioconcentration, **7:**48, 60
    BOD emissions by country/industry, **7:**278–81
    emerging, **7:**48
    health effects, **7:**60
    from hydraulic fracturing for natural gas, **8:**72
    industrial wastewater, **7:**50, 51
    See also Persistent organic pollutants; Pesticides
Organisation for Economic Cooperation and Development (OECD):
    description of, **2:**56, **3:**90, 91, 164, **4:**6
    water tariffs, **6:**312–19
    See also Overseas Development Assistance
Orion Nebula, **3:**220
Outer space, search for water in:
    clouds, interstellar, **3:**219–20
    Earth's water, origin of, **1:**93–98, **3:**209–12, **6:**5
    exploration plans, **3:**216–17
    Jupiter's moons, **3:**217–18
    Mars, **3:**214–17, **5:**175–80
    moon, the, **3:**212–14

solar system, beyond our, **3:**218–19
summary/conclusions, **3:**221–22
universe, on the other side of the, **3:**220–21
Overseas Development Assistance (ODA), **4:**6, 278–83, **5:**262–72
water supply and sanitation
by donating country, **7:**273–74
2004-2011, **8:**317–19
by subsector, **7:**275–77
2007-2011, **8:**320–22
Oxfam Adaptation Financing Index, **6:**52
Ozguc, Nimet, **3:**185
Ozone, for bottled water, **7:**159, 161

P
Pacific Environment, **8:**10
Pacific Institute:
bottled water, **4:**24
climate change, **4:**232, 234, 236
Colorado River and the Basin Study, **8:**141
desalination, **8:**93
hydraulic fracturing, **8:**65, 77
privatization, **4:**45–46, 193–94, **5:**132–33
sustainable water jobs research, **8:**35–36
Water Conflict Chronology (*See* Water Conflict Chronology)
water use in California, **4:**105, **8:**38, 89
Pacific Island developing countries (PIDCs):
climate change
overview, **3:**xix
precipitation, **3:**115–16, 124–25
projections for 21st century, **3:**121–23
science overview, **3:**119–21
sea-level rise, **3:**124
severe impacts of, **5:**136
storms and temperatures, **3:**125
freshwater resources
description and status of, **3:**115–18
overview, **3:**113–14
threats to, **3:**118–19
profile of, **3:**115
summary/conclusions, **3:**125–27
terrain of, **3:**116
See also Oceania
Pacific Region, Western:
mortality rate, under-5, **7:**263
Packard Humanities Institute, **3:**187
Pakistan:
agriculture, **4:**88, 89
bottled water, **4:**40
conflicts concerning freshwater, 2011-2012, **8:**163
conflicts/cooperation concerning freshwater, **5:**10, 13, 15
dams
with Chinese financiers/developers/builders, **7:**331–33
World Commission on Dams, **3:**160
Wullar, construction by India, **8:**163
dracunculiasis, **1:**53
groundwater, **4:**82, 88, 89
Orangi Pilot Project, **4:**71–72
sanitation services, **4:**71–72
Palau, **3:**118
Palestine:
conflicts concerning freshwater, 2011-2012, **8:**165
Gaza Strip, **8:**154
West Bank, **8:**165
Palestinian National Authority:
Red Sea-Dead Sea projects, **8:**144, 153–57
Palestinians, **1:**109, 118, **5:**6, 10, 13–15
Panama, **3:**160
Papua New Guinea, **7:**333
Paraguay, **3:**13

Parasites. *See under* Diseases, water-related
Partnerships:
with community-based organizations, **8:**59
public-private (*See* Public-private partnerships)
strategic corporate, **6:**31–32
Pathogenic organisms. *See* Diseases, water-related
Peak oil, **6:**1–3, 14
Peak water:
description of, **6:**8, **8:**159
ecological, **6:**10–12, 15
fossil groundwater, **6:**9–10
limitations of term, **6:**15
summary of, **6:**15
utility of, **6:**9–14
Pennsylvania, **2:**118
Dimock Township, methane contamination from hydraulic fracturing, **8:**69, 70–71
disposal of hydraulic fracturing wastewater, **8:**74
illegal wastewater disposal, **8:**76
Philadelphia, **8:**44
Susquehanna River Basin, **8:**67, 68, 70
Pepsi and PepsiCo, **4:**21, **5:**146, **6:**27, **7:**26, 36, 161
Permitting:
desalination plant, **8:**101, 112
groundwater wells, **8:**137, 138
wastewater, **4:**150
Perrier bottled water, **4:**21, 38, 40, 41, **5:**151
Persian Gulf War, **1:**110, 111
Persians, **3:**184
Persistent organic pollutants (POP), **7:**48, 60. *See also* Organic contaminants
Peru, **1:**46, 59–60, **2:**178–79
cholera, **8:**359, 373
conflicts concerning freshwater and mining, **8:**166
Pesticides, **5:**20, **7:**48, 60. *See also* Runoff, agricultural
PET. *See* Polyethylene terephthalate
Petroleum and fossil fuels:
carbon footprint, **8:**85, 86
case studies, **7:**90–92
climate change caused by burning of, **6:**9, 40–43, 53, **7:**80, 84
energy content, **7:**75
impacts of contamination
drinking water, **7:**88–89, **8:**63
economic, **7:**64–65
freshwater ecosystems, **7:**84–87
health effects, **7:**51–52, 57
human communities, **7:**87–88
overview, **7:**92–93
water quality, **7:**51, 73–74, 76–84
mining process (*See* Mining, fossil fuels)
oil spills, **7:**73, 74, 77, 87, 90
origins of, **6:**4
spills from hydraulic fracturing for natural gas, **8:**65, 66, 75, 76
water consumption (water footprint), **7:**25, 26, 31, 73, 74–75
Pets, purchased food going to feed, **2:**77
Pharmaceutical contaminants, **7:**49, 51
Philippines:
bottled water, **4:**40
cholera, **1:**63
conflict/cooperation concerning freshwater, **5:**11
dams, **1:**71
with Chinese financiers/developers/builders, **7:**333
World Commission on Dams, **3:**161
environmental concerns, top, **7:**287
loss of tourism revenue due to water pollution, **7:**64
mining spill, **7:**62, 65
prices, water, **3:**69
privatization, **3:**60, 61, 66, **4:**46

Pinchot, Gifford, **1:**80–81
Pluto, **3:**220
Pneumonia, **7:**259–63
Poland, **3:**66, 160–61, **4:**38
 fossil-fuel production, **7:**75
Poland Spring bottled water, **4:**21, **7:**162
Polar satellite, **3:**209–10
Poliomyelitis, **7:**272
Politics. *See* Government/politics
Pollution. *See* Environmental issues; Water quality, contaminants
Pollution prevention, **7:**65, **8:**24–25
Polychlorinated biphenyls (PCBs), **7:**48, 60
Polycyclic aromatic hydrocarbons (PAHs), **7:**89
Polyethylene terephthalate (PET), **4:**39, 41, **7:**158–59
Pomona (CA), **2:**138
Population issues:
 by continent, **2:**213–14
 diseases, projected deaths from water-related, **4:**12–13
 displaced people, **6:**145–46, **8:**148, 164 (*See also* Dams, social impacts, displaced people)
 drinking water access, **6:**63, 68
 expanding water-resources infrastructure, **1:**6
 food needs, **2:**63–64, 66–67
 growth
  0–2050, **2:**212
  2000–2020, **4:**10
  effects on water quality, **7:**53
 Millennium Development Goals (MDGs) affected by, **6:**63
 sanitation, **6:**63
 total and urban population data, **1:**246–50
 withdrawals, water, **1:**10, 12, 13, **3:**308–9
 See also Developing countries
Portugal, **1:**71
Poseidon Resources Corp., **2:**108, **8:**110–16
Poseidon Water Resources, **5:**61
Postel, Sandra, **1:**111
Potash, **8:**155
Poverty, **4:**40–41, **5:**123–24, **6:**58, **7:**61–62. *See also* Developing countries; Environmental justice
Power, measuring, **2:**308, **3:**326, **4:**333, **7:**349
Power generation. *See* Hydroelectric production; Nuclear power; Solar power; Wind power
Precipitation:
 acid rain, **7:**87
 Atlanta (GA), **6:**104
 China, **6:**87
 climate change, **1:**140–41, 146–47, **4:**159, 166, **6:**40, 87
 as "green water" in water footprint concept, **8:**85–86
 Las Vegas (NV), **6:**104
 Pacific Island developing countries, **3:**115–16, 124–25
 rainwater catchment/harvesting, **7:**120–21, **8:**25, 42, 48
 Seattle (WA), **6:**104
 snowfall/snowmelt, **1:**142, 147, **4:**160–61
 Standard Precipitation Index, **5:**93
 stocks and flows of freshwater, **2:**20, 22
 and use of term "withdrawal," **7:**222
 volume, and stormwater runoff production, **8:**39
Precision Fabrics Group, **1:**52
Pressure, measuring, **2:**307, **3:**325, **4:**332, **7:**348
Preston, Guy, **4:**50
Pricing, water:
 agricultural irrigation, **7:**111–12
 Australia, **7:**111–12, 118
 block, **1:**26, **4:**56
 bottled water, **4:**22–23
 climate change, **4:**180–81
 desalinated seawater, **8:**102, 106, 110–11, 113, 115
 households in different/cities/countries, **3:**304
 for hydraulic fracturing for natural gas, **8:**68
 Jordan, **1:**117
 from major water projects, **8:**104
 market approach, **1:**27, **7:**111
 off-take agreement, **8:**106
 peak-load, **1:**26
 privatization, **3:**69–71, 73, **4:**53–55
 rate structures (*See* Water rate structures)
 San Diego (CA), **8:**100
 seasonal, **1:**26
 take-or-pay contract, **8:**106, 112, 115, 117, 119
 tier, **1:**26
 twentieth-century water-resources development, **1:**24–28
 urban areas, **1:**25–27, **4:**124–25, 150
 and water policy reform, **7:**153
 See also Economy/economic issues, subsidies
Private goods, **3:**34
Privatization:
 business/industry, water risks that face, **5:**152–53
 conflict/cooperation concerning freshwater, **3:**xviii, 70–71, 79, **4:**54, 67
 defining terms, **3:**35
 drivers behind, **3:**58–59
 economic issues, **3:**70–72, **4:**50, 53–60
 environmental justice, **5:**131–33
 failed, **3:**70
 forms of, **3:**63–67, **4:**47, 48
 history, **3:**59–61
 opposition to, **3:**58
 overview, **3:**57–58, **4:**xvi, 45–46
 players involved, **3:**61–63
 principles and standards
  can the principles be met, **4:**48–49
  economics, use sound, **3:**80–81, **4:**53–60
  overview, **3:**79, **4:**47–48, **7:**106
  regulation and public oversight, government, **3:**81–82, **4:**60–73
  social good, manage water as a, **3:**80, **4:**49–53
 risks involved
  affordability questions, pricing and, **3:**69–73
  dispute-resolution process, weak, **3:**79
  ecosystems and downstream water users, **3:**77
  efficiency, water, **3:**77–78
  irreversible, privatization may be, **3:**79
  local communities, transferring assets out of, **3:**79
  monitoring, lack of, **3:**75–77
  overview, **3:**67–68, **8:**19–31
  public ownership, failing to protect, **3:**74–75
  underrepresented communities, bypassing, **3:**68
  water quality, **3:**78
 sanitation services, **5:**273–75
 summary/conclusions, **3:**82–83, **4:**73–74
 update on, **4:**46–47
 World Water Forum (2003), **4:**192, 193–94
Procter & Gamble, **5:**157, **6:**27
Productivity:
 agricultural, **2:**74–76
 land, **8:**86
 water, **3:**17–19
Progressive Habitat Development Alternative, **6:**135
Projections, review of global water resources:
 Alcamo et al. (1997), **2:**56–57
 analysis and conclusions, **2:**58–59
 data constraints, **2:**40–42
 defining terms, **2:**41
 Falkenmark and Lindh (1974), **2:**47–49
 Gleick (1997), **2:**54–55
 inaccuracy of past projections, **2:**43–44

Kalinin and Shiklomanov (1974) and De Mare
(1976), **2:**46–47
L'vovich (1974), **2:**44–47
Nikitopoulos (1962, 1967), **2:**44
overview, **2:**39–40
Raskin et al. (1997, 1998), **2:**55–56
Seckler et al. (1988), **2:**57–58
Shiklomanov (1993, 1998), **2:**50–53
in 2002, **3:**xvii–xviii
World Resources Institute (1990) and Belyaev
(1990), **2:**49–50
See also Sustainable vision for the Earth's
freshwater
Public Citizen, **4:**69
Public goods, **3:**34
Public Limited Companies (PLC), **4:**72–73
Public participation:
    business/industry water management, **7:**37–38
    climate change adaptation, **6:**49
    drought management, **7:**116–17
    Great Lakes–St. Lawrence River Basin Water
        Resources Compact, **7:**169
    sustainable vision, **1:**82, **7:**24–25
    water decision making, **5:**150–51, **6:**96–97, **8:**11
    See also Education
Public perception:
    company public relations, **8:**23, 29–30
    environmental concerns around the world, top,
        7:285–88
    environmental concerns of U.S. public, top, **7:**282–
        84
    hydraulic fracturing for natural gas, **8:**66
    North American Water and Power Alliance, **8:**129
    satisfaction with air quality (*See* Air quality,
        satisfaction)
    satisfaction with water quality (*See* Water quality,
        satisfaction)
    terrorism, **5:**2–3
    water costs and desalination, **8:**99
    water risks of business/industry, **7:**26–27
Public-private partnerships, **3:**74–75, **4:**60–73, 193–94,
    6:92, **7:**40
    for desalination projects, **8:**110, 111–12, 119
    lessons learned, **8:**110, 111–12, 119
    risk allocation, **8:**105–6
    shared risk, **8:**19–31
Public Trust Doctrine, **5:**37
Puerto Rico, **3:**46, **5:**34
Pupfish, desert, **3:**142
Pure Life, **4:**40
Purified water, **4:**29

Q
Qatar, **7:**75
Qinghai-Tibetan Plateau, **6:**88
Quality of life (QOL), **3:**88–96, **6:**61
Quantitative measures of water availability/use, **2:**25

R
Race and environmental discrimination, **5:**118. *See
    also* Environmental justice
Radiative forcing, **3:**120
Radioactive contaminants, **7:**52, 73, **8:**72
Rail, Yuma clapper, **3:**142
Rainfall. *See under* Precipitation
Ralph M. Parsons Company, **8:**124, 130
Rand Water, **1:**95, 96
Rates:
    wastewater (*See* Wastewater, rates for)
    water (*See* Pricing, water; Water rate structures)
Raw or value-added resource, water traded as a,
    3:42–47
Reagan, Ronald, **2:**95

Rebates, for water conservation, **6:**110–12, **7:**119, 121
Reber Plan (San Francisco Bay Project), **8:**131–34
Reclaimed water:
    agricultural water use, **1:**28, 29, **2:**139, 142, 145–46,
        7:54, **8:**74
    Australia, **7:**114, 120–21
    blackwater, **7:**120
    California (*See* California, reclaimed water)
    costs, **2:**159, **8:**49
    defining terms, **2:**139
    environmental and ecosystem restoration, **2:**149–
        50
    food needs for current/future populations, **2:**87
    graywater (greywater), **7:**120, **8:**38, 48–49
    groundwater recharge, **2:**150–51
    health issues, **2:**152–56
    from hydraulic fracturing, **8:**74
    Israel, **1:**25, 29, **2:**138, 142
    Japan, **2:**139, 140, 158–59
    Namibia, **2:**152, 156–58
    overview, **1:**28–29, **2:**137–38, **7:**60, **8:**48
    potable water reuse, direct/indirect, **2:**151–52
    primary/secondary/tertiary treatment, **2:**138
    processes involved, **2:**140
    on roads, **8:**74
    summary/conclusions, **2:**159–61
    urban areas, **1:**25, **2:**146–49, **4:**151, **7:**114
    uses, wastewater, **2:**139, 141–42, **8:**49
Recreation:
    costs of poor water quality, **7:**64
    effects of water restriction policies, **7:**115–16
    tourism, **7:**64, **8:**26, 128, 155, 164
Red List of Threatened Species (IUCN), **7:**292–302
Red Sea:
    Dead Sea projects, **8:**144, 153–57
    desalination, proposed, **8:**154, 155–56
Refugees. *See* Displaced people
Regulatory bodies. *See* Law/legal instruments/
    regulatory bodies
Rehydration therapy, cholera and, **1:**57
Reliability, desalination and water-supply, **5:**73–74
Religious importance of water, **3:**40
Remediation, **8:**45, 46, 48
Renewable freshwater supply:
    by continent, **2:**215–17
    by country, **1:**235–40, **2:**197–202, **3:**237–42, **4:**261–
        66, **5:**221–27, **6:**195–201
    2011 update, **7:**215–20
    2013 update, **8:**221–26
    developing, **1:**28, **8:**48–50, 51, 52
    fossil groundwater, **6:**9–10
    globalization and intl. trade of water, **3:**39
    Overseas Development Assistance, **7:**273–77
        by donating country
            2004-2011, **8:**317–19
        by subsector
            2007-2011, **8:**320–22
Renewable resources, **6:**6–7
Reptiles:
    crocodiles, **7:**56
    extinct or extinct in the wild species, **7:**298
    threatened, **1:**291–96, **7:**56
    turtles, **3:**49, 50
Reservoirs:
    built per year, number, **2:**116
    climate change, **1:**142, 144–45
    environmental issues, **1:**75, 77, 91, **8:**128
    Martian ice, **3:**215–16
    number larger than 0.1 km by continent/time
        series, **2:**271–72
    orbit of Earth affected by, **1:**70
    sediment, **1:**91, **2:**127, **3:**136, 139, **4:**169
    seismic activity induced by, **1:**77, 97, **6:**144–45

Reservoirs (*continued*)
    total number by continent/volume, **2:**270
    twentieth-century water-resources development, **1:**6
    U.S. capacity, **1:**70
    U.S. volume, **2:**116
    volume larger than 0.1km by continent/time series, **2:**273–74
    See also Dams
Reservoirs, specific:
    Diamond Valley, **3:**13
    Imperial, **3:**136
    Itaipú, **1:**75
    Mesohora, **1:**144
    Occoquan, **2:**152
Restrooms and CII water use, **4:**135, 136
Reuse, water. *See* Reclaimed water
Revelle, Roger, **1:**149
Reverse osmosis (RO):
    bottled water treatment, **7:**159–60, 161
    and desalination, **2:**96, 102–3, **5:**51, 55, 57, 58, 60, 72
        costs, **8:**94
    energy requirements, **7:**161
    for industrial water treatment, **8:**28
"Rey" water (water footprint concept), **8:**86, 323, 332, 341, 350
Rhodesia, **5:**7
Rice, **2:**74–79
Right to water. *See* Human right to water; Nature's right to water
Risk assessment:
    dams and, **3:**153, **7:**137
    desalination projects, **8:**101–6, 118–19
    hydraulic fracturing for natural gas, **8:**77–78
Risk communication, **8:**11
Risk management:
    droughts and, **5:**99
    policy engagement to reduce, **8:**20
    and regulatory uncertainty, **8:**24
    risk allocation with desalination projects, **8:**115–16
    "shared risk" and the private sector, **8:**19, 25–27
Rivers:
    climate change, **1:**142, 143, 145, 148, **7:**2–10
    consequences of poor water quality, **7:**54–55
    dams' ecological impact on, **1:**77, 91
    deltas, **3:**xix
    diversion projects, large-scale, **1:**74, **8:**124–31, 136–44
    Federal Wild and Scenic Rivers Act of 1968, **1:**15, **8:**129
    floods, **5:**104, **7:**14–15
    flow rates, **2:**305–6, **3:**104, 324, **4:**168, 172, 331
    National Wild and Scenic Rivers Act of 1997, **2:**120, **5:**36
    Overseas Development Assistance, **7:**277
        by donating country
            2004-2011, **8:**317–19
        by subsector
            2007-2011, **8:**320–22
    pollution and large-scale engineering projects, **1:**6
    restoration, **2:**xix, 127, **3:**143–44, **8:**45–46, 48
    runoff, **1:**142, 143, 148, **2:**222–24, **4:**163–67, **5:**29
        Composite Runoff V1.0 database, **8:**274, 292
        effects of climate change, **7:**10
        limitations in data, **8:**274, 292
        monthly for major river basins, by basin name, **8:**292–309
        monthly for major river basins, by flow volume, **8:**274–91
        "natural," **8:**274, 292
        Rhine River Basin, **7:**10
    transboundary (*See* International river basins)

    wastewater dumping into, **6:**82
    See also Environmental flow; Stocks and flows of freshwater; Sustainable vision for the Earth's freshwater
Rivers, specific:
    Abang Xi, **1:**69
    Agrio, **7:**65
    Allier, **2:**125
    Amazon, **1:**75, 111, **2:**32, **7:**56, **8:**165
    American, **1:**16
    Amu Darya, **3:**3, 39–40, **7:**52
    Amur, **7:**131
    Apple, **2:**118
    Athabasca, **7:**87, 91–92
    AuSable, **2:**119
    Beilun, **7:**131
    Bhagirathi, **1:**16
    Boac, **7:**65
    Brahmaputra, **1:**107, 111, 118–19, 206–9, **6:**290
        watershed within China, **7:**131
    Butte Creek, **2:**120–23
    Cache, **8:**46, 48
    Carmel, **5:**74
    Cauvery, **1:**109, **2:**27, **7:**3, **8:**163
    Clyde, **2:**125–26
    Colorado (*See* Colorado River)
    Columbia, **3:**3, **7:**9, 18
        water transfer, proposed, **8:**126, 128, 129, 130–31
    Congo, **1:**75, 111, 156, **2:**31, 32–33
    Crocodile, **1:**123
    Danube, **1:**109, **2:**31, **5:**33, 111, **7:**5
    Elwha, **2:**117, 127–28
    Emory, **7:**84
    Euphrates, **1:**109–11, 118, **2:**33, **3:**182, 183–87, **6:**290
    Ganges, **1:**107, 111, 118–19, 206–9, **4:**81, **6:**290
        threatened/at risk species, **7:**56
        watershed within China, **7:**131
    Gila, **3:**139, 140
    Gordon, **2:**126–27
    Hai He, **6:**82, 90
    Han, **1:**109–10
    Har Us Nur, **7:**131
    Hsi/Bei Jiang, **7:**131
    Ili/Junes He, **7:**131
    Incomati, **1:**120–23
    Indus, **1:**16, 77, 111, **7:**131
    Irrawaddy, **7:**131, 132–33
    Jinsha, **6:**92
    Jordan, **1:**107, 109, 111, 115–16, **2:**31, **5:**33
        contamination, **8:**155
        diversion, effects on the Dead Sea, **8:**154–55
    Juma, **6:**90
    Kennebec, **2:**117, 123–25, **5:**34
    Kettle, **2:**119
    Kissimmee, **5:**32
    Kosi, **7:**6
    Kromme, **5:**32
    Laguna Salada, **3:**139
    Lamoille, **2:**128
    Lancang, **6:**88, **7:**130–32 (*See also* Rivers, specific, Mekong)
    Lerma, **3:**77
    Letaba, **1:**123
    Limpopo, **1:**120–23
    Logone, **5:**32
    Loire, **2:**117, 125
    Lower Snake, **2:**131–34
    Luvuvhu, **1:**123
    Mahaweli Ganga, **1:**16, **5:**32
    Malibamats'o, **1:**93
    Manavgat, **1:**203, 204–5, **3:**45, 47
    Manitowoc, **2:**119

Maputo, **1:**121
McCloud, **5:**123
Meghna, **6:**290, **7:**131
Mekong, **1:**111, **7:**9, 13–15, 14, 130–32
Merrimack, **2:**118
Meuse, **5:**15
Milwaukee, **2:**117–19
Mississippi, **2:**32, 33, **3:**13, **5:**110, 111
    water transfer, proposed, **8:**127, 140, 141
Missouri, **3:**13, **8:**140–44
Mooi, **4:**51
Murray-Darling, **5:**33, 42
    decline in water flow due to drought, **7:**102
    ecological effects of drought, **7:**100–101
    location, **7:**99
    water management, **7:**110–11, 146–47
    water markets, **7:**110–14, 146
Murrumbidgee, **7:**113
Narmada, **5:**133
Naryn, **8:**166
Neuse, **2:**126
Niger, **1:**55, 111, **2:**31, 85, **7:**88
Nile, **1:**77, 111, **2:**26, 32–33, **3:**10–11, **5:**111
    canal in Sudan, **8:**144
    dams, **8:**164
    effects of water contamination on fisheries, **7:**62
    hydrology, **7:**10
    intl. agreements, **7:**5–6, 10–13, **8:**91
    oil spills, **7:**77
Nujiang, **6:**89
Ob, **7:**131
Okavango, **1:**111, 119, 121–24
Olifants, **1:**123, **7:**64
Orange, **1:**93, 98–99, 111, **7:**52
Orontes, **1:**111, 115
Pamehac, **5:**34
Paran, **1:**111, **3:**13
Patauxent, **5:**34
Po, **5:**111
Prairie, **2:**118
Puerco, **7:**52
Pu-Lun-To, **7:**131
Red/Song Hong, **7:**131
Rhine, **5:**111, **7:**5, 10
Rhone, **3:**45, 47
Rio Grande, **1:**111, **5:**41, **7:**4, 8, **8:**125
Rogue, **2:**119, 128
Sabie, **5:**32
Sacramento, **2:**120–23, **4:**164, 167, 169, **5:**34, 111
    delta region, **8:**133
    effects of dams on fish, **8:**128
    and the Reber Plan, **8:**132
St. Lawrence, **7:**165, 167–69
Salween (Nu), **7:**131, 132
San Joaquin, **4:**164, 169, **8:**132, 133
Senegal, **1:**111, **2:**85
Shingwedzi, **1:**123
Sierra Nevada, **4:**164
Silala/Siloli, **8:**165–66
Snake, **2:**131–34, **3:**3, **8:**129
Songhua, **6:**83
Spöl, **5:**33
Sujfun, **7:**131
Susquehanna, **8:**67, 68, 70
Suzhou, **5:**32
Syr Darya, **3:**3, 39–40, **7:**52
Tarim, **5:**32, **7:**131
Temuka, **5:**33
Theodosia, **5:**34
Tigris, **1:**69, 111, 118, **2:**33, **3:**182, 187–90, **6:**290
Tumen, **7:**131
Vaal, **1:**95, **7:**52, **8:**27–28
Vakhsh, **8:**166
Volga, **1:**77
Wadi Mujib, **5:**33
Waitaki, **5:**33
White Salmon, **2:**119
Wind, **8:**73
Xingu, **8:**165
Yahara, **2:**118
Yalu, **7:**131
Yangtze, **5:**133, **6:**81, 86, 88, 91, 143–44 (*See also* Dams, specific, Yangtze River)
    threatened/at risk species, **7:**56
Yarlung Sangpo/Siang, **7:**134
Yarmouk, **1:**109, 115–16
Yellow, **5:**5, 15–16, 111, **6:**86, 91
Zambezi, **1:**111
Zhang, **6:**90
Zhujiang, **6:**86
See also China, dams, construction overseas; Dams, by continent and country; Lesotho Highlands project
The Road not Taken (Frost), **3:**1
Roads:
    permeable pavement, **8:**42
    reduction of water percolation due to, **7:**53
    runoff from, **7:**53, 77
    use of reclaimed water on, **8:**74
Roaring Springs/Global Beverage Systems, **4:**39
Rodenticides, **5:**20
Rome, ancient, **1:**40, **2:**137, **3:**184
Roome, John, **1:**99
Roosevelt, Franklin, **1:**69
Roundworm, **7:**61–62
Runoff:
    agricultural, **5:**128, 305–7, **7:**46, 48, 49–50, **8:**50, 155
    construction, **8:**76
    effects of climate change, **6:**43, **7:**10
    from hydraulic fracturing for natural gas, **8:**76–77
    "natural," **8:**274, 292
    river (*See* Rivers, runoff)
    from roads and parking lots, **7:**53, 77
    stormwater, **7:**53, 76, 77–78, **8:**39, 41–44
Rural areas:
    development and the World Water Forum, **4:**203
    drinking water, **6:**70–71 (*See also* Drinking water, access, by country)
    sanitation services, **6:**70–71 (*See also* Sanitation services, access by country)
Russell, James M., III, **1:**196
Russia:
    dams, **1:**75, 77
    environmental concerns, top, **7:**287, 288
    fossil-fuel production, **7:**75
    groundwater, **5:**125
    hydroelectric production, **1:**71, **7:**129
    irrigation, **4:**296
    and the Kyrgyzstan-Tajikistan conflict over Rogun Dam, **8:**166
    Siberia, reversal of rivers' flows, proposed, **8:**129
    threatened/at risk species, **1:**77
    water policy reform, **7:**149
    See also Soviet Union, former
Rwanda, **1:**62, **7:**11, 264
RWE/Thames, **5:**162

S
SABMiller, **7:**36
Safeway Water, **4:**39
Saint Lucia, **7:**264
St. Petersburg (FL), **2:**146–47
Salinization:
    climate change, **4:**168, 169–70, **7:**8, 54
    continental distribution, **2:**268

Salinization (*continued*)
:   by country, **2:**269
    ecological effects, **7:**47
    from fossil-fuel production, **7:**76
    groundwater, **4:**87, **7:**55, **8:**148–49
    salt concentrations of different waters, **2:**21, 94
    soil fertility, **2:**73–74
    See also Desalination
Salmon, **1:**77, **2:**117, 120–21, 123, 128, 132, 133, **3:**3
:   at risk due to dams, **8:**128–29
Salt. *See* Brine; Salt water
Salton Sea:
:   air-quality monitoring, **6:**137
    background of, **6:**129
    Bureau of Reclamation, **6:**135
    California water transfers, **6:**129–31
    Colorado River inflows, **6:**129, 132
    Imperial Irrigation District, **6:**130
    inflows, **6:**129–33
    location of, **6:**127–28
    restoration of, **6:**131–37, **8:**144
    salinity of, **6:**128, 137
    seismic activity, **6:**135
Salton Sea Authority (SSA), **6:**134
Salt water, **6:**5, **8:**72
Samoa, **3:**118
Samosata, **3:**184–85
Samsat, **3:**184–85
San Francisco Bay, **3:**77, **4:**169, 183, **5:**73, **8:**132
San Francisco Bay Project (Reber Plan), **8:**131–34
San Francisco Chronicle, **4:**24
Sanitation services:
:   access by country, **1:**256–60, **3:**261–69, **5:**247–55, **6:**221–29, **7:**241–50
    :   urban and rural, 1970-2008, **8:**252–53, 255–62
        urban and rural, 2011, **8:**252–53, 263–67
    access by region, MDG progress on, **3:**270–72, **5:**256–61, **6:**67–70, 233–35, **7:**254–56
    :   2013 update, **8:**271–73
    childhood mortality and, **6:**58, **7:**57–58, 61–62
    costs of, **6:**73
    developing countries, **1:**263–64, **7:**62
    diarrhea reduction through, **6:**75, **7:**58
    economic return on investments in, **7:**63–64
    education services affected by, **6:**58, **7:**61
    environmental justice, **5:**127–31
    falling behind, **1:**39–42, **5:**117, 124
    funding of, **6:**73
    as a human right, **8:**271
    importance of, **6:**58
    "improved," use of term, **7:**241, 254, **8:**252–53, 271
    inadequate, **6:**58
    intl. organizations, recommendations by, **2:**10–11
    investment in infrastructure projects with private participation, **5:**273–75
    lack of, statistics, **7:**52, **8:**1, 3
    limitations in data and reporting, **6:**61–62, **7:**242, 254–55, **8:**253, 271–72
    maternal health affected by, **6:**58
    nongovernmental organization resources for, **6:**74
    Overseas Development Assistance, **4:**282–83, **7:**273–77
    :   by donating country 2004-2011, **8:**317–19
        by subsector 2007-2011, **8:**320–22
    poverty eradication and, **6:**58
    prioritizing of, **6:**75
    rural areas, **6:**70–71
    twentieth-century water-resources development, **3:**2
    urban areas, **6:**70–71
    well-being, measuring water scarcity and, **3:**96–98
    within-country disparities in, **6:**77
    women and access to water, **5:**126 (*See also* Women, responsibility for water collection)
    World Health Organization, **4:**208, **6:**62
    World Water Forum (2003), **4:**202, 205
    See also Health, water issues; Human right to water; Millennium Development Goals; Soft path for water; Well-being, measuring water scarcity and
San Jose/Santa Clara Wastewater Pollution Control Plant, **2:**149–50
San Pellegrino bottled water, **4:**21
Santa Barbara (CA), **5:**63–64, **8:**102–3, 119
Santa Rosa (CA), **2:**145–46
Sapir, Eddie, **4:**69
Sargon of Assyria, **1:**110
Sasol, **7:**39, **8:**27–28
Saudi Arabia:
:   desalination, **2:**94, 97, **8:**93
    Disi Aquifer, **8:**157
    dracunculiasis, **1:**52
    fossil-fuel production, **7:**75
    groundwater, **3:**50
    intl. river basin, **2:**33
    pricing, water, **1:**24
Save the Children Fund, **4:**63
Save Water and Energy Education Program (SWEEP), **4:**114
Saving Water Partnership (SWP), **6:**109
SCA, **6:**27
Schistosomiasis, **1:**48, 49, **7:**58, 272
School of Environmental Studies (SOES), **2:**167, 171
Scientific American, **3:**89
Seagram Company, **3:**61
Sea-level rise, **3:**124, **4:**169–70, **7:**8, 54
Seattle (WA):
:   description, **6:**103
    per-capita water demand, **6:**104–6
    population growth, **6:**104
    precipitation, **6:**104
    temperature, **6:**104
    wastewater rate structure, **6:**118–19
    water conservation, **6:**109, 110–12
    water rate structures, **6:**115–17
    water-use efficiency, **6:**109, **8:**38
Sedimentation:
:   dams/reservoirs and, **1:**91, **2:**127, **3:**136, 139, **4:**169
    due to deforestation, **8:**26
    ecological effects, **7:**46, 53, 85
    of wetlands, **7:**56
Seismic activity:
:   caused by filling reservoirs, **1:**77, 97, **6:**144–45
    and floods, **5:**106
    Salton Sea, **6:**135
    San Andreas fault, **6:**135
    Three Gorges Dam, **6:**144–45
Seljuk Turks, **3:**184, 188
Senegal, **1:**55
Serageldin, Ismail, **1:**166
Serbia, **7:**333
Services, basic water. *See* Drinking water, access; Employment; Health, water issues; Human right to water; Municipal water; Sanitation services
Servicio Nacional de Meterologia e Hidrologia, **2:**179
Sewer systems, condominial, **3:**6. *See also* Sanitation services
Shad, **2:**123
Shady, Aly M., **1:**174
Shale, **7:**75, 78–79, **8:**63, 67, 68
Shaping the 21st Century project, **3:**91
Shellfish. *See under* Invertebrates

Shigella, **7:**57
Shoemaker, Eugene, **1:**196
Showerheads, **4:**109, 114, **6:**106, **7:**28
Shrimp, **3:**49, 50, 141, 142
Siemens, **7:**133
Sierra Club, **1:**81
Singapore:
    access to water, strict, **2:**41
    conflict/cooperation concerning freshwater, **1:**110
    desalination, **2:**108, **5:**51
    disputes with Malaysia, **1:**22
    toilets, energy-efficient, **1:**22
    water-use efficiency, **4:**58–60
Sinkholes, **8:**155
Skanska, **3:**167
Slovakia, **1:**109, 120
Slovenia, **1:**71
SNC, **1:**85
Snow. *See under* Precipitation
Snow, John, **1:**56–57
Social goods and services, **3:**36–37, 80, **4:**49–53
Société de distribution d'eau de la Côte d'Ivoire (SODECI), **4:**66
Société pour l'aménagement urbain et rural (SAUR), **4:**66
Socioeconomic issues, **5:**94, **7:**29–30
Soft Energy Paths , **3:**xiii
Soft path for water:
    definition of, **6:**13, 101, **8:**37
    description of, **6:**12–14, **7:**150
    economies of scale in collection/distribution, **3:**8
    efficiency of use, definitions/concepts
        agriculture, **3:**19–20
        businesses, **3:**22–24
        conservation and water-use efficiency, **3:**17
        maximum practical/cost-effective savings, **3:**23
        municipal scale, **3:**20–22
        overview, **3:**16–17
        poem, **5:**219
        productivity and intensity, water, **3:**17–19
        social objectives, establishing, **3:**17
    emerging technologies, **5:**23–24
    end-use technology, simple, **3:**8–9
    how much water is really needed, **3:**4
    and integrated water resource management, **8:**27
    moving forward, **3:**25–29
    myths about
        cost-effective, efficiency improvements are not, **3:**12–15
        demand management is too complicated, **3:**15–16
        market forces, water demand is unaffected by, **3:**9
        opportunities are small, efficiency, **3:**9
        real, conserved water is not, **3:**10–11
        risky, efficiency improvements are, **3:**11–12
    overview, **3:**30, xviii
    redefining the energy problem, **3:**xiii
    sewer systems, condominial, **3:**6
    user participation, **3:**5, 6
    vs. hard path, **3:**3, 5–7, **6:**13–14
    See also Sustainable vision for the Earth's freshwater
Soil:
    changes, **1:**141–42, 148, **4:**167
    climate change and moisture, **4:**167
    compaction, **7:**83
    degradation by type/cause, **2:**266–67
    dust storms, **7:**105–6
    erosion, **7:**46, 79, 105–6, **8:**75
    food needs for current/future populations, **2:**71, 73–74
    hard path for meeting water-related needs, **3:**2
    moisture in, as "green water" in water footprint concept, **8:**85–86
Solar power:
    carbon footprint, **8:**85
    Dead Sea project, proposed, **8:**154
    desalination and, **2:**105–6
    as flow-limited resource, **6:**6–7
    water consumption (water footprint), **7:**25
Solar radiation powering climate, **1:**138
Solomon Islands, **8:**168
Solon, **5:**5
Somalia, **5:**106, **8:**167
Sonoran Desert, **3:**142
South Africa:
    bottled water, **4:**22
    conflict/cooperation concerning freshwater, **1:**107, 119–21, 123–24, **5:**7, 9
    conflicts concerning freshwater 2011-2012, **8:**168
    dams, **1:**81, **3:**161, **7:**52
    Development Bank of South Africa, **1:**95, 96
    drinking water, access to, **7:**145
    environmental flow, **5:**32, 35, 37, 42
    fossil-fuel production, **7:**75
    human right to water, **2:**9, **4:**211
    hydrology, **1:**156–58
    introduced/invasive species, **7:**48
    legislation and policy
        Apartheid Equal Rights Amendment (ERA), **1:**158–59
        Constitution and Bill of Rights, **1:**159–60, **2:**9
        General Agreement on Tariffs and Trade, **3:**49
        National Water Conservation Campaign, **1:**164–65
        review process for, **1:**160–64
        water policy reform, **7:**145–46
        White Paper on Water Supply, **1:**160
    loss of tourism revenue due to water pollution, **7:**64
    mining, **7:**65
    privatization, **3:**60, **4:**49–51
    sanitation services, **7:**145
    South African Department of Water Affairs and Forestry, **1:**96
    threatened/at risk species, **7:**56
    water conservation, **8:**27–28
    See also Lesotho Highlands project
South America:
    aquifers, transboundary, **7:**3
    bottled water, **4:**18, 289, 291, **5:**163, 281, 283
    cholera, **1:**266, 270, 271
    dams, **1:**75, **3:**293
    drinking water, **1:**253, **3:**257, **5:**243, **7:**236–37
    environmental flow, **5:**34
    groundwater, **4:**86
    hydroelectric production, **1:**278
    irrigation, **1:**299, **2:**80, 259, 265, **3:**289, **4:**296, **5:**299, **6:**327, 333
    mortality rate, under-5, **7:**260–61
    population data, total/urban, **1:**248
    privatization, **3:**60
    renewable freshwater supply, **1:**238, **2:**200–201, 217, **3:**240, **4:**264, **5:**224
        2011 update, **7:**218
        2013 update, **8:**224
    reservoirs, **2:**270, 272, 274
    river basins, **6:**289, 308–11
    rivers, transboundary, **7:**3
    runoff, **2:**23
    salinization, **2:**268
    sanitation services, **1:**258, **3:**266, **5:**252, **7:**246–47
    threatened/at risk species, **1:**293
    water availability, **2:**217

South America (*continued*)
  withdrawals, water, **1:**243, **2:**207–8, **3:**247–48, **4:**271–72, **5:**232–33
    by country and sector, **7:**225–26, **8:**231–32
  See also Latin America
Southeastern Anatolia Project (GAP):
  archaeology in the region, **3:**183
  Euphrates River, developments on the, **3:**183–87
  overview, **3:**181–83
  summary/conclusions, **3:**190–91
  Tigris River, developments along the, **3:**187–90
Southern Bottled Water Company, **4:**39
South Sudan, **8:**164
Soviet Union, former:
  air quality, satisfaction, 2012, **8:**391
  cholera, **1:**58
  climate change, **1:**147
  dams, **1:**70, **3:**293
  environmental movement, **1:**15
  intl. river basin, **2:**29, 31
  irrigation, **1:**301, **2:**263, 265, **3:**289, **4:**296
  renewable freshwater supply, **4:**266
    2011 update, **7:**220
    2013 update, **8:**226
  water quality, satisfaction, 2012, **8:**391
  withdrawals, water, **1:**244, **2:**211, **3:**250–51
  See also Russia
Spain:
  agriculture, **4:**89
  conflict/cooperation concerning freshwater, **5:**5
  dams
    hydroelectric production, **1:**71
    Three Gorges Dam, **1:**89
    World Commission on Dams, **3:**161
  environmental flow, **5:**33
  globalization and intl. trade of water, **3:**45, 47
  groundwater, **4:**89
  mining, **7:**65
Sparkling water, **4:**30, **7:**158. *See also* Bottled water
Special Climate Change Fund (SCCF), **6:**51–52
Spectrometer, neutron, **3:**213
Spectroscopy, telescopic, **5:**175
Spiritual issues. *See* Religious importance of water
Spragg, Terry, **1:**203–5, **8:**135
Spring water, **4:**30, **7:**161–62, 163. *See also* Bottled water
Sri Lanka, **1:**69, **2:**86, **3:**161–62
  dams, **5:**134, **7:**334
  environmental flow, **5:**32
  floods, **5:**106
Starbucks, **5:**163, **7:**26
State Environmental Protection Administration, Chile (SEPA), **6:**80–81, 94
Stationarity, **6:**45
Statoil, **6:**23
Stewardship, water, **8:**22, 26
Stock-limited resources, **6:**6–7
Stocks and flows of freshwater:
  flows of freshwater, **2:**22–24
  hydrologic cycle, **2:**20–27
  major stocks of water on Earth, **2:**21–22
  overview, **2:**19–20
  rivers (*See* Rivers, runoff)
  summary/conclusions, **2:**36–37
  transboundary agreement strategies, **7:**8
  See also International river basins
Stone & Webster Company, **2:**108, **5:**61
Storage volume relative to renewable supply (S/O), **3:**102
Storm frequency/intensity, changes in, **1:**142–43, **4:**161–63
Stormwater runoff, **7:**53, 76, 77–78
  and hydraulic fracturing, **8:**76–77
  sustainable practices, **8:**39, 41–44

Streams, **5:**305–7
  effects of poor water quality on, **7:**54–55
  impacts of fossil-fuel extraction/processing, **7:**85, 87
  restoration, **8:**45–46
Strong, Maurice, **1:**88
Structure of Scientific Revolutions (Kuhn), **1:**193
Stunting, **6:**58
Sturgeon, **1:**77, 90, **2:**123
S2C Global Systems, **8:**135–36
Submillimeter Wave Astronomy Satellite (SWAS), **3:**219–20
Sub-Saharan Africa. *See* Africa, sub-Saharan
Subsidies. *See* Economy/economic issues
Substitutes, **6:**8–9
Sudan, **1:**55, **2:**26, **5:**7, 13
  dams with Chinese financiers/developers/builders, **7:**133, 334–35
  and the Nile River Basin, **7:**11, **8:**144, 164
  schistosomiasis and dam construction, **7:**58
  water shortages in refugee camps, **8:**164
Suez Canal, **8:**153
Suez Lyonnaise des Eaux, **3:**61–63, **4:**46
Supervisory Control and Data Acquisition (SCADA), **5:**16
Supply. *See* Renewable freshwater supply
Supply-chain management policies and programs, **6:**24
Supply-side development. *See* Twentieth-century water-resources development
Surface water:
  as "blue water" in water footprint concept, **8:**86, 89–90, 323, 332, 341, 350
  in China, **6:**81
  effects of climate change, **6:**43
  "produced" water from hydraulic fracturing, **8:**72
  See also Lakes; Rivers; Streams
Sustainability reports. *See* Global Reporting Initiative
Sustainable Asset Management (SAM) Group, **3:**167
Sustainable vision for the Earth's freshwater:
  agriculture, **1:**187–88
  business/industry, **7:**23, **8:**19–22, 30, 31
  climate change, **1:**191
  conflict/cooperation concerning freshwater, **1:**190
  criteria, **1:**17–18, **8:**37
  diseases, water-related, **1:**186–87
  ecosystems water needs identified and met, **1:**188–90
  funding, need for sustainable, **8:**8, 9–10
  human needs, basic, **1:**185–86
  overview, **1:**183–84, **8:**37
  public participation/perception, **1:**82, **7:**24–25
  stormwater management, **8:**41–44
  sustainable water jobs, **8:**35–59
  See also Soft path for water; Twenty-first century water-resources development
Swaziland, **1:**121
Sweden, **1:**52, 96, **3:**162
Switzerland, **1:**89, 171, **3:**162, **5:**33, 35
S&W Water, LLC, **5:**61
Sydney Morning Herald, **5:**66
Synthesis Report (2001), **5:**136
Syria:
  conflicts and the role of water, **1:**109, 110–11, 116, 118
  climate change, **8:**149–51
  overview and context, **8:**147–49, 165
  strategic attacks on infrastructure, **8:**149
  2011-2012, **8:**165
  dams, **7:**335
Systems Research, **2:**56

T

Taenia solium, **4**:8
Tahoe-Truckee Sanitation Agency, **2**:152
Tajikistan, **5**:9, **7**:335, **8**:166
Tampa Bay (FL). *See under* Desalination, plants
Tanzania, **1**:63, **7**:11, 335, **8**:168
Tapeworm, pork, **4**:8
Target Corporation, **6**:27
Tar sands, **7**:75, 78, 85, 87, 91–92
Tear Fund, **5**:131
Technical efficiency, **4**:103–4
Technology development, **5**:219, 220, **6**:60, **7**:67
Technology transfer, **8**:10–11
Temperature, measuring, **2**:307, **3**:325, **4**:332, **7**:348
Temperature rise, global, **1**:138, 145, **3**:120–23, **4**:159, 166. *See also* Climate change *listings*; Greenhouse effect; Greenhouse gases
Tennant Method and environmental flow, **5**:38
Tennessee Valley Authority, **1**:69–70, 145, **7**:307
Terrorism, **2**:35, **5**:1–3
   chemical/biologic attacks, vulnerability to, **5**:16–22
   cyberterrorism, **5**:16, **6**:152, **7**:176
   defining terms, **5**:3–5
   detection and protection challenges, **5**:23
   early warning systems, **5**:23–24
   environmental terrorism, **5**:3–5
   overview, **5**:1–2, **7**:176
   physical access, protection by denying, **5**:22–23
   poisoning of water supply, **8**:163, 165, 167
   policy in the U.S., security, **5**:23, 25
   public perception/response, **5**:2–3
   response plans, emergency, **5**:24–25
   summary/conclusions, **5**:25–26
   water infrastructure attacks, **5**:15–16, **8**:149, 167
   in water-related conflict, **5**:5–15, **7**:176, **8**:165 (*See also* Water Conflict Chronology)
   and water treatment, reducing vulnerability, **5**:2
Texas, **3**:74–75
   Austin, **1**:22
   hydraulic fracturing for natural gas, **8**:67–68
   North American Water and Power Alliance, **8**:126
Thailand, **4**:40, **5**:33, 106, 134
   arsenic in groundwater, **7**:59
   dams with Chinese financiers/developers/builders, **7**:335–36
   drought, **7**:132
   and the Mekong River, **7**:14, 130
   and the Salween River, **7**:132
Thames Water, **3**:63, **5**:62
Thatcher, Margaret, **1**:106, **3**:61
Thirsty for Justice, **5**:122
Threatened/at risk species:
   Colorado River, **3**:134, 142
   by country, **2**:291–97
   dams, **1**:77, 83, 90, **2**:120, 123
   extinct in the wild, freshwater animal species, **7**:292–302
   fish (*See* Fish, threatened/at risk species)
   proportion of species at risk in U.S., **4**:313–16
   Red List, **7**:292
   by region, **1**:291–96, **7**:56
   twentieth-century water-resources development, **3**:3
   water transfers, **3**:39–40
   See also Extinct species
Three Affiliated Tribes, **5**:123
Three Gorges Dam, **6**:139–49
   chronology of events, **1**:85–87, **6**:147–48
   climatic change caused by, **6**:146–47
   costs of, **6**:141–42
   dimensions of, **6**:140–41
   displaced people, **1**:78, 85, 90, **5**:134, 151
   economic issues, **1**:16, **6**:141–42
   financial costs of, **6**:141–42
   fisheries, **6**:142–43
   flood protection benefits, **6**:144
   funding of, **1**:86–89, **6**:141–42
   geological instability caused by, **6**:144–45
   history, **6**:140, **7**:133
   hydroelectric production, **1**:84, **6**:140
   impacts of, **1**:89–92, **6**:142–43
   largest most powerful ever built, **1**:84
   military targeting of, **6**:146
   opposition to, **1**:91–93
   overview, **6**:148–49, **7**:129
   population relocation and resettlement caused by, **6**:145–46
   river sediment flow effects, **6**:143–44
   seismicity caused by, **6**:144–45
   shipping benefits of, **6**:144
   size of, **6**:140
   storage capacity of, **6**:140
   threats to, **6**:139
Time, measuring, **2**:304, **4**:329, **7**:345
Timor Leste, **5**:9, **7**:264
Togo, **1**:55, **7**:336
Toilets, **1**:21–22, **3**:4, 118, **4**:104, 109, 113–14, **6**:106, 110
   low-flush, employment to install, **8**:39
Tonga, **3**:46, 118
Touré, A. T., **1**:53
Tourism, **7**:64, **8**:26, 128, 155, 164
Toxic waste dumps, **5**:119, 124
Toxic Wastes and Race in the United States, **5**:119
Trachoma, **1**:48, **7**:272
Trade. *See* Globalization and international trade of water
Traditional planning approaches, **1**:5. *See also* Projections, review of global water resources; Twentieth-century water-resources development
Transfers, water, **1**:27–28, 74–75, **3**:39–40
   Alaska water shipments, proposed, **8**:133, 135–36
   bag technology, **1**:200–205, **8**:135
   Disi Water Conveyance Project, **8**:157
   Garrison Diversion Project, **8**:140
   from Gulf of California to the Salton Sea, **8**:144
   Las Vegas Pipeline Project, **8**:136–40
   Missouri River diversion project, **8**:140–44
   North American Water and Power Alliance (NAWAPA), **1**:74, **8**:124–31
   Reber Plan (San Francisco Bay Project), **8**:131–34
   Red Sea-Dead Sea projects, **8**:144, 153–57
   by tanker, **8**:130, 135, 136
   See also Dams
Transparency International, **8**:12
Transparency issues, **8**:12–14, 112
Transpiration loss of water into atmosphere, **1**:141, **2**:83, **4**:159–60
Transportability, **6**:7–8
Transportation:
   canals, **8**:128, 144, 153, 154
   energy costs, **7**:161–62, 163
   by tanker, **8**:130, 135, 136
Treaties. *See* Law/legal instruments/regulatory bodies; United Nations
Trichuriasis, **1**:48, **7**:272
Trinidad and Tobago, **2**:108, **5**:72, **7**:264
Trout Unlimited, **2**:118, 123, 128
True Alaska Bottling Company, **8**:135–36
Trypanosomiasis, **7**:272
Tuna, **3**:49, 50
Tunisia, **2**:142, **5**:33, **7**:336
Turkey:
   bag technology, water, **1**:202–5, **8**:135
   conflict/cooperation concerning freshwater, **1**:110, 118, **5**:8

Turkey (*continued*)
  conflicts concerning freshwater, and Syria, **8:**148
  dams with Chinese financiers/developers/builders, **7:**336
  environmental concerns, top, **7:**287
  globalization and intl. trade of water, **3:**45–47
  terrorism, **5:**22
Turkish Antiquity Service, **3:**183. *See also* Southeastern Anatolia Project
Turtles, **3:**49, 50
Tuvalu, **5:**136
Twentieth-century water-resources development:
  Army Corps of Engineers and Bureau of Reclamation, U.S., **1:**7–8
  benefits of, **3:**2
  capital investment, **1:**6–7
  drivers of, three major, **1:**6
  end of
    alternatives to new infrastructure, **1:**17–18
    demand, changing nature of, **1:**10–14
    economics of water projects, **1:**16–17
    environmental movement, **1:**12, 15–16
    opposition to projects financed by intl. organizations, **1:**17
    overview, **1:**9–10
    shift in paradigm of human water use, **1:**5–6
      government, reliance on, **1:**7–8
      limitations to, **1:**8–9, **3:**2–3
      problems/disturbing characteristics of current situation, **1:**1–2
    summary/conclusions, **1:**32
    supply-side solutions, **1:**6
Twenty-first century water-resources development:
  agriculture, **1:**23–24
  alternative supplies, **1:**28, **8:**48–50, 51, 52
  desalination, **1:**29–32
  efficient use of water, **1:**19–20
  governance issues, **8:**6–16
  industrial water use, **1:**20–21
  overview, **1:**18–19
  Pacific Island developing countries, **3:**121–23
  paradigm shifts, **1:**5–6, **8:**19, 22
  pricing, water, **1:**24–28
  reclaimed water, **1:**28–29
  residential water use, **1:**21–23
  summary/conclusions, **1:**32–33
  See also Soft path for water; Sustainable vision for the Earth's freshwater
Typhoid, **1:**48, **7:**57, 58
Typhus, **1:**48

U
Uganda, **1:**55, **4:**211
  conflicts concerning freshwater, 2011-2012, **8:**167
  dams with Chinese financiers/developers/builders, **7:**336
  and the Nile River Basin, **7:**11
  wastewater treatment by the Akivubo Swamp, **7:**63
Ultraviolet radiation, bottled water treatment, **7:**159, 161
Unaccounted for water, **3:**305, 307, **4:**59
Underground storage tanks (UST), **7:**77
Undiminished principle and the human right to water, **5:**37
Unilever, **5:**149, **6:**24, **7:**35
United Arab Emirates (UAE), **5:**68–69, **7:**75
United Kingdom:
  environmental concerns, top, **7:**287, 288
  See also specific countries
United Nations:
  Agenda 21, **1:**18, 44, **3:**90
  arsenic in groundwater, **2:**167, 172
  Children's Fund (UNICEF), **1:**52, 55, **2:**167, 172, 173, **6:**60, 62
    collaboration with UN-Water, **8:**7
    data collection by, **7:**230, 241, **8:**236, 252
    Group DANONE aid, **7:**41
  Commission on Human Rights, **2:**5
  Commission on Sustainable Development, **2:**10, **3:**90
  Committee on Economic, Social, and Cultural Rights, **5:**117, 137
  Comprehensive Assessment of the Freshwater Resources of the World (1997), **1:**42–43
  Conference on International Organization (1945), **2:**5
  conflict/cooperation concerning freshwater, **1:**107, 114, 118–19, 124, 210–30, **2:**36, **8:**7–8
  data, strict access to water, **2:**41–42
  Declaration on the Right to Development (1986), **2:**8–10
  Development Programme (UNDP), **1:**52, 82, 171, **2:**172, 173, **3:**90, **4:**7, **5:**100, **8:**7
  diseases, water-related, **5:**117
  dracunculiasis, **1:**52, 55
  drinking water, **1:**40, 251
  droughts, **5:**100
  Earth Summit (1992), **3:**38, 88, 101
  Economic Commission for Asia and the Far East (UNECAFE), **7:**13
  Educational, Scientific, and Cultural Organization (UNESCO), **8:**7, 11, 14
  environmental justice, **5:**137–38
  Environment Programme (UNEP), **1:**137, **3:**127, 164, **7:**34, **8:**7
  Food and Agriculture Organization (FAO), **2:**64, 67, **5:**126
    AQUASTAT database, **4:**81–82, **7:**215, 221, **8:**221, 310
    collaboration with UN-Water, **8:**7
    Syria, **8:**148
  food needs for current/future populations, **2:**64, 66, 67
  Framework Convention on Climate Change (UNFCC), **3:**126, **6:**48–51
  Global Water Partnership, **1:**165, 166, 171, 175, **5:**183, **6:**73
  greenhouse gases, **3:**126
  groundwater, **2:**167, 172, **4:**80–81
  Human Poverty Index, **3:**87, 89, 90, 109–11
  human right to water, **2:**3, 5–9, 14, **4:**208, 214, **5:**117
    formal recognition, **7:**251, **8:**268
  Industrial Development Organization, **7:**278–81
  Inter-agency Group for Child Mortality Estimation, **7:**264
  intl. river basins assessment, **7:**2–3
  intl. water transfer policy, **8:**130
  public participation and sustainable water planning, **1:**82
  Summit for Children (1990), **1:**52, **2:**14
  Universal Declaration of Human Rights, **2:**4–10, **4:**208
  UN-Water, **8:**7, 8, 12
  watercourses, uses of (*See* Convention of the Law of the Non-Navigational Uses of International Watercourses)
  well-being, measuring water scarcity and, **3:**90, 96, 109–11
  World Water Council, **1:**172, 173, 175
  See also Law/legal instruments/regulatory bodies, International Covenant on Economic, Social, and Cultural Rights; League of Nations; Millennium Development Goals
United States:
  Alaska water shipments, proposed, **8:**133, 135–36
  bottled water (*See under* Bottled water)
  budgets, U.S. federal agency water-related, **7:**303–7

Index 471

business/industry, water risks, **5:**162–63
cholera, **1:**56, 266, 270, 271
climate change, **1:**148, **7:**153
Colorado River Basin, intl. agreements, **7:**6, 8–9, 15–16, **8:**91
conflict/cooperation concerning freshwater, **1:**110, 111, **5:**6–9, 11–12, 24
dams, **1:**69–70, **3:**293, **7:**52
desalination, **2:**94–95, 97, **5:**58–63
diseases, water-related, **4:**308–12
dracunculiasis, **1:**52
drinking water, **1:**253, **3:**257, **5:**242
    access, **3:**280–88, **7:**236, **8:**36–37
droughts, **5:**93, **6:**44, **8:**130
economic productivity of water, **4:**321–24
environmental concerns, **6:**339–41
environmental concerns of the public, top, **7:**282–84, 287, 288
environmental flow, **5:**34, 36–37
environmental justice, **5:**119–20, 122–23
floods, **4:**305–7
food needs for current/future populations, **2:**68–69
fossil-fuel production, **7:**75, 78–79, 82, 85, 88, 90
General Agreement on Tariffs and Trade, **3:**50
Great Lakes Basin, intl. agreements, **7:**165–69
groundwater, **3:**2, **4:**82, 86, 96, **5:**125
    arsenic in, **7:**59
human right to water, **4:**213
hydroelectric production, **1:**71, 278, **7:**129
introduced/invasive species, **7:**48
irrigation, **1:**299, **2:**265, **3:**289, **4:**296, **5:**299, **7:**16
Las Vegas Pipeline Project, **8:**136–40
meat consumption, **2:**79–80
Missouri River diversion project, **8:**140–44
mortality rate, under-5, **7:**261
North American Free Trade Agreement, **3:**47–48, 51–54, **8:**130
North American Water and Power Alliance, **8:**124–31
pesticides, **5:**305–7
population data/issues, **1:**248, **2:**214
precipitation changes, **1:**146–47
privatization, **3:**58–60
radioactive contaminants, **7:**52
renewable freshwater supply, **1:**238, **2:**200, 217, **3:**240, **4:**264, **5:**224, **6:**83–84
    2011 update, **7:**218
    2013 update, **8:**224
reservoirs, **2:**270, 272, 274
runoff, **2:**23
salinization, **2:**268
sanitation services, **1:**258, **3:**266, 272, **5:**251, **7:**246
terrorism, **5:**21–23, 25
threatened/at risk species, **1:**293, **4:**313–16, **7:**56
transboundary waters, **7:**3, **8:**124–31
unemployment statistics, **8:**35
usage estimates, **1:**245, **8:**125
water availability, **2:**217, **8:**36–37
water footprint, **8:**87–88, 89, 90, 91
water industry revenue/growth, **5:**303–4
water policy reform
    background, **7:**143–44
    key steps to, **7:**151–54
    need for, **7:**143, 150–51, 154
well-being, measuring water scarcity and, **3:**92
withdrawals, water, **1:**10, 11, 12, 13, 243, **2:**207, **3:**247, 308–12, **4:**271, 317–20, **5:**232
    1900-2005, **8:**142
    2005, **7:**225, **8:**231
United Utilities, **3:**63
United Water Resources, **3:**61, 63
United Water Services Atlanta, **3:**62, **4:**46
Universidad de San Augustin, **2:**179

University of California at Santa Barbara (UCSB), **3:**20–22
University of Kassel, **2:**56
University of Michigan, **3:**183
Upper Atmosphere Research Satellite (UARS), **1:**196
Uranium, **7:**74
Urban areas:
    drinking water access in, **6:**70–71, **7:**97, 115–17 (*See also* Drinking water, access, by country)
    droughts, **5:**98, **7:**114–21
    floods, **5:**104
    future demands in, **6:**102–4
    municipal water, **1:**29, **4:**29, **5:**73, **6:**101
    pricing, water, **1:**25–27, **4:**124–25, **7:**118
    privatization, **3:**76
    reclaimed water, **1:**25, **2:**146–49, **7:**114
    sanitation services in, **6:**70–71 (*See also* Sanitation services, access by country)
    soft path for meeting water-related needs, **3:**20–22, **7:**150
    stormwater runoff management techniques, **8:**42
    water rate structure (*See* Water rate structures)
    water use in, **6:**101–2, **7:**115–17
Urbanization, **5:**98, **7:**53
Urfa, **3:**185
Urlama, **5:**5
U.S. Filter Company, **3:**63
U.S. National Water Assessment, **5:**112
User fees and environmental flows, **5:**41–42
Utah:
    North American Water and Power Alliance, **8:**124–31
    Snake Valley, **8:**139
Utilities and risks that face business/industry, **5:**162–63
Uzbekistan, **1:**52, **4:**40, **7:**336, **8:**166

V
van Ardenne, Agnes, **4:**196
Varieties of Environmentalism (Guha & Martinez-Alier), **5:**123
Vegetation:
    Colorado River, **3:**134, 139–42
    gardens, **1:**23, **4:**122–23, **7:**115, 116, **8:**42
    green roofs, **8:**42
    lawns, **1:**23, **4:**122–23, **7:**115, 116
    planter boxes, **8:**42
    rain gardens, **8:**42
Vehicles, impacts of truck traffic, **8:**75
Velocity, measuring, **2:**305, **3:**323, **4:**330, **7:**346
Venezuela, **7:**75
Veolia, **6:**93
Veolia Environnement, **5:**162
Vermont Natural Resources Council, **2:**128
Vibrio, **7:**47. *See also* Cholera
Vibrio cholerae, **1:**56, 57, 58, **7:**57. *See also* Cholera
Vietnam, **4:**18, **5:**134, 163
    arsenic in groundwater, **7:**59
    dams with Chinese financiers/developers/builders, **7:**130, 336–37
    and the Mekong River Basin, **7:**14, 130
Viking, **3:**214, **5:**177
Virginia, **2:**152, **8:**54
Virgin Islands, U.S., **3:**46
Visalia (CA), **1:**29
Vision 21 process, **2:**3
Vivendi, **3:**61–64, 70, **4:**47
Voith and Siemens, **1:**85
Volume, measuring, **2:**303–4, **3:**321–22, **4:**328–29, **7:**344–45. *See also* Stocks and flows of freshwater

W

Waggoner, Paul, **1:**149
Waimiri-Atroari people, **5:**134
Wales, **7:**63. *See also* United Kingdom
Wall Street Journal, **1:**89
Warfare, **5:**4–5. *See also* Conflict/cooperation concerning freshwater; Terrorism; Water Conflict Chronology
Warming, global, **1:**138. *See also* Climate change *listings;* Greenhouse effect; Greenhouse gases
Washing machines, **1:**23, **4:**114–16, **5:**219, 220
Washington. *See* Seattle
Washington, D. C., green roofs initiative, **8:**44
Waste management:
    hazardous/toxic waste landfills, **5:**119, 124
    Overseas Development Assistance, **7:**277
        by donating country 2004-2011, **8:**317–19
        by subsector 2007-2011, **8:**320–22
    See also Wastewater
Wastewater:
    business/industry effluent, **7:**26, 28–29, 51, 55–56, 73–74
    disposal associated with hydraulic fracturing, **8:**72, 74–75, 76
    dumped into rivers, **6:**82
    human waste disposal, **7:**52–53
    from hydraulic fracturing for natural gas, **8:**64, 65, 66, 71–75
    pits for storage, **8:**72
    rates for, **6:**118–19
    treatment of, **1:**6, **2:**138, **5:**153, 159–60, **6:**27, 92
        energy considerations, **8:**74
        expansion and improvement, **7:**65
        inappropriate, in municipal treatment plants, **8:**74
        overwhelmed by stormwater runoff, **7:**53, **8:**39
        for reuse, **8:**49
        reverse osmosis, **8:**28
    volume produced per day, U.S., **8:**49
    See also Reclaimed water; Sanitation services
Wasting, **6:**58
Water:
    "blue," in water footprint concept, **8:**86, 89–90, 323, 332, 341, 350
    bottled (*See* Bottled water)
    embedded/embodied/virtual/indirect, **3:**18–19, **6:**7, **8:**83, 91 (*See also* Water footprint)
    footprint (*See* Water footprint)
    in goods, **6:**335–38
    "green," in water footprint concept, **8:**85–86, 90, 323, 332, 341, 350
    "grey," in water footprint concept, **8:**86, 323, 332, 341, 350
    lack of substitutes for, **6:**8–9, 13
    origins of, **1:**93–98, **3:**209–12, **6:**5
    pricing (*See* Pricing, water)
    produced, **6:**23, **7:**73, 76, 79
    rights (*See* Water rights)
    right to (*See* Human right to water; Nature's right to water)
    running out of, **6:**4–5
    stocks of, **6:**6
    units/data conversions/constants, **2:**300–309, **3:**318–27, **4:**325–34, **5:**319–28, **7:**341–50, **8:**395–404
    vs. oil, **6:**3–9, 14
Water (Vizcaino, et al.), **5:**219
WaterAid, **5:**131
Water allocation:
    instream, preserving/restoring, **5:**29–30 (*See also* Environmental flow)
    transboundary water agreements, **8:**14–15, 166
    transboundary waters and climate change, **7:**7, 13
    volumetric systems, **4:**95–97
Water-based diseases, **7:**58. *See also* Diseases, water-related
Waterborne diseases, **1:**47–49, 274–75, **4:**8, **7:**57–58. *See also* Diseases, water-related
Water Conflict Chronology, **1:**108–9, 125–30, **2:**35, 182–89, **3:**194–206, **4:**xvii–xviii, 238–56, **5:**5–15, 190–213, **6:**151–93
    cited by the media, **8:**173
    overview and use of terms, **8:**160
    2011 update, **7:**1, 175–205
    2013 update, **8:**173–208
    website, **7:**175, **8:**147, 160
Water conservation:
    Atlanta (GA), **6:**108–9
    Australia, **7:**106, 109, 114–21
    California commercial/industrial water use
        background to CII water use, **4:**132–33
        calculating water conservation potential, methods for, **4:**143–47
        current water use in CII sectors, **4:**133–38
        data challenges, **4:**139–40, 148–50, 152–53
        defining CII water conservation, **4:**132
        evolution of conservation technologies, **4:**149
        overview, **4:**131–32
        potential savings, **4:**140–43
        recommendations for CII water conservation, **4:**150–53
        summary/conclusions, **4:**153–54
        water use by end use, **4:**138–39
    California residential water use
        abbreviations and acronyms, **4:**126–27
        agricultural water use, **4:**107
        current water use, **4:**105–6, **8:**38
        data and information gaps, **4:**108–9
        debate over California's water, **4:**102–3
        defining conservation and efficiency, **4:**103–5
        economics of water savings, **4:**107–8
        indoor water use
            dishwashers, **4:**116
            end uses of water, **4:**112–13
            faucets, **4:**117
            leaks, **4:**117–18
            overview, **4:**109
            potential savings by end use, **4:**111–12
            showers and baths, **4:**114
            summary/conclusions, **4:**118
            toilets, **4:**113–14, **8:**39
            total use without conservation efforts, **4:**110
            washing machines, **4:**114–16
        outdoor water use
            current use, **4:**119–20
            existing efforts/approaches, **4:**120–21
            hardware improvements, **4:**122–23
            landscape design, **4:**122–23, **8:**38
            management practices, **4:**121–22
            overview, **4:**118–19
            rate structures, **4:**124–25
            summary/conclusions, **4:**125–26
        overview, **4:**101–2
    cost analysis, **8:**100
    description of, **6:**106
    indoor, **6:**110–12, **8:**38
    Las Vegas (NV), **6:**107–8, 110–12
    outdoor, **6:**112, **8:**38
    rainwater catchment, **7:**120–21, **8:**25, 42, 48
    rebates and incentives for, **6:**110–12, **7:**119, 121
    South Africa, **8:**27–28
    Tampa Bay (FL), **8:**107
    technology to monitor, **8:**39
    U.S., **8:**142
Water Efficient Technologies, **6:**107

Index 473

Water footprint:
 calculation, **8:**85–86
 California *vs.* U.S., **8:**88–90, 91
 concept, **8:**83–85, 323, 332, 341, 350
 conclusion, **8:**90–92
 findings, **8:**87–90
 limitations in data and reporting, **8:**323–24, 332–33, 341–42, 350–51
 and nexus thinking, **8:**85–86
 per-capita, of national consumption, by country 1996-2005, **8:**323–31
 per-capita, of national consumption, by sector and country 1996-2005, **8:**332–40
 total, of national consumption, by country 1996-2005, **8:**341–49
 total, of national consumption, by sector and country 1996-2005, **8:**350–58
Water Footprint Network (WFN), **8:**85, 87
Water in Crisis: A Guide to the World's Fresh Water Resources (Gleick), **2:**300, **3:**318
Water industry. *See* Business/industry, water risks; Economy/economic issues
Water Integrity Network, **8:**12
Water landscape, **6:**36
Water market and water trading, **3:**47–48, **7:**110, 111–14, 146, **8:**102. *See also* Pricing, water
Water performance reporting, **6:**28–31
A Water Policy for the American People, **3:**16, **4:**103
Water & Process Technologies, **5:**159
Water quality:
 acidification (*See* Acidification)
 bottled water, **4:**17, 25–26, 31–32, 37–40, **7:**159–60, 161
 business/industry, water risks that face, **5:**146–49, **7:**26
 China, **6:**80–82
 climate change, **4:**167–68, **6:**44, **7:**7
 community-level consequences of poor, **7:**61–62
 contaminants
  assimilation, and "grey water" in water footprint concept, **8:**86, 323, 332, 341, 350
  emerging, **7:**48–49
  fecal, **7:**52–53, 57–58, **8:**155
  from fossil-fuel extraction/processing, **7:**88–89, **8:**63–64
  methane, **8:**69, 70–71, 73
  organic, **7:**278–81, **8:**155
  overview, **7:**46–47, **8:**3
  pathogenic organisms, **7:**47
  See also specific contaminants
 droughts, **5:**98, 102
 ecological consequences of poor, **7:**54–57
 economic/social consequences of poor, **7:**62–65
 environmental justice, **5:**127–29
 floods, **5:**109
 groundwater, **4:**83, 87, **8:**68
 Guidelines for Drinking-Water Quality, **4:**26–27, 31
 human health consequences of poor, **7:**57–60
 impacts of fossil-fuel extraction/processing, **7:**51, 73–74, 76–93, **8:**66
 monitoring technology, **8:**49
 overview, **7:**45
 pollution prevention, **7:**65, **8:**24–25, 52
 privatization, **3:**78
 salinity issues (*See* Desalination; Salinization)
 satisfaction
  countries with most and least, **8:**392–93
  by country, **7:**289–91
  limitations on data collection by poll, **8:**387, 389, 392
 by region
  2012, **8:**389–91
  sub-Saharan Africa, by country, **7:**290
  2012, **8:**387–88
 sedimentation (*See* Sedimentation)
 temperature/thermal pollution, **7:**46–47, 51, 53, 85
 three-tier classification of impacts, **7:**52
 transfer of water by tanker, **8:**136
 water quantity consequences of poor, **7:**60
 See also Desalination; Drinking water, access; Environmental flow; Salinization
Water rate structures, **3:**303
 Atlanta (GA), **6:**115–16
 average price, **6:**117
 benefits of, **6:**112, 114
 consumption charges, **6:**117
 flat, **6:**114
 inclining block, **6:**114
 Las Vegas (NV), **6:**115–16
 seasonal, **1:**26, **6:**114
 Seattle (WA), **6:**115–17
 summary of, **6:**119
 uniform, **6:**114
 wastewater, **6:**118–19
Water reporting:
 by companies, **6:**18, 20 (*See also* Corporate reporting)
 inconsistency in, **6:**33
 performance, **6:**28–31
 recommendations for, **6:**37–38
 by sector, **6:**32–35
Water Resources Policy Committee, **3:**16
Water rights, **7:**111, 112, 113, **8:**138, 156
Water risk, corporate. *See* Business/industry, water risks
Watersheds, **6:**10–11. *See also* International river basins
 impact of hydraulic fracturing for natural gas, **8:**67–68
 restoration, **8:**45
Water Supply and Sanitation Collaboration Council, **7:**230, 241, **8:**236, 252
Water use:
 company reporting on, **6:**18, 20 (*See also* Corporate reporting)
 consumptive/nonconsumptive, **6:**7, **8:**84, 87 (*See also* Water footprint)
 definitions and use of term, **1:**12, **8:**84
 direct *vs.* indirect, **3:**18–19, **6:**7, **8:**83 (*See also* Water footprint)
 domestic (*See* Water footprint, per capita, of national consumption, by sector and country)
 efficiency (*See* Water-use efficiency)
 estimates, **1:**46, 246
 fossil-fuel extraction/processing and energy production, **7:**25, 26, 31, 73–84
 global, statistics, **8:**87
 increases in, **6:**1
 industrial, percent used for, **7:**74
 institutional, **6:**102
 measurement, **2:**25, **7:**30–31
 process water use and CII water use, **4:**134–36
 restriction policy in urban areas, **7:**115–17
 vs. Gross Domestic Product, **8:**142
 See also Withdrawals, water
Water-use efficiency, **1:**19–20, **3:**77–78, **4:**xvi–xvii, 58–60, **5:**153, 157, **6:**106–9
 age of homes and, **6:**112
 agriculture, **3:**4, 19–20, **7:**109–10, 146, **8:**50–54
 Atlanta (GA), **6:**108–9
 Australia, **7:**118–21
 business/industry, **7:**35, 37, 39, **8:**28

Water-use efficiency (*continued*)
    jobs created by, **8:**39
    Las Vegas (NV), **6:**107–8
    legislation and policy, **7:**28, 118–19, 153–54
    measurement, **7:**30–31
    overview of techniques to increase, **8:**37–38
    Seattle (WA), **6:**109, **8:**38
    Tampa Bay (FL), **8:**107
    U.S. policy, **7:**153–54
    See also Soft path for water; Sustainable vision for the Earth's freshwater; Twenty-first century water-resources development
Waynilad Water, **4:**46
WCD. *See* World Commission on Dams
Weather Underground group, **5:**20
Websites, water-related, **1:**231–34, **2:**192–96, **3:**225–35
    short documentaries and films, **7:**174
    Water Conflict Chronology, **7:**175, **8:**147, 160
Weight of water, measuring, **2:**309, **3:**327, **4:**334, **7:**350
Well-being, measuring water scarcity and:
    Falkenmark Water Stress Index, **3:**98–100
    multifactor indicators
        Human Poverty Index, **3:**87, 89, 90, 109–11
        Index of Human Insecurity, **3:**107, 109
        International Water Management Institute, **3:**108, 197
        overview, **3:**101
        vulnerability of water systems, **3:**101–4
        Water Poverty Index, **3:**110–11
        Water Resources Vulnerability Index, **3:**105–6
    overview, **3:**xviii–xix, 87–88
    quality-of-life indicators, **3:**88–96, **6:**61
    single-factor measures, **3:**96–98, 101–3
    summary/conclusions, **3:**111
Well water, **4:**28, 30. *See also* Groundwater
Western Pacific Region. *See* Pacific Region, Western
Wetlands, **1:**6, **3:**141–43, **5:**111, **6:**86, 88
    degradation, **7:**56, 63
    ecosystem services by, **7:**56, 63
    loss, **8:**4, 132
    restoration, **8:**48
Wetlands, specific:
    Amazon River, **7:**56
    Ciénega de Santa Clara, **3:**141–43
    El Doctor, **3:**141
    El Indio, **3:**141, 142
    Nakivubo, **7:**63
    Rio Hardy, **3:**141
    San Francisco Bay, **8:**132
Wheat, **2:**75, **4:**89
Whipworm, **7:**61
White, Gilbert, **3:**16
Wildlife:
    Colorado River, **1:**77, **3:**134
    National Fish and Wildlife Foundation, **2:**124
    reserves, **7:**47–48
    World Wildlife Fund International, **3:**157
    See also Birds; Mammals
Williams, Ted, **2:**119
Wind power, **2:**105, **7:**25
Wintu people, **5:**123
Wisconsin, **2:**117–18
Withdrawals, water:
    conflict/cooperation concerning, **1:**112, **8:**24
    by country and sector, **1:**241–44, **2:**203–11, **3:**243–51, **4:**267–75, **5:**228–36, **6:**202–10, **7:**223–29
        2013 update, **8:**227–33
    definitions and use of term, **1:**12, **7:**221–22, **8:**84, 227–28
    Great Lakes Basin restrictions, **7:**168, 169
    gross national product
        China, **3:**316–17
        Hong Kong, **3:**313–15
        U.S., **3:**310–12
    hydraulic fracturing for natural gas, **8:**65, 66, 67–69
    Las Vegas area, proposed, **8:**137, 138, 139
    population in the U.S., **1:**10, 12, 13
    soft path for meeting water-related needs, **3:**23–24
    threatened/at risk species, **3:**3
    total/per-capita, **1:**10, 11
    See also Groundwater, monitoring/management problems; Projections, review of global water resources
Wolf, Aaron, **2:**28
Wolff, Gary, **3:**xiv
Women:
    effects of poor water quality on, **7:**61
    and environmental justice, **5:**126, 134
    increased exposure to contaminated water, **7:**89
    responsibility for water collection, **7:**61, 89
    violence to, during water collection, **8:**168
World Bank:
    arsenic in groundwater, **2:**172–73
    business/industry, water risks that face, **5:**147
    dams, **1:**82–83, **7:**133, 134, 138
    Development Research Group, **7:**278
    diseases, water-related, **4:**9
    displaced people, dams and, **1:**78
    dracunculiasis, **1:**52
    effects of climate change on Middle East and Northern Africa, **8:**149
    Global Burden of Disease assessment, **7:**270
    Global Water Partnership, **1:**165, 171, **5:**183
    human needs, basic, **1:**44, 47
    human right to water, **2:**10–11
    Lesotho Highlands project, **1:**96, 99
    opposition to projects financed by, **1:**17
    overruns, water-supply projects, **3:**13
    privatization, **3:**59, 70, **4:**46
    Red Sea–Dead Sea projects, **8:**154, 155
    sanitation services, **5:**273
    self-review of dams funded by, **1:**175–76
    Southeastern Anatolia Project, **3:**190–91
    Three Gorges Dam, **1:**85, 88
    World Water Council, **1:**173
World Business Council for Sustainable Development (WBCSD), Global Water Tool, **7:**33, 34
World Climate Conference (1991), **1:**149
World Commission on Dams (WCD):
    data and feedback from five major sources, **3:**150
    environmental flow, **5:**30
    environmental justice, **5:**134, 135
    findings and recommendations, **3:**151–53
    goals, **3:**150–51, **7:**136
    organizational structure, **3:**149–50, **7:**136
    origins of, **1:**83, 177–79, **7:**136
    overview, **3:**xix, **7:**136–37
    priorities/criteria/guidelines, **3:**153–58, **7:**137–38, 139
    reaction to the report
        conventions, intl., **3:**166
        development organizations, intl., **3:**164–65
        funding organizations, **3:**158, 162–64, 167–69, **7:**138–39
        governments, **3:**170–71, **7:**139–40
        industry/trade associations, intl., **3:**169–70, **7:**139
        national responses, **3:**159–62, **7:**138–39
        nongovernmental organizations, **3:**157, **7:**138
        overview, **3:**155–56, **7:**138
        private sector, **3:**166
        regional groups, **3:**165
    rights and risk assessment, **3:**153, **7:**137
    Southeastern Anatolia Project, **3:**191
    summary/conclusions, **3:**171–72

Index

World Conservation Union (IUCN), **1:**82–83, 121, 177, **3:**164. *See also* International Union for Conservation of Nature
World Council on Sustainable Development, **5:**158
World Court, **1:**109, 120
World Food Council, **2:**14
World Fund for Water (proposed), **1:**174–75
World Health Assembly, **1:**52
World Health Organization (WHO):
    arsenic in groundwater, **2:**166, 167, 172
    bottled water, **4:**26–27
    childhood mortality, data, **7:**257
    cholera, **1:**61, 271 (*See also* Cholera, cases reported to the WHO; Cholera, deaths reported to the WHO)
    collaboration with UN-Water, **8:**7
    desalination, **5:**75
    diseases, water-related, **4:**9, **5:**117
    dracunculiasis, **1:**52, 55
    drinking water, **3:**2, **4:**2, 208, **6:**211
        access to, data, **7:**230, **8:**236
    Global Burden of Disease assessment, **7:**270
    human needs, basic, **1:**44
    human right to water, **2:**10–11
    Joint Monitoring Programme, **6:**60, 73, **7:**230, 241, **8:**236, 252
    reclaimed water, **2:**154, 155
    sanitation services, **1:**256, **3:**2, **4:**2, 208, **6:**62
        access to, data, **7:**241, **8:**252
    unaccounted for water, **3:**305
    well-being, measuring water scarcity and, **3:**90, 91
World Health Reports, **4:**8, 9
World Meteorological Organization (WMO), **1:**137, **5:**100, **8:**7
World Resources Institute, **2:**27–28, 49–50, **8:**2
World Trade Organization (WTO), **3:**48–50
Worldwatch Institute, **2:**28
World Water Council (WWC), **1:**172–76, **3:**165, **4:**192–93, **5:**183
World Water Development Report, **8:**6
World Water Forum:
    2000, **3:**xviii, 58, 59, 90, 173
    2003
        background to, **4:**192–94
        Camdessus Report, **4:**195–96, 206
        efficiency and privatization, lack of attention given to, **4:**192
        focus of, **4:**191
        human right to water, **4:**212
        Millennium Development Goals, **4:**6, 7
        Ministerial Statement, **4:**194–95, 200–204
        NGO Statement, **4:**192, 198, 205–6
        overview, **4:**xv
        successes of, **4:**191–92
        Summary Forum Statement, **4:**196–97
        value of future forums, **4:**192
    2006, **5:**186–88
World Wildlife Fund International, **3:**157
Wyoming, Pavillion gas field, **8:**71, 72, 73

X
Xeriscaping, **1:**23, **4:**123–24
Xylenes, **8:**72

Y
Yangtze! Yangtze!, **1:**92
Yeates, Clayne, **1:**194–95
Yemen, **1:**53, 55, **8:**164–65

Z
Zambia, **1:**63, **4:**211, **5:**9, 32
    dams with Chinese financiers/developers/builders, **7:**133, 337–38
Zimbabwe, **1:**107, **5:**7, 140, **7:**338
Zombie water projects:
    Alaskan water shipments, **8:**133, 135–36
    Las Vegas Pipeline Project, **8:**136–40
    Missouri River diversion project, **8:**140–44
    North American Water and Power Alliance (NAWAPA), **1:**74, **8:**124–31
    Reber Plan (San Francisco Bay Project), **8:**131–34
    water transfer to the Salton Sea, **8:**144
Zuari Agro-Chemical, **1:**21

Island Press | Board of Directors

Decker Anstrom
(Chair)
Board of Directors
Comcast Corporation

Katie Dolan
(Vice-Chair)
Conservationist

Pamela B. Murphy
(Treasurer)

Merloyd Ludington Lawrence
(Secretary)
Merloyd Lawrence, Inc.
and Perseus Books

Stephen Badger
Board Member
Mars, Inc.

Terry Gamble Boyer
Author

Melissa (Shackleton) Dann
Managing Director, Endurance
Consulting

Margot Paul Ernst

Russell Faucett
General Partner,
Barrington Partners

William H. Meadows
Counselor and Past President,
The Wilderness Society

Alexis G. Sant
Managing Director
Persimmon Tree Capital

Charles C. Savitt
President
Island Press

Ron Sims

Sarah Slusser
Executive Vice President
GeoGlobal Energy LLC